PRIMER ON MECHANICS OF MATERIALS
VOLUME 1

PRIMER ON MECHANICS OF MATERIALS
VOLUME 1

N.J. MASSON

ISBN 9781482374896

TABLE OF CONTENTS

DEDICATION

****

This book is dedicated to the Shell International Petroleum Company Limited, London who, through their local subsidiary the United British Oilfields Limited, Point Fortin, awarded me a scholarship in 1954 to pursue tertiary technical education and training in the United Kingdom.

Whatever my subsequent brief service with Shell Trinidad Limited, and by comparison much longer service with the Government of Trinidad and Tobago might have contributed to national development, is attributed wholly to my benefactor.

N J Masson

This page is intentionally left blank.

PREFACE

True education is self-education, is application. It has been said that most men fail not because of a lack of capacity but rather because of failure to apply themselves; a shortcoming of which the distinguished jurist Edward Abbott Parry inferred was a failure "... to read and learn and digest beneath the lamp of industry."

The lectures, symposia and whatnots of formal education are but mere adjuncts to the self education in which every student aspiring to professional status in engineering no less than in any other field of endeavour, must earnestly and fervently engage.

This book is offered as a supplementary text to students of Mechanics of Materials. It was designed as an aid to self-study of the topics which would normally be included in any undergraduate curriculum. However, it was not written to satisfy the requirements of any syllabus in particular but hopefully in such a style as to make the subject delightful for any reader and thereby to facilitate the process of digestion while in the refulgent rays of Judge Parry's third lamp.

The reader may wonder why in the face of an astounding number of books on the subject at hand, consisting as they do of numerous ideas and manner of expressing them, the author found it necessary to add another. The answer is simply that it was felt one was needed which consisted of his own selection of topics, the order of their arrangement and method of teaching; and in doing so to point out some of the dodges and pitfalls along the way.

Much of the science of Mechanics of Materials is expressed in the language of the Queen and Handmaiden of the Sciences, viz. Mathematics[1], a subject with a distinct hierarchical structure. Consequently the topics dealt with in this text are presented in a logical, 'vertical sequence' akin to the arrangement of the floors of a ziggurat: each one relying on the strength and support provided by preceding constructive effort. Inclusion of computer software for the solution of problems was avoided on the ground that while computer programs "... are capable of quite refined analysis" "... these have not led to an improvement in the reliability of engineering design." I have however drawn attention to the importance of the need for skill in the use of the electronic digital computer and for proficiency in computer programming in applications involving the Finite Element Method. This is so because invariably the solution of engineering problems using this technique literally involves an enormous number of calculations.

---

[1] "One reason", said Albert Einstein, "why Mathematics enjoys special esteem above all other sciences is that its propositions are absolutely certain and indisputable while those of the other sciences are to some extent debatable and in constant danger of being overthrown by newly discovered facts."

In a work such as this, total originality in respect of the treatment of subject matter is well nigh impossible. Having myself ploughed through and quarried hard in an extraordinarily large number of volumes on Strength of Materials, Mechanics, Applied Physics and other books devoted to special topics related to the main subject, it is inevitable that much of what was retained would probably have unconsciously manifested itself here and there on the written page. That notwithstanding, it is hoped that the book survives what Isaac Asimov called the "Road-to-Xanadu Test". Readers with a knowledge of the laws of statical equilibrium, elementary calculus, and of matrix algebra should have no difficulty in following the text which is replete with drawings, illustrated examples and solved problems, some based on questions set at engineering examinations of external authorities. I shall be most grateful for notifications (nojose@tstt.net.tt) of any errors in the text and for constructive criticism of the work that would aid in its improvement. The bibliography in the Appendix lists the authors of the referenced texts upon whose shoulders I stand in sincere acknowledgement of their kind assistance.

As a consequence of the large number of drawings and detailed solutions of many illustrated examples and solved problems, the text had of necessity to be divided into three parts: **Volume 1** containing Chapters 1-5; **Volume 2**, Chapters 6-12; and **Volume 3**, Chapters 13-20.

All units used in the text except those in which some quantities are expressed in so-called permitted units e.g. angular measurement in degrees and pressure in bars, are based on the Systeme Internationale (SI) Metric System.

I end this Preface in a manner similar to that of its beginning with a quotation attributed this time to the great French aphorist Francois de Rochefoucauld: "Our capacity exceeds our will power and it is often only to excuse ourselves that we hug the belief that things are impossible." I urge, implore, beg, beseech you to let your will power be the equal of your capacity which is limitless.

CHAPTER 1

INTRODUCTION

The subject variously called Strength of Materials, Mechanics of Materials, Mechanics and Elasticity of Solids and Strength and Elasticity of Materials, just four of the names by which it is perhaps most commonly known, is one of the great disciplines of modern-day Applied Science. It has widespread application throughout the entire spectrum of the engineering profession. No bridge, no building, no petrochemical plant, no piece of machinery, no pressure vessel, no drilling rig, no overpass, no rapid-rail, no spacecraft, no computer, no aircraft, no development of new materials, almost nothing in engineering practice, can be designed without application of the theories, methods and practices of mechanics of materials.

Its birth dates back to the seventeenth century at the time when Galileo Galilei the famous engineering scientist who, having studied several inexplicable structural failures in shipbuilding and construction practice, debunked the extraordinary hypothesis then prevailing "...that geometry alone was sufficient to analyse and design structures and machines". Galileo proposed the revolutionary postulate that supplementary information was needed which required empirical knowledge of the physical and other properties of materials "...if failures were to be avoided".   Thus was born the new science of Strength of Materials.

But if Galileo is credited with the fatherhood of the subject, then his own confirmation of a false hypothesis concerning the problem of the cantilever beam, attributes to him a similar if not more exalted status in respect of "...a paradigm of error that has been repeated throughout the history of engineering". We need not get into the details of this here, but in a nutshell, Galileo's erroneous analysis of the maximum load–carrying capacity of a cantilever, overstated by a factor of 3, the maximum load given by the now–known correct equation derived on the basis of modern beam theory. Accordingly, without a factor of safety, perhaps I should say factor of ignorance, to mask the error, a cantilever beam would have failed long before Galileo's predicted load was reached.

The cautionary tale in Galileo's paradigm of error exemplifies the vital and immortal dictum which is of such fundamental importance in engineering education and practice that it can never be repeated often enough, namely, that it is of utmost paramountcy to know and understand thoroughly the assumptions and limitations of an underlying theory; in other words to 'ignore the process whereby formulas are derived and used can be as wanting as those that ignore the rules of calculus and the laws of nature'.

Sophisticated theoretical analyses, aided and abetted by dazzling computer software do not absolve professional engineers and students alike, of the need for such knowledge and understanding. In these circumstances, the need is perhaps greater.

Incidentally, the correct treatment of the cantilever–beam problem is attributed to the French mathematician A. Parent whose work was published in 1713, 71 years after Galileo's death. Sadly, Parent's work was not recognized earlier partly because of Galileo's engineering – scientific credibility and partly because Parent was unpopular with his contemporaries and maintained what I call a low visibility coefficient. Galileo, it would appear was human after all; and not God.

Today, Strength of Materials has grown many special branches, some having names which reflect an emphasis on particular aspects or phenomena such as for example, elasticity, plasticity, metal fatigue, or which deal with special matters such as mechanical–failure analysis, fracture mechanics and mechanical testing of metallic and plastic materials, to name but a few. The discipline rests upon a vast and solid foundation fashioned from practical knowledge and experience since the Renaissance, and nurtured by the findings of an almost continuous infusion of research and development activity involving theoretical and experimental studies on a worldwide basis, some of it quite intensely mathematical and abstruse. With the advent of the digital computer has come novel procedures for the solution of stress and strain problems in Strength of Materials, one method of comparatively recent vintage being that of the Finite Element Method (FEM), concerning which the first textbook was published in 1967; and another being a new branch of mechanical engineering called Fracture Mechanics; previously mentioned. Some of the most brilliant and original thinkers in the field of the physical sciences and engineering have made seminal contributions in the work of development of the discipline, perhaps the best known among the pioneers being Galileo, Parent, Hooke, Young, Poisson, St. Venant, Cauchy, Coulomb, Timoshenko, Airy, Love, Southwell and Allen to name but a small coterie.

Methodology of Problem Solving

Strength of Materials is principally about determination of stresses and strains, and matters related thereto. For example, the component members of a sports pavilion super–structure or, of say, a mechanical device, are subjected in service to various loading conditions which give rise to internal forces in the members. These forces cause stresses and strains in them. But the materials of which the members are made must have the capacity to endure the maximum values of these effects without impairing the safe functioning of the structure or mechanical device, in any way whatsoever.

The discipline then provides the ways by which such stresses and strains are determined. And notwithstanding the apparent complexity of any problem, there are very broadly speaking, and in an extremely general sense, remarkably few basic steps to be followed in working out a solution. Before getting to these steps I proffer a few words of advice: Do not attempt to learn Strength of Materials by memorizing formulae unless of course you want to be alzheimic in old age. Even if you feel you can do it, do not try; perish the thought at once. You should instead start by making every effort to understand the basics. It is not always easy; you may have to quarry hard in some texts; this is the nature of the undertaking. Upon graduation, a fledgling or apprentice engineer would probably have to spend upwards of anything like 4000 hours to 5000 hours in some kind of work/training mode in his or her employer's establishment before either he or she is anywhere near proficient. Yet so many expect to master Strength of Materials after only 1% or less of this time; such great utopian dreams hardly ever materialise unless of course one is a genius. The Dadourian practice I found useful: whenever I came to a new topic in the field, I wrote out in full, the analytical work given in the textbooks for proofs of laws, theorems, whatever, always paying particular attention as was mentioned earlier to the underlying assumptions. If you try this technique then sometimes you may find it necessary to fill in yourself any intermediate steps omitted by the author in moving from one result to another. Having said that however, it should be emphasized that it is not the intention here to convey the notion that memory is not important. By tackling many problems you will use certain fundamental formulae so frequently that they will become embedded in your mind; they become second nature so to speak. And it cannot be overstressed that Strength of Materials, as indeed all mathematically–based subjects can be learned only by solving problems; that is by practice; the more the merrier.

Now to the basic steps and illustrations.

Step 1: Preparation of a Free-Body Diagram

A Free–Body Diagram, FBD for short, is simply a diagrammatic representation of a body in equilibrium showing all the external forces acting on it, and if dissected the internal reactions, and where they act. To say that FBDs are indispensable aids in the solution of problems in Engineering Mechanics (Statics and Dynamics) is to grossly underestimate their value. Many students have difficulty in solving such problems precisely because they do not apply the methodology of the FBD.

Consider the simple example of the bell–crank at Fig. 1. The crank is pivoted at 'C' and carries a horizontal force 'P' and force 'F' inclined at $\alpha°$ to the horizontal at B and A, respectively. The weight of the crank is deemed negligible.

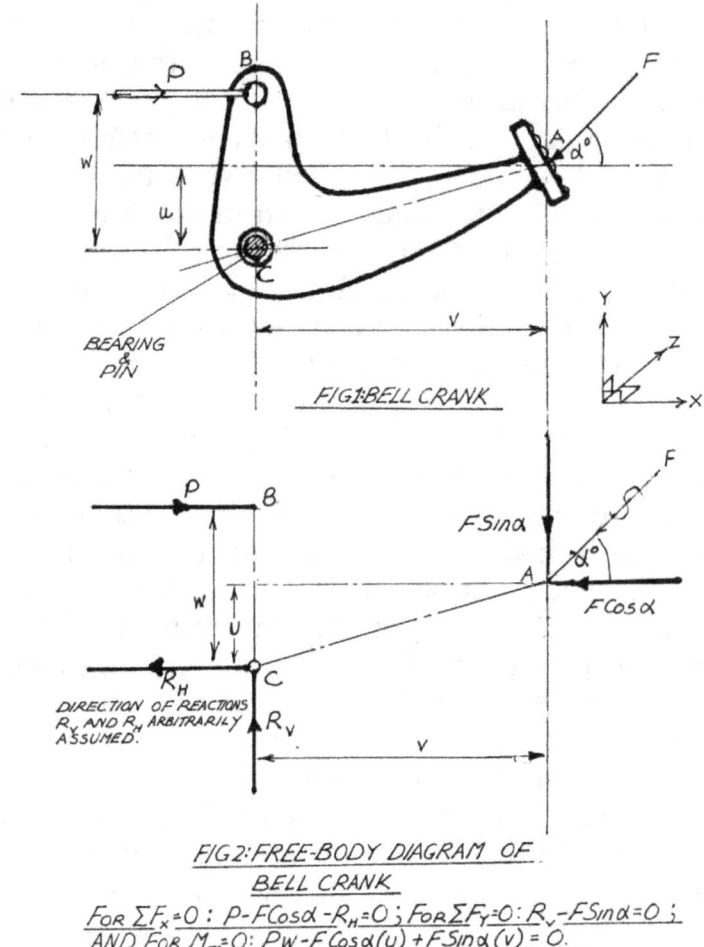

FIG 1: BELL CRANK

FIG 2: FREE-BODY DIAGRAM OF
BELL CRANK

For $\Sigma F_x = 0$: $P - F\cos\alpha - R_H = 0$; For $\Sigma F_y = 0$: $R_v - F\sin\alpha = 0$;
And For $M_z = 0$: $Pw - F\cos\alpha(u) + F\sin\alpha(v) = 0$.

Question: Are 'P' and 'F' the only external forces acting on the crank? In order to draw the FBD for the bell–crank, we separate it from the pivot at 'C' and make the FBD shown at Fig.2. The answer to the question is "No"!

In the FBD all the forces acting on the crank are shown. We do not know the direction of the reaction at the pivot 'C', so we substitute respectively its vertical and horizontal components 'R'$_V$ and 'R'$_H$ as shown. The crank is 'free', stable, unshackled and at rest. Thus the laws of static equilibrium are applicable, i.e. $\Sigma F_y = 0$; $\Sigma F_x = 0$; $\Sigma M_z = 0$. Normally the X – Y plane is the 2 – dimensional surface or plane on which a component is drawn, and the Z – axis is perpendicular to this plane. Therefore when we write $\Sigma M_z = 0$, it means the sum of moments about the Z axis through some specific point is zero. Thus, we may say moments about the Z – axis through a centre of mass, for example. It should also be noted that the directions of R_H and R_V were arbitrarily chosen. If it turns out in the supporting calculations that a positive sign is associated with any of these reactions then the arbitrarily chosen direction is correct; if otherwise, then the correct direction is the reverse of that assumed. Next take the case of the six reinforced–concrete cylinders all of the same weight 'W' stacked on a truck; a side view of the arrangement is shown as Fig.3. It is also assumed purely for the purposes of illustration that all the contact surfaces of the cylinders and of the tray of the truck are smooth.

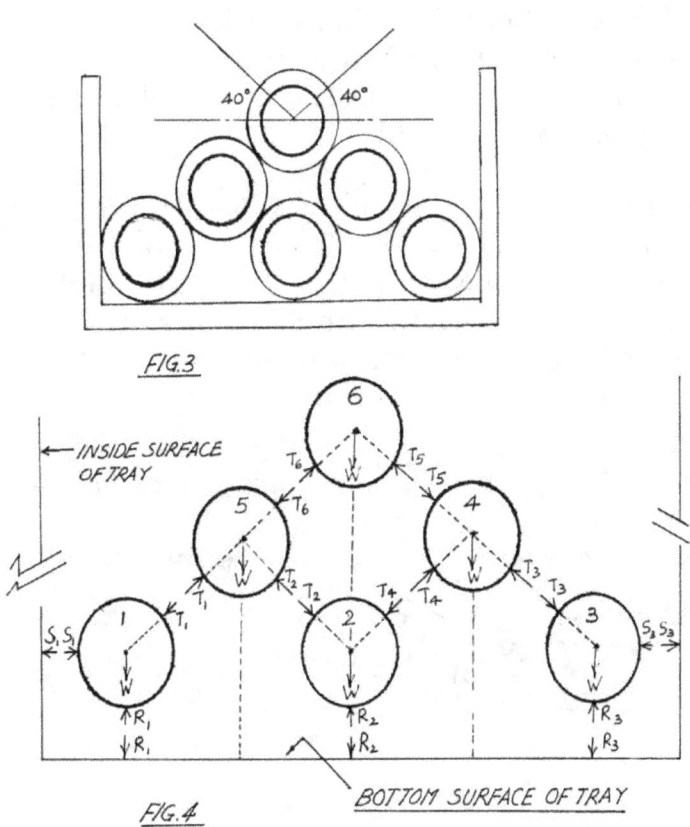

FIG.3

FIG.4

The FBD for each cylinder is shown in Fig.4. Cylinder 1 has three forces acting on it, viz. its weight 'W' and the reactions R_1 and S_1, at the bottom and side surfaces of the tray of the truck respectively. Cylinder 2 is acted upon by its own 'W; and by 'R_2', the reaction of the tray of the trunk and also by T_4 and T_2 the reactions at the point of contact with cylinders 4 and 5 respectively. Similarly, cylinder 3 is acted upon by its own weight 'W' and 'R_3' the reaction at the bottom surface of the tray of the truck and also by 'S_3' the reaction at the side surface of the tray of the truck. Now you describe the FBDs for cylinders 4, 5 and 6. In a practical situation the effect of friction between the contact surfaces of cylinders and between the contact surfaces of cylinders and the tray of the truck would have to be taken into account, but we need not concern ourselves with that here.

En passant': a word of caution. The stacking, storage and transport of concrete cylinders and other 'cylindricals' such as ductile–iron pipes are not to be done "vie-ki-vie". There are recorded deaths caused by improper stacking of pipes and failure to strap bundles of pipe properly for storage and for transport. Therefore, safety is of the utmost importance. For example, in order to ensure stability of a certain stack, certain diameter pipes and cylinders must not be stacked above a certain height. Pipe manufacturers consider the matter to be so important that when they fulfil orders for large volumes of their products they invariably send technicians to the delivery sites to oversee off-loading, transportation and stacking of the materials. There is method to the business; a technology in its own right.

Let us now consider an example in Dynamics. The motor vehicle shown in Fig.5 is of mass 'M' and propelled by a force 'F' which gives it an acceleration 'a' in the same direction as 'F'. It is assumed that the combined resistance to motion due to road–surface friction and to wind drag is 'R'. The FBD for such a dynamic situation is reduced to that shown by Fig.5.

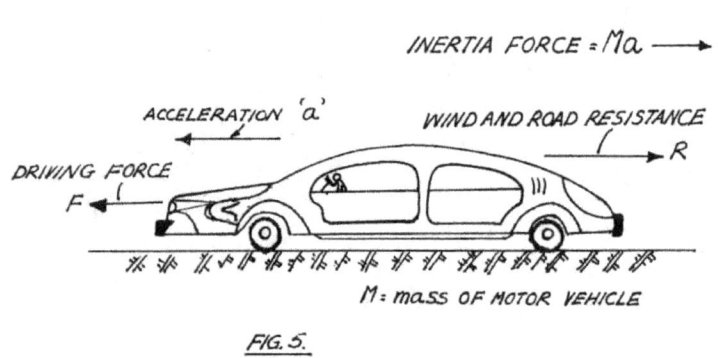

FIG. 5.

Observe that the reactive force, i.e. the inertia force, as its description connotes, opposes the motion. This inertia force = M(a). Thus for dynamic equilibrium we may write the following expression, $F - Ma - R = 0$, treating the problem as one in statics, i.e. $\Sigma F_x = 0$. When the vehicle decelerates, the inertia force Ma acts in the opposite direction to that shown in the figure and $F + Ma - R = 0$. Also when R = 0 and the vehicle is accelerating, $F = Ma$ which as every engineering schoolboy knows is a particular case of Newton's second law of motion. The term 'body' in the context of an FBD can mean different things. For example it could be used in reference to a single object such as the bell-crank just considered or it could be a complete structure comprising many components such as a six–panel Howe truss shown as Fig.6; or it could be part of the same truss.

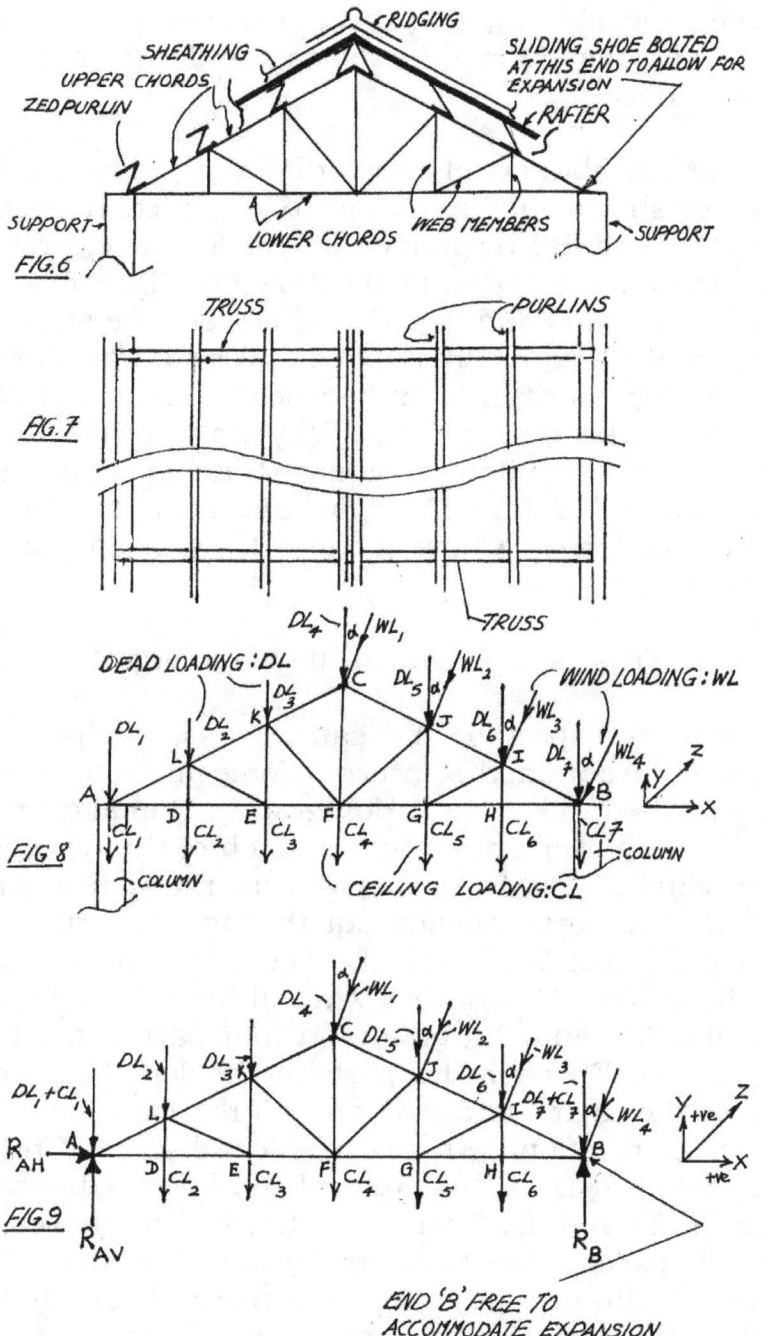

FIG.6

FIG.7

FIG 8

FIG 9

END 'B' FREE TO
ACCOMMODATE EXPANSION

Referring to Fig. 6, it is seen that the truss itself is made up of components called chords of which there are upper and lower types. The loading on the truss shown in Fig. 8 consists of the roof covering which may be either standing seam – or galvanized–iron corrugated sheeting or clay–or slate–tiles or other cladding material, purlins, rafters, braces, ceiling – and other suspended loads, and of course the weight of the truss itself (estimated at the design stage); these comprise the so–called dead loads. Then there are the live loads such as those due to wind, which in the Caribbean may get up to hurricane force, and to rain. Fortunately in the tropics, we do not have to consider live loads due to snow and ice.

Let us suppose the distribution of these different loads and the truss is as shown in Fig. 9. As shown in Fig. 8, the fully loaded truss is supported by columns located at 'A' and 'B'. To prepare a FBD for the entire truss, the entire frame complete with all the external loads shown in Fig. 8 is separated or freed from the supporting columns and in order to prevent the structure from falling to the ground, external reactions are applied at A and B as shown in Fig. 9: R_{AH} and R_{AV}, respectively, the horizontal and vertical components of the reaction at 'A'; and 'R_B' the vertical reaction at B. Why only a vertical reaction at B? Because in order to accommodate expansion due to temperature changes a 'sliding shoe' is bolted at end 'B' to allow for this. Thus there is no resistance to horizontal motion of the truss. Consequently $R_{BH} = 0$. For large spans, rollers are substituted for shoes.

The FBD of the Howe Truss as specified is shown in Fig. 9.

A 'body' could also in respect of the same truss, comprise part of it. For example, if we wish to obtain the internal force in web member KF and in chords KC and EF, then one way of doing so is to make an 'imaginary' cut through the truss as represented in Fig. 10. Each of the two separated parts of the truss in represents a Free–Body diagram for that part. Consider the left – hand portion of the two parts. Having cut through the truss and, as it were, unleased the internal forces T_1, T_2 and T_3. Therefore, these internal forces and the external reactions R_{AH}, R_{AV} together with all the external loads, in this case DL_1,+ CL_1, DL_2, DL_3, CL_2 and CL_3 acting on that part of the truss must be in equilibrium. Hence the FBD for the part on the left. Similarly, the internal forces T_1, T_2, T_3 for the right–hand side and all the external loads on that side including the external reaction R_B must be in equilibrium. The diagram on the right is accordingly the FBD for that part of the truss. Observe the arrows on the force vectors T_1, T_2 and T_3. These directions were chosen arbitrarily. But notice when the two parts of the truss are brought together, T_1, T_2 and T_3 are respectively equated with each other and the truss is back to an entire whole. In Fig. 11 I have separated again the truss into 2 parts. Your exercise is to write down for each part, the forces that must be equilibrium to constitute the relevant FBD.

Finally consider as a body the single member of the truss LK. In Fig. 12. LK is shown with all the internal forces T_1, T_2 T_4, T_6, T_7 and T_8 acting on it. If we wish to determine the force T_{LK}, then we can cut through the member and produce two FBDs. On the left: External load DL_3 must be in equilibrium with internal forces T_4, T_6, T_7 and T_{LK} ; and, on the right external load DL_3 must be in equilibrium with internal forces T_1, T_2, T_{LK} and T_8. When the two FBDs are brought together, the two forces T_{LK} neutralize one another. Do not concern yourself at this stage with arrow directions. Directions will come out in the wash when actual calculations are made. Once a certain direction is assumed you have to perform your calculations on that basis. If your assumption was correct then your answer will have a positive sign; if negative then the direction you assumed is the reverse.

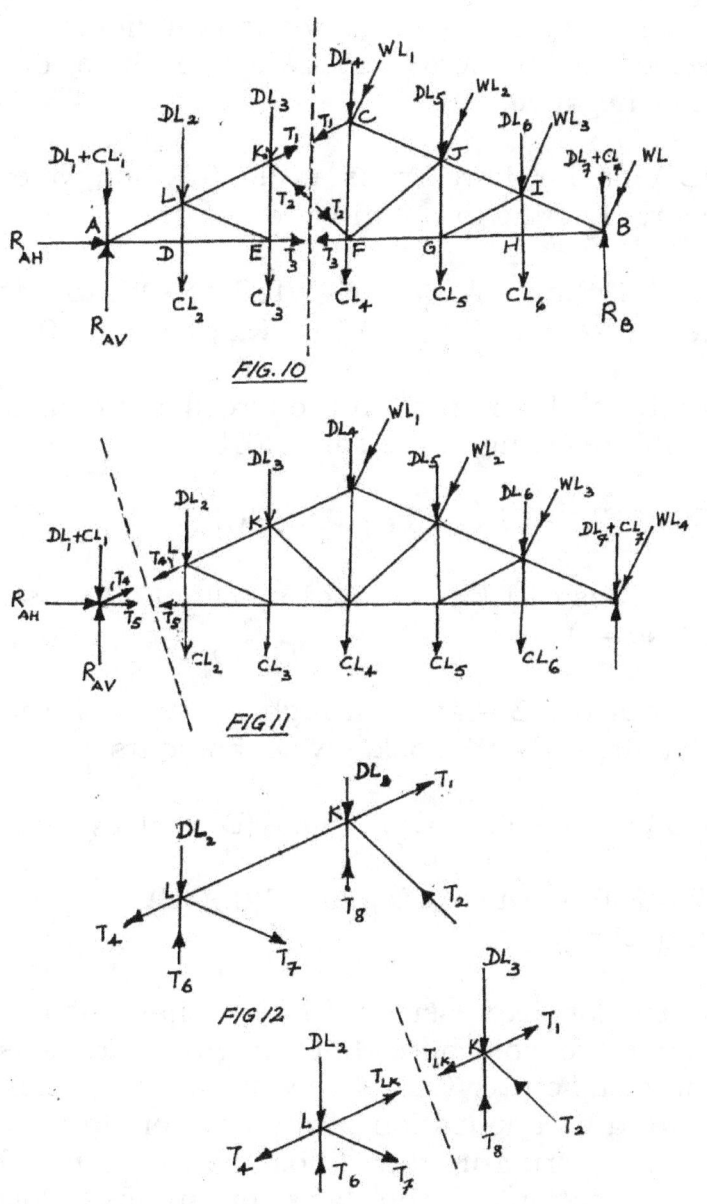

FIG. 10

FIG 11

FIG 12

Step 2: Application of the Laws of Statics to the FBDs

These laws are symbolized as follows:

$\Sigma F_x = 0$; $\Sigma F_y = 0$; $\Sigma F_z = 0$; and
$\Sigma M_x = 0$; $\Sigma M_y = 0$; $\Sigma M_z = 0$

Evidently for any particular configuration the orientation of the orthogonal axes must be known, positive and negative directions indicated. Similarly we must decide on the sense of moment: clockwise: positive or negative; anti – clockwise: positive or negative.

Referring to the FBD for the Howe Truss at Fig. 9, when we employ $\Sigma F_x = 0$, i.e. the sum of all the forces acting in the direction of the X – axis = 0, we obtain

$$R_{AH} - WL_1 Sin\alpha - WL_2 Sin\alpha - WL_3 Sin\alpha - WL_4 Sin\alpha = 0$$
$$\text{i.e. } R_{AH} - Sin\alpha (WL_1 + WL_2 + WL_3 + WL_4) = 0$$

Similarly, for $\Sigma F_y = 0$ i.e. the sum of all the forces acting in the direction of the y – axis, we may write, referring to the same FBD

$$R_{AV} + R_B - CL_2 - CL_3 - CL_4 - CL_5 - CL_6 - (DL_1 + CL_1)$$

$$DL_2 - DL_3 - DL_4 - WL_1 Cos\alpha - DL_5 - WL_2 Cos\alpha - DL_6 - WL_3 Cos\alpha$$
$$(DL_7 + CL_7) - WL_4 Cos\alpha = 0$$

Taking moments about the Z – axis through B, we obtain for $\Sigma M_z = 0$, taking each lower chord length = a, and clock – wise moments, positive

$$R_{AV} (6a) - (DL_1 + CL_1) (6a) - (DL_2 + CL2) (5a) - (DL_3 + CL_3) (4a)$$

$$(DL_4 + CL_4)(3a) - WL_1(CB) - (DL_5 + CL_5)(2a) - WL_2(JB)$$
$$(DL_6 + CL_6)(a) - WL_3(IB) = 0$$

We shall explore similar expressions in the chapter on "Forces in Plane Frameworks". It is well to note here that the Howe Truss is what is called a statically determinate structure. This is because all the forces in the bars of the structure can be found by application of the laws of statics alone. By contrast statically indeterminate structures are those in which bar forces may not be found solely by application of the laws of statics. Compatibility of bar deflections or deformation must also be taken into account. Let us consider an example. The structure is a rigid square platform of side 's' supported by 4 identical cables each of length 'L' and carrying a load of 'W' Newton as shown in Fig. 13. Some such platforms are seen in churches with spotlights affixed to them.

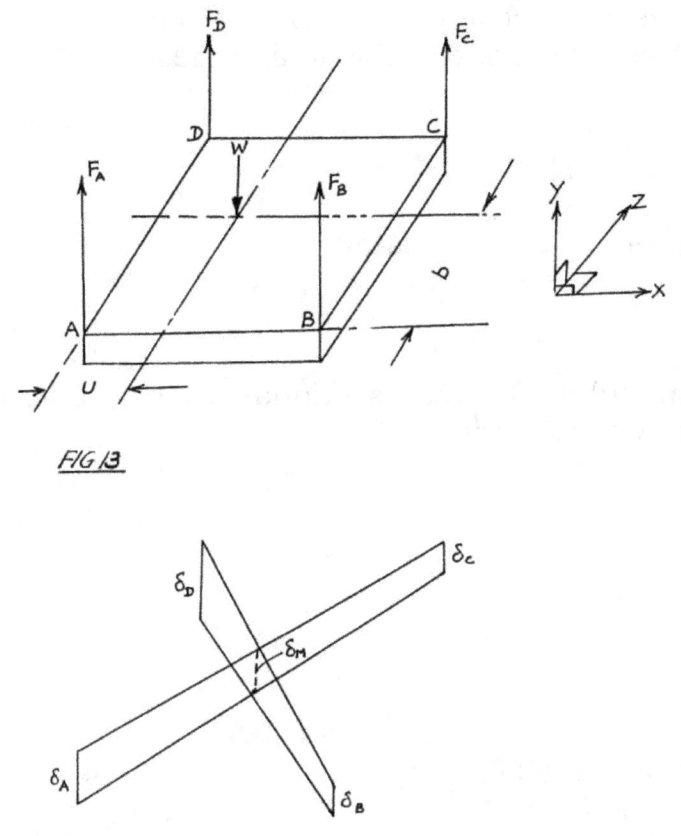

FIG 13

FIG 14

The force in each cable is to be determined.

For $\Sigma F_y = 0$
$$F_A + F_B + F_C + F_C + F_D - W = 0 \qquad \text{............... (i)}$$

There are no forces in the direction of the X – axis. Therefore $\Sigma F_x = 0$

For $\Sigma M_x = 0$, anti–clockwise moments, positive, i.e. moments about 'AB'
$$F_D(s) + F_C(s) - Wb = 0 \qquad \text{.......................... (ii)}$$

For $\Sigma M_z = 0$ anti–clockwise moments positive, i.e. moments about 'AD'
$$F_B(s) + F_C(s) - Wu = 0 \qquad \text{.......................... (iii)}$$

This is as far as we can get by applying the laws of statics. There are 4 unknowns and only 3 equations.

In order to move forward, account must be taken of the deflections δ_A, δ_B, δ_C and δ_D of the cables respectively at A, B, C, and D. In Fig. 14 'δ_M' is the deflection at the intersections of the diagonal planes.

By geometry,
$$\delta_M = (\delta_A + \delta_C) / 2 = (\delta_B + \delta_D) / 2 \qquad \ldots\ldots\ldots\ (iv)$$

and by Hooke's law, of which much later,
$$E = \frac{F}{a} \cdot \frac{L}{\delta}$$

Because Young's modulus E, cross–sectional area 'a' and length 'L' are the same for each cable, we may write

$$\delta = F \cdot \frac{L}{aE} = F\ (k)$$

k, being a constant $= \dfrac{L}{AE}$

Accordingly,
$$\delta_A = kF_A\ ;\ \delta_B = kF_B\ ;\ \delta_C = kF_C\ ;\ \delta_D = kF_D$$

so that from (iv)
$$F_A + F_C = F_B + F_D \qquad \ldots\ldots\ldots\ldots\ldots\ldots\ldots\ldots\ (v)$$

a result which enables each cable tension to be determined.

Step 3: Application of the appropriate analytical tools of the engineering science of mechanics of materials

Like the physician or surgeon who having examined a patient and arrived at a diagnosis and then sets about administering the relevant medication or decides to perform a particular surgical technique, the engineer having isolated the components as a Free–Body or Free Bodies, selects from his Mechanics of Materials' tool kit the necessary analytical instruments to determine forces, stresses and deflections in the component. An example should clarify the matter.

Let us suppose that the maximum stress in the leg 'BC' of the portal frame A-B-C-D shown in Fig. 15 is to be determined. It is assumed that because of its slenderness ratio BC, is not a strut. There are rigid joints at C and D; Pin joints are at A and B. When point load 'W' is on CD, the feet of the frame tend to splay or move apart; hence 'H' tending to prevent such spreading.

Referring to Fig. 15, application of $\Sigma F_Y = 0$, produces: $R_A + R_B - W = 0$. Also we may take moments about 'B' to obtain: $R_A (a + b) - W (b) = 0$

Applying $\Sigma F_x = 0$, gives: $H - H = 0$. Evidently statics alone cannot help to determine reaction 'H'. Thus, we may say that 'H' is an indeterminate force. There are ways to find the value of 'H'. But for the present let us assume that we know the value of 'H'. Let us draw the FBD for member BC.

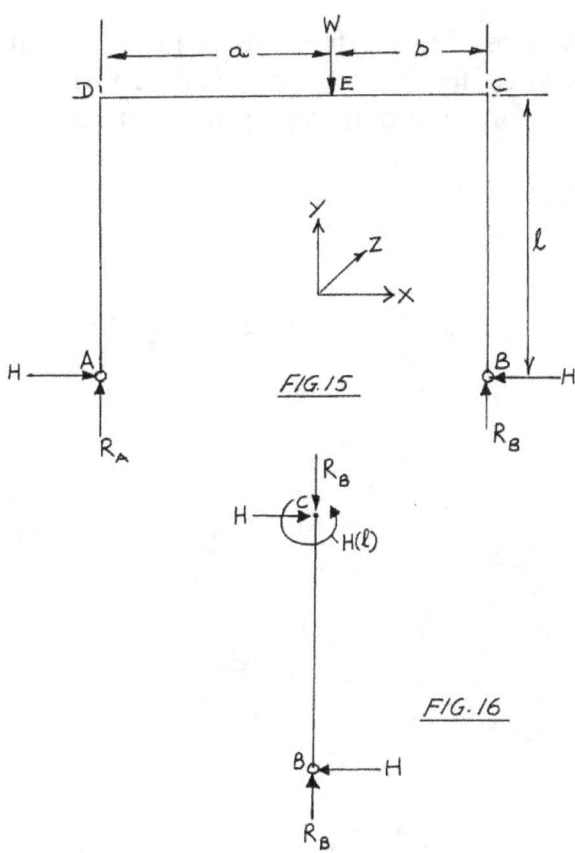

FIG. 15

FIG. 16

In Fig. 16, the right to left force H at B is equalized by left to right force H and at C; the upward reaction R_B at B is equilibrated by a downward force R_B at C. Also, H at B causes a clockwise moment H times 'l' which must be countered by an anti – clockwise moment of the same magnitude at 'C'.

Our analysis by FBD for member BC, therefore reveals that BC is subjected to (i) compression due to R_B; (ii) shear due to H and (iii) bending moment = Hl, the nature of which need not concern us here. Accordingly 'BC' is under the influence of direct compressive stress due to R_B, shear stress due to 'H' and bending stress due to the bending moment. A state of complex stress therefore exists; and the engineer goes to his tool kit for the necessary analytical equipment in order to design the member. In the exercises following I have included one (Exercise 10) on the same portal frame shown in Fig. 15.

However, you have to draw the FBDs for AD and CD the other two members of the frame in order to complete that exercise.

EXERCISES

The principal objective of the following exercises is the preparation of Free-Body Diagrams and not the production of mathematical results. In some cases answers are given. Not all diagrams are fully dimensioned.

Some of the exercises are drawn from the field of kinetics, the branch of dynamics which deals with forces that produce motion. One of the exercises requires a field trip to an oil-well drilling rig in operation.

Exercise 1

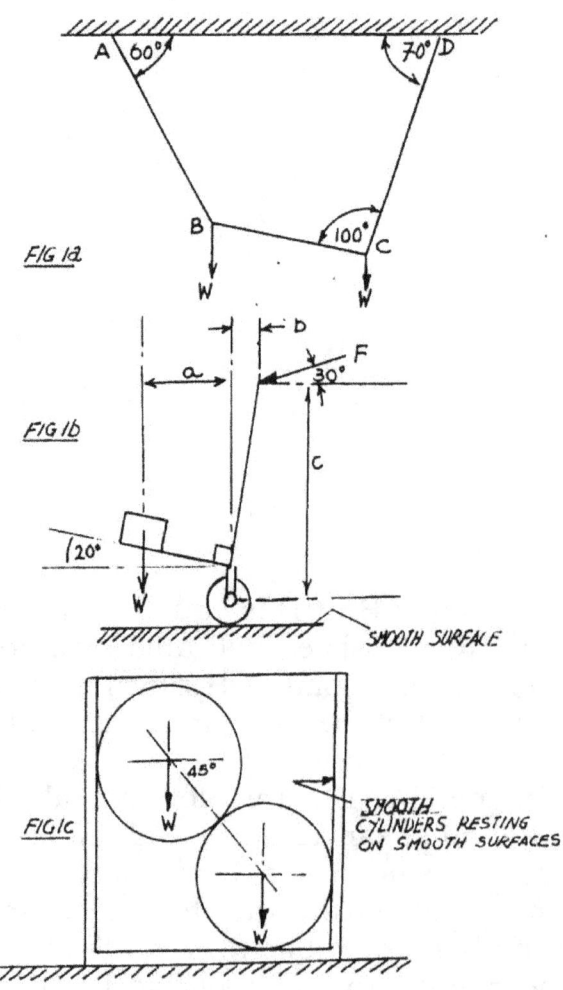

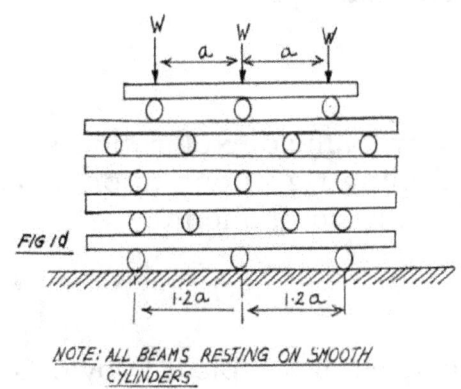

FIG 1d

NOTE: ALL BEAMS RESTING ON SMOOTH
CYLINDERS

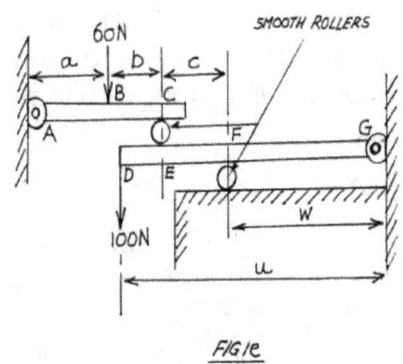

FIG 1e

Exercise 2

Determine the magnitude of force 'P' in Fig. 2 such that the pulleys 'A' and 'B' are equivalent to 120N and 60N respectively. Draw the FBD for each pulley. What is the magnitude of the reaction at each bearing?

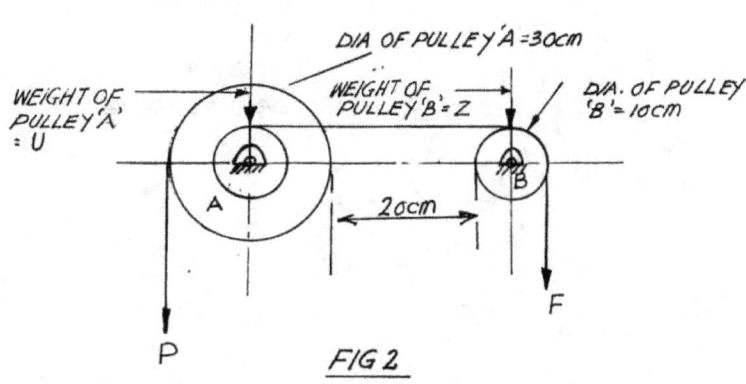

FIG 2

Exercise 3 (Part assignment; part question)

(i)   <u>Assignment</u>

Make a visit to an active oilfield exploration site (with your hard hat) and witness an oil-well drilling rig in operation. Make a neat drawing of the derrick and label its main components.

(ii)   <u>Question</u>

In Fig. 3, the five sheaves of a typical Crown Block and the four sheaves of its companion Travelling Block, both fitted with drilling cable, all of which make up part of the hoisting equipment of an oil derrick, are shown. Draw the FBD for the arrangement: what is the total load on the derrick in terms of 'W' the total load on the rotary hook: i.e. the drilling string comprising, kelly (or grief stem), drill collar/s, drill pipe, drill bit, etc.)? How might the load on the derrick be reduced?

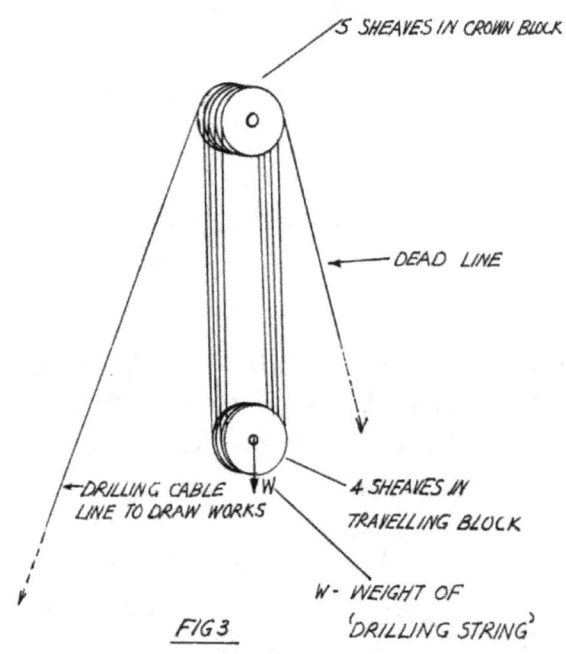

FIG 3

Exercise 4

Note: Unlike a truss which supports loads only at the extremities of its members, a frame is capable of supporting loads at any point on the structure.

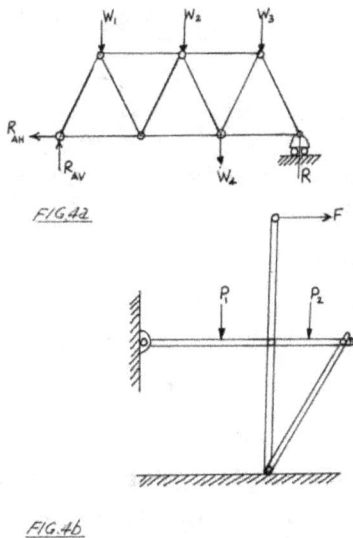

Fig. 4a is a drawing of a typical truss; and, Fig. 4b is a drawing of a frame.

(i)   Draw FBDs for the two parts of the truss shown in Fig. 4c, separated by MM′ the trace of the plane dividing it.

(ii) Draw FBDs for the three members of the frame shown in Fig. 4d.

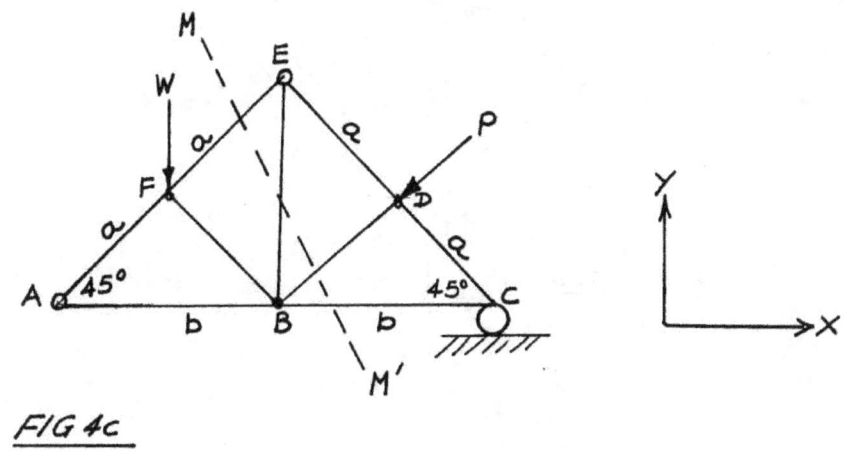

FIG 4c

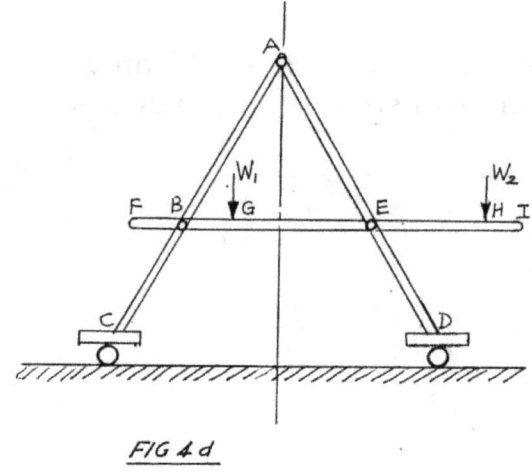

FIG 4 d

Exercise 5

The four uniform rods of the frame shown in Fig. 5 are connected by pin joints at A, C, E, and G. Each rod rests on a smooth peg at B, D, F, and H. Draw the FBD for each rod and write down the equations of equilibrium for one-half of the frame. Determine therefrom the horizontal and vertical components of the pin reactions at A and C. Observe that the weights of the rods were treated as negligible and therefore excluded.

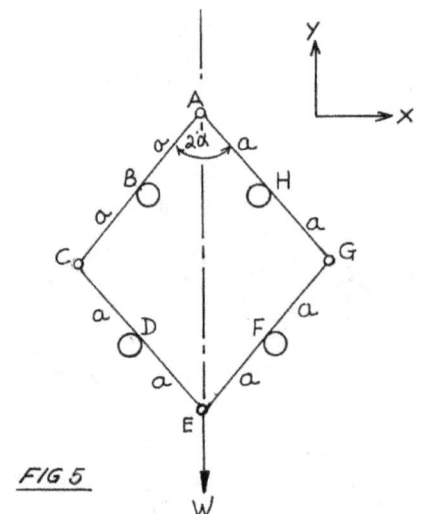

FIG 5

Exercise 6

Draw the FBD for the ladder shown in Fig. 6 and using it write down the equations of equilibrium from which the minimum angle, say, α° at which the ladder can be placed against the smooth wall without slipping, can be determined. The mass of the woman on the ladder is 60 kg and that of the ladder 30 kg shown acting through its centre of mass. Assume a coefficient of friction = 0.65 and take g = 9.81 m/s².

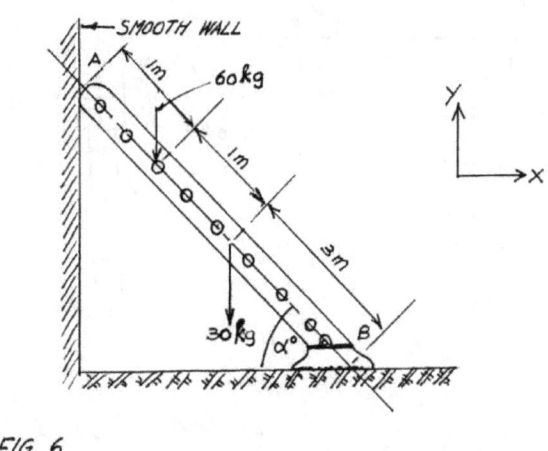

FIG 6

Exercise 7

Draw the FBDs for the pulleys A, B, and C and for the frame of the hydraulic boom shown in Fig. 7.

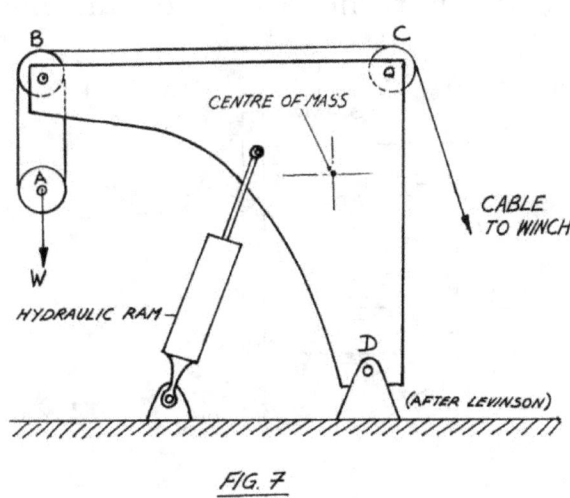

FIG. 7

Exercise 8

Draw FBDs for each portion of the three compound bars shown in Figs. 8a, 8b, and 8c. Note that the bar in Fig. 8c is fixed to rigid supports at both ends.

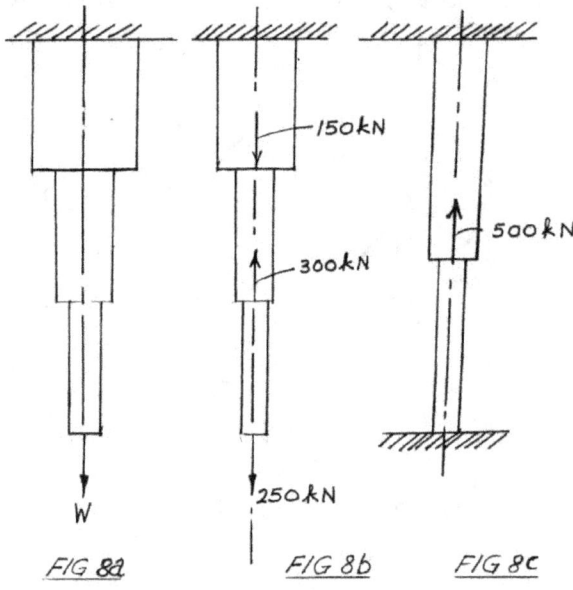

FIG 8a   FIG 8b   FIG 8c

Exercise 9

As part of an irrigation system a farmer constructs a retaining concrete structure across a channel. As a means of supporting the wall against hydrostatic pressure, the farmer erects a strut inclined at $\alpha°$ to the vertical and pin-jointed at its lower end. Draw the FBD for the arrangement. See Fig. 9.

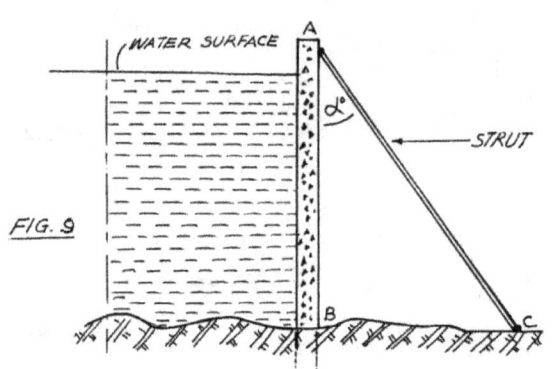

FIG. 9

Exercise 10

A portal frame ABCD is shown in Fig. 10. It is pin-jointed at supports A and B, and has rigid joints at C and D. The force 'H' is statically indeterminate but its magnitude need not concern you now. Draw the FBD for each member of the frame.

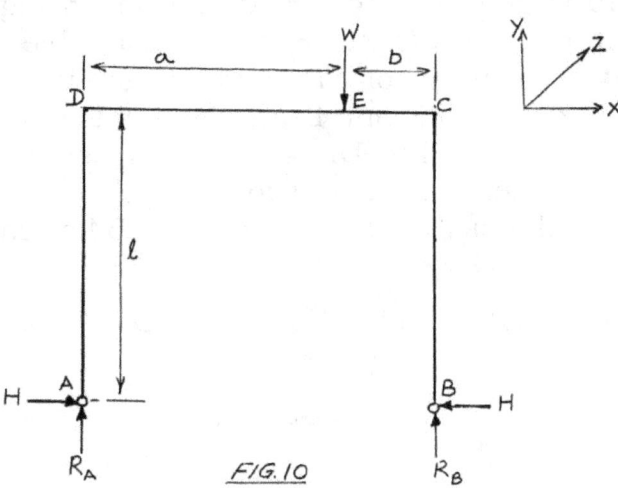

FIG. 10

Exercise 11

The two concrete blocks of mass 2000 kg and 1000 kg shown in Fig. 11 are coupled together by an inextensible steel chain and pulled along the ground by a tractor exerting a tractive force of 30 kN on the heavier block. The coefficient of kinetic friction between the blocks and the ground is 0.65. Draw the FBD reflecting dynamic equilibrium for each block and write down the equations of motion for such equilibrium in each case. Evaluate the tension in the chain and the acceleration of the motion.

Take $g = 9.81$ m/s^2.

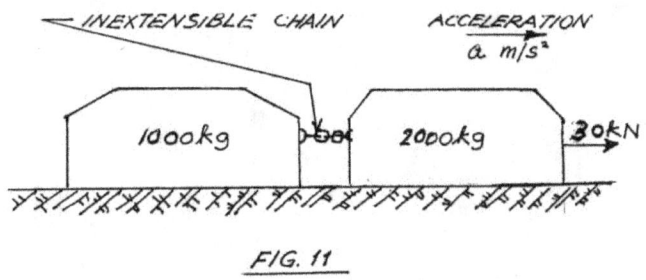

FIG. 11

Exercise 12

(i)   A vessel has a displacement of 20,000 metric tonnes in salt water. What is the displacement volume?

(ii)   This same vessel's metacentric height is 0.8m above the ship's centre of mass. The ship is heeled 3° to port as shown in Fig. 12. It has to take on board 350 tonnes of cargo. Space is available in a particular hold where the centres of mass of the containers with the on-coming cargo will be located 2.5m and 5m from the centre line on the port and starboard sides respectively. Draw a diagram showing the forces acting in such an arrangement and use it to determine how the loading is to be distributed. Take density of salt water = 1,025 kg/(m)³.

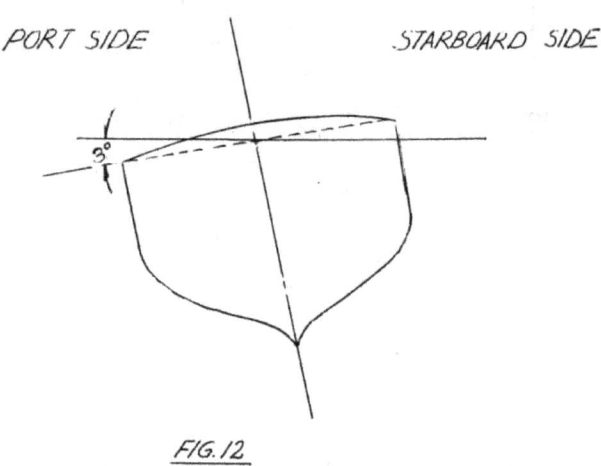

PORT SIDE   STARBOARD SIDE

3°

FIG. 12

Exercise 13

A four-wheel drive "Hill Rover" of mass 'M' kg is travelling up a slope of angle '$\alpha°$' to the horizontal with an acceleration of 'a' metres/sec². The tractive efforts at its rear and front wheels are T_1 newton and T_2 newton respectively. The total resistance to motion is 'K' newton. See Fig. 13. On a labelled diagram show the various forces acting on the vehicle and write down the equation of dynamic equilibrium. Designate the reactions at the front and back axles R_1 and R_2 respectively. Take 'g' as acceleration due to gravity. What is the maximum value of tractive effort when both front and rear wheels will slip? Take μ = coefficient of adhesion between the wheels and the ground.

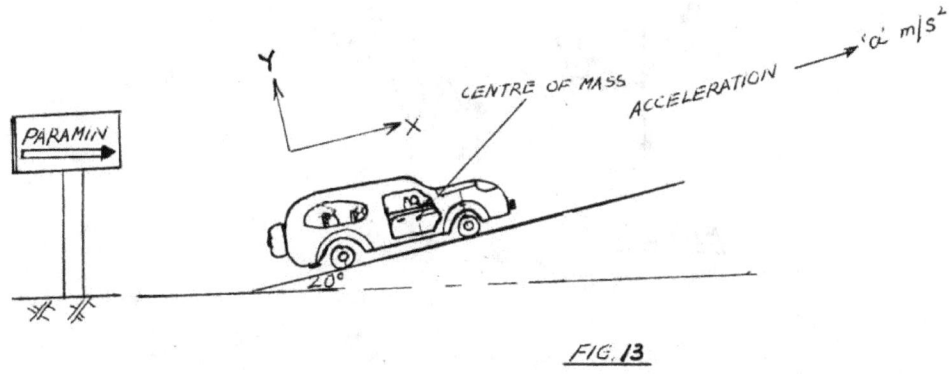

FIG. 13

Exercise 14

A two-gear speed reducer is shown in Fig. 14. The smaller gear 'A' is the driver and it is its torque T_A that gives 'A' an acceleration of α_A rad/s². An acceleration α_B rad/s² is transmitted to gear A. Draw the FBD for each gear.

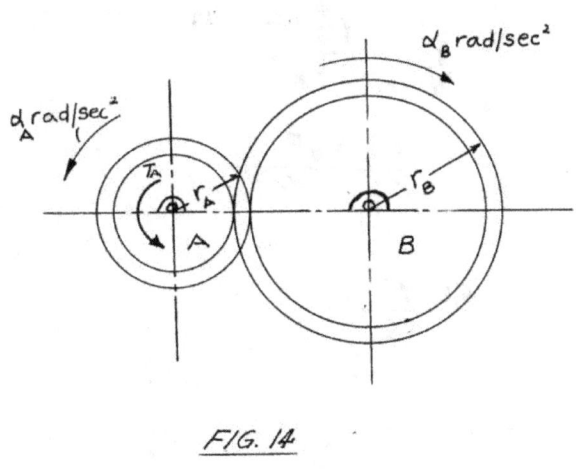

FIG. 14

~ 23 ~

SOLUTIONS TO EXERCISES

Exercise 1

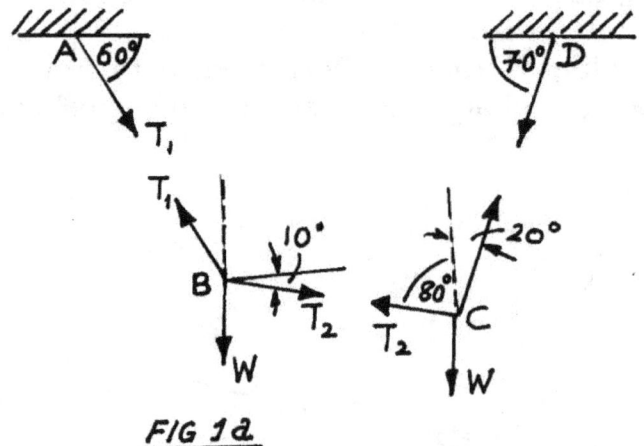

FIG 1a

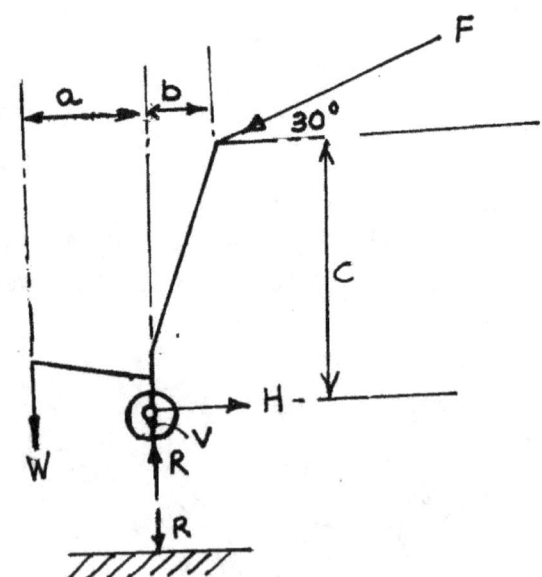

FIG:1b

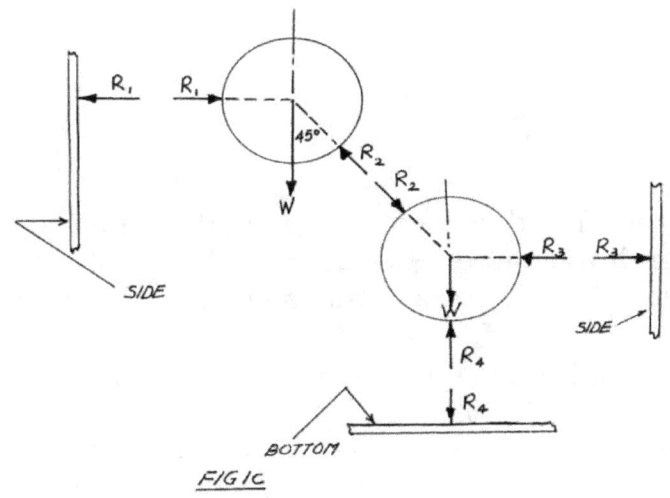

FIG 1c

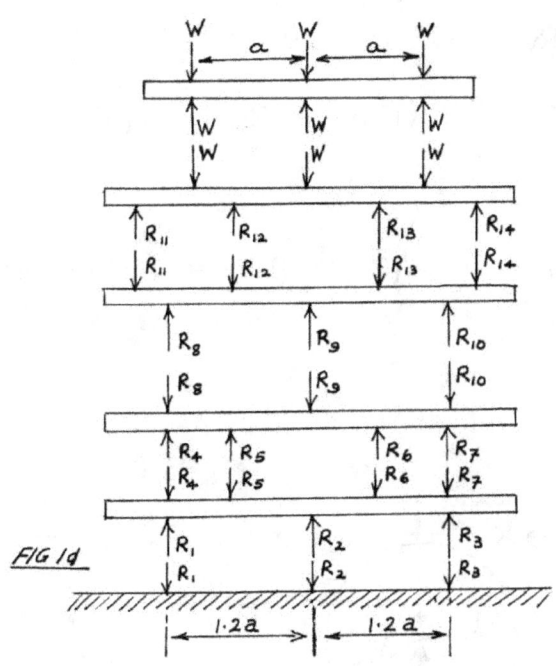

FIG 1d

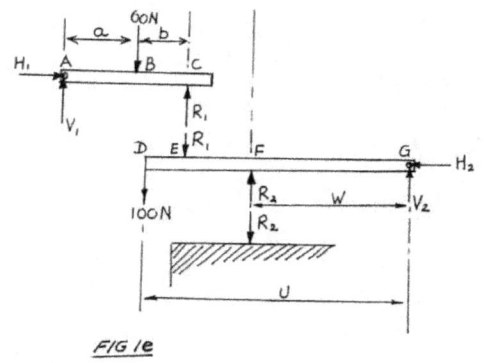

FIG 1e

~ 25 ~

Exercise 2

FBD labeled Fig. 2.

R_H' and R_V' are horizontal and vertical components of bearing reaction at 'A'; R_H and R_V are corresponding components at 'B'. 'U' and 'Z' are the weights of the pulleys acting through their centres of mass viz the centres of the pulleys.

For $\Sigma F_Y = 0$; $P+U-R'_V = 0$; $R_V-Z-F = 0$
For $\Sigma F_{YX} = 0$; $F+R'_H = 0$; $R_H-F = 0$
Moments about centre of pulley 'A':
$Px10-FX5 = 0$
Or $F = 2P$

Accordingly, $R_H'= 2P$; $R_V' = P+U$; $R_H = 2P$, $R_V = 2P+Z$

$\therefore$ Reaction at pulley 'A' = $\quad \sqrt{(R_H')^2 + (R_V')^2} = \sqrt{5P^2 + 2PU + U^2}$

and

$\quad$ reaction at pulley 'B' = $\quad \sqrt{(R_H)^2 + (R_V)^2} = \sqrt{8P^2 + 4PZ + Z^2}$

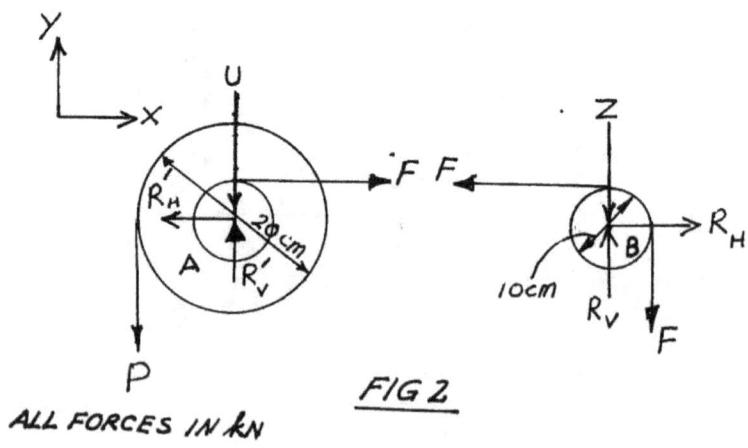

FIG 2

ALL FORCES IN kN

Exercise 3

The drawing of a typical oil derrick is shown in Fig 3. The FBD for the Crown Block/Travelling Block arrangement is shown in Fig. 3a. Considering the vertical equilibrium of the lower half of the diagram it is evident that the load on each cable is W/8. When the load on the drilling line to the draw-works and the dead line are included we have ten times W/8. The load on the derrick is therefore $\frac{10}{8}$W or 1.25W.

If, for example, the dead line is attached to the Travelling Block, then each of the nine lines must carry W/9. When the line to the cat head of the draw works is included we have ten times W/9. The total load on the derrick is now $\frac{10}{9}$W or 1.11W.

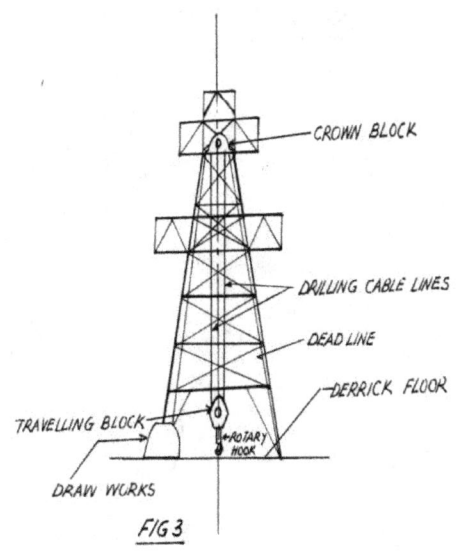

FIG 3

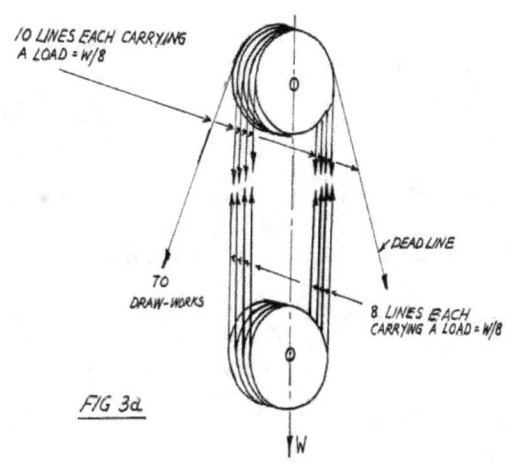

FIG 3a

Exercise 4(i)

FBD is shown as Fig. 4c.

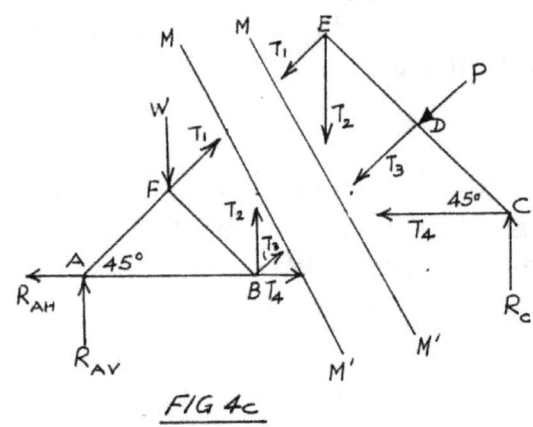

FIG 4c

Exercise 4(ii)

FBD is shown as Fig. 4d.

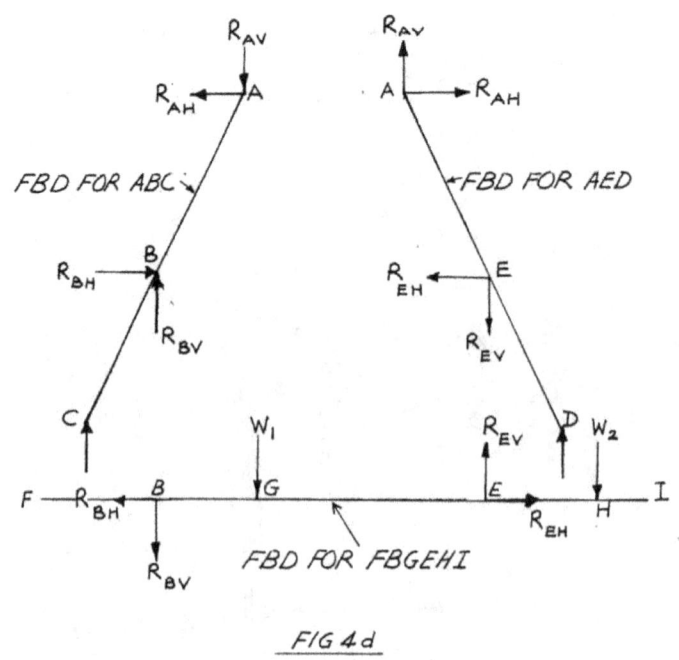

FIG 4d

Exercise 5

The FBD is shown in Fig. 5.

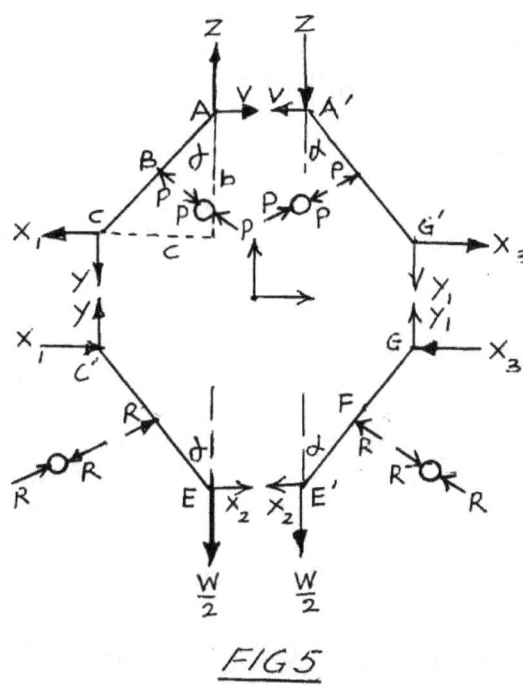

$$FIG\ 5$$

The pins being smooth the reactions 'P' and 'R' are perpendicular to the rods. Commencing the analysis with the lower rod, we have for:

$\Sigma F_Y = 0$

$R Sin\alpha + Y - \dfrac{W}{2} = 0$ .... (i)

$\Sigma F_Y = 0$

$X_1 - X_2 - R\ Cos\ \theta = 0$ .... (ii)

Moments about C

$X_2(2a\ Cos\ \alpha) + \dfrac{W}{2}\ (2a\ Sin\ \alpha) - Pa = 0$ .... (iii)

From (iii) $X_2 = \dfrac{Pa - Wa\ Sin\alpha}{2a\ Cos\alpha}$

From (ii) $X_1 = X_2 + R\ Cos\ \alpha$

$= \dfrac{Pa - Wa\ Sin\alpha}{2a\ Cos\alpha} + R Cos\ \alpha$

$$X_1 = \frac{Pa - Wa\,Sin\alpha + 2aR\,Cos^2\alpha}{2a\,Cos\alpha}$$

From (i) $\quad Y = \frac{W}{2} - R\,Sin\alpha$

Proceeding now to the top bar ABC, we have for:

$$\Sigma F_y \quad = \quad 0$$

$$Z + PSin\alpha - Y \quad = \quad 0 \qquad\qquad 1\ldots\ldots\ldots\ldots \quad (iv)$$

$$\Sigma F_x \quad = \quad 0$$

$$X_1 + PCos\alpha - V \quad = \quad 0 \qquad\qquad 1\ldots\ldots\ldots\ldots \quad (v)$$

From (iv) $\quad Z + PSin\alpha - \frac{W}{2} + RSin\alpha \quad = \quad 0$

$$\therefore \quad Z \quad = \quad \frac{W}{2} - Sin\alpha\,(P + R)$$

From (v) $\quad \dfrac{Pa - WaSin\alpha + 2aR\,Cos^2\alpha + PCos\alpha - V = 0}{2aCos\alpha}$

$$V \quad = \quad \frac{2a^2P - WaSin\alpha + 2aRCos^2\alpha + 2aPCos^2\alpha}{2aCos\alpha}$$

Exercise 6

The FBD is shown as Fig. 6.

For $\quad\quad\quad\quad\quad\quad \Sigma X \quad = \quad 0$
$$R_1 - \mu R_2 \quad = \quad 0 \qquad\qquad \ldots\ldots\ldots\ldots (i)$$

For $\quad\quad\quad\quad\quad\quad \Sigma Y \quad = \quad 0$
$$60(9.81) + 30(9.81) - R_2 = \quad 0 \qquad\qquad \ldots\ldots\ldots\ldots (ii)$$

Moments about A
$$\mu R_2(5\,Sin\alpha) + 60(9.81)(Cos\alpha) + 30(9.81)(2Cos\alpha) - R_2 5\,Sin\alpha = 0\ldots\ldots\ldots\ldots (iii)$$

From (ii) $\quad\quad\quad R_2 \quad = \quad 883N$
From (i) $\quad\quad\quad\; R_1 \quad = \quad 574N$

Accordingly (iii) may be written as:

2870 Sinα + 589 Cosα + 589 Cosα - 4415 Sinα
i.e. Tanα = 1178/1545

∴ α = 37°

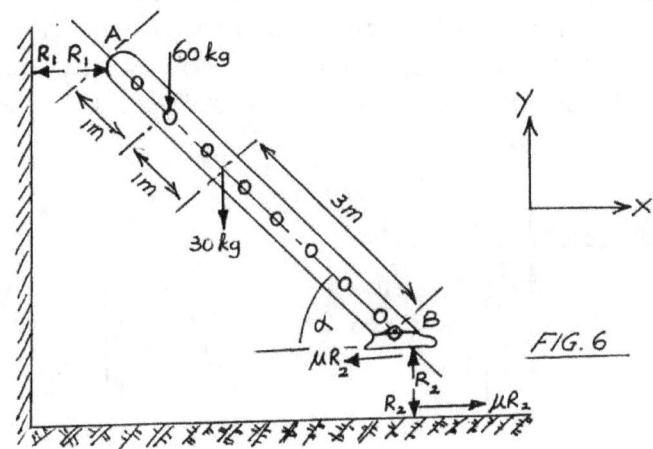

FIG. 6

Exercise 7

The FBD is shown in Fig. 7.

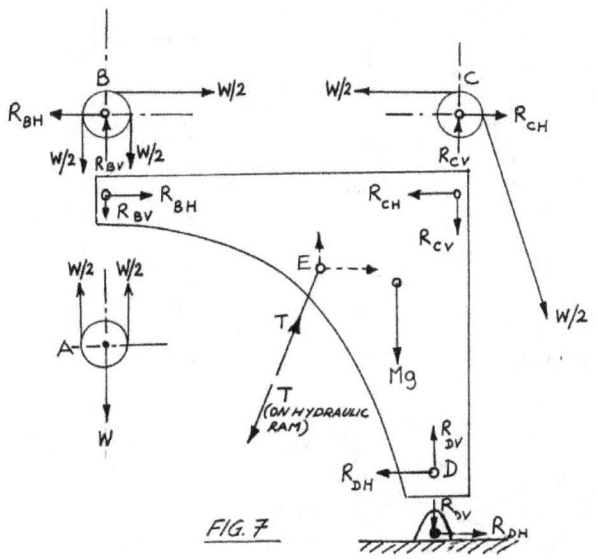

FIG. 7

Exercise 8

The FBDs are shown in Figs. 8a, 8b, 8c(i) and 8c(ii). Observe that each uniform section is subjected to a constant stress. For example in Fig. each portion of the compound bar is under a constant load 'W'. Accordingly each portion is under a constant stress = W/relevant cross-sectional area. Observe also that the merging of the different sections at junctions has the effect of producing the original loading configuration. For example in Fig. 8b when the lower two portions of the bar are brought together, the nett effect is 300 kn upwards.

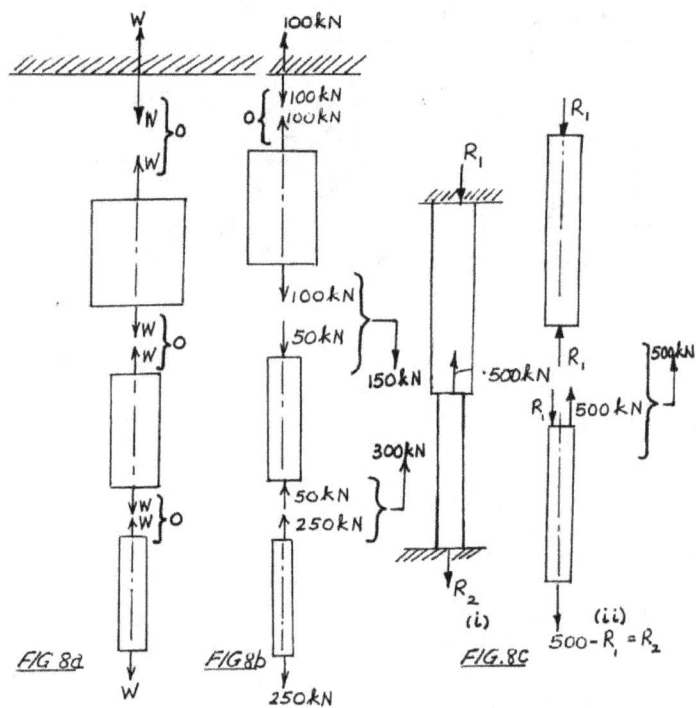

FIG 8a   FIG 8b   FIG. 8c

Exercise 9

The FBD is shown in Fig. 9.

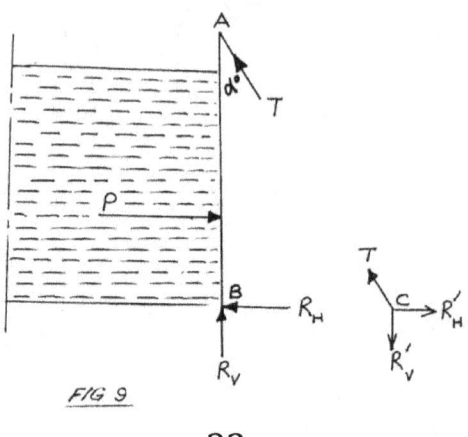

FIG 9

~ 32 ~

Exercise 10

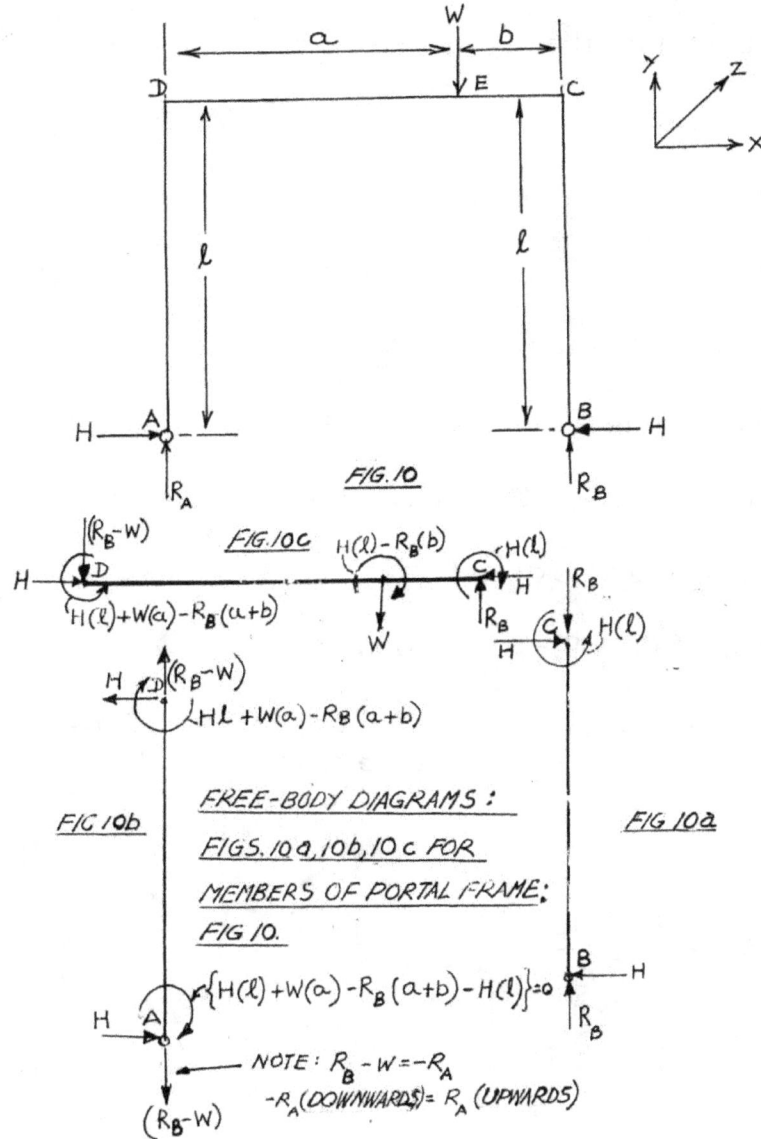

FIG. 10

FIG. 10c

FREE-BODY DIAGRAMS:
FIGS. 10a, 10b, 10c FOR
MEMBERS OF PORTAL FRAME:
FIG 10.

FIG 10b   FIG 10a

$\{H(\ell) + W(a) - R_B(a+b) - H(\ell)\} = 0$

NOTE: $R_B - W = -R_A$
$-R_A$ (DOWNWARDS) = R_A (UPWARDS)

The FBDs are given above.

For BC: Fig. 10a; For CD: Fig. 10c; and For AD: Fig. 10b.

Note that with reference to Fig. 10, when moments are taken about 'A', $R_B(a+b)$ = $W(a)$; also, in Fig. 10b, R_B-W, downwards, which is equivalent to R_A, upwards.

Exercise 11

The FBD for each block is shown in Fig. 11

The equation of motion for the 2000kg block is:

$$3,000 - \mu R_1 - T - M_1 a \quad = \quad 0 \qquad [M_1 = 400 \text{ kg}]$$

And for the other block:

$$T - \mu R_2 - M_2 a \qquad = \qquad 0 \qquad [M_2 = 200 \text{ kg}]$$

Now $\quad \mu R_1 = \mu M_1 g$ and $\mu R_2 = \mu M_2 g$

$$\therefore \qquad 3,000 - \mu_1 M_1 g - T - 2,000a = 0$$

i.e. $\qquad 150,000 - 0.65\,(2000)(9.81) - T - 2,000a = 0 \qquad \ldots\ldots\ldots\ldots\ldots$ (i)

Similarly for the 1,000kg block,

$$T - 0.65(1,000)(9.81) - 1,000a = 0 \qquad \ldots\ldots\ldots\ldots\ldots \text{(ii)}$$

The magnitude of 'T' and 'a' may be obtained by solving the simultaneous equations (i) and (ii). Ans: T = 10kn; a = 3.6m/s²

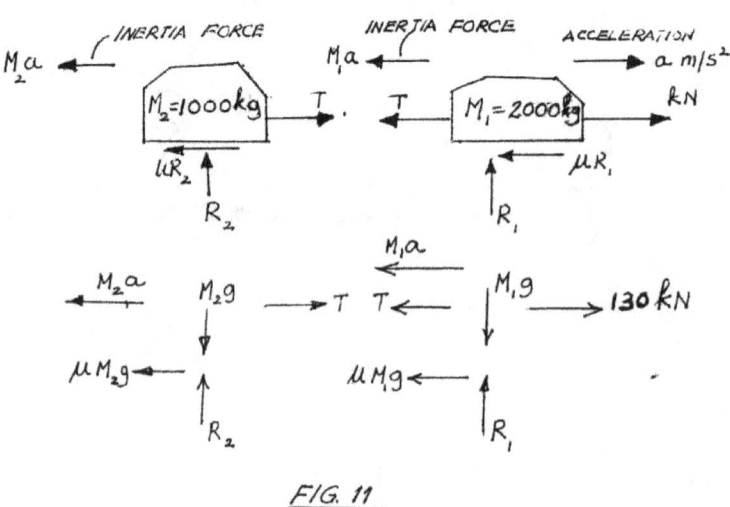

FIG. 11

Exercise 12

(i)   The volume of salt water displaced $= \dfrac{20{,}000 \times 1{,}000}{1{,}024}$

$$= 19{,}531\text{m}^3$$

(ii)   Fig. 12a is a diagram showing the vessel heeling to the port. The ship's centre of mass is at G. When the vessel heels, G moves to G_1.

In Fig. 12b, the force due to buoyancy is shown acting upwards on the same centreline of the ship's centre of mass. For the vessel to be upright after loading:

Heeling moment to port = heeling moment to starboard

From geometrical considerations:

$$\frac{GG_1}{GM} = \text{Tan } 3° = 0.05241$$

$\therefore$ $GG_1 = 0.8(0.5241) = 0.041928$

Working in newtons and metres :

$$C_1(1{,}000)g \times 2.5 + 20{,}000(1{,}000)g \times 0.041928 = (350 - C_1)(1{,}000)\, g \times 5$$

$\therefore$ $2.5C_1 + 838.6 \quad = \quad (350 - C_1)5$

                            $= \quad 1{,}750 - 5C_1$

i.e. $7.5C \quad = \quad 911.4$

or $C_1 \quad = \quad 121/5$

Therefore the mass of the container with its contents on the port side is 121.5 tonnes. Accordingly, the container and its contents on the starboard side = 228.5 tonnes.

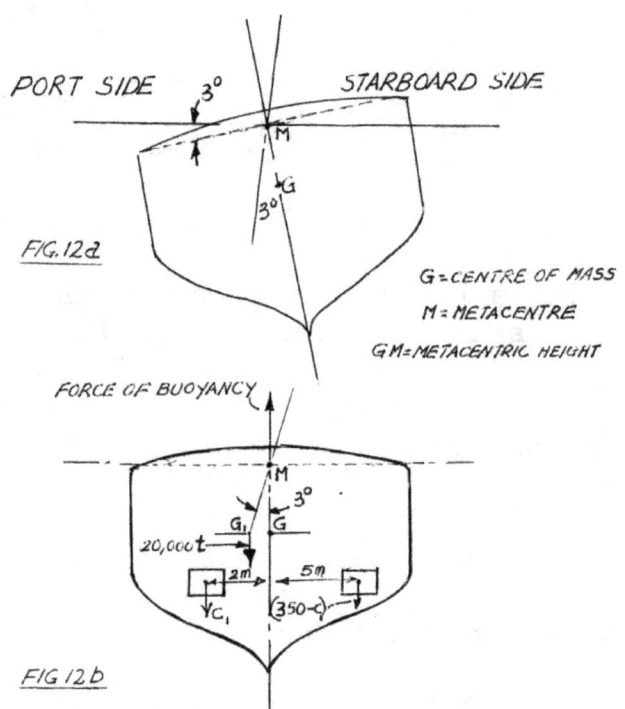

PORT SIDE   3°   STARBOARD SIDE

M

3° ᴳG

FIG. 12a

G = CENTRE OF MASS

M = METACENTRE

GM = METACENTRIC HEIGHT

FORCE OF BUOYANCY

M

3°

G_1 G

20,000 t

2m   5m

C_1 (350-c)

FIG 12 b

Exercise 13

Equation of dynamic equilibrium:

$$T_1 + T_2 - Mg\sin\alpha - K - Ma = 0$$

Maximum value of Tractive Effort when the vehicle's front and rear wheels will slip is:

$$T_m = \mu R_1 + \mu R_2$$

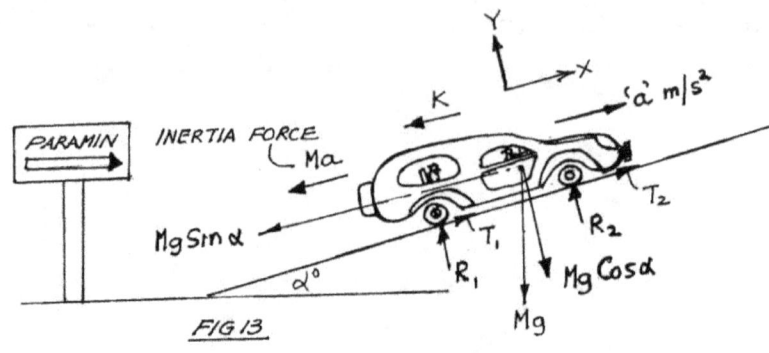

PARAMIN   INERTIA FORCE
                 Ma

$Mg\sin\alpha$

$\alpha°$

R_1 T_1

$Mg\cos\alpha$

Mg

R_2

T_2

K

Y

X

'a' m/s²

FIG 13

Exercise 14

The FBD is shown in Fig. 14.

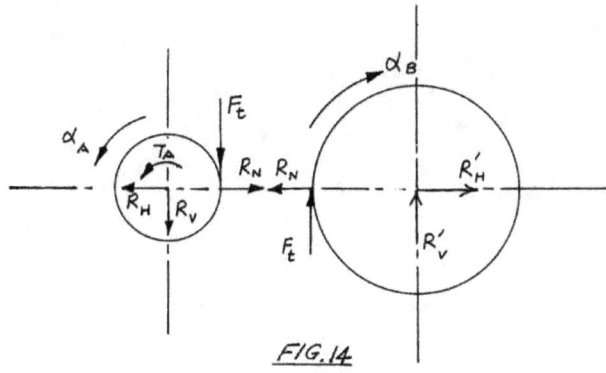

FIG.14

This page is intentionally left blank.

CHAPTER 2

FORCES IN PLANE FRAMEWORKS

Plane framework is a generic term used to describe a class of engineering structures such as roof trusses which are formed from various assemblies of bars and other components for the purpose of resisting geometric distortion; such assemblies, their applied loading and external reactions being in one and the same plane. Similarly, but by contrast, the term "Space Framework" refers to those framed structures in which the bars and other component parts, applied loading and external reactions are in more than one plane, i.e. in 3 dimensions.

It is necessary to distinguish between (i) plane frameworks having external loads which are assumed to be acting only at smooth pin-joints, and in respect of which theoretically the forces induced in its members are wholly of either a direct tensile or direct compressive nature, and, (ii) plane frames in which external loads are applied at places in the structure other than pin-joints, in which case shearing forces and bending moments would induce additional stresses in components apart from those due to direct tensile and compressive forces. At Fig. 1a, a pin-jointed plane framework is shown loaded by external forces at points 'F' and 'E', whereas at Fig. 1b a plane frame pin-jointed at C,D,E is shown carrying a load positioned between pin-joints at 'D' and 'E'. At Fig. 1c, I have shown the computed forces acting in all members together with the external loads of Fig. 1a. At Fig. 1d, the results of resolution of forces at pin-joints E and F of the framework at Fig. 1a are shown.

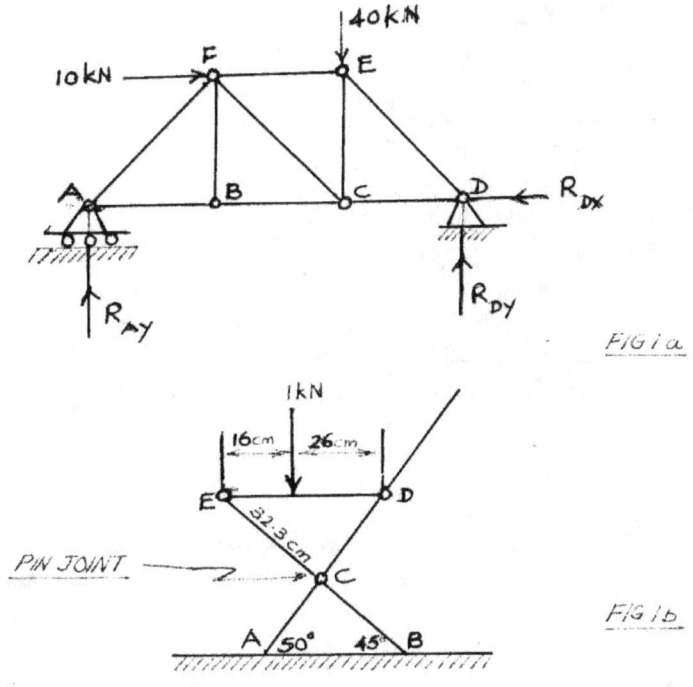

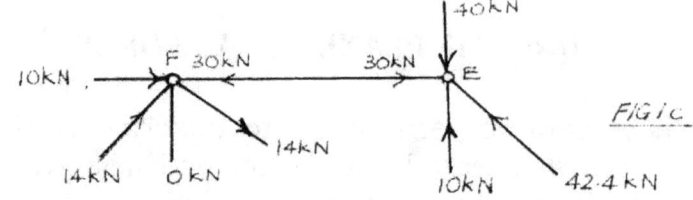

FIG 1c

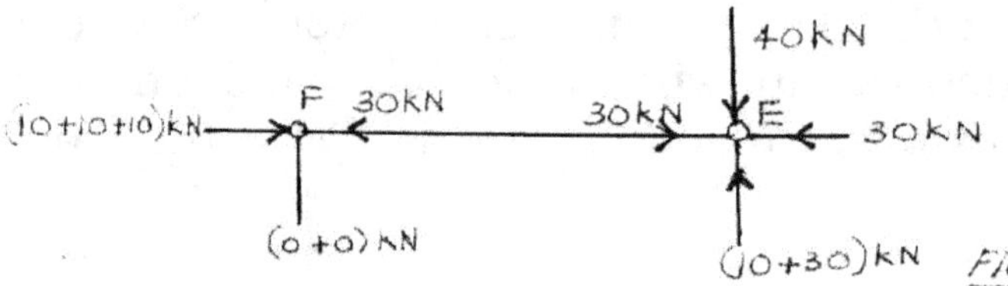

FIG 1d

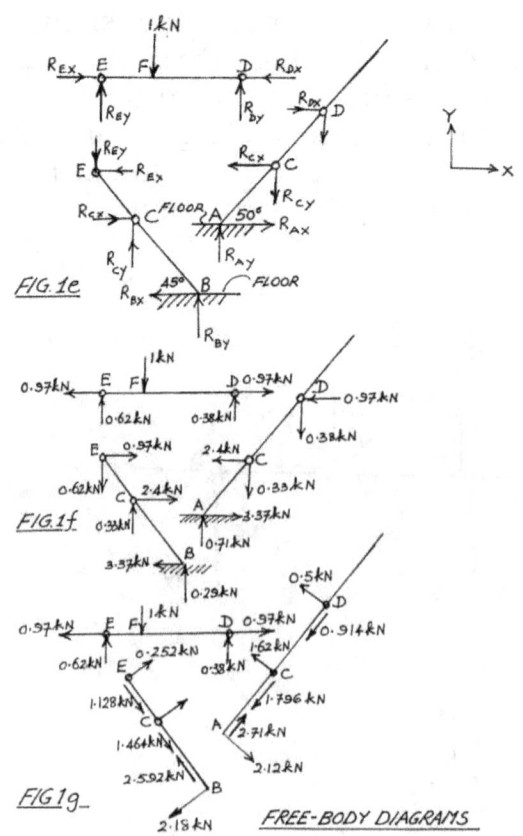

FIG. 1e

FIG. 1f

FIG 1g

FREE-BODY DIAGRAMS

Consider next the case of the pin-connected frame of the chair in Fig. 1b. Member ED supports a load of 1kN. Free-body diagrams for each member of the frame are shown in Figs. 1e, 1f and 1g. It must be pointed out that the directions of the horizontal reactions at the pins at 'E' and 'D' in Fig. 1e are arbitrarily chosen. Evidently, the vertical components at 'E' and 'D' had to be pointing upwards. In Fig. 1f, the computed values of all reactions are indicated in the diagram. See the solution of problem No. 6 at the end of this Chapter for the necessary calculations. It will be observed there that the directions which were arbitrarily chosen for R_{EX} and R_{DX} in Fig. 1e were incorrect: calculation proved they were opposite. They are shown correctly in Fig. 1f. Lastly in Fig. 1g, all the horizontal and vertical pin-reactions at pin-connections E, C and B of member ECB, and at A, C and D of member ACD, have been resolved into components parallel and perpendicular to these members. As may be seen all members are in a state of equilibrium. Observe however that at pin-connection E for member ED there is an upwards force of 0.62 kN between 'E' and the 1kN at 'F' tending to push that part of the member upwards: a shearing force in fact. Likewise, there is a clockwise moment of 0.62kN times 16/100 cm equivalent to approximately 0.1 kNm about point F: a bending moment. By observation, it is seen that there are similar shearing forces and bending moments acting on the other members of the frame. Referring again to all the pin joints of Fig. 1d it is observed that all vertical and horizontal forces equalize each other. Hence in Fig. 1d there are no shearing forces and bending moments acting on any of the members of the framework.

Plane and space frameworks are analysed in order to determine the magnitude and nature of the forces and deflections caused by the loads imposed on them, on the basis of which determinations, component parts may then be appropriately sized to provide the necessary resistance in a manner consistent with safety and economy. Frameworks may also be divided into two categories viz. (i) statically-determinate and (ii) statically- indeterminate. Generally speaking, a framework or frame if you will, is subjected to loading of some kind whether static or dynamic, and supported by reactions as in the structure at Fig. 1a. Provided the reactions can be found by applying the equations of statical equilibrium as they relate to one plane, i.e. $\sum F_x = 0$; $F_y = 0$; $\sum M = 0$, then the frame is said to be statically determinate.

By way of illustrating the condition of statical indeterminacy, if an additional support were to be introduced at say pin-joint 'B' of the structure at Fig. 1a, then there would be more reactions than that which could be determined by the application of $\sum F_x = 0$; $\sum F_y = 0$; and $\sum M = 0$. In such a case the structure would be regarded as statically-indeterminate. Another word for the term statically-indeterminate is hyperstatic. Whereas for statically determinate plane frames there are 3 equations of equilibrium viz. $\sum F_x = 0$; $\sum F_y = 0$; $\sum M = 0$, in the case of statically-determinate space frames there are six equations of equilibrium to be satisfied : $\sum F_x = 0$; $\sum F_y = 0$; $\sum F_z = 0$; $\sum M_x = 0$; $\sum M_y = 0$; $\sum M_z = 0$.

Forces in plane frameworks

The methods which are most commonly applied to the solution of forces in statically-determinate plane frames are:

Graphical determination by means of funicular or line polygon and force diagram;
Method of Joints;
Method of Sections or Ritter's method; and
Method of Tension Coefficients.

Each of these methods is best understood by following its application to a specific example, for which purpose consider the pin-jointed cantilever frame shown at Fig. 2.

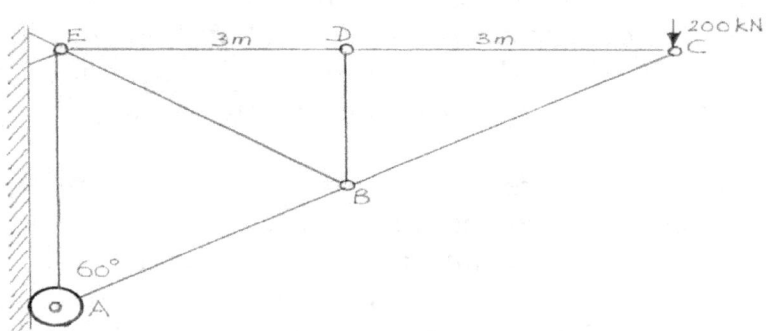

FIG. 2

Before doing so however it is well to recall the nature of actions or reactions for the different types of supports commonly encountered in frame analysis, in which connection refer to Fig. 3.

In Figs 3a through to 3f there is a single reaction R acting in a direction normal to the surface on which either the rollers or the rocker rests at the point of contact; there is no horizontal component in the case of Figs. 3a, 3b and 3c and none parallel to the inclined surface in the case of Figs. 3d, 3e and 3f.

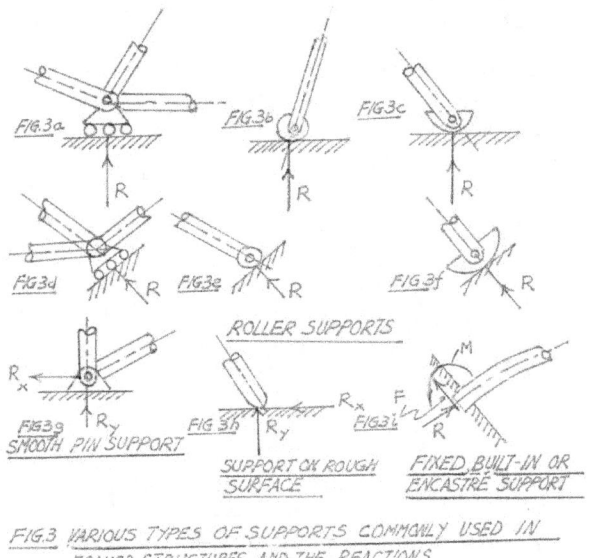

FIG.3 VARIOUS TYPES OF SUPPORTS COMMONLY USED IN
FRAMED STRUCTURES AND THE REACTIONS

As illustrated at Fig. 3g smooth-pin supports generally have a single inclined reaction of which it is normal practice to represent by means of a vertical component and horizontal component, exemplified by R_Y and R_X respectively at Fig. 3g; so too in the case of a member resting on a rough surface as at Fig. 3h. In both these cases, Figs. 3g and 3h, the reaction $R = \sqrt{R_X^2 + R_Y^2}$.

Where a component is fixed in direction as at Fig. 3i, in addition to the component reactions 'F' and 'R', there is also a fixing moment 'M'.

Referring again to Fig. 2, a roller being at A, it is known immediately that the single reaction there, R_{AX} is horizontal; its direction could be either leftwards or rightwards. The imposed load is 200kN at C and there is a smooth-pin support at E where the reaction R_E has two components viz. R_{EX} horizontally; and R_{EY} vertically. The directions shown for R_{AX}, R_{EY} and R_{EX} are assumed.

Let us now proceed to determine the reactions and internal forces graphically.

(i)   **Graphical determination**

The reactions will first be obtained. Refer to Fig. 4a where the external forces R_{AX}, R_E and 200kN are shown. These constitute a system of forces in equilibrium. Consequently where the lines of action of the 200kN force and R_{AX} meet defines a point through which the third force R_E must pass.

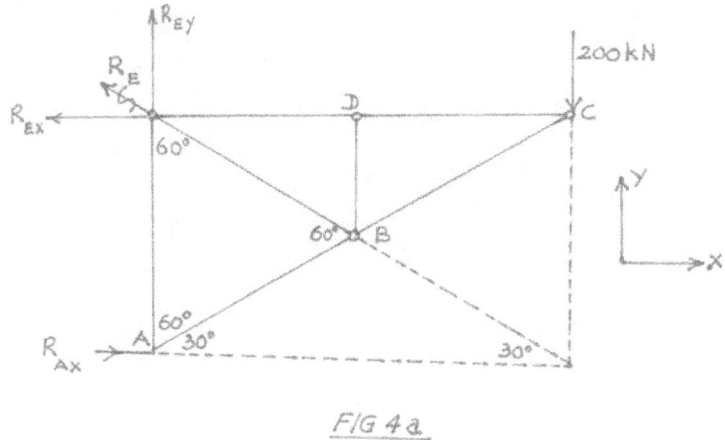

FIG 4a

Thus the triangle of forces shown at Fig. 4b represents, to an appropriate scale, the forces 200kN, RAX and RE in a state of equilibrium. Note the direction of the arrows of the force vectors in Fig. 4b. Magnitude-wise REX = RECos30° = RAX and REY= RESin30°. From Fig 4b according to our force scale which need not concern us at this juncture RAX=346kN; RE=400kN. The next step is to label a drawing done to scale of the frame using Bow's notation as in Fig. 5.

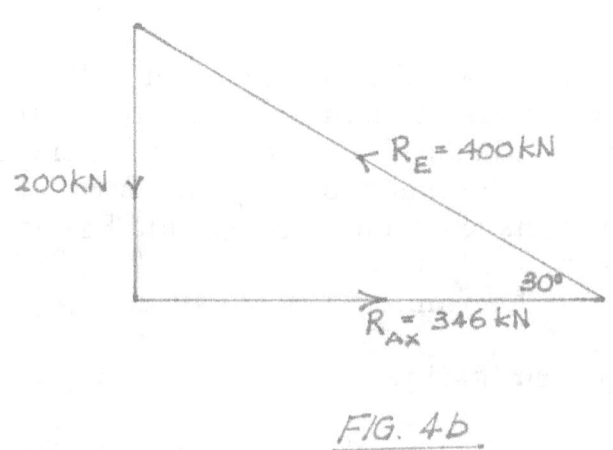

FIG. 4b

~ 44 ~

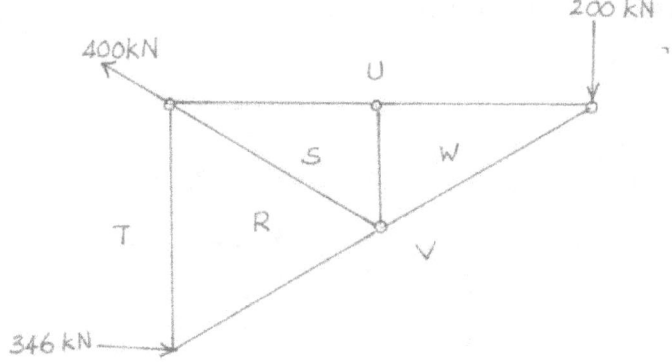

FIG. 5 : *STRUCTURE LABELLED*
USING BOW'S NOTATION

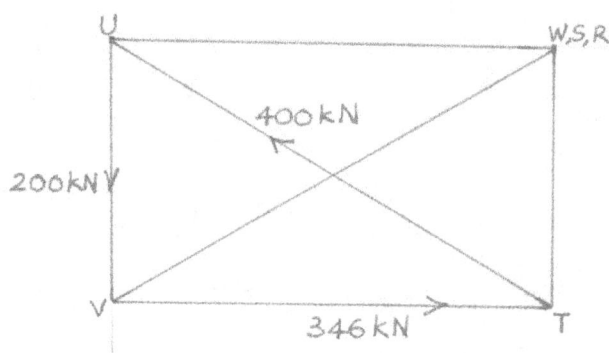

FIG. 6 a *FORCE DIAGRAM*

Using Fig. 4b as its foundation, the force diagram is constructed by drawing constituent forces parallel to corresponding structural members as shown. Evidently, for geometrical consistency to be preserved, w,s,r must be coincident points which means in effect that there are no forces in members WS, i.e. BD; and SR, i.e. BE. The magnitude of each force is therefore determined by measurement to scale. To find whether a force is a strut or a tie, consider equilibrium at each joint. For example at joint 'A' where the forces 346kN, 'tr' and 'rv' meet. The vector order of these forces in Fig. 6a is : 'vt' to 'tr', to 'rv'. The arrow in 'vt' is fixed and inviolable. Therefore the arrows in 'tr' and 'rv' must proceed in an anti-clockwise manner.

~ 45 ~

See diagrams labelled Fig. 6b and Fig. 6c.

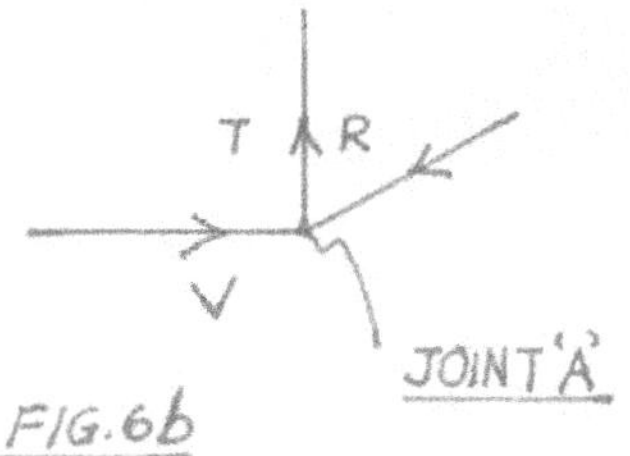

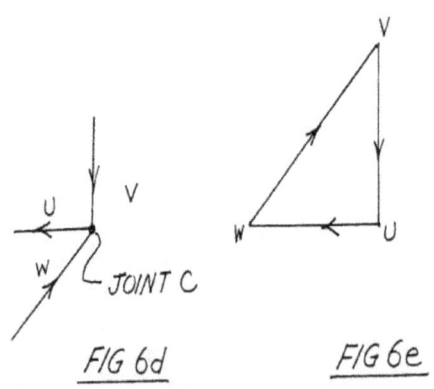

FIG.6b   JOINT 'A'

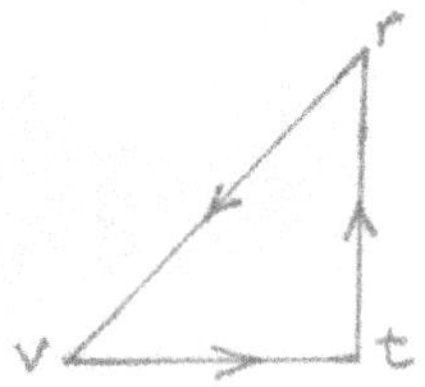

FIG. 6C
JOINT 'C'

Accordingly member TR, i.e. AE is a tie, the arrow in Fig. 6b pointing away from joint 'A' and member RV, i.e. AB is a strut, the arrow in Fig. 6b pointing towards joint 'A'. Similarly by considering joint 'C' where forces 200kN, 'vw' and 'wu' meet, the relevant portion of the force diagram is triangle 'uvw'. The direction of uv is fixed. Therefore for equilibrium, the arrows must proceed around this triangle in a clockwise manner. Refer to Fig. 6d and 6e.

JOINT C

FIG 6d   FIG 6e

Hence WV, i.e. BC is a strut; and WU, i.e. CD is a tie

~ 46 ~

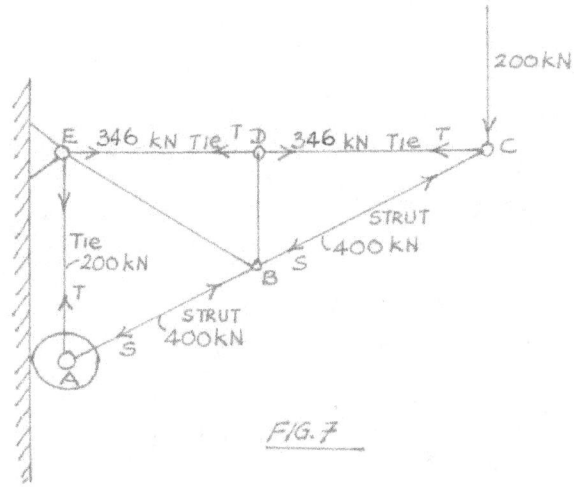

FIG. 7

The magnitude and nature of the forces in the framework are entered in the diagram labelled Fig. 7. S = strut; T = tie. In summary therefore the 4 steps in a graphical determination of forces in a pin-jointed plane framework are:

✓ obtain reactions by means of link or funicular polygon or triangle of forces;
✓ label scale drawing of framework using Bow's notation;
✓ construct force diagram by considering each joint of frame; and
✓ obtain magnitude of each force by measurement to scale and determine nature of forces i.e. whether strut or tie, by examining equilibrium at each joint, using the force diagram for this purpose.

(ii) Method of Joints

In a sense the method of joints may be said to be an analytical derivative of the method of graphical determination, because each relies on a consideration of each joint of a framework in equilibrium considerations.

Because the directions of the forces are known it is clear that in order to complete a force polygon there must be one known force e.g. the downward vertical 200kN load at C of the truss at Fig. 2 and a maximum of two unknown forces e.g. R_{AX} and R_E of the same truss. All the analytical tools required are the following equations of statics:

$$\sum F_y = 0 \quad ; \sum F_x = 0; \ \sum M = 0$$

or equations giving a summation of forces on any two axes perpendicular to one another. Let us use these tools to determine the forces in the truss at Fig. 2. As with the graphical determination, the first order of business is finding the reactions.

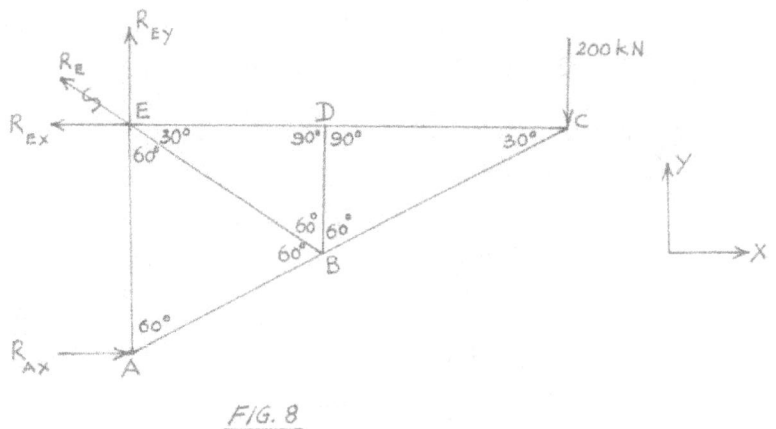

FIG. 8

The truss is again shown at Fig. 8, but R_E is resolved into its 2 components viz R_{EX} and R_{EY}.

Considering equilibrium of external forces:

For $\sum F_y = 0$

$$R_{EY} - 200 = 0$$
$$\therefore \quad R_{EY} = 200kN$$

For $\sum F_x = 0$

$$R_{AX} - R_{EX} = 0 \qquad \therefore \ R_{AX} = R_{EX}$$

For $\sum M_E = 0$

$$R_{AX}(AE) - 200(CE) = 0$$
$$\frac{AE}{6} = Tan30^o = \frac{1}{\sqrt{3}}$$

$$\therefore \qquad AE = 2\sqrt{3}$$

so that

$$R_{AX}(2\sqrt{3}) = 200(6)$$

$$\therefore \quad R_{AX} = 200\sqrt{3}$$

$$R_{AX} = 346kN$$

It follows therefore that R_{EX} = 346.4kN. The values of these external forces are entered in Fig. 9a.

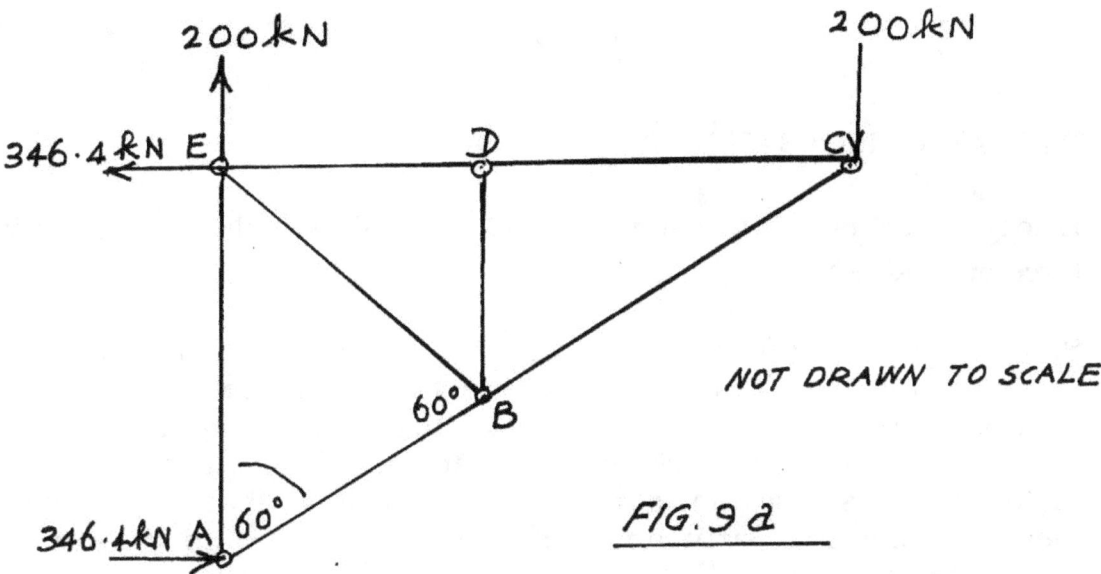

200kN

200kN

346.4 kN E

NOT DRAWN TO SCALE

60° B

346.4kN A 60°

FIG. 9a

Let us now proceed to determine the internal forces of the truss.

Consider joint 'C' first. Fig. 9b refers.

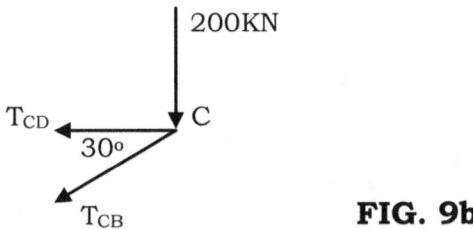

200KN

T_{CD}

30°

C

T_{CB}

FIG. 9b

Assume forces in CD and BC as ties and enter arrows to show this as indicated in the diagram. Apply the tools from your kitbag now.

For $\sum F_y = 0$

$$200 + T_{CB} \operatorname{Sin}30 = 0$$

$$\therefore \quad T_{CB} = -400kN$$

For $\sum F_x = 0$

$$T_{CD} + T_{CB} \cos 30 = 0$$

i.e. $T_{CD} = -T_{CB} \cdot \dfrac{\sqrt{3}}{2}$

$$= +400 \cdot \dfrac{\sqrt{3}}{2}$$

$T_{CD} = 200\sqrt{3} \text{ kN} \equiv 346\text{kN}$

Evidently, CB is a strut carrying a force of 400kN and CD a tie carrying a force of $200\sqrt{3}$ kN.

Some students confuse themselves by commencing to change arrows in the members at this stage of the analysis. It is better to assume all members are ties i.e. all arrows pointing away from the joint under consideration and to work with the arithmetical values as calculated including the signs; and, simply to bear in mind that a negative value means a force is a strut and a positive value a tie (in the convention being followed here).

Next consider joint 'D'. Fig. 9c refers.

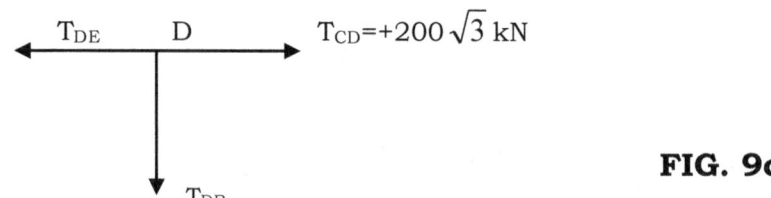

FIG. 9c

For $\sum F_x = 0$, it can straightway be seen that

$$T_{CD} - T_{DE} = 0$$

or $T_{DE} = T_{CD}$

$\therefore \quad T_{DE} = +200\sqrt{3} \text{ kN} \equiv 346\text{kN}$ (as in Fig. 7)

For $\sum F_y = 0;$ $\quad T_{DB} = 0$

Next consider joint 'A', as in Fig. 9d.

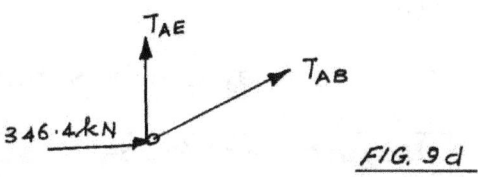

FIG. 9 d

For $\sum F_x = 0$,

$$346.4 + T_{AB}Sin60° = 0$$

i.e. $\quad T_{AB} = -346.4\dfrac{(2)}{\sqrt{3}}$

$$T_{AB} = -400kN$$

For $\sum F_y = 0$

$$T_{AE} + T_{AB}Cos60° = 0$$

$$T_{AE} = -T_{AB}Cos60°$$

$$= +400\left(\dfrac{1}{2}\right)$$

$$T_{AE} = 200kN$$

and finally joint 'B' as in Fig. 9e

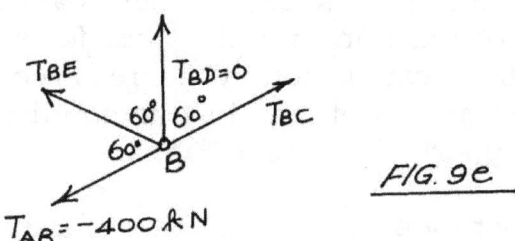

FIG. 9e

for which in the case of $\sum F_x = 0$

$$T_{AB}Cos30° - T_{BC}Cos30° + T_{BE}Cos30° = 0$$

i.e. $\quad T_{AB} - T_{BC} + T_{BE} = 0$

or $-400 - (-400) + T_{BE} = 0$

The results of the calculations may now be summarized thus:

Member	Magnitude of Force (kN)	Strut(S) or Tie (T)
AB	-400	S
BC	-400	S
CD	$+200\sqrt{3} = 346.4$	T
DE	$+200\sqrt{3} = 346.4$	T
AE	+200	T
BD	0	-
BE	0	-

To summarise, the principal steps in solving a pin-jointed plane framework are:

(a)   determine reactions analytically by considering the entire frame to be in equilibrium and by applying the conditions : $\sum F_z = 0$; $\sum F_y = 0$ and $\sum M = 0$.

Remember that rollers and 'shoes' cannot support any lateral component of reaction and also that for smooth-pin supports the reactions there can generally be resolved into 2 components at right angles to each other. In this condition review the examples shown of support reactions in Fig. 3.

(b)   start consideration of equilibrium of forces at each joint, starting with joints which have no more than 2 unknowns. Remember that in considering such equilibrium, all the forces including any external load, whether a reaction or imposed, must enter into the analysis; prepare free-body diagrams for each joint; and

(c)   apply $\sum F_y = 0$; $\sum F_x = 0$.

(iii)   Method of Sections

This method also relies on the laws of statical equilibrium : $\sum F_x = 0$; $\sum F_y = 0$ and $\sum M = 0$. The truss which was labelled Fig. 2, being in equilibrium, implies that any applied or imposed loading and the external reactions are balanced by the internal forces induced in its members. If one were to cut through the truss with an imaginary plane S_1S_1 as shown at Fig. 10a, then the members "exposed" as it were are : CD, BD, BE and AB. Now re-establish equilibrium by bringing each section separately to a state of equilibrium, i.e. the internal forces exposed on either part of the 2 sections resulting from the 'cut' remain in equilibrium with all the external reactions and any imposed loadings acting on either part.

Thus, for the section to the right of S_1S_1 internal forces T_{CD}, T_{BD}, T_{BE}, T_{AB} and the external load of 200kN at C comprise a system of forces in equilibrium to which the laws of statics may be applied. Similarly, for the section to the left of S_1S_1 the internal forces exposed, viz. T_{BD}, T_BE and T_{AB} must be in equilibrium with the external reactions: R_{EX} = 346kN and R_{EY} = 200kN at E; and R_{AX} = 346.4kN at A.

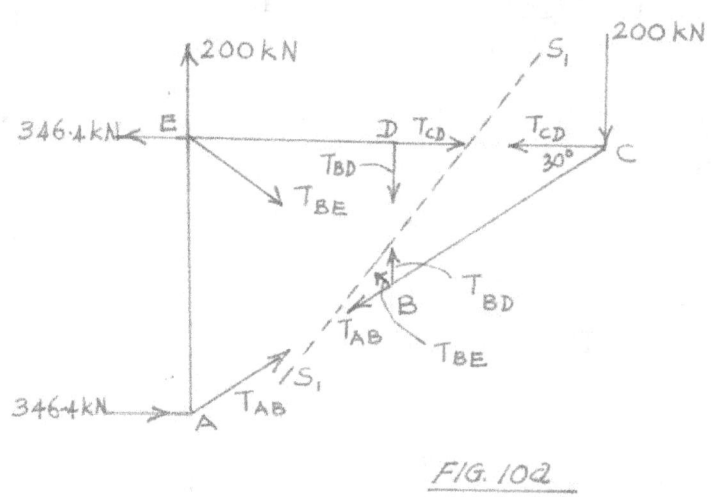

FIG. 10a

~ 53 ~

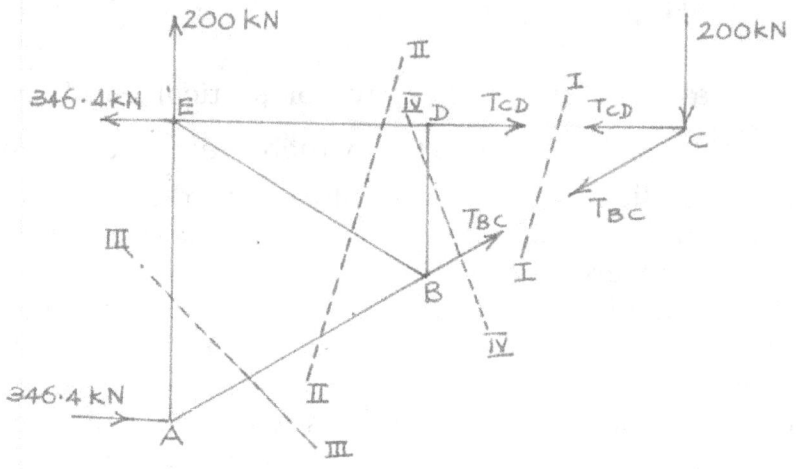

FIG. 10 b

That in a nutshell is the *raison d'etre* of the method of sections; and it could be said to include the method of joints as a sub-set. In applying this method to the solution of our truss let us start by taking a section (I)-(I) as shown at Fig. 10b.

The section to the right of the imaginary 'cut' is in equilibrium; as also the section to the left of it.

Considering equilibrium of the section to the right and applying: $\sum F_y = 0$; and $\sum F_x = 0$, we have:

For $\sum F_y = 0$

$$200 + T_{BC}\text{Sin}30° = 0$$
$$\therefore \quad T_{BC} = -400\text{kN}$$

For $\sum F_x = 0$

$$T_{CD} + T_{BC}\text{Cos}30° = 0$$

$$\therefore \quad T_{CD} = -T_{BC}\frac{\sqrt{3}}{2}$$

$$= -(-400)\frac{\sqrt{3}}{2}$$

i.e. $T_{CD} = +200\sqrt{3}\text{ kN}$

Next, cut the frame with (II)-(II). The internal forces exposed are : T_{DE}, T_{BE} and T_{AB}. Consider equilibrium of the section of the frame to the right of (II)-(II). See Fig. 10c.

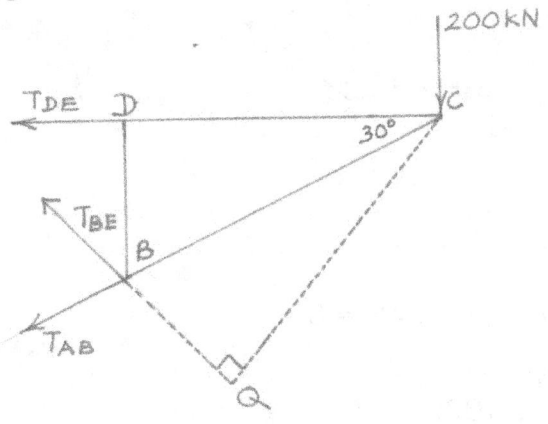

FIG. 10c

For $\quad \sum M_B = 0$

$\qquad T_{DE}(DB) - 200(CD) = 0$

Referring to, say, Fig. 7: $\dfrac{DB}{AE} = \dfrac{CD}{CE} = \dfrac{3}{6} = \dfrac{1}{2}$

and because $\qquad AE = 2\sqrt{3}$

$\qquad\qquad\qquad DB = \sqrt{3}$

$\therefore \qquad T_{DE}(\sqrt{3}) - 200(3) = 0$

$\qquad \therefore \qquad\qquad T_{DE} = 200\dfrac{(3)}{\sqrt{3}}$

$\qquad\qquad\qquad\qquad T_{DE} = +200\sqrt{3}\ kN$

It is also evident that $T_{BE} = 0$, because $\sum M_C$ the moment of T_{BE} about C, i.e. $T_{BE}(CQ) = 0$

Thus $\quad T_{BE} = 0$

For $\quad \sum F_x = 0$

$\qquad T_{DE} + T_{BE}\cos 30^\circ + T_{AB}\cos 30^\circ = 0$

$\qquad \therefore \qquad T_{DE} + T_{AB}\cos 30^\circ = 0$

$\qquad$ or $\qquad 200\sqrt{3} + T_{AB} \cdot \dfrac{\sqrt{3}}{2} = 0$

$\qquad\qquad\qquad T_{AB} - 400kN$

Take section (III)-(III) and consider equilibrium of the frame to the left of it. See Fig. 10d.

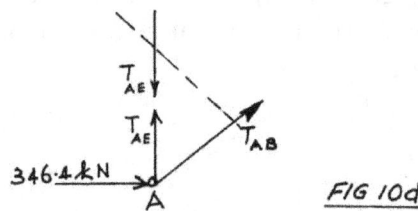

346·4kN

FIG 10d

For $\sum F_y = 0$

$$T_{AE} + T_{AB}\cos 60° = 0$$

i.e. $T_{AE} + (-400) \cdot \dfrac{1}{2} = 0$

∴ $T_{AE} = +200\text{kN}$

Finally, cut the frame with section (IV)-(IV) and consider equilibrium of the part of the frame to the right of it. See Fig. 10e.

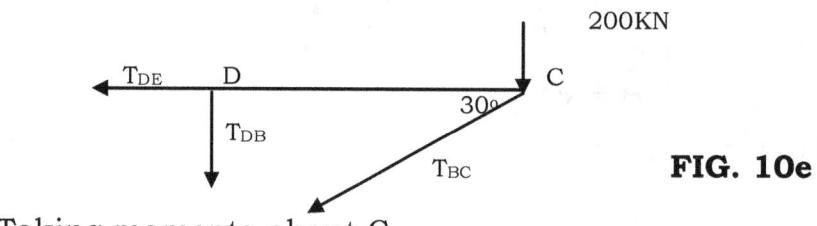

200KN

FIG. 10e

Taking moments about C,

$$T_{DB}(DC) = 0$$

∴ $T_{DB} = 0$

The results obtained in this analysis correspond exactly with those obtained by using the Method of Joints.

In summary therefore, the steps to be followed in conducting an analysis by the Method of Sections are similar to those stated in the case of determination by the Method of Joints, viz

(i) determine reactions acting on the framework; include any external loads;

(ii) inspect diagram of framework and select sections which would give no more than two unknown forces; isolate sections to be analysed,

~ 56 ~

one at a time and make free-body diagrams for each. Remember always that for any one of the two parts of the structure produced by a cut, the internal forces exposed must be in equilibrium with all the applied loads and external reactions acting on that part; and,

(iii)   apply the laws of statical equilibrium as they relate to plane members (i.e. $\sum F_y = 0$; $F_x = 0$; and $\sum M = 0$) to each section.

(iv)   Method of tension coefficients

This is yet another analytical method and like those considered previously is based on the fundamental laws of statical equilibrium. However, the method of tension coefficients has the advantage over the method of joints and the method of sections in that it is more easily applied to space frames than the other two. In fact it would be true to say that it is among the simplest and most accurate methods available for analysis of space frames.

In order to illustrate the method of tension coefficients, consider again equilibrium at joint 'C' of the truss at Fig. 2. The forces are shown in the free-body diagram at Fig. 11a.

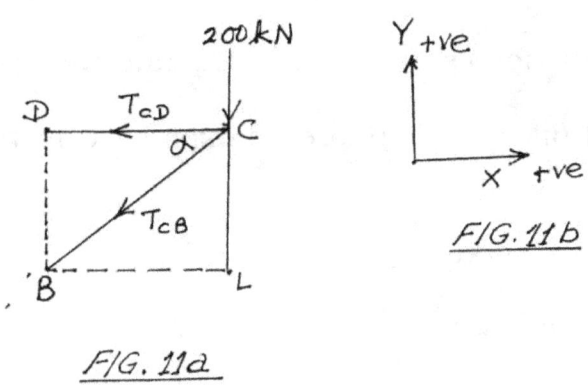

F/G. 11a

F/G. 11b

Analysing, we have:

For $\sum F_x = 0$

-T$_{CD}$ - T$_{CB}$Cosα = 0   . . . . (i)

Note that we have defined the positive directions of the co-ordinate axes as at Fig. 11b and followed this convention in writing equation (i) ; T_{CD} is in the negative direction of the X-axis; also, T_{CB} is in the negative direction when projected on the Y-axis.

For
$$\sum F_y = 0$$
$$-T_{CB}\operatorname{Sin}\alpha - 200 = 0 \qquad\qquad \dots\dots \text{(ii)}$$

[Note the 200kN external load is in the negative direction of the Y-axis] Referring to Fig. 11a and considering trigonometrical relations

$$Cos\,\alpha = \frac{Pr\,ojected \ \ length \ \ of \ \ CB \ \ on \ \ X - axis}{CB}$$

$$Sin\,\alpha = \frac{Pr\,ojected \ \ length \ \ of \ \ CB \ \ on \ \ Y - axis}{CB}$$

Now, the tension coefficient of a member is defined as the tension (positive sign) or compression (i.e. negative tension; negative sign) divided by the length of the member. Accordingly, the tension coefficients of concern to us here, viz t_{CD} and t_{cB} are defined as follows:

$$t_{CD} = \frac{T_{CD}}{CD}; \qquad t_{BC} = = \frac{T_{CB}}{CB}$$

T_{CD} and T_{CB} are the forces in members CD and CB respectively.

Equations (i) and (ii) may therefore be expressed in the form. Taking (i) first

$$-T_{CD} - T_{CB}\frac{\begin{pmatrix} Pr\,ojected \ \ length \ \ of \ \ CB \\ on \ \ X - axis \end{pmatrix}}{CB} = 0$$

i.e.
$$-t_{CD}(CD) - t_{CB}\begin{pmatrix} projected \ \ length \ \ of \ \ CB \\ on \ \ X - axis \end{pmatrix} = 0$$

CD is the projection of CB on the X-axis; CD being on the X-axis. See Fig. 11a

$$\therefore \qquad -t_{CD}\begin{pmatrix} projected \ \ length \\ of \ \ CB \ \ on \ \ X - axis \end{pmatrix} - t_{CB}\begin{pmatrix} projected \ \ length \\ of \ \ CB \ \ on \ \ X - axis \end{pmatrix} = 0$$

[Note CD = projected length of CB on X-axis]

so that $\quad$ -t$_{CD}$(CD) – t$_{CB}$(CD) = 0

or $\hspace{6em}$ t$_{CD}$ = -t$_{CB}$ $\hspace{6em}$ . . . . . (iii)

Equation (ii) may therefore be rewritten as:

$$-T_{CB}\frac{\left(\begin{array}{c}Projected\ length\ of\ CB\\ on\ Y-axis\end{array}\right)}{CB}-200=0$$

Putting tension coefficient $t_{CB}=\dfrac{T_{CB}}{CB}$

$$-t_{CB}\left(\begin{array}{c}Projected\ length\ of\\ CB\ on\ Y-axis\end{array}\right)-200=0$$

i.e. $\hspace{4em}$ $-t_{CB}(DB)-200=0$ $\hspace{5em}$ . . . . . (iv)

To obtain the tension coefficients we have to solve equations (iii) and (iv).

Noting that $DB=\sqrt{3}$

$$\therefore\hspace{3em}-t_{CB}(\sqrt{3})=200$$

$$t_{CB}=-\frac{200}{\sqrt{3}}$$

and $\hspace{4em}$ $t_{CD}=-\left(\dfrac{-200}{\sqrt{3}}\right)=+\dfrac{200}{\sqrt{3}}$

The units of tension coefficient are force per unit length, in this case kN/m. Now we have to multiply the values of t$_{CB}$ and t$_{CD}$ by the length of their respective members in order to obtain the forces in these members. Now, CD = 3m and $CB=2\sqrt{3}$

$$\therefore\hspace{3em}T_{CB}=\frac{-200\times 2\sqrt{3}}{\sqrt{3}}=-400kN$$

and $\quad$ $T_{CD}=\dfrac{200}{\sqrt{3}}\times 3$

$T_{CD}=+200\sqrt{3}$

These results match the values obtained by the other methods.

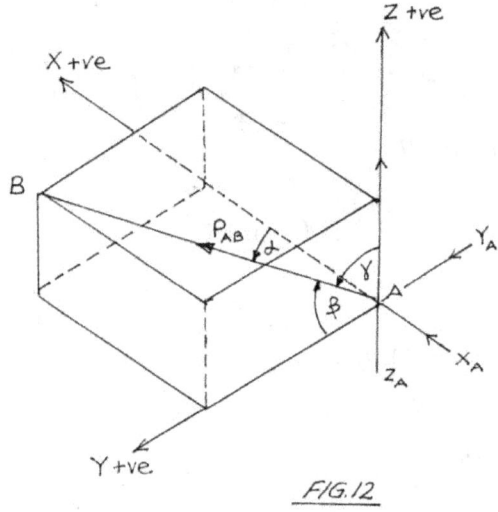

FIG.12

In general, if, say, an internal force P_{AB} of a member of a framework is acting at a joint 'A' as at Fig. 12 where there are also applied loading: X_A acting in the positive direction of the X-axis, Y_A along the positive direction of the Y-axis, and Z_A along the positive direction of the Z-axis as shown in the diagram, then for the equilibrium condition: $\sum F_Y = 0$; $\sum F_X = 0$; $\sum Fz = 0$, we have for:

$$\sum F_X = 0 \qquad P_{AB} Cos\alpha + X_A = 0 \qquad \ldots\ldots \text{(v)}$$
$$\sum F_y = 0 \qquad P_{AB} Cos\beta + Y_A = 0 \qquad \ldots\ldots \text{(vi)}$$
$$\sum F_z = 0 \qquad P_{AB} Cos\gamma + Z_A = 0 \qquad \ldots\ldots \text{(vii)}$$

Now,

$$Cos\alpha = \frac{Projected\ length\ of\ AB\ on\ X-axis}{AB}$$

$$Cos\beta = \frac{Projected\ length\ of\ AB\ on\ Y-axis}{AB}$$

$$Cos\gamma = \frac{Projected\ length\ of\ AB\ on\ Z-axis}{AB}$$

[Note the positive directions of the X-, Y- and Z-axes]

~ 60 ~

Writing

Tension coefficient, $t_{AB} = \dfrac{P_{AB}}{AB}$

Equations (v), (vi) and (vii) may be expressed as follows :

$$t_{AB}\left(\begin{array}{l}Projected\ \ length \\ of\ member\ on\ X-axis\end{array}\right) + X_A = 0$$

$$t_{AB}\left(\begin{array}{l}Projected\ \ length \\ of\ member\ on\ Y-axis\end{array}\right) + Y_A = 0$$

$$t_{AB}\left(\begin{array}{l}Projected\ \ length \\ of\ member\ on\ Z-axis\end{array}\right) + Z_B = 0$$

Evidently if there are say, 'n' members of the framework at joint 'A' then considering equilibrium in relation to the three axes you would have for

X-direction : $\displaystyle\sum_{k-1}^{n} t_k(x_k) + X_A = 0,$ . . . . . . (viii) ;

Y-direction : $\displaystyle\sum_{k-1}^{n} t_k(y_k) + Y_A = 0,$ . . . . . . (ix) ; and

Z-direction : $\displaystyle\sum_{k-1}^{n} t_k(z_k) + Z_A = 0,$ . . . . . . (x)

These equations all relate to the single joint 'A' with 'n' members at the joint. For each joint in the frame a similar triad of equations may be written thus giving 3m equations if there are 'm' joints.

If one were to apply these equations to a statically-determinate plane frame, then there would be 3 more equations than three are unknowns; whereas in the case of a statically-indeterminate space frame there would be six more equations than there are unknowns. Do not be too concerned about these superfluities; they can be used to check reactions and the values of Tension Coefficients in other equations of equilibrium; for an internal check if you will.

In Fig. 11a, we highlighted the forces T_CB and T_CD at joint 'C' of our original plane frame shown at Fig. 2. When joint 'B' is being analysed, notwithstanding the other 'new' members that may constitute the joint, there must be a tension coefficient t_BA for BA. If therefore T_AB was considered as a tie, then its projections on each of the axes X,Y and Z would be in the positive direction of these axes directions of these axes. Conversely when we move to joint 'B' and consider T_{BA} as a tie, all the projections are in the negative directions of the X-, Y-, and Z-axes.

Accordingly it should be noted that in the practical application of the method of tension coefficients in the solution of frameworks, both plane and space, it is assumed at the outset that the forces at all joints are ties; and, further that the terms of the equations arising from expansion of expressions (viii), (ix) and (x) are positive or negative as the relevant forces would move the joint being considered in the positive or negative direction of the axis being considered.

When therefore a tension coefficient is positive this means the member is in fact a tie, and when negative a strut (in our convention).

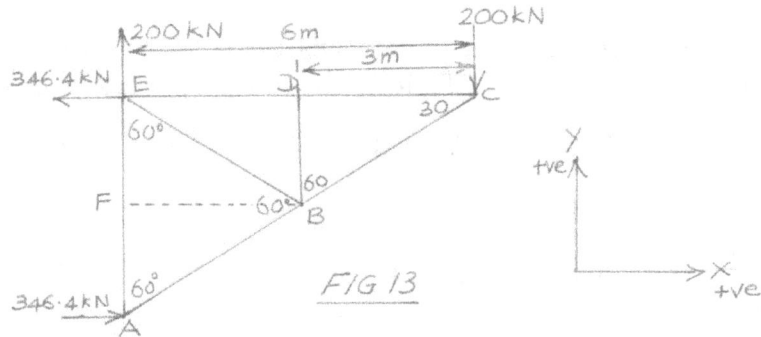

FIG 13

Let us now use the method of tension coefficients to find all the forces in our cantilever truss shown again at Fig. 13. Considering joint C first in the diagram below.

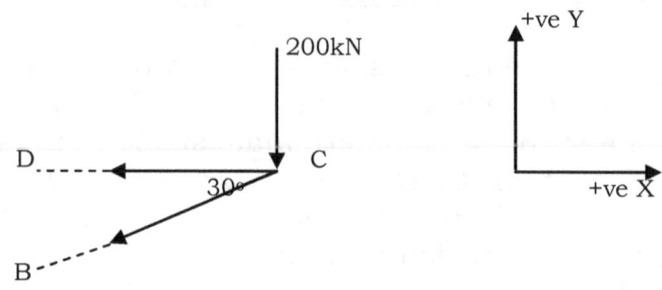

Let t_{CD} = tension coefficient of CD

  t_{CB} "   "   " CB

~ 62 ~

Considering the equilibrium equations for the X- and Y-axes, we may straightway write

For X-direction
$$-t_{CD}(CD) + t_{CB}(-CD) = 0$$
i.e.
$$t_{CD} = -t_{CB}$$
and for Y-direction :
$$-t_{CD}(0) + t_{CB}(-BD) - 200 = 0$$
$$\therefore \quad -t_{CB}(\sqrt{3}) = 200$$
$$t_{CB} = \frac{-200}{\sqrt{3}}$$

and for Y-direction:

so that
$$t_{CD} = \frac{+200}{\sqrt{3}}$$

Next, take joint D and let t_{DB}, t_{DE} be the tension coefficients of DB and DE respectively.

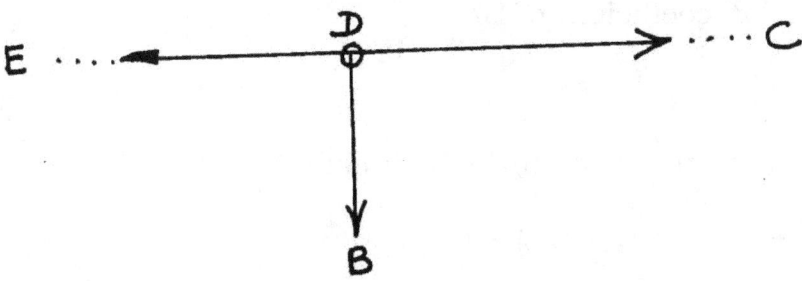

For equilibrium in the X-direction

$$t_{DC}(+DC) + t_{DE}(-DE) = 0$$
$$t_{DC}(3) - t_{DE}(3) = 20$$
i.e. $\quad t_{DC}(3) - t_{DE}(3) = 0$
$$\therefore \quad t_{DC} = t_{DE}$$
But $\quad t_{DC} = t_{CD}$
$$\therefore \quad t_{DE} = \frac{200}{\sqrt{3}}$$

For Y-direction

$$t_{DC}(0) + t_{DE}(0) + T_{DB}(-DB) = 0$$
$$\text{i.e.} \qquad t_{DB} = 0$$

Next, take joint 'B' in diagram immediately below:

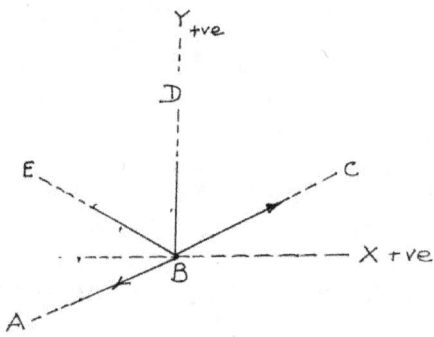

As before, let

t_{BA} = tension coefficient of BA
t_{BE} =   "   "   "   BE

Considering equilibrium in the X-direction

$$T_{BC}(+DC) + t_{BD}(0) + t_{BE}(-DE) + t_{BA}(-DE) = 0$$

i.e. $t_{BC}(3) + 0 - t_{BE}(3) - t_{BA}(3) = 0$

or $t_{BC} - t_{BE} - t_{BA} = 0$

Putting $t_{BC} = t_{CB} = \dfrac{-200}{\sqrt{3}}$

$$-t_{BE} - t_{BA} = -t_{BC} = \dfrac{+200}{\sqrt{3}}$$

$$\therefore \qquad t_{BE} + t_{BA} = \dfrac{-200}{\sqrt{3}}$$

For equilibrium in the Y-direction

$t_{BC}(+BD) + t_{BD}(+BD) + t_{BE}(+BD) + t_{BA}(-AF) = 0$

i.e.

$t_{BC}(\sqrt{3}) + 0(\sqrt{3}) + t_{BE}(\sqrt{3}) - t_{BA}(\sqrt{3}) = 0$

or $t_{BC} + t_{BE} - t_{BA} = 0$

or $t_{BE} - t_{BA} = -t_{BC} = \dfrac{200}{\sqrt{3}}$

It follows therefore that

$2t_{BE} = 0$

i.e. $t_{BE} = 0$

and $t_{BA} = \dfrac{-200}{\sqrt{3}}$

Finally, consider joint A

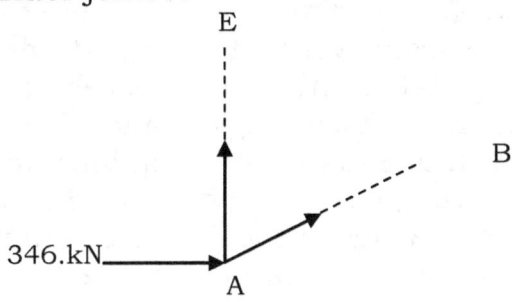

Considering equilibrium in Y-direction and designating t_{AE} as tension coefficient of AE, we have

$t_{AB}(-AF) + t_{AE}(+AE) = 0$

Note that with reference to Fig. 13, the projection of BA (which is assumed a tie) on Y-axis is (-AF), so that

$t_{AB}(-\sqrt{3}) + t_{AE}(2\sqrt{3}) = 0$

$\therefore$ $t_{AE} = \dfrac{-t_{AB}}{2} = \dfrac{+100}{\sqrt{3}}$

Now, tabulate the results.

Member	Tension Coefficient kN/m	Length of Member M	Force in member kN	Strut(S) or Tie (T)
AB	$-200/\sqrt{3}$	$2\sqrt{3}$	-400	S
AE	$+100/\sqrt{3}$	$2\sqrt{3}$	+200	T
BC	$-200/\sqrt{3}$	$2\sqrt{3}$	-400	S
BD	0	$\sqrt{3}$	0	-
BE	0	$\sqrt{3}$	0	-
CD	$200/\sqrt{3}$	3	$200\sqrt{3}$	T
DE	$200/\sqrt{3}$	3	$200\sqrt{3}$	T

The tabulated values correspond exactly with those obtained by the other analytical methods considered previously. Clearly, the method of tension coefficients is a powerful instrument for determining forces in frameworks, even in plane frameworks as was just demonstrated. Many of the steps which were taken here were solely for the purpose of illustrating its application. Once angles and distances are known, expansion of the fundamental equations for equilibrium at each joint can be done on the basis of inspection. Because the method of tension coefficients is based on knowledge of the projected lengths of members of the framework on an orthogonal-axes system : 2 for plane frameworks and 3 for space frameworks, it is important to have drawings which either provide such data or permit of such determination.

In summary, the steps to be followed in performing analysis by the method of tension coefficients are:

(a)   define the co-ordinate-axis system to be followed, labelling the positive direction of each;

(b)   determine angles and lengths of members which would permit finding horizontal and vertical projections of members on the co-ordinate-axes system;

(c)   consider each joint of framework assuming all forces are ties; select joints having no more than two unknown tension coefficients; tension coefficient = t;

(d)   expand equilibrium equations : $\sum t$ (projected length of member on X-axis) + component/s of any external force at joint in X-direction = 0; $\sum t$ (projected length of member on Y-axis) + component/s of any external force at joint in Y-direction = 0; $\sum t$ (projected length of member on Z-axis) + component/s of any external force at joint in Z-direction = 0. Note especially that terms are positive when the projected 'tie' on an axis is in the positive direction of that axis, and negative when in the negative direction of that axis;

(e)   solve the resulting expanded equilibrium equations to obtain values of the tension coefficients;

(f)   multiply tension coefficient of member by its length to obtain value of internal force on member. A positive tension coefficient indicates a tie; a negative tension coefficient, a strut. Note too that a tension coefficient has the units : force per unit length so that if 't' is say kN/metre, then the length of a member must be expressed in metres, m, to obtain force in Kn; and

(g)   tabulate the results.

We consider in Chapter 17, Elastic Deflection in Statically Determinate Frameworks.

SOLVED AND OTHER PROBLEMS

Q1. (a) Identify by means of a well labelled diagram the following terms as applied to a roof truss : (a) rise; (b) span; (c) panel; (d) panel point; (e) upper chord; (f) lower chord; (g) brace; (h) stanchion; (i) web member; (j) ridging; (k) purlin; (l) rafter; (m) upper chord bracing. Explain what is meant by "the pitch of a roof".

(b) By means of sketches show three types of frames employed in the design of roof trusses. Provide a brief explanation for each chosen.

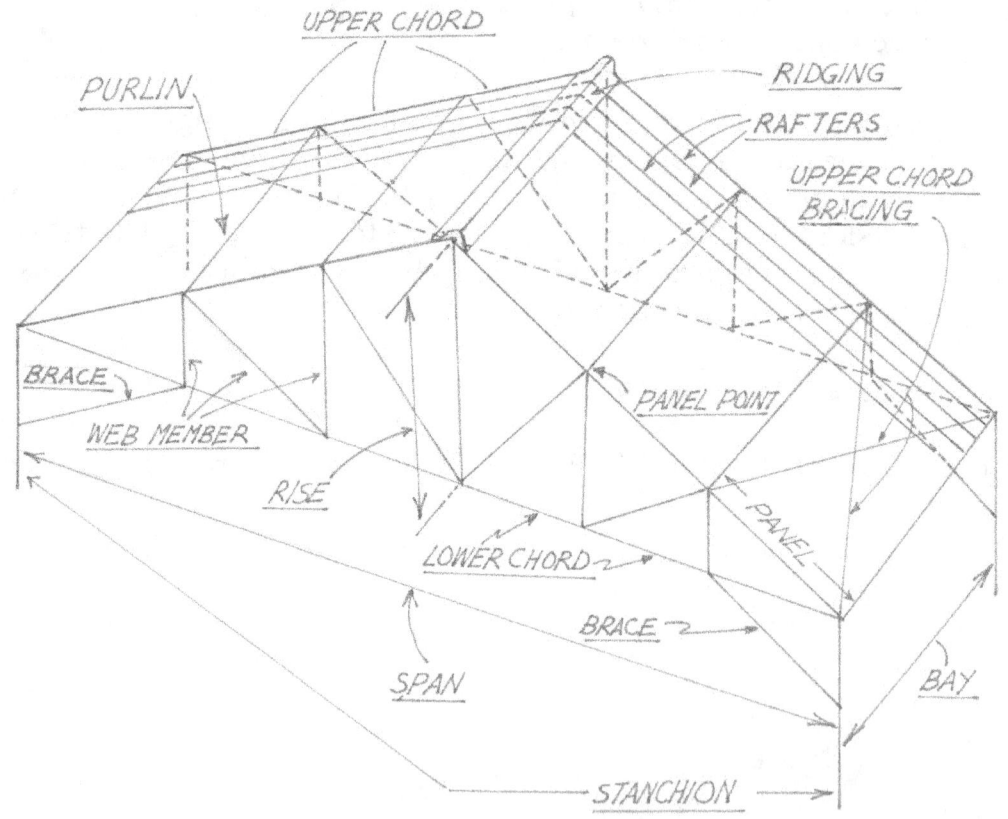

FIG 1

THE PITCH OF A ROOF MAY BE EXPRESSED IN SEVERAL WAYS. HERE IT IS EXPRESSED AS : RISE/SPAN

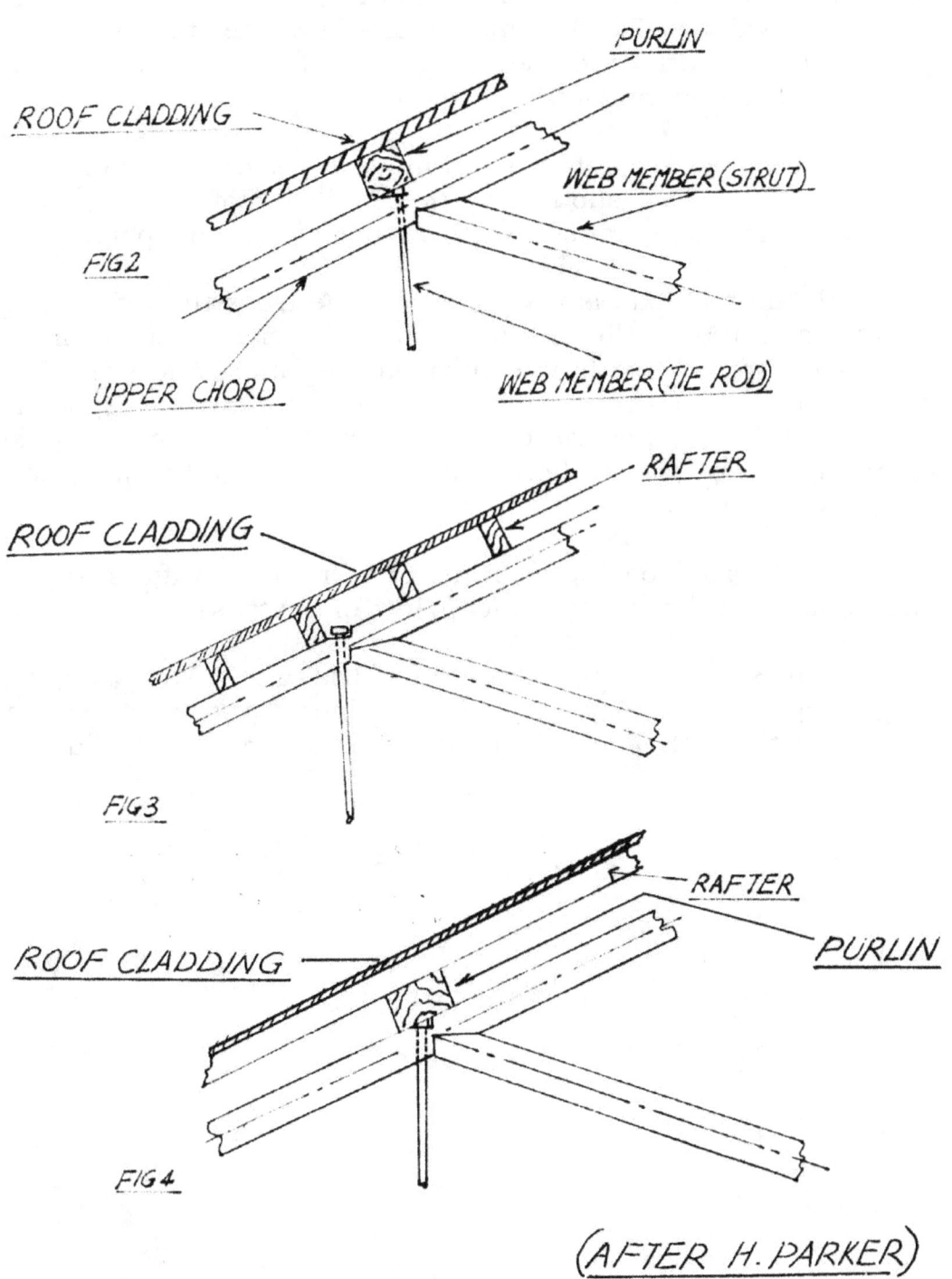

PURLIN

ROOF CLADDING

WEB MEMBER (STRUT)

FIG 2

UPPER CHORD

WEB MEMBER (TIE ROD)

RAFTER

ROOF CLADDING

FIG 3

RAFTER

ROOF CLADDING

PURLIN

FIG 4

(AFTER H. PARKER)

Answer:

(a)   A drawing of two trusses accounting for one bay of a roof structure is provided in Fig. 1. This should allow the student to become *'au courant'* with the terms commonly used in the building-construction industry.

(b)   Three types of roof truss framing are shown in Figs 2, 3 and 4. First of all it should be noted that roof loading is generally assumed to be transferred to a truss at its panel points.

In Fig. 2, the roof load is transferred to the truss through the agency of the purlins : from purlin to purlin and to the panel points. In Fig 3, the roof load is transferred to the truss via the rafters and from them to the upper chords, via the purlins. In such circumstances, loads act at points which do not necessarily coincide with panel points and accordingly the upper chords are subjected to bending, shear and compressive stresses.

In Fig. 4, the roof loading is transferred from the rafters to the purlins and from the purlins to the panel points of the trusses.

Why are trusses made up of system of triangles? Well, simply because the triangle is the only geometrical figure whose shape cannot be changed without changing the length of one or more of its sides.

Q2. A cantilever truss is part of the framework of a bridge, the supports being a hinge at E and rollers at A. The load at B is 200kN. See Fig. 1. Determine graphically the magnitude and nature of the forces of the members of the framework.

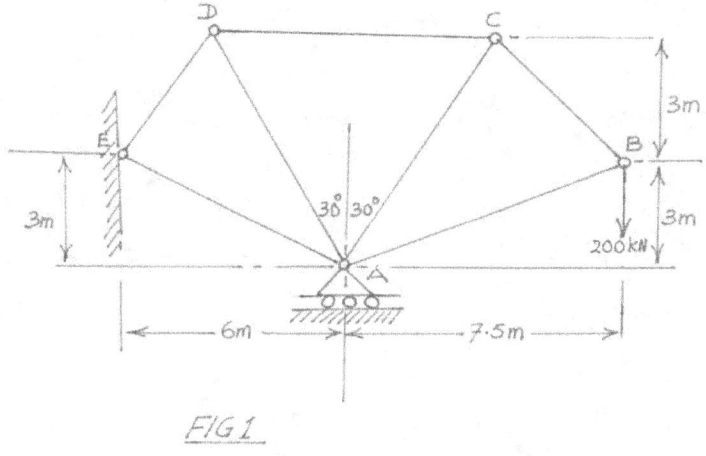

FIG 1

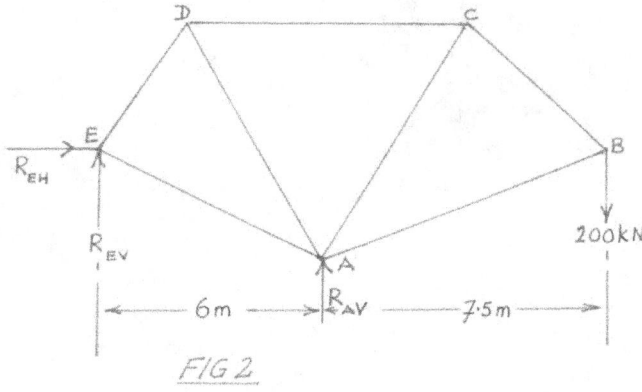

FIG 2

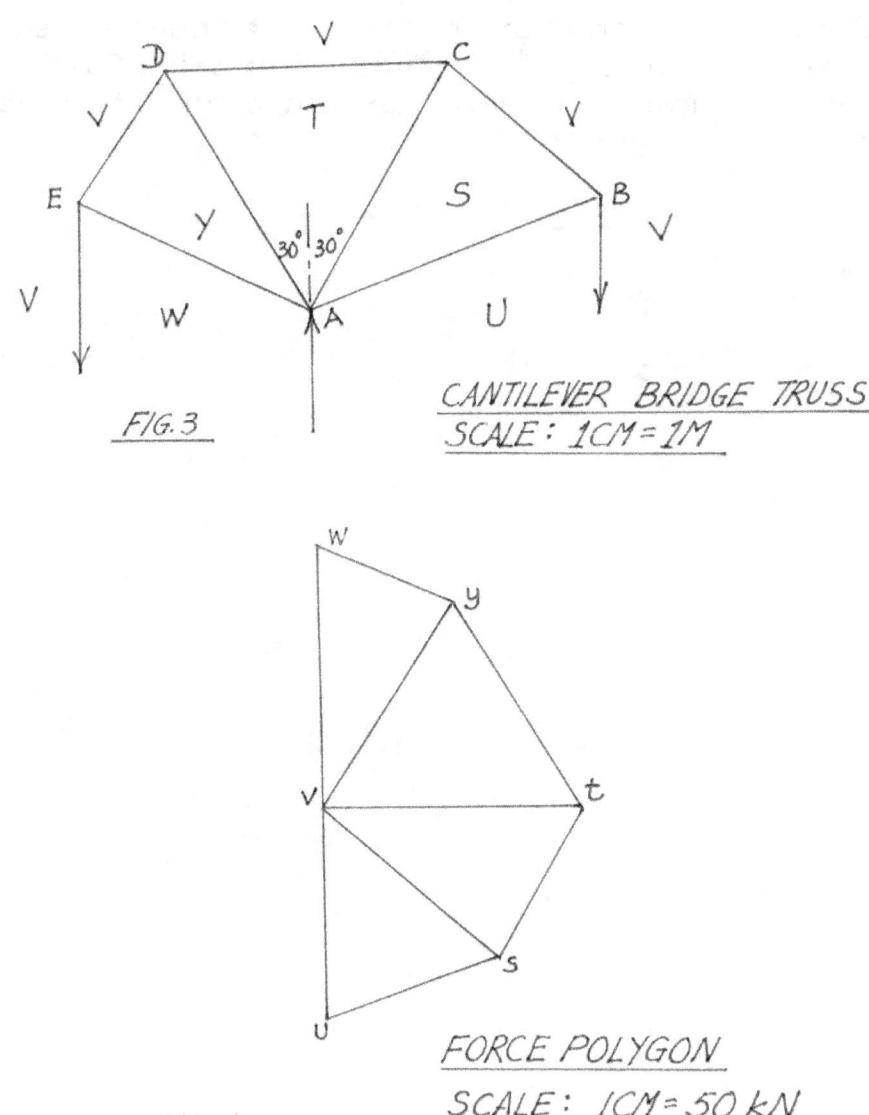

FIG. 3

CANTILEVER BRIDGE TRUSS
SCALE : 1CM = 1M

FORCE POLYGON
SCALE : 1CM = 50 kN

FIG 4

The first requirement is determination of the reaction R_{EV} at E. Because of the rollers at 'A', the reaction there can have only one component, viz. R_{AV} acting vertically as shown in Fig. 2. If one assumes two components for the reaction at E, say, R_{EH} and R_{EV}, then clearly since there are no horizontal external forces on the frame, other than the one we assumed, R_{EH} must be zero.

Secondly, we take moments about 'A'. Therefore

$$R_{EV}(6) + 200(7.5) = 0$$

$$\therefore \quad R_{EV} = -250\text{kN, i.e. } R_{EV} \text{ acting downwards;}$$

evidently, $R_{AV} = 450$ kN, upwards

~ 72 ~

The next step is preparation of the force polygon, for which purpose, a line diagram of the truss is labeled appropriately in Bow's notation as in Fig. 3. After that, the force vectors $\overrightarrow{wv}$ and $\overrightarrow{vu}$ are set out vertically in accordance with their directions on the frame as in Fig. 3. The force polygon is completed in the usual manner, that is to say 'wy' on the polygon is drawn parallel to members WY on the Bow-notated diagram; 'vy' to VY; and where 'wy' and 'vy' intersect locates point 'y' on the polygon, and so on. The magnitude of the forces are obtained by measuring the length of the component members comprising the polygon and converting to forces on the basis of the force-scale employed.

In order to determine which component members of the truss are ties or struts, we consider each joint in turn on the space diagram with reference to the forces meeting there in the force polygon. When therefore we consider joint 'E', the vector arrow in bow-notated member 'VY' represented by 'vy in the force polygon, points from 'v' to 'y' . When we apply this direction to member 'ED' at joint 'e' the arrow points from 'E' to 'D', i.e. away from 'E'. Hence member 'ED' is a tie. Similarly, when the direction 'y' to 'w' in the force polygon is applied at joint 'E', because the arrow points towards the joint e, AE is a strut. Similarly when we consider joint 'B', the force polygon 'VUS' (a triangle really) is in equilibrium, with directions 'V' to 'U' to 'S' to 'V'. Applying these directions to members AB is a strut, the arrow being directed at 'B'; and 'BC' is a tie, the arrow moving away from B. For members AD and AC, consider 'UW' in the force polygon – direction upwards. Continuing in this direction around the polygon, applying the directions 'y' to 't' and 't' to 's' to members AD and AC, at joint 'A', we deduce that both these members are struts.

The results are stated in the following table:

Member	Magnitude kN	Strut (S) or Tie (T)
AE/wy	-135	S
AB/us	-170	S
BC/vs	210	T
AC/st	-160	S
CD/tv	230	T
AD/ty	-225	S
DE/vy	225	T

Q3. Fig. 1 shows a small space support carrying a load of 5kN as shown. Determine the tension in the rods PQ, QR and the pin reaction at the boom support S.

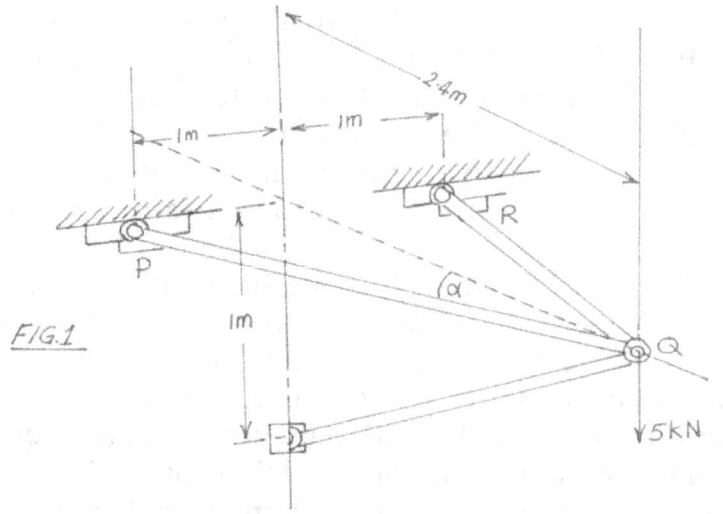

FIG.1

Fig 2 is the free-body diagram of the loaded space support.

Distance PQ is obtained from :

$$PQ^2 = 1^2 + (2.4)^2 = 6.76$$

i.e. $PQ = 2.6m$

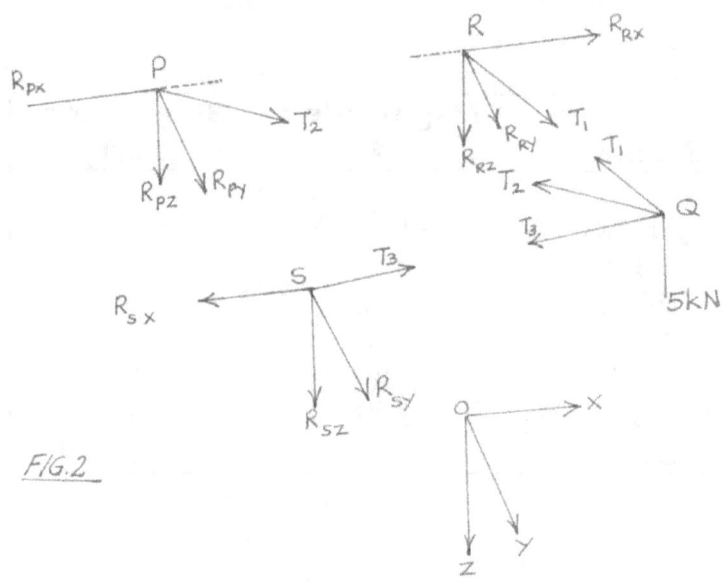

FIG.2

Forces acting at Q : See Fig. 3

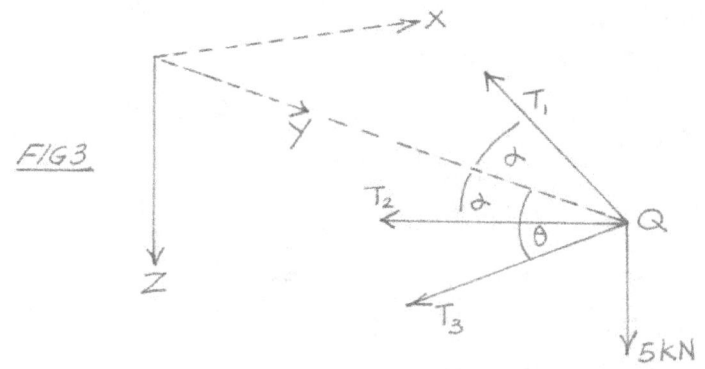

Let T_1, T_2 be the tensions in QR and PQ respectively
Let T_3 = tension in boom QS

Resolving forces in the Z-direction at Q

$T_3 \, Sin\,\theta = 5$

Note that T_1 and T_2 have no component in the Z-direction.

$$Sin\,\theta = \frac{1}{2.6}$$

$\therefore$ T_3 = 5(2.6)

i.e T_3 = 13kN

Still at Q, but resolving along the Y-axis

$T_1 Cos\alpha + T_2 Cos\alpha + T_3 Cos\alpha = 0$

From symmetry $T_1 = T_2$

$\therefore$ $2T_1 Cos\alpha + 13 Cos\theta = 0$

$Cos\alpha = \dfrac{2.4}{2.6}$; $Cos\theta = \dfrac{2.4}{2.6}$

so that

$2T_1 + 13 = 0$

from which $T_1 = T_2 = -6.5kN$

Forces acting at S

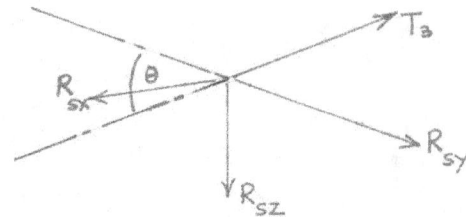

FIG 4

The components of the reaction at S, R_s are as indicated in Fig. 4; R_{sx} in the direction of the X-axis; R_{sy} in the direction of the Y-axis; and R_{sz} in the direction of the Z-axis. These are in equilibrium with the internal resisting force in the boom, T_3.

Considering equilibrium in the Y-direction

$$R_{sy} + T_3 Cos\theta = 0$$

$$\therefore \quad R_{sy} = -13 \cdot Cos\,\theta$$

$$= -13\left(\frac{2.4}{2.6}\right)$$

$$\therefore \quad R_{sy} = -12kN$$

Considering equilibrium in the Z-direction

$$R_{sz} - T_3 Sin\theta = 0$$

$$\therefore \quad R_{sz} = 13 Sin\theta$$

$$\therefore \quad R_{sz} = 13 \cdot \frac{1}{2.6} = 5kN$$

Evidently, $R_{sx} = 0$

$$\therefore \quad R_s^2 = \sqrt{(R_{sy})^2 + (R_{sz})^2 + R_{sx})^2} = \sqrt{169}$$

i.e. $R_s = 13kN$

Q4. The plan view of a space framework made out of aluminium tubing is as shown in Fig. 1. The ridge bar EF is completely horizontal and is 4 metres above the ground to which surface, supports at A, B, C and D are pinned. Loads of 2.5kN and 4kN acting in a horizontal plane are applied as indicated at E and F which are also pin-jointed. Use the method of tension coefficients or any other method to determine the magnitude and nature of the forces in the framework.

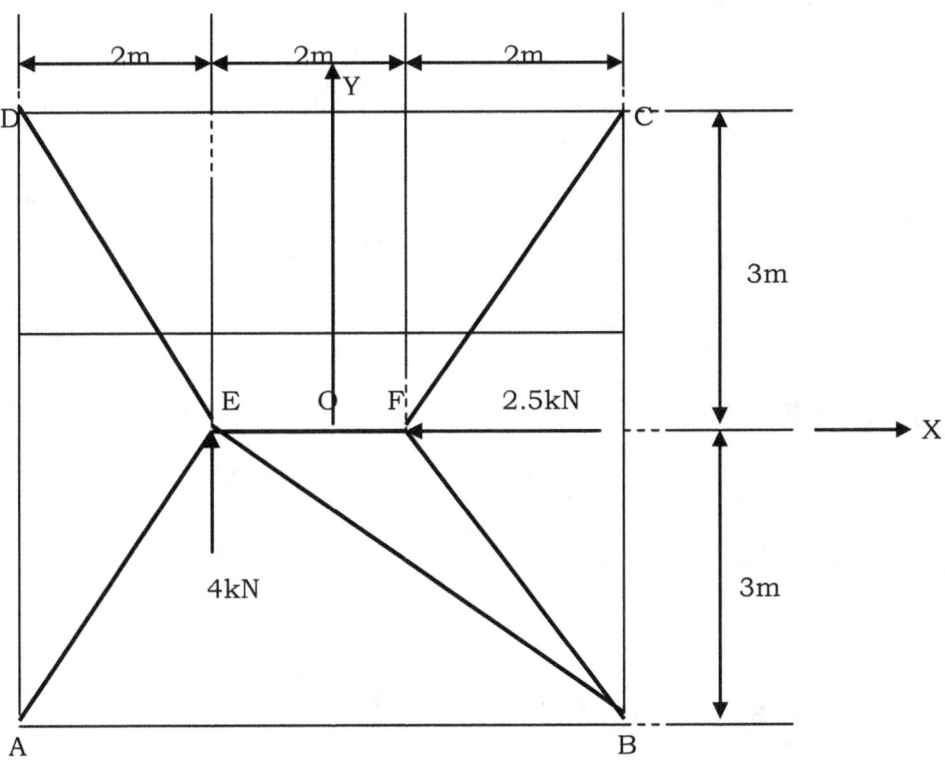

FIG. 1

The positive directions of the X- and Y-axes are as indicated at Fig. 1. The Z-axis is perpendicular to the plane of this sheet, its positive direction being upwards.

The plan length of ED = $\sqrt{(3)^2 + (2)^2} = \sqrt{13}$ m and its true length =

$\left\{ \left(\sqrt{13}\right)^2 + (4)^2 \right\}^{1/2} = \sqrt{29}$. Similarly the plan length of $EB = \sqrt{(4^2) + (3)^2} = 5$m

and its true length = $\sqrt{(5)^2 + (4)^2} = \sqrt{41}m.$ The next step is to consider each joint and to write down the equilibrium equations for the X-, Y-, and Z-axis.

First, take joint E

Remember to assume all members are ties: This is done by having all arrows on the force vectors excluding those of any external loads point away from the joint as shown in the accompanying diagram.

Expanding the equilibrium equations we have, <u>denoting tension coefficients of members by the letters defining them,</u>

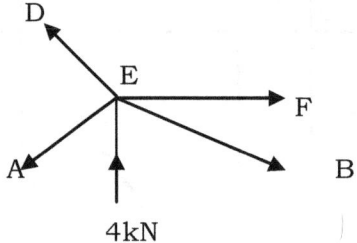

Considering joint E

In X-direction

$$-ED(2) - EA(2) + EF(2) + EB(4) = 0 \qquad \ldots \ldots \text{(i)}$$

In Y-direction

$$+ED(3) - EA(3) - EB(3) + 4 = 0 \qquad \ldots \ldots \text{(ii)}$$

In Z-direction

$$-ED(4) - EA(4) - EB(4) = 0 \qquad \ldots \ldots \text{(iii)}$$

(Note: EF has no projection on Z-axis)

Considering joint F next and following the same procedure we have

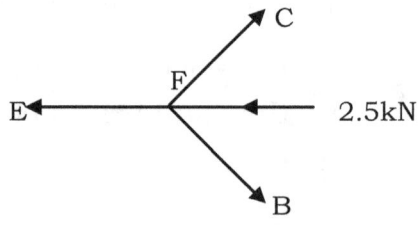

In X-direction

$$-FE(2) + FC(2) + FB(2) - 2.5 = 0 \qquad \qquad \dots \dots \text{(iv)}$$

In Y-direction

$$+FC(3) - FB(3) = 0 \qquad \qquad \dots \dots \text{(v)}$$

In Z-direction

$$-FC(4) - FB(4) = 0 \qquad \qquad \dots \dots \text{(vi)}$$

From (vi)   FC = -FB

Which result when substituted in (v) gives

$$FB = 0 \quad ; \quad FC = 0$$

so that (iv) becomes

$$-2FE = 2.5$$

or $FE = -1.25$

When this value is substituted in (i) we have

$$-2ED - 2EA + 4EB = 2.50 \qquad \qquad \dots \dots \text{(vii)}$$

the other two relevant equations at this juncture being (i) and (iii).

Multiplying (vii) by 2 and subtracting equation (iii) from the result

$$-4ED - 4EA + 8EB = 5$$

$$-4ED - 4EA - 4EB = 0$$

from which

$$12EB = 5$$

or $EB = 5/12$

Putting this value of EB in (vii) and multiplying the resulting equation by 3 gives

$$-6ED - 6EA = 2.50 \qquad \qquad \dots \dots \text{(viii)}$$

Also, putting EB = $\frac{5}{12}$ in (ii) and multiplying by 2, we have

6ED – 6EA = -5.5   . . . . . (ix)

From (viii) and (ix)

   -12EA = -3

or   EA = +0.25

Then from $6ED - 6\left(\frac{1}{4}\right) = -5.5$

   ED $= \frac{-2}{3}$

Collecting results we have :

Tension coefficient of   FB = 0

"   "   "   FC = 0

"   "   "   FE = -1.25

"   "   "   EB = 5/12

"   "   "   EA = 0.25

"   "   "   ED = -2/3

The lengths of FE, EB, EA and ED are respectively : 2m; $\sqrt{41}$ m; EA = ED $= \sqrt{29}$. Accordingly the forces in members are as follows :

t_{FE} = -1.25 x 2 = -2.50kN,   strut

t_{EB} = $\frac{5}{12}$ x $\sqrt{41}$ = 2.67kN,   tie

t_{EA} = 0.25 x $\sqrt{23}$ = 1.35kN,   tie

t_{ED} = -2/3 x $\sqrt{29}$ = -3.59kN,   strut

Q5. Determine entirely by graphical means the force in each member and the reactions at F and A of the pin-jointed framework shown in Fig. 1.

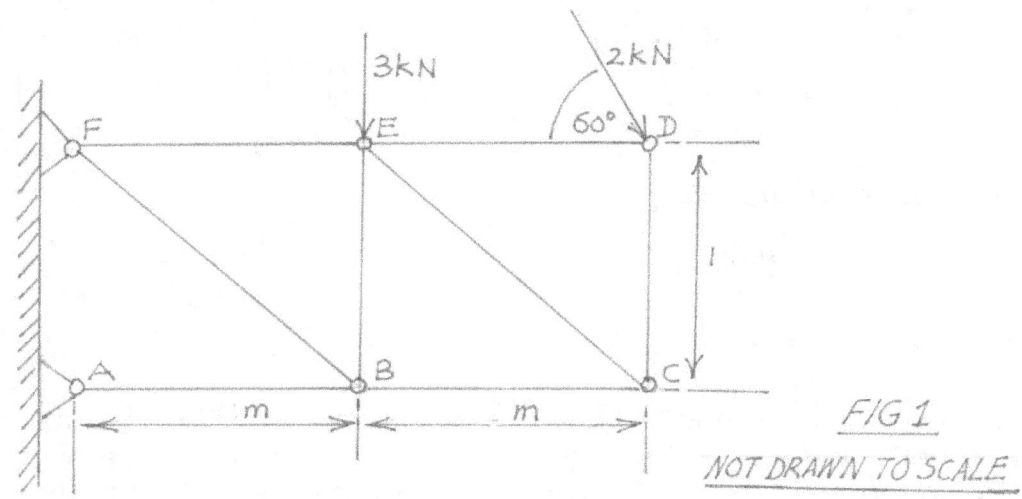

First, we redraw Fig. 1 showing the reactions at A and F; and the other external forces 3kN and 2kN, as indicated in Fig. 2.

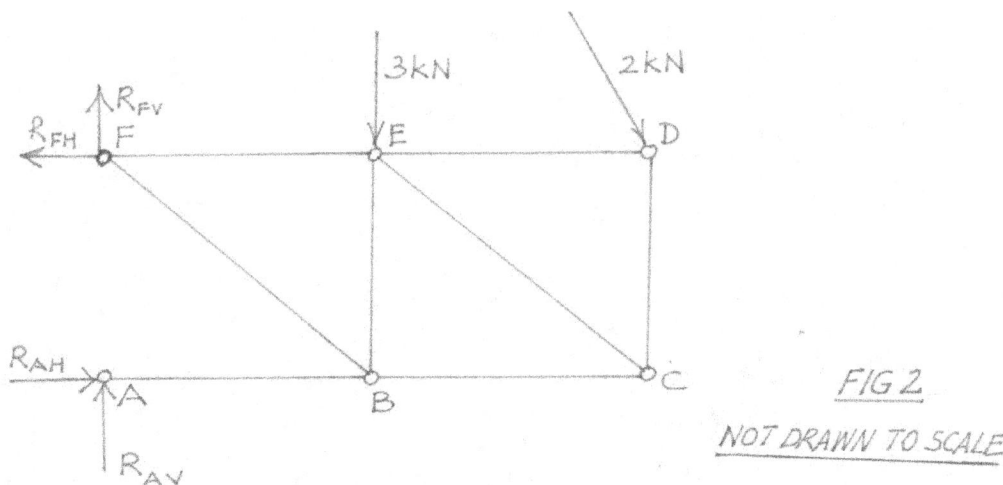

It should be observed that if we were to cut AB, the pin-joint at A will have the 3 forces acting as shown in Fig. 3 : the components R_{AH} and R_{AV} of the reaction; and the internal resisting force F of member AB. Let us assume F to be tensile force.

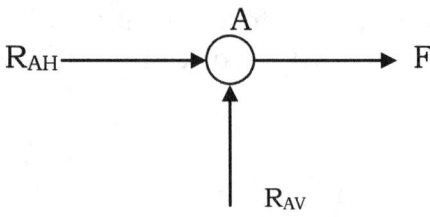

FIG. 3

For equilibrium at A,

$$R_{AH} + F = 0$$

and $\qquad R_{AV} = 0$

Clearly therefore the external force or external reaction, if you will, at A must act along AB; and so we have the direction of the reaction at A. If next we find the resultant of the two external forces : (i) 2kN acting at 60° to the horizontal at D, and (ii) 3kN acting vertically at 'E', then, the point where this resultant meets the line of action of the reaction at A is the same point at which the line of action of the resultant at F must pass. This is illustrated in Fig. 4. (See overpage).

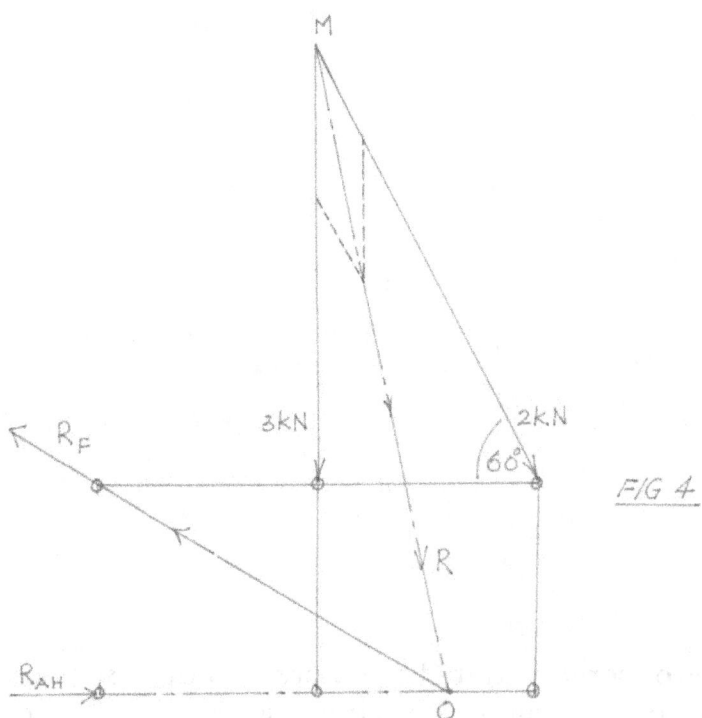

The lines of action of the 2kN and 3kN forces meet at M. The parallelogram of forces is completed to produce resultant R the line of

~ 82 ~

action of which meets the line of action of the reaction through A at point O. The direction of the resultant through F, designated R_F may now be drawn as shown.

Because the magnitude and direction of R are known, and also because we know the line of action of R_{AH} and R_F, we can quite easily construct the vector polygon as in Fig. 5.

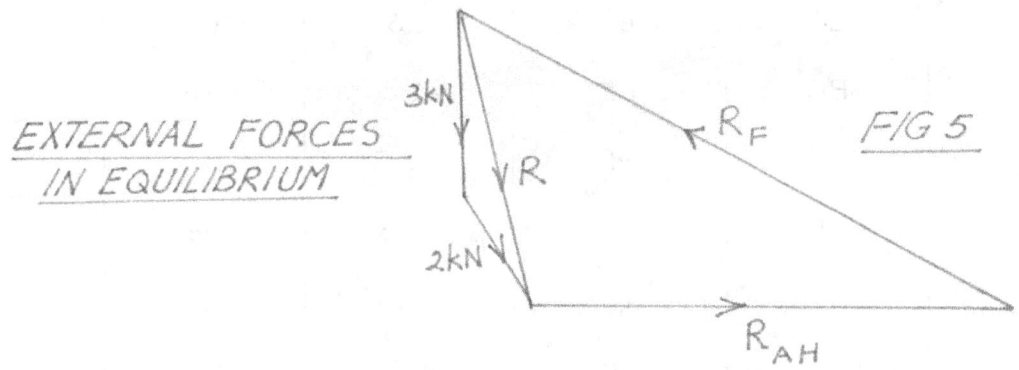

EXTERNAL FORCES
IN EQUILIBRIUM
FIG 5

The construction illustrated in Fig. 4 is drawn to scale in Fig. 6. The Bow-notated framework is presented as Fig. 7, and Bow's diagram or force polygon is at Fig. 8.

In Table 1 I have stated the magnitude of the forces as scaled off from Fig. 8 and indicated their nature as well.

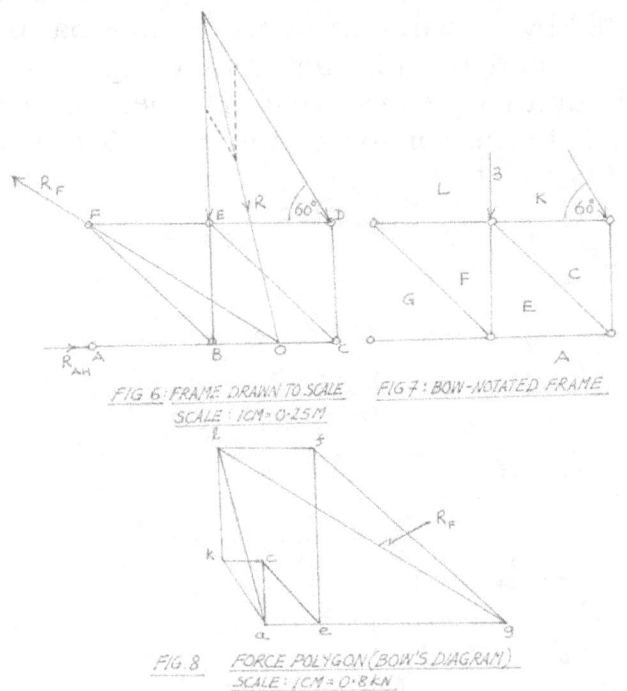

FIG 6 : FRAME DRAWN TO SCALE
SCALE : 1CM = 0.25M

FIG 7 : BOW-NOTATED FRAME

FIG 8   FORCE POLYGON (BOW'S DIAGRAM)
SCALE : 1CM = 0.8 kN

Member as in Fig. 1	Designation in Fig. 8	Magnitude of Force, kN	Nature
DE	kc	1	tie
DC	ac	-1.75	strut
EC	ec	2.5	tie
BC	ae	-1.8	strut
BE	ef	-4.75	strut
EF	lf	2.85	tie
BF	fg	6.6	tie
AG	ag	-6.5	strut

The student should note that in drawing Bow's diagram, 'lk' and 'ak' representing the external forces of 3kN vertically downwards and 2kN at 60° to the horizontal, are first set out. The rest follows by say locating point 'c' the intersection of lines parallel to CK and AC. Next 'ce' is drawn parallel to CE and point 'f' is located at the intersection of 'lf' and 'ef'. Point 'g' is located by drawing, through 'f' a line parallel to FG. Hence we obtain the lines of action and magnitude of 'lg' and 'ag' which represent the reactions R_F and R_{AH} respectively. By measurements taken from Fig. 8, R_F = 9kN at 32° to the horizontal and R_{AH} = 6.4kN acting horizontally.

Q6 A manufacturer of furniture is considering using the frame shown in Fig. 1 as part of the design for a chair. In order to size the pins to be used at joints C, D and E the manufacturer must first determine the reactions there. What are the reactions at C, D and E? Take AC = 35cm; CB = 5cm; CD = 30cm and DE = 42cm.

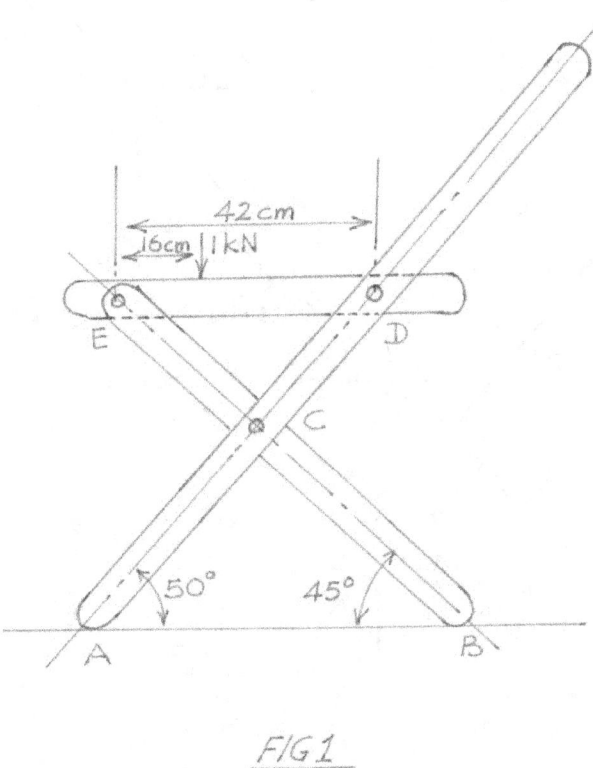

FIG 1

The first thing to be done is preparation of the free-body diagrams for the three members of the frame. These are presented at Figs. 2a, 2b and 2c.

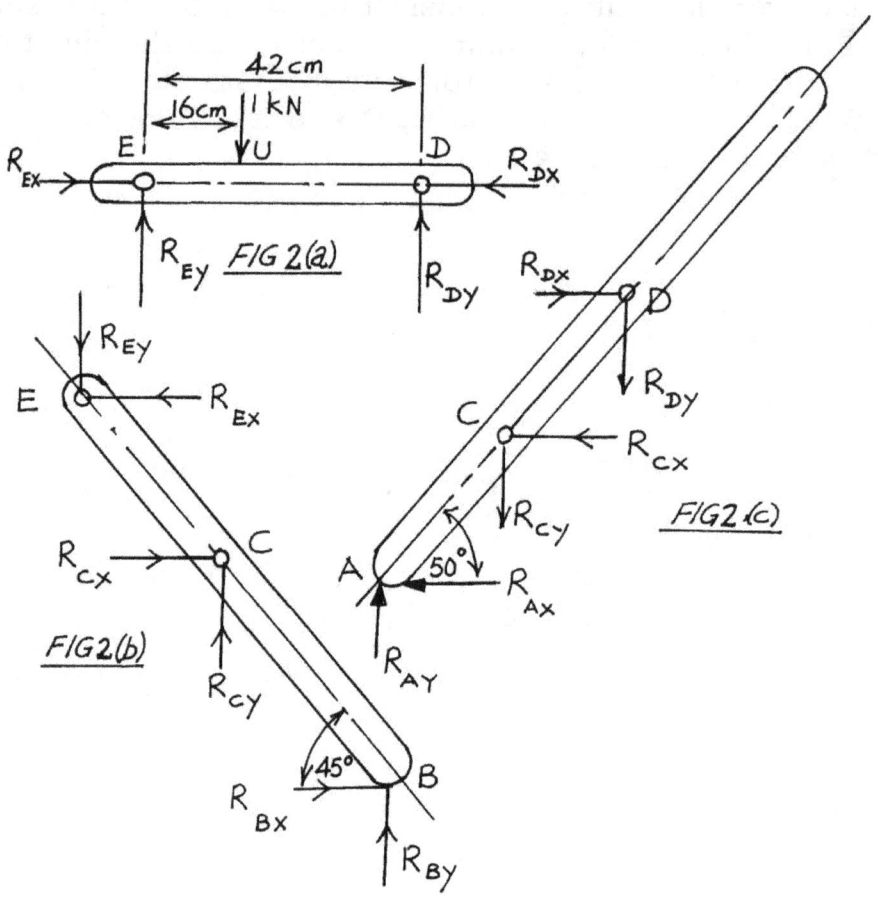

Because the frame is in equilibrium, each component part must also be in equilibrium. Accordingly we apply the conditions for static equilibrium to each part.

<u>For Fig. 2a</u>

For $F_Y = 0$; $R_{EY} + R_{DY} - 1 = 0$ . . . . . (i)

For $F_X = 0$; $R_{EX} - R_{DX} = 0$ . . . . . (ii)

For   moments about 'U': $R_{EY}(16) - R_{DY}(26) = 0$ . . . . . (iii)

From (iii) $R_{EY} = 1.62R_{DY}$, say, 1.63kN

so that by (i) $R_{DY} = 0.381$kN, say, 0.38kN

and $R_{EY} = 0.619$kN, say, 0.62kN

Before we go any further let us determine the length of EC. Applying the cosine law to triangle CDE

$$(EC)^2 = (CD)^2 + (42)^2 - 2(CD)(42)Cos50°$$

$$= (30)^2 + (42)^2 - 2(30)(42)Cos50°$$

$$= 2664 - 1620 = 1044$$

$$\therefore \quad EC = 32.3cm$$

so that EB = 32.3 + 45 = 77.3cm

Returning to Fig. 2(b)

For $\quad \sum F_Y = 0$; $\quad R_{EY} - R_{CY} - R_{BY} = 0$ $\hspace{3cm}$ . . . . . (iv)

For $\quad \sum F_X = 0$; $\quad R_{EX} - R_{CX} - R_{BX} = 0$ $\hspace{3cm}$ . . . . . (v)

For $\quad \sum M_B = 0$; $\quad R_{EY}(EBCos45°) + R_{EX}(EBSin45°)$

$-R_{CX}(CBSin45°) - R_{CY}(CBCos45°) = 0$

i.e. $\quad R_{EY}\{77.3(0.707)\} + R_{EX}\{77.3(0.707)\}$

$R_{CX}\{45(0.707)\} - R_{CY}\{45(0.707)\} = 0$

or $\quad 54.7R_{EY} + 54.7R_{EX} - 31.8R_{CX}$
$31.8R_{CY} = 0$

$54.7R_{EY} + 54.7R_{EX} - 31.8R_{CX} - 31.8R_{CY} = 0$ $\hspace{3cm}$ . . . . . (vi)

Next, we consider the free-body diagram Fig. 2c.

For $\sum F_Y = 0$; $\quad R_{AY} - R_{CY} - R_{DY} = 0$ $\hspace{3cm}$ . . . . . (vii)

For $\sum F_X = 0$; $\quad R_{DX} - R_{CX} - R_{AX} = 0$ $\hspace{3cm}$ . . . . . (viii)

For $\sum M_A = 0$; $\quad R_{DX}(ADSin50) + R_{DY}(ADCos50°)$
$$- R_{CX}(35Sin50) + R_{CY}(35Cos50°) = 0$$

i.e. $\quad R_{DX}(65Sin50) + R_{DY}(65Cos50)$
$-R_{CX}(35Sin50) + R_{CY}(35Cos50) = 0$

or $R_{DX}\{65(0.766)\} + R_{DY}\{65(0.643)\}$
           $-R_{CX}\{35(0.766)\} + R_{CY}\{35(0.643)\} = 0$

or $49.8R_{DX} + 41.8R_{DY} - 26.8R_{CX} + 22.5R_{CY} = 0$ . . . . . (ix)

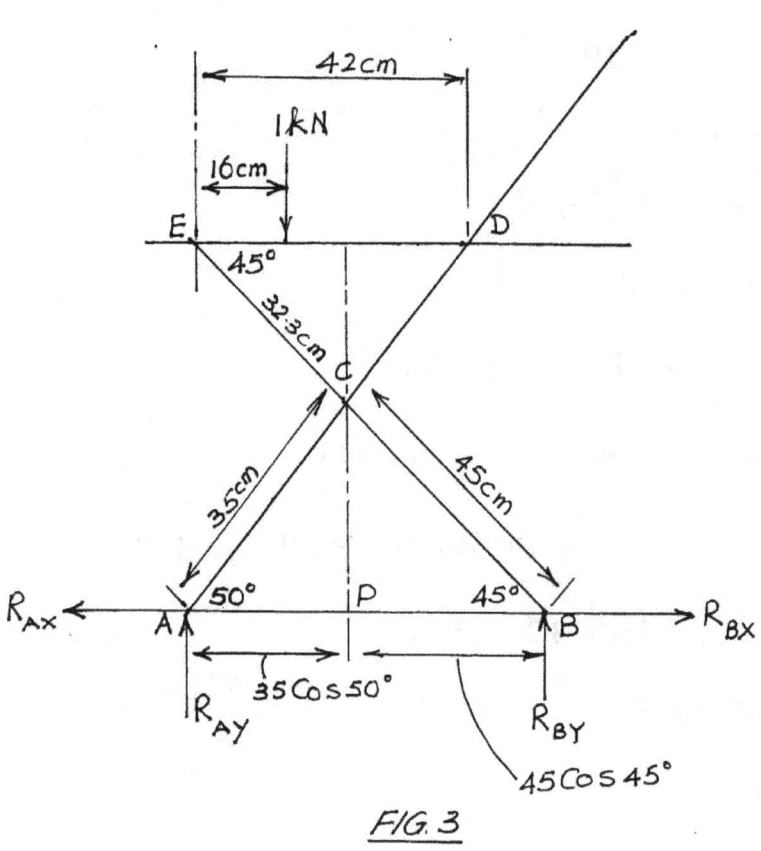

FIG. 3

Lastly, by considering the external forces acting on the frame, that is to say the reactions at A and B and the force on ED, we can determine R_{AY}, R_{BY}, R_{AX} and R_{BX} by applying the conditions for equilibrium.

Referring to Fig. 3.

For $\sum F_Y = 0$; $R_{AY} + R_{BY} - 1 = 0$ . . . . . (x)

For $\sum F_X = 0$; $R_{AX} - R_{BX} = 0$ . . . . . (xi)

and for $\sum M_P = 0$; $R_{AY}(35\text{Cos}50^\circ) - R_{BY}(45\text{Cos}45^\circ)$

          $- 1(32\text{Cos}45 - 16) = 0$

or $22.5R_{AY} - 31.8R_{BY} - 1(22.8 - 16) = 0$

i.e. $22.5R_{AY} - 31.8R_{BY} - 6.8$ $= 0$ . . . . . (xii)

Multiplying equation (x) by 22.5 and subtracting the result from equation(xii) we get

$$22.5R_{AY} - 31.8R_{BY} = 6.8$$

$$22.5R_{AY} + 22.5R_{BY} = 22.5$$

$$\therefore \quad 54.3R_{BY} = 15.7$$

$$\text{or} \quad R_{BY} = 0.29\text{kN}$$

$$\therefore \quad R_{AY} = 0.71\text{kN}$$

From (xi), $R_{AX} = R_{BX}$

Putting $R_{EY} = 0.62\text{N}$, and $R_{BY} = 0.29\text{kN}$ in equation (iv) produces

$$0.62 - R_{CY} - 0.29 = 0$$

from which

$$R_{CY} = 0.33\text{kN}$$

Also, putting $R_{EY} = 0.619\text{kN}$ and $R_{CY} = 0.33\text{kN}$ in equation (vi) produces

$$54.7(0.619) + 54.7R_{EX} - 31.8R_{CX} - 31.8(0.33) = 0$$

i.e. $33.9 + 54.7R_{EX} - 31.8R_{CX} - 10.5 = 0$

or $54.7R_{EX} - 31.8R_{CX} = -23.4$ . . . . . (xiii)

Further, putting $R_{DX} = 0.381\text{kN}$ and $R_{CY} = 0.33\text{kN}$ in equation (ix) produces

$$49.8R_{DX} + 41.8(0.381) - 26.8R_{CX} + 22.5(0.33) = 0$$

or $49.8R_{DX} + 15.9 - 26.8R_{CX} + 7.4 = 0$

i.e. $49.8R_{DX} - 26.8R_{CX} = -23.3$ . . . . . (xiv)

According to equation (ii), $R_{DX} = R_{EX}$. Therefore, equation (xiv) may be rewritten as :

$49.8R_{EX} - 26.8R_{CX} = -23.3$ . . . . . (xv)

Multiplying (xiii) by 26.8 and (xv) by 31.8, the following are obtained :

$1465.96R_{EX} - 852.24R_{CX} = -627.12$

and

$\qquad 1583.64R_{EX} - 852.24R_{CX} = -740.94$

$\qquad \therefore \;\; 117.68R_{EX} = -113.82$

from which

$\qquad R_{EX} = -0.97kN$

and $R_{DX} = -0.97kN,$

the minus sign merely indicating that the directions assumed for these forces in the free-body diagrams of Fig. 2 are the reverse of what are shown there.

Continuing, if we put $R_{EX} = -0.97kN$ in equation (xv)

$54.7(-0.97) - 31.8R_{CX} = -23.4$

i.e. $53.1 - 31.8R_{CX} = -23.4$

$$\therefore \qquad R_{CX} = \frac{76.5}{31.8}$$

or $R_{CX} = 2.4kN$

From equation (v) which states that

$R_{EX} - R_{CX} - R_{BX} = 0$

we obtain R_{BX}

i.e. $-0.97 - 2.4 - R_{BX} = 0$

$\qquad \therefore \qquad R_{BX} = -3.37kN$ and $R_{AX} = -3.37kN$

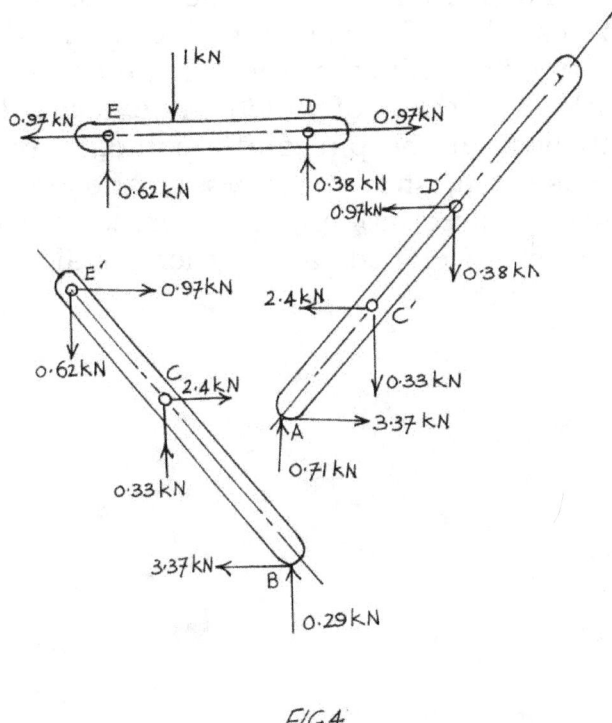

FIG 4

In Fig. 4, I have shown the values of the various components of forces shown in the free-body diagram/s presented as Fig. 2. Directions of those forces which were evaluated with a negative sign have been changed as appropriate.

Reaction at pin E $= \sqrt{(0.62)^2 + (0.97)^2}$

$= 1.15\text{kN}$

Reaction at pin D $= \sqrt{(0.38)^2 + (0.97)^2}$

$= 1.04\text{kN}$

Reaction at pin C $= \sqrt{(0.33)^2 + (2.4)^2}$

$= 2.42\text{kN}$

The student should realize that in sizing member ED, the reality that the load (in this case 1kN) is acting at a position between the pin-joints at 'E' and 'D' means that bending moments, resulting in bending stresses and shearing forces giving rise to shearing stresses will be induced in this member. Consequently these must be taken into account in the design of 'ED'. I make this point here to reinforce the distinction between (i) frames which are capable of supporting external loads at points between the pin-joints at the extremities of their members such as 'ED' in Fig. 1 and (ii) frameworks in respect of which the external loads and reactions are applied exclusively at the pin joints, sometimes called panel points. Whereas in the case of (i) shearing stresses and bending stresses, apart from direct tensile or direct compressive forces, are induced, in (ii), the internal resistances developed are either straightforward tensile or compressive forces.

Q7 Two views of a tripod typical of those used for lifting loads on the building sites of small construction projects, are shown at Figs 1a and 1b. C is pin-jointed and 3m above the ground. The back stay AC is 45° the horizontal. For a load of 10kN determine the forces in the shear legs BC as CD and backstay AC.

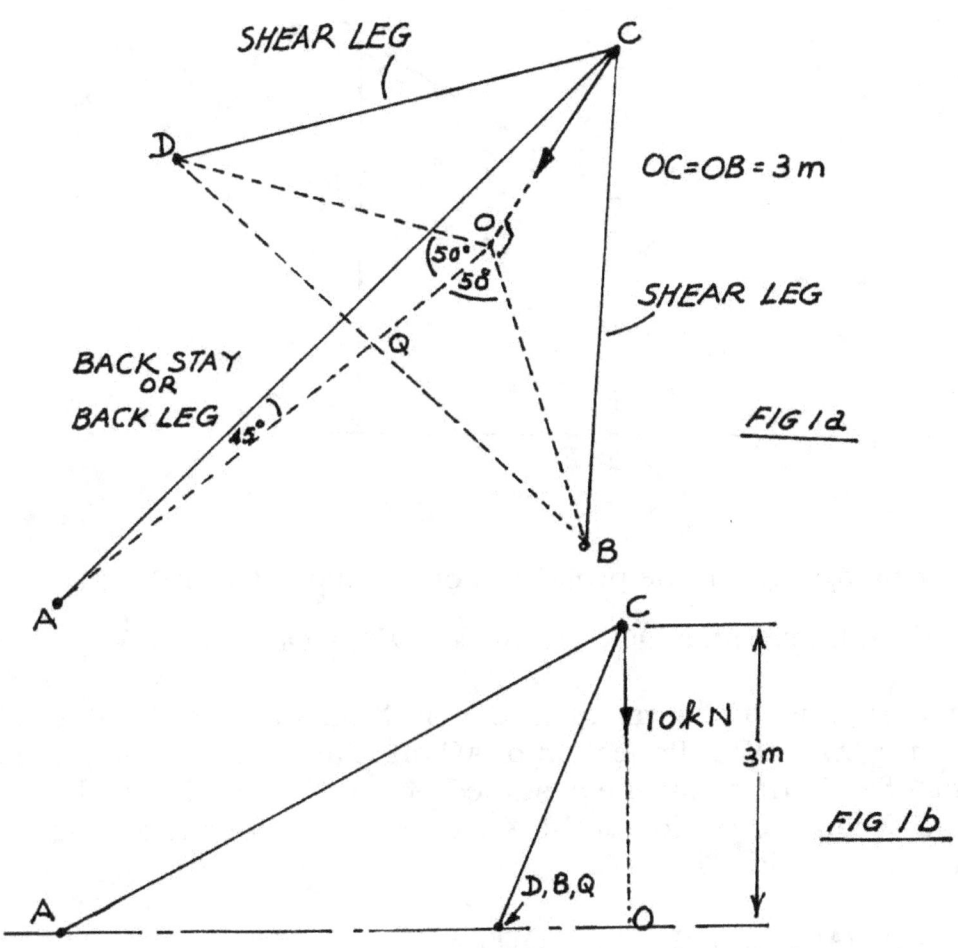

SHEAR LEG

BACK STAY
OR
BACK LEG

OC=OB = 3 m

SHEAR LEG

FIG 1a

10kN

3m

D,B,Q

FIG 1b

Let the tension coefficients of AC, BC and CD be denoted by AC, BC and CD respectively. Consider first, equilibrium in the X-direction remembering that we treat all members as ties and further that terms in the equilibrium equation are positive if the force moves the joint in the positive direction of whatever axis is being considered and likewise negative if the force moves the joint in the negative direction

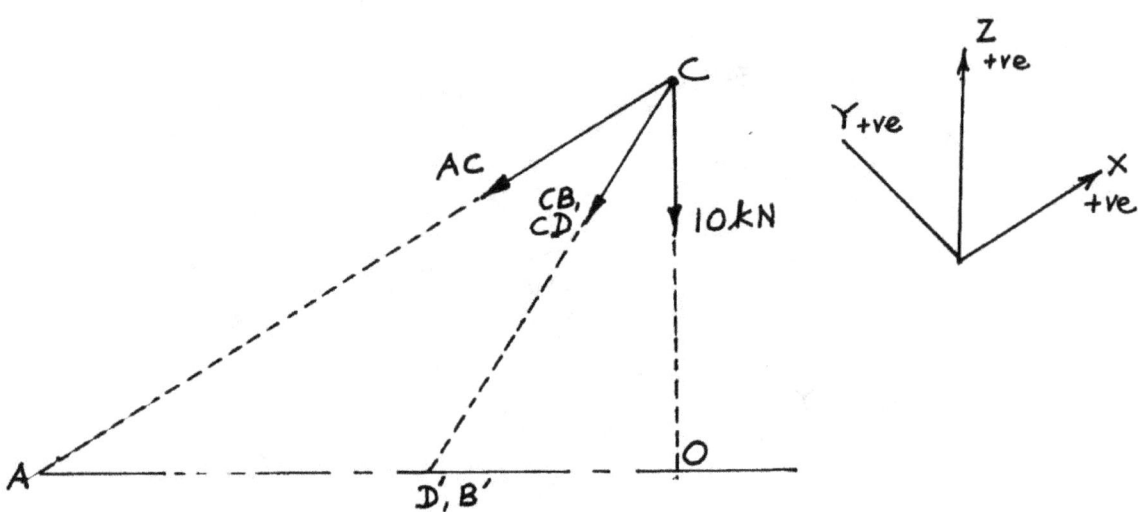

Referring to Fig. 1b, the projection of member 'BC' on X-axis = OQ; $\dfrac{OQ}{3}$ = Cos50°, from which OQ = 1.93m³ = OB' = OD'

Similarly projection of member 'CD' on X-axis is also equal to OQ = 1.93m = OB' - OD'. Projection of AC on X-axis = AO = 3m, angle OAC being 45°. Noting that the assumed ties AC, BC and CD all pull joint C in the negative direction of the X-axis, we have for the equilibrium equation for this axis :

$$-AC(AO)-BC\ (OB') - CD(OD')\ =\ 0$$

i.e.

$$-AC(3) - BC(1.93) - CD(1.93) = 0 \qquad\qquad \cdots\cdots \text{(i)}$$

Consider the equilibrium equation for the Y-axis.

The projection of CD on the Y-axis = QD; $\dfrac{QD}{3}$ = Sin50°

∴   QD = 2.30m.

Similarly, the projection of BC on the Y-axis is QB = QD.

Observe however that with reference to FIG 1(a) whereas the force in CD will tend to move joint 'C' in the positive direction of the Y-axis, the opposite occurs in the case of the force in BC.

Accordingly,

+CD(2.3) – BC(2.3) = 0 . . . . . (ii)

Consider finally the equilibrium equation for the Z-axis.

The projection of each of the lengths AC, BC, and CD on the Z-axis is 3m. Each of the assumed ties tends to pull joint 'C' in the negative direction of the Z-axis. Therefore

-AC(3) – BC(3) – CD(3) – 10 = 0 . . . . . (iii)

Solving equations (i), (ii) and (iii) we have from (ii)

CD = BC

Substituting this result in (i) and (iii) gives

-3AC – 1.93BC – 1.93BC = 0

i.e.

$\qquad$ -3AC – 3.86BC = 0 . . . . . (iv)

and

$\qquad$ -3AC – 3BC – 3BC – 10 = 0

or

$\qquad$ -3AC – 6BC = 10 . . . . . (v)

Subtracting (v) from (iv)

$\qquad$ +2.14BC = -10

$\qquad\qquad\therefore$ BC = -4.67

and CD = -4.67

$$-3AC + 3.86(4.67) = 0$$

$$\text{i.e.} \quad AC = 6$$

Remember that these values for AC, BC and CD are tension coefficients having units of force/unit length. These now have to be multiplied by the lengths of the relevant members in order to obtain the forces in each. Evidently the length of each member of the tripod is $3\sqrt{2}$ metres i.e. 4.242m. Accordingly,

Force in AC, F_{AC} = 6 x length of AC

$$= + 6 \text{ x } 4.242kN$$

F_{AC} = 25.5kN, tie

Force in BC, F_{BC} = -4.67 x length of BC

$$= -4.67 \text{ x } 4.242$$

$$\therefore \quad F_{BC} = -19.8kN$$

Likewise F_{CD} = -19.8kN

F_{BC} and F_{CD} being of negative sign are struts.

Q8. The pitch-pine truss shown in Fig. 1 is to be used in a building project. Assuming that the uniformly distributed load of 'w' kilonewtons per metre is distributed uniformly among the supports determine what the maximum value of 'w' is. You may assume that compression members can carry a maximum load of 30kN and tension members 15kN

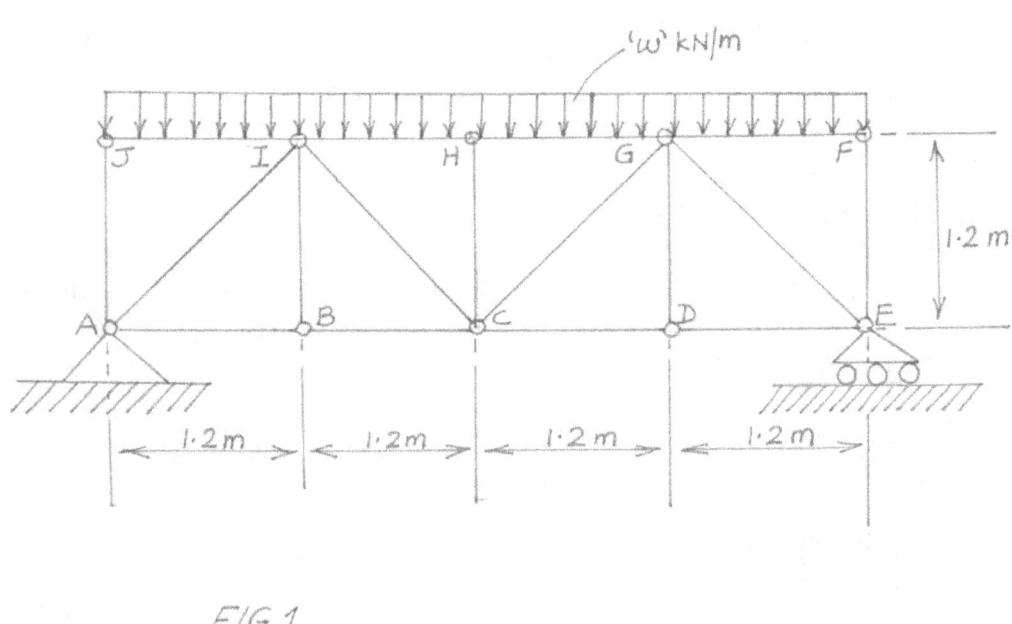

FIG 1

It is not considered out of place to state here for general information that a truss is a special kind of framed structure comprising straight members joined to each other such that the stresses induced by the applied loadings are wholly axial and either tensile or compressive in nature in each member.

Theoretically, members of a truss are assumed to be connected together at joints by frictionless pins. In practice however, as most of you student engineers would have observed, gusset plates, rivets and welds are not uncommon connectors used in the fabrication of trusses. But it is essential that the axes of all members at all joints should meet at a common point there so that none of the forces induced in the member can produce a moment about the joint.

The truss is 4.8 metres long and the uniformly distributed load is 'w' kN/m. Therefore the total load is 4.8wkN. There are 5 panel points so that each panel point will carry a load 0.96wkN, as shown in Fig. 2.

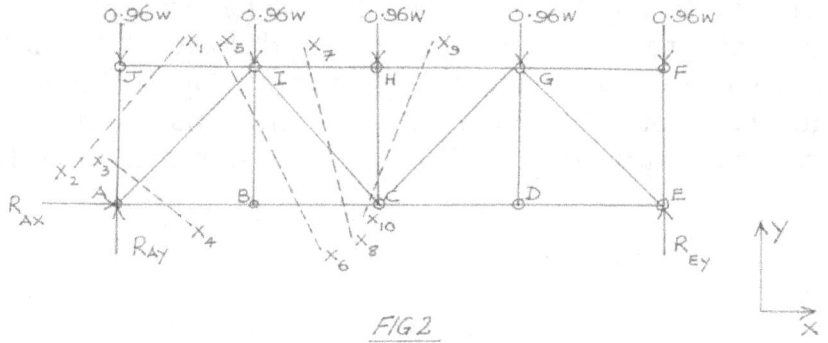

FIG 2

Evidently, $R_{AX} = 0$; $R_{AY} = R_{EY} = 2.4w$

The truss is symmetrical and so we need only determine by calculation the forces for one-half its members. Let us do so using the Method of Sections. All forces are in kN

Considering section X_1X_2

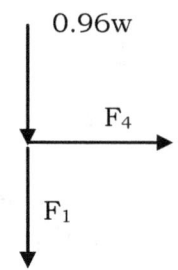

For $\sum F_y = 0$; $0.96w + F_1 = 0$ $\quad \therefore \quad F_1 = -0.96w$; strut

By inspection, $F_4 = 0$

Considering section X_3X_4

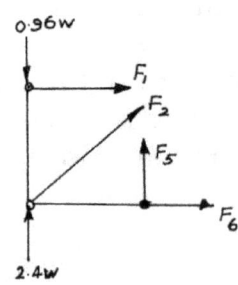

For $\sum F_y = 0$; $F_1 + F_2\cos45 + 2.4w = 0$

$\therefore$ $-0.96w + F_2 \cdot \dfrac{1}{\sqrt{2}} + 2.4w = 0$

$\therefore$ $\dfrac{F_2}{\sqrt{2}} = = -1.4w$

i.e. $F_2 = -(1.44\sqrt{2})w$;   strut

For $\sum F_x = 0$; $F_2\sin45 + F_3 = 0$

i.e. $-144\dfrac{\sqrt{2}}{\sqrt{2}}w + F_3 = 0$

$\therefore$ $F_3 = +1.44w$;   tie

Considering section X_5X_6

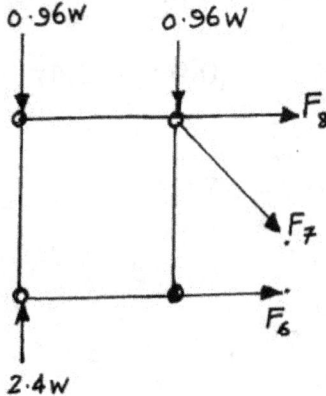

For $\sum F_y = 0$: $F_5 + F_2\cos45 + 2.4w - 0.9w = 0$

i.e. $F_5 - 1.44\dfrac{\sqrt{2}}{\sqrt{2}} \cdot w + 2.4w - 0.96w = 0$

$\therefore$ $F_5 = 0$

For $\quad \sum F_x = 0: \quad F_1 + F_2 Sin45 + F_6 = 0$

$$\therefore \quad = -0.96w - 1.44 \frac{\sqrt{2}}{\sqrt{2}} w + F_6 = 0$$

$$\therefore \quad F_6 = 2.4w; \quad tie$$

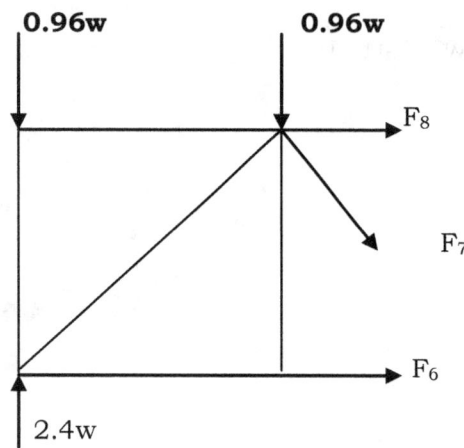

Considering section $X_7 X_8$ and equilibrium of the truss to the left of it, just as we did in the case of the previous sections, we have :

For $\quad \sum F_y = 0; \quad F_7 Cos45 + 2(0.9w) - 2.4w = 0$

$$\text{i.e.} \quad \frac{F_7}{\sqrt{2}} + 1.92w - 2.4w = 0$$

$$\frac{F_7}{\sqrt{2}} = 0.48w$$

$$\therefore \quad F_7 = (0.48\sqrt{2})w, \quad tie$$

For $\quad \sum F_x = 0 \quad F_8 + F_7 Sin45 + F_6 = 0$

$$\text{i.e.} \quad F_8 + 0.48\sqrt{2}w \cdot \frac{1}{\sqrt{2}} + 2.4w = 0$$

$$\therefore \quad F_8 = -2.88w; \quad strut$$

Finally, considering $X_9 X_{10}$

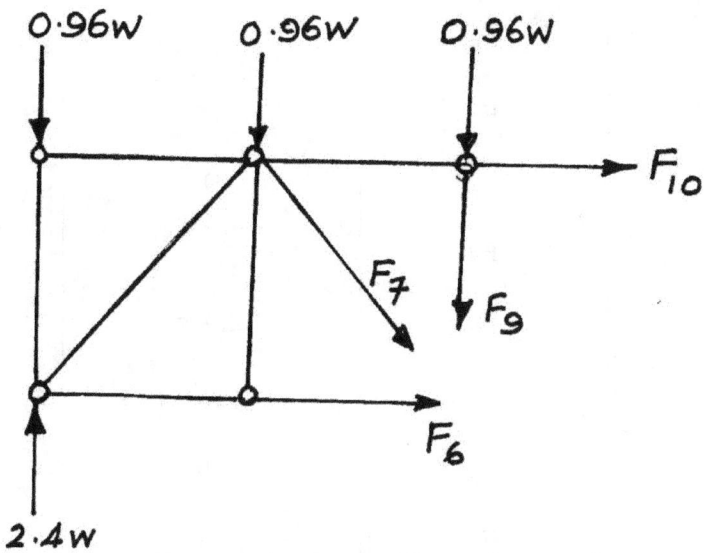

For $\sum F_y = 0$

$F_9 + F_7 Cos45 + 3(0.96w) - 2.4w = 0$

i.e. $F_9 + 0.48\sqrt{2} \cdot w \cdot \dfrac{1}{\sqrt{2}} + 2.88w - 2.4w = 0$

$F_9 + 3.36w - 2.4w = 0$

∴ $F_9 = -0.96w$; strut

The results of the calculations are summarized in the table following.

Truss Member	Force Designation	Magnitude in kN	Strut(S) or Tie(T)
AJ	F_1	-0.96w	S
AI	F_2	-2.04w	S
AB	F_3	1.44w	T
JI	F_4	0	-
BI	F_5	0	-
BC	F_6	2.4w	T
CI	F_7	$0.48\sqrt{2}w$	T
HI	F_8	-2.88w	S
CH	F_9	-0.96w	S

Inspection of the table reveals that the maximum force carried by a compression member occurs in member HI and is equal to 2.88w; also that the maximum force carried by a tension member, i.e. a tie, occurs in member BC, and is equal to 2.4w.

Having regard to the limitations placed on the maximum loads which may be carried by struts and ties: 30kN in the case of a strut and 15kN in the case of a tie, we have:

$$2.88w \ = \ 30kN$$

$$\therefore \quad w \ = \ 10.4kN/metre$$

and   2.4w = 15kN

so that   w = 6.25kN/metre

Evidently the maximum value of 'w' which would satisfy both conditions is w = 6¼ kN/m.

Q9. The truss shown in Fig. 1 carries 2 pulleys at D and F as shown, with a cable supporting a load equivalent to 100kN. The cable is anchored at P and the truss is pinned at A and supported by rollers at C. Determine the forces in EF, BF and AB.

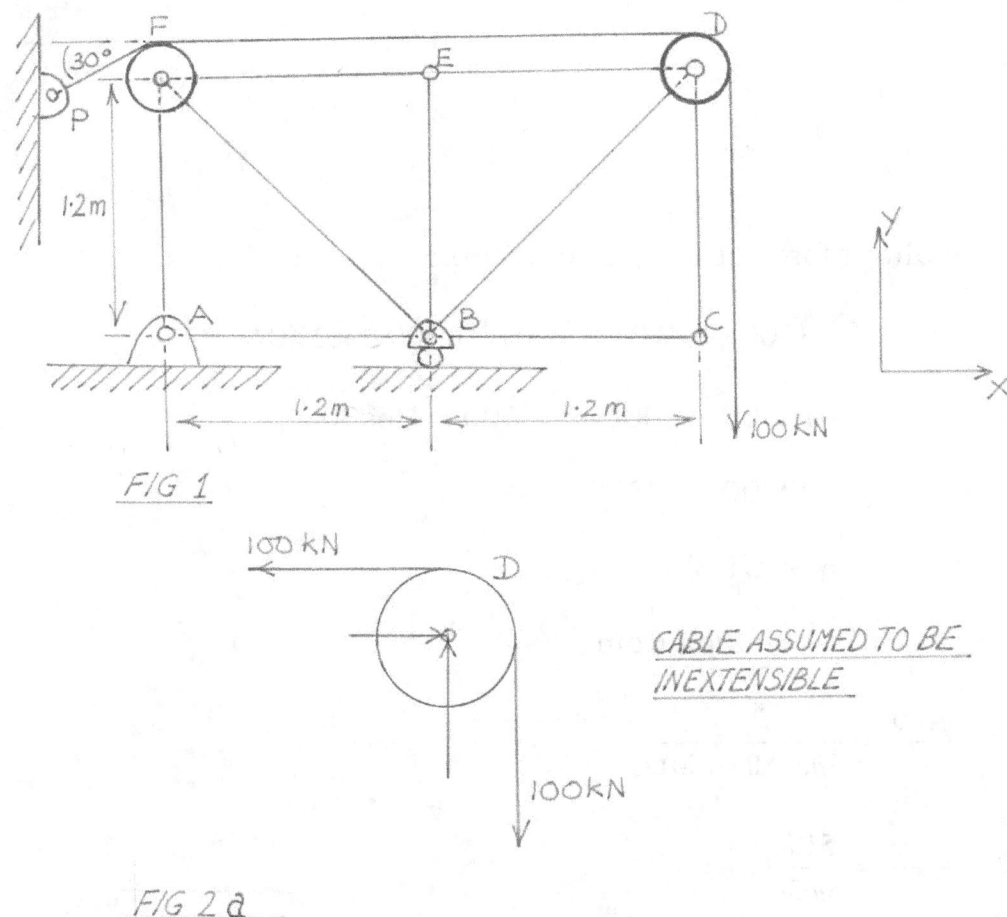

FIG 1

FIG 2 a

CABLE ASSUMED TO BE INEXTENSIBLE

In Fig. 2a which is the free-body for the pulley at D, the resultant force at D, R_D is given by :

$$R_D^2 = (100)^2 + (100)^2$$

∴ $R_D = 100\sqrt{2} = 141.4\text{kN}$

R_D acts at 45° to the horizontal, evidently.

In Fig. 2b I have shown the free-body diagram for the pulley at F.

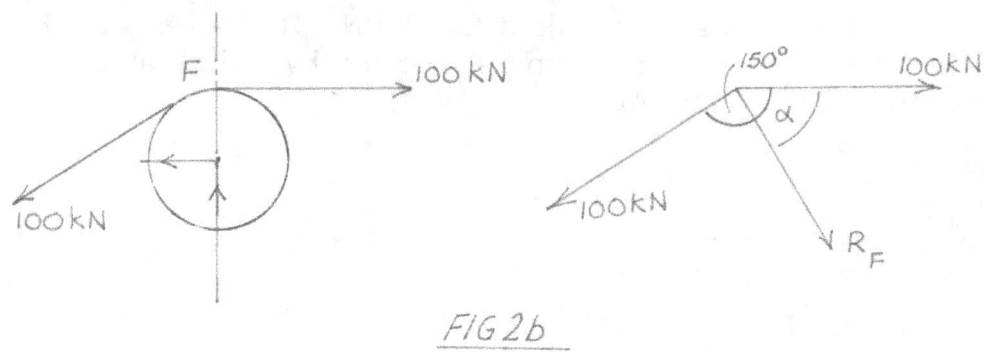

FIG 2b

The resultant force at F, R_F due to cable forces is obtained from

$$R_F^2 = (100)^2 + (100)^2 + 2(100)100Cos150°$$

$$= 10000 + 10000 + 20000(-Cos30°)$$

$$= 20000 - 17320$$

$$\therefore \quad R_F = 51.8kN$$

Direction of R_F obtained from

$$\frac{100}{Sin\alpha} = \frac{R_F}{Sin(180-150)}$$

$$\therefore \quad \frac{100}{Sin\alpha} = \frac{51.8}{Sin30}$$

$$\therefore \quad Sin\alpha = \frac{100}{103.6} = 0.9652$$

$$\alpha, = 74.85, \text{ say } 75°$$

The external loads on the truss including the reactions are shown in Fig. 3.

For $\quad \sum F_y = 0$: $\quad R_{AY} + R_B = 51.8Sin75 + 100\sqrt{2}\cdot Sin45°$

i.e. $\quad R_{AY} + R_B = 50 + 100 = 150$

For $\quad \sum F_x = 0$: $\quad R_{AX} - 51.8Cos75° + 100\sqrt{2}\cdot Cos45° = 0$

~ 104 ~

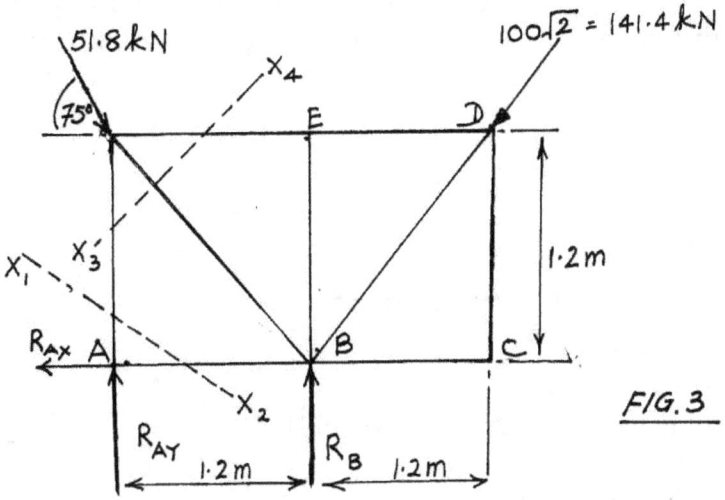

FIG. 3

∴ $R_{AX} - 13.4 + 100 = 0$

∴ $R_{AX} = -86.6kN$

Evidently the load in AB, $F_{AB} = R_{AX}$

i.e. $F_{AB} = -86.6kN$

Taking moments about A, for $\sum M_A = 0$

$R_B(1.2) - 51.8Cos75(1.2) + 100\sqrt{2}\,Cos45(1.2)$

$- 100\sqrt{2}\,Sin45(2.4)=0$

i.e. $R_B - 13.4 + 100 - 200 = 0$

or $R_B = 113.4kN$

and because $R_{AY} + R_B = 150$

$R_{AY} = 36.6kN$

Evidently the load in member AF, $F_{AF} = -R_{AY}$

i.e. F_{AF} = -36.6kN

For those to whom the conclusion about F_{AB} and F_{AF} are not so evident, consider equilibrium of the framework to the left of section X_1X_2

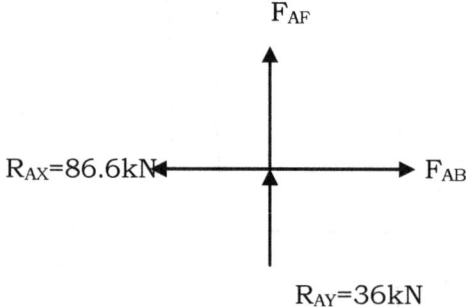

For $\sum F_y = 0$: $R_{AY} + F_{AF} = 0$

i.e.   36.6 + F_{AF} = 0

∴ F_{AF} = -36.6 kN; strut

Likewise, for $\sum F_x = 0$: $F_{AB} - R_{AX} = 0$

i.e. F_{AB} - (-86.6) = 0

∴ F_{AB} = -86.6kN; strut

Considering section X_3X_4

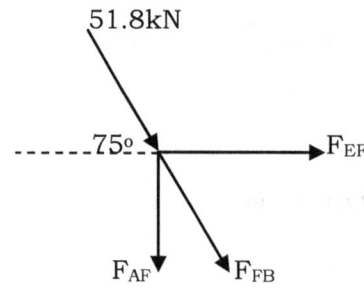

For $\sum F_y = 0$: F_{AF} + 51.8Sin75° + F_{FB} Cos45 = 0

i.e.   -36.6 + 50 + F_{FB} $\cdot \dfrac{1}{\sqrt{2}}$ = 0

$$\therefore \quad F_{FB} = -13.4\sqrt{2} \text{ kN}$$

or $\quad F_{FB} = -18.9 \text{KN};$ strut

For $\quad \sum F_X = 0 \quad : \quad F_{EF} + F_{FB}\text{Sin}45 + 51.8\text{Cos}75 = 0$

i.e. $\quad F_{EF} - 18.9\text{Sin}45 + 13.4 = 0$

$\qquad F_{EF} - 13.4 + 13.4 = 0$

$\qquad \therefore \quad F_{EF} = 0$

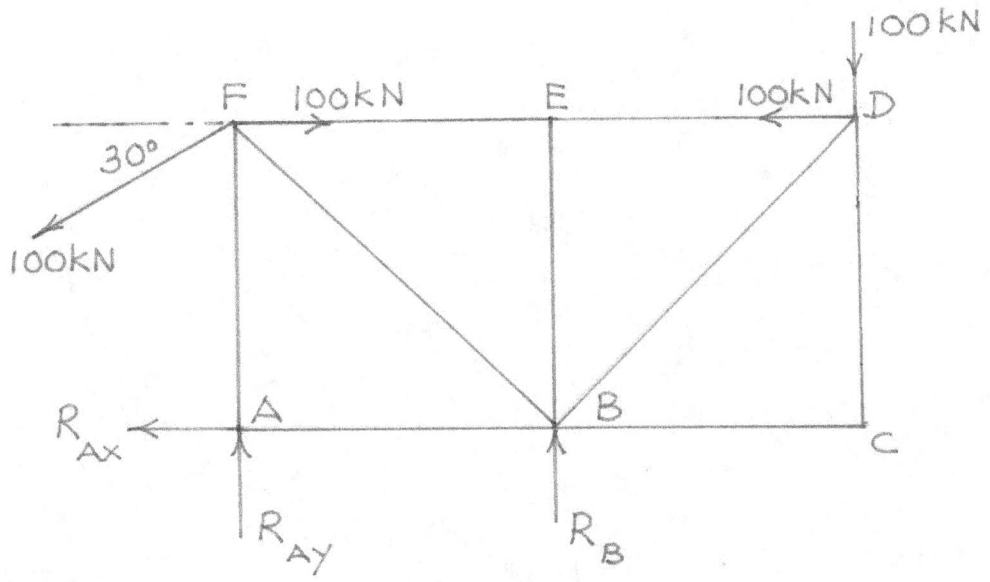

FIG 4

The student should note that the external loading shown in Fig. 3 is equivalent to that of Fig. 4 and that the problem may be solved using the latter as an alternative configuration.

Q10. Calculate the reactions at the pin-joints at A, B and C of the frame shown in Fig. 1. A horizontal force of 10kN acts at the mid-point of BC and a vertical load of 20kN acts at mid-point of AC.

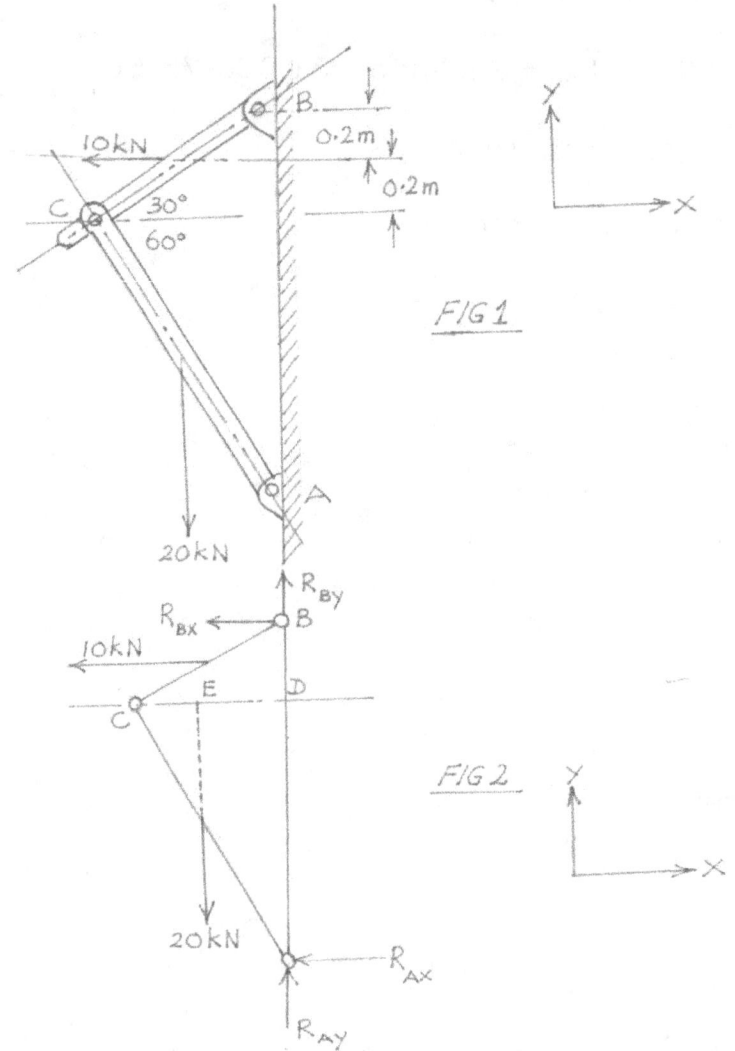

In Fig. 2, I have shown the components of the reactions at the wall. Taking moments about B for $\sum M_B = 0$

$$R_{AX}(AB) - 20(ED) + 10(0.2) = 0 \qquad\qquad \cdots\cdots \text{ (i)}$$

From the geometry of the configuration

$$\frac{CD}{0.4} = \text{Tan}60° = \sqrt{3}$$

$$\therefore \qquad\qquad CD = 0.4\sqrt{3}\,\text{m}$$

Also $\dfrac{CD}{AD} = \text{Tan}30^\circ = \dfrac{1}{\sqrt{3}}$

$\therefore \quad AD = \sqrt{3} \cdot CD = 0.4\sqrt{3} \cdot \sqrt{3} = 3 \times 0.4$

or $\quad AD = 1.2\text{m}$

and $\quad AB = AD + DB = 1.2 + 0.4 = 1.6\text{m}$

Further $\quad ED = \dfrac{1}{2} CD = 0.2\sqrt{3}\,\text{m}$

Equation (i) therefore becomes

$$R_{AX}(1.6) - 20(0.2\sqrt{3}) + 10(0.2) = 0$$

i.e. $\quad 1.6R_{AX} - 6.9 + 2 = 0$

from which $\quad R_{ax} \approx 3.1 \text{ kN}$

Freebody diagrams for CB and AC are shown in Figs 3a and 3b.

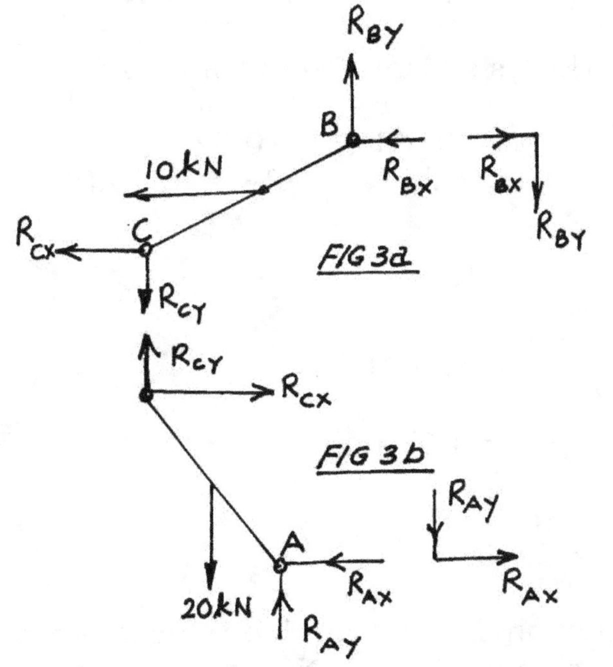

FIG 3a

FIG 3b

Considering equilibrium at CB and with reference to Fig. 3a

For $\sum F_x = 0$: $\quad R_{BX} + R_{CX} + 10 = 0$

$\quad\quad\quad \therefore \quad R_{BX} = -R_{CX} - 10$ $\quad\quad\quad\quad\quad\quad\quad\quad\quad\quad$ . . . . . (i)

For $\sum F_y = 0$: $\quad R_{BY} - R_{CY} = 0$

$\quad\quad\quad \therefore \quad R_{BY} = R_{CY}$ $\quad\quad\quad\quad\quad\quad\quad\quad\quad\quad\quad\quad$ . . . . . (ii)

For $\sum M_B = 0$: $\quad 10(0.2) + R_{CX}(0.4) - R_{CY}(0.4\sqrt{3}) = 0$

$\quad\quad\quad \therefore \quad 0.4R_{CX} - 0.69R_{CY} = -2$ $\quad\quad\quad\quad\quad\quad$ . . . . . (iii)

Considering equilibrium of AC and with reference to Fig. 3b

For $\sum F_X = 0$: $\quad R_{CX} - R_{AX} = 0$

$\quad\quad\quad \therefore \quad R_{CX} = R_{AX}$ $\quad\quad\quad\quad\quad\quad\quad\quad\quad\quad\quad\quad$ . . . . . (iv)

For $\sum F_y = 0$: $\quad R_{CY} - 20 + R_{AY} = 0$

$\quad\quad\quad \therefore \quad R_{CY} + R_{AY} = 20$ $\quad\quad\quad\quad\quad\quad\quad\quad\quad$ . . . . . (v)

For $\sum M_C = 0$ $\quad 20(0.2\sqrt{3}) - R_{AY}(0.4\sqrt{3}) + R_{AX}(1.2) = 0$

$\quad\quad\quad\quad\quad \therefore \quad 1.2R_{AX} - 0.69R_{AY} = -6.9$ $\quad\quad\quad$ . . . . . (vi)

When $R_{AX} = 3.1\text{kN}$ is substituted in (vi) we get

$$0.69R_{AY} = 10.62$$
$$\therefore \quad\quad R_{AY} = 15.4\text{kN}$$

From (v)

$\quad\quad R_{CY} = 20 - 15.4$

$\quad\quad\quad \therefore \quad R_{CY} = 4.6\text{kN}$

From (ii) $\quad\quad R_{BY} = R_{CY} = 4.6\text{kN}$

From (iv) $\quad\quad R_{CX} = R_{AX}$

$\quad\quad \therefore \quad\quad R_{CX} = 3.1\text{kN}$

We may use equation (iii) as a check on this latter result. Substituting $R_{CY} = 4.6\text{kN}$, we obtain $R_{CX} = 3\text{kN}$, near enough.

From (i)

$$R_{BX} = -R_{CX} - 10$$
$$\therefore \quad R_{BX} = -3.1 - 10$$

$$R_{BX} = -13.1 \text{kN}$$

Collecting our results, I summarise for ease of reference

$$R_{AX} = 3.1 \text{kN} \quad ; \quad R_{AY} = 15.4 \text{kN}$$

$$R_{BX} = -13.1 \text{kN} \quad ; \quad R_{BY} = 4.6 \text{kN}$$

$$R_{CX} = 3.1 \text{kN} \quad ; \quad R_{CY} = 4.6 \text{Kn}$$

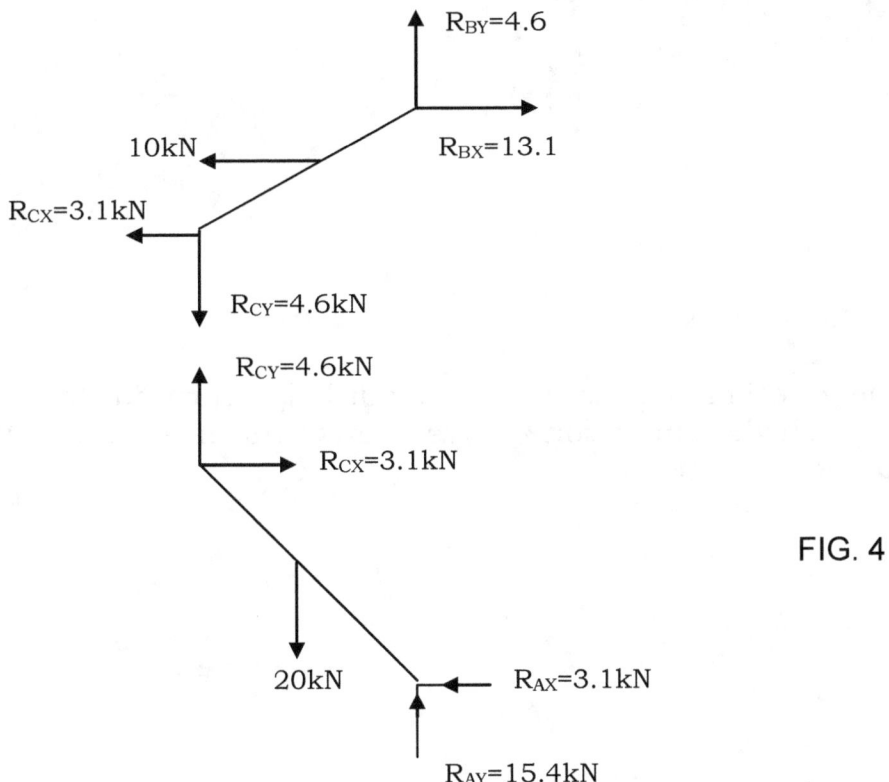

FIG. 4

As an overall check on the calculations, I have redrawn the free-body diagrams in Fig. 4 with the values of the components of the reactions and the mid-point forces indicated. Note that for the direction of R_{BX} in Fig 3(a) its value was determined as –13.1kN. This meant that the direction assumed in Fig. 3(a) was incorrect. I have shown its correct direction in Fig. 4; its value in that direction being +13.1kN.

So far we have only determined the components of the pin reactions. Designating the pin reactions at A, B and C as R_A, R_B and R_C respectively, we have

$$R_A = \sqrt{(3.1)^2 + (15.4)^2}.$$

$$= \sqrt{246.77}$$

$$\therefore \quad R_A = 15.7 \text{kN}$$

$$R_B = \sqrt{(-13.1)^2 + (4.6)^2}$$

$$\sqrt{192.77}$$

$$\therefore \quad R_B = 13.9 \text{kN}$$

and

$$R_C = \sqrt{(3.1)^2 + (4.6)^2}$$

$$\sqrt{30.77}$$

$$\therefore \quad R_C = 5.5 \text{kN}$$

Why did the 2 decimal places under the surd sign turn out to be a point 77 (.77) in each determination? Mysterious? As an exercise, find the direction of these reactions.

Q11. Fig. 1 shows the line drawing of a jib crane used on a sugar estate. Compute the reactions at the supports A, B, and E and also find the force in each member indicating whether it is a strut or a tie,

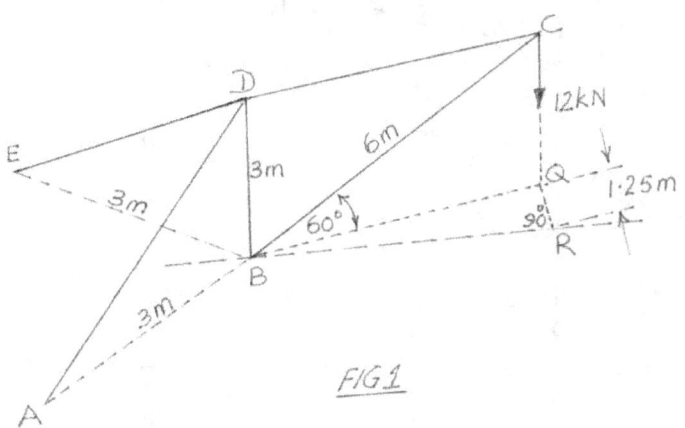

FIG 1

DIAGRAM NOT DRAWN TO SCALE

From trigometrical considerations, BQ = 3m, angle BCQ being 30°. Also, $(BR)^2 = (3)-(1.25)^2$, i.e. BR = 2.73m

[See overpage]

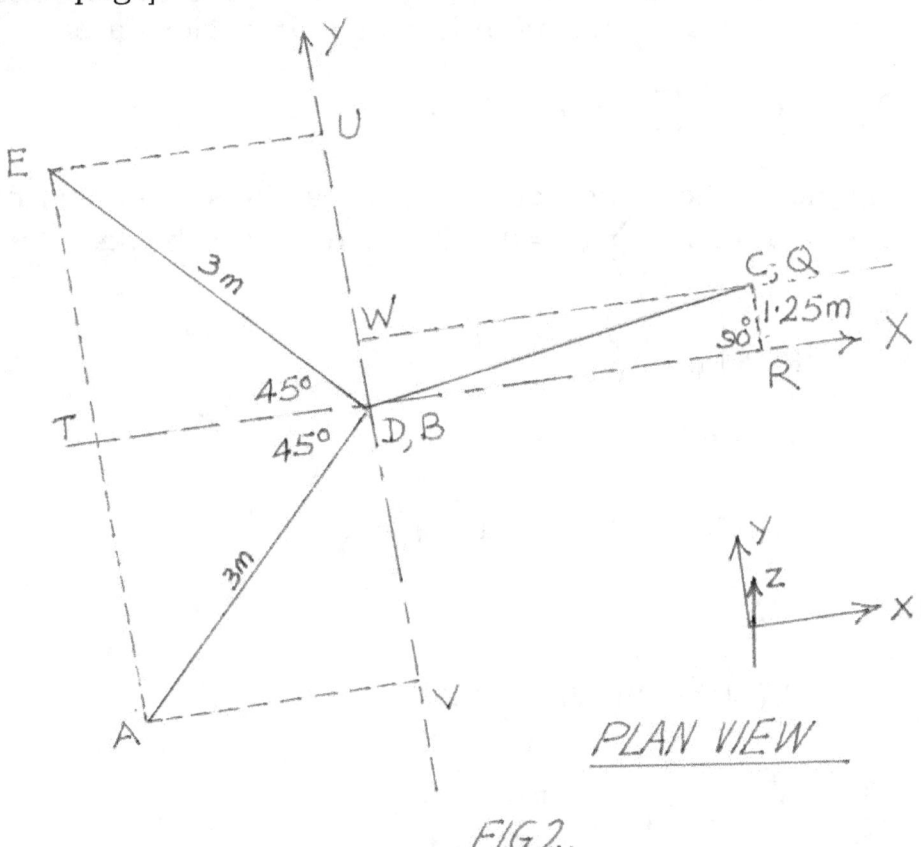

PLAN VIEW

FIG 2.

DIAGRAM NOT DRAWN TO SCALE

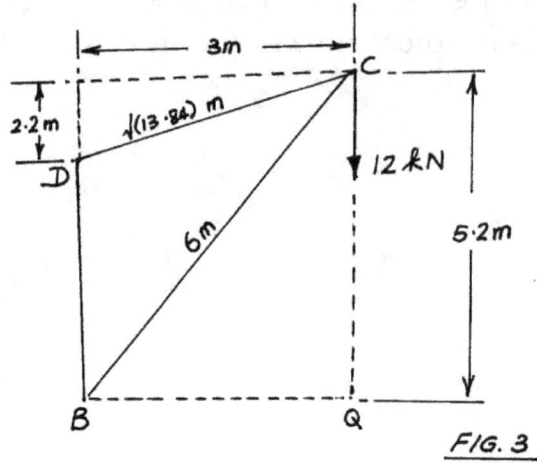

FIG. 3

Fig. 2 shows the plan view of the jib crane and Fig. 3 a view on a plane containing 'BQCD'.

In order to determine the reactions at A, E and B, respectively R_A, R_E and R_B, consider equilibrium of the applied loading and these reactions. The basic relations for statical equilibrium in 3 dimensions are :

$$\sum F_X = 0 \quad ; \quad \sum F_Y = 0 \quad ; \quad \sum F_Z = 0$$

There are no applied loads or external reactions in either the X- or Y-directions and so for $\sum F_Z = 0$, all the reactions being assumed to be vertical

$$R_A + R_B + R_E = 0 \qquad \qquad \ldots \ldots \text{(i)}$$

Also, for $\sum M_X = 0$

$$R_E(ET) - R_A(AT) - 12(1.25) = 0 \qquad \qquad \ldots \ldots \text{(ii)}$$

and for $\sum M_Y = 0$

$$R_E(EU) + R_A(AV) - 12(CW) = 0$$

Now $ET = AT = \dfrac{3}{\sqrt{2}} m = EU = AV$

$$\therefore \quad R_E \cdot \frac{3}{\sqrt{2}} - R_A \cdot \frac{3}{\sqrt{2}} - 15 = 0$$

$$3R_E - 3R_A = 15\sqrt{2} = 21.21 \qquad \qquad \ldots\ldots \text{(iii)}$$

and

$$R_E \frac{3}{\sqrt{2}} + R_A \frac{3}{\sqrt{2}} - 12(CW) = 0$$

But $CW = \sqrt{(3)^2 - (1.25)^2} = \sqrt{(7.4375)} = 2.73\text{m}$

$$\therefore \quad R_E \frac{3}{\sqrt{2}} + R_A \frac{3}{\sqrt{2}} - 12(2.73) = 0$$

$$\therefore \quad 3R_E + 3R_A = 46.3 \qquad \qquad \ldots\ldots \text{(iv)}$$

Adding (iii) and (iv)

$$6R_E = 67.5$$
$$\text{i.e.} \quad R_E = 11.25\text{kN}$$

and $\qquad 33.75 - 3R_A = 21.21$

$$\therefore \quad R_A = 4.18\text{kN}$$

Substituting these results in (i)

$$4.18 + R_B + 11.25 = 12$$

i.e. $\quad R_B = 12 - 15.43\text{kN}$

$$R_B = -3.43\text{kN}$$

Magnitude and direction of the reactions are :

$\qquad R_A = 4.18\text{kN, vertically upwards}$

$\qquad R_B = 3.43\text{kN, vertically downwards}$

$\qquad R_E = 11.25\text{kN, vertically upwards}$

The method of tension coefficients is ideally suited to determination of the forces in this space frame.

First, consider joint 'D'. Remember to assume all members as being in tension and that the terms of the equations are positive or negative if the forces move a joint in the positive or negative direction, respectively.

In order to avoid confusion, tension coefficients shall be identified by the letters specifying their members. Thus the tension coefficients of DE and AD are merely referred to as DE and AD, respectively.

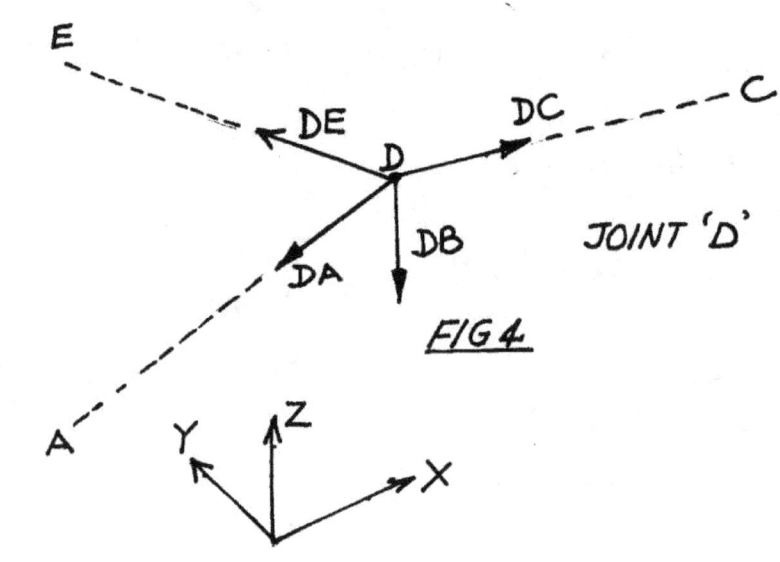

Accordingly,

Joint 'D'

For X-direction : $-DE\left(\dfrac{3}{\sqrt{2}}\right) - DA\left(\dfrac{3}{\sqrt{2}}\right) + DC(2.73) = 0$ . . . . . (v)

For Y-direction : $+DE\left(\dfrac{3}{\sqrt{2}}\right) - DA\left(\dfrac{3}{\sqrt{2}}\right) + DC(1.25) = 0$ . . . . . (vi)

For Z-direction : $-DE(3) + DA(3) + DC(2.2) - DB(3) = 0$ . . . ... (vii)

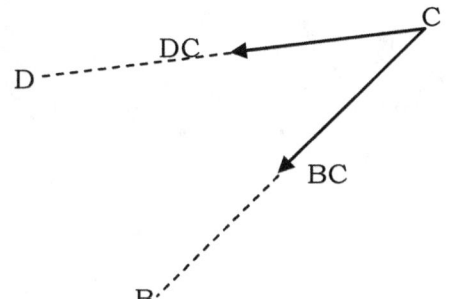

Joint C in Plane 'BQCD'

FIG. 5

Joint 'C'

For X-direction : -DC(2.73) – BC(2.73) = 0   . . . . . (viii)

For Y-direction : -DC(1.25) – BC(1.25) = 0   . . . . . (ix)

For Z-direction   : -DC(2.2) – BC(5.2) – 12 = 0   . . . . . (x)

From (ix)   DC = -BC

Substituting this result in equation (x)

-2.2(-BC) – 5.2BC –12 = 0

$$-3BC = 12$$

$$\therefore \quad BC = -4$$

so that   DC = +4

Putting DC = 4 in (v)

$$-\frac{3}{\sqrt{2}} DE - \frac{3}{\sqrt{2}} DA + 10.92 - 0$$

i.e.   -3DE – 3DA = -15.44   . . . . . (xi)

and doing the same for (vi) yields

$$\frac{3}{\sqrt{2}} DE - \frac{3}{\sqrt{2}} DA + 5 = 0$$

i.e. $3DE - 3DA = -7.07$ . . . . . (xii)

From (xi) and (xii)

$$-6DA = -22.51$$

$$\therefore \quad DA = 3.75$$

and $3DE - 11.26 = -7.07$

or $DE = 1.40$

Solving for DB in (vii)

$$-140(3) + 3.75(3) + 4(2.2) - DB(3) = 0$$
i.e. $-4.20 + 11.25 + 8.8 - 3DB = 0$

$$-3DB = -15.85$$

$$\therefore DB = 5.28$$

The results are tabulated below, the forces in each member being obtained on the basis of the product of tension coefficient and length of member.

Member	Length of Member	Tension Coefficient (kN/m)	Force in Member (kN)
BC	6m	-4	-24, Strut
BD	3m	5.28	16.84, Tie
DA	$(3\sqrt{2})m$	3.75	15.9, Tie
DC	$\sqrt{(13.84)}m$	4	14.9, Tie
DE	$(3\sqrt{2})m$	1.40	5.94, Tie

Q12. A mechanical engineering technician is proposing the scissors truss shown in Fig. 1 as part of a stand for a water tank. The loading may be regarded as consisting of two parts : a uniformly-distributed load of 30kN/m and a concentrated load of 60kN as shown. Show how he might compute the forces in all the members of the truss and determine which are struts and which are ties.

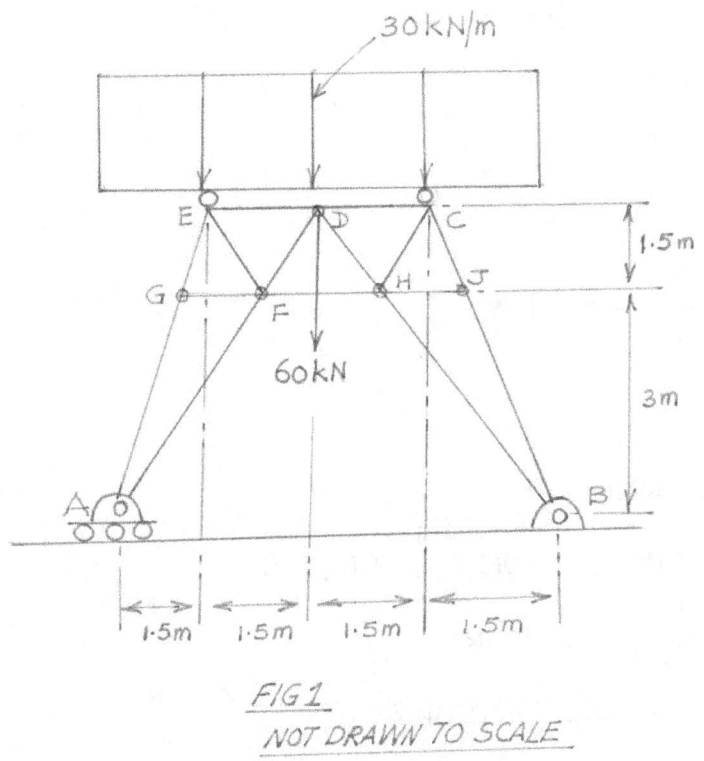

FIG 1

NOT DRAWN TO SCALE

The uniformly distributed load of 30kN/m over a distance of 6m translates into panel-point loads of 90kN at E and C. The load at panel point D is 60kN.

The truss is redrawn in Fig. 2 to show the vertical reaction at A; the components of reaction R_B at B and the panel point loads.

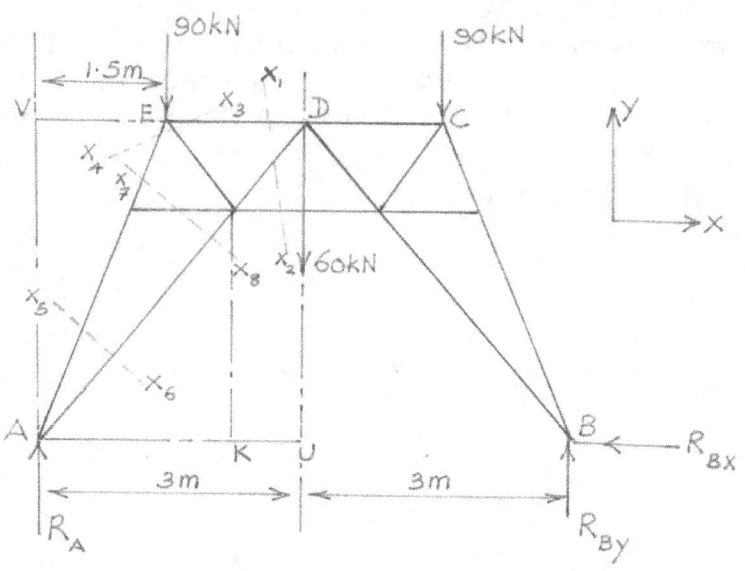

$$FIG\ 2$$

Taking moments about B

$$R_A(6) - 90(4.5) - 90(1.5) - 60(3) = 0$$

$$\therefore \quad 6R_A = 720$$

$$\text{i.e.} \quad R_A = 120kN$$

Evidently, $\quad R_{BY} = 120kN$

From the geometry of the truss

$$\text{Tan ADU} = \frac{4.5}{3} = 1.5$$

$$\therefore \quad \angle ADU = 56.3° \quad ; \quad \angle EDF = \angle DFW = 33.7°$$

Using the Method of Sections, let us consider section X_1X_2, and equilibrium of that section of the truss to the left of it. See Fig. 3

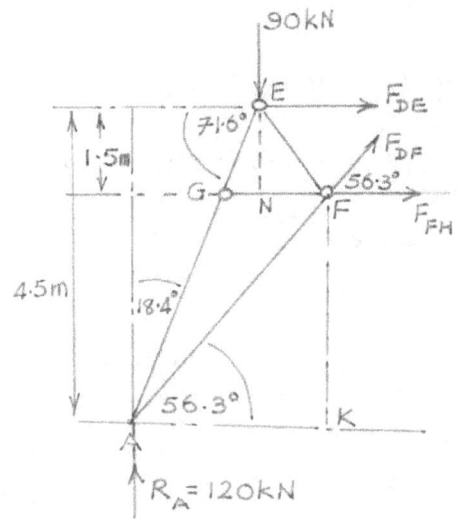

FIG 3

For vertical equilibrium $\quad \sum F_y = 0$

$F_{DF}\text{Sin}56.3° + 120 - 90 = 0$ $\qquad\qquad\qquad\qquad$ . . . . . (i)

$\qquad \therefore \qquad F_{DF}(0.832) = -30$

$\qquad\qquad \therefore \quad F_{DF} = -36.1\text{kN}$

Taking moments about T for $\sum M_T = 0$

$F_{DE}(SF) - 90(NF) + 120(1.5 + NF) = 0$ $\qquad\qquad$ . . . . . (ii)

Now, with reference to Fig. 2

$$\text{Cos}18.4° = \frac{3}{AG}$$

$\qquad \therefore \qquad AG = 3.2\text{m}$

Also,

$$\text{Sin}56.3° = \frac{EK}{AF}$$

$$= \frac{3}{AF}$$

$\qquad \therefore \qquad AF = 3.6\text{m}$

Using Cosine rule

$$GF^2 = AG^2 + AF^2 - 2(AG)(AF)\text{Cos } GAF$$

$$= (3.2)^2 + (3.6)^2 - 2(3.2)(3.6)\text{Cos } 15.3°$$

$$= 10.24 + 12.96 - 22.2$$

i.e.   GF = 1m

Now, $\text{Tan}18.4° = \dfrac{GN}{1.5}$

∴   GN = 1.5 Tan 18.4°

= 0.4989,   say 0.5m

and because NF = GF – GN

NF = 1 – 0.5 = 0.5m

Substituting this result in (ii), noting that SF = 1.5m

$$F_{DE} (1.5) - 90(0.5) + 120(2) = 0$$

i.e. $1.5F_{DE} = -240 + 45$

∴ $F_{DE} = -130\text{kN}$

Also for horizontal equilibrium

$$F_{DF}\text{Cos } 56.3° + F_{FH} + F_{DE} = 0 \qquad \qquad \ldots \ldots \text{(iii)}$$

∴ $-36.1\text{Cos } 56.3° + F_{FH} - 130 = 0$

or $-20 + F_{FH} - 130 = 0$

∴ $F_{FH} = 150\text{kN}$

Considering section X_3X_4

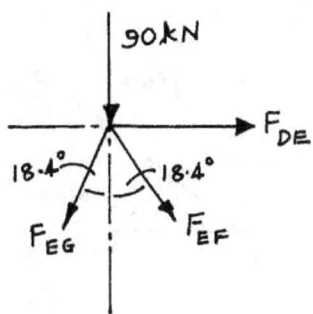

For $\sum F_y = 0$

$F_{EG}\cos 18.4° + F_{EF}\cos 18.4° + 90 = 0$ . . . . . (iv)

For $\sum F_x = 0$

$F_{EG}\sin 18.4° - F_{EF}\sin 18.4° - F_{DE} = 0$ . . . . . (v)

$\therefore$ $0.95F_{EG} + 0.95F_{EF} = -90$

and

$0.32F_{EG} - 0.32F_{EF} - (-130) = 0$

Solving for F_{EG} and F_{EF}

$0.95F_{EG} + 0.95F_{EF} = -90$

$0.95F_{EG} - 0.95F_{EF} = -386$

Adding

$1.90F_{EG} = -476$

$\therefore$ $F_{EG} = -251kN$

and, $0.95(-251) + 0.95F_{EF} = -90$

$-238.5 + 0.95F_{EF} = -90$

$0.95F_{EF} = +148.5$

$\therefore$ $F_{EF} = +156.3kN$

Considering section X_5X_6

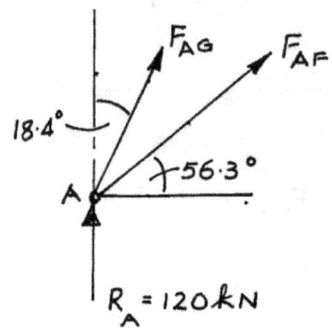

For $\sum F_y = 0$: $F_{AG}\cos 18.4° + F_{AF}\sin 56.3° + 120 = 0$

For $\sum F_X = 0$: $F_{AG}\sin 18.4° + F_{AF}\cos 56.3° = 0$

i.e. $0.95F_{AG} + 0.83F_{AF} = -120$

$0.32F_{AG} + 0.55F_{AF} = 0$

$0.95F_{AG} + 0.83F_{AF} = -120$

$0.95F_{AG} + 1.633F_{AF} = 0$

∴ $0.803F_{AF} = +120$

∴ $F_{AF} = 149.4$kN

From

$0.32F_{AG} + 0.55(149.4) = 0$

$0.32F_{AG} = -82.2$

∴ $F_{AG} = -256.8$kN

Considering section X₇X₈

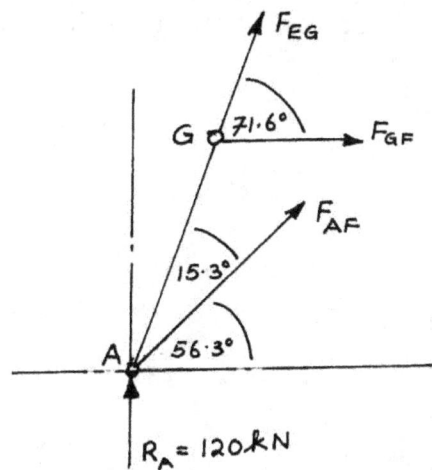

For $\sum F_x = 0$

$$F_{GF} + F_{AF}\cos 56.3° + F_{EG}\cos 71.6° = 0$$

i.e. $F_{GF} + 149.4(0.55) + (-251)(0.32) = 0$

or $F_{GF} + 82.2 - 80.3 = 0$

$$F_{GF} = -1.9\text{kN}$$

The results of our calculations are shown in the table below.

Member	Force (in kN)	Strut(S) or Tie(T)
AG	-256.8	S
AF	149.4	T
EG	-251	S
EF	156.3	T
DE	-130	S
DF	-36.1	S
FH	150	T
GF	-1.9	S

Q13. The pin-jointed framework shown in Fig. 1 is hinged at A and supported by a cable at E. A vertical load of 30kN is supported at C. Draw the force diagram for the framework, determining the magnitude of the load in each member and distinguishing between struts and ties. Also, what is the magnitude and direction of the reaction at A?

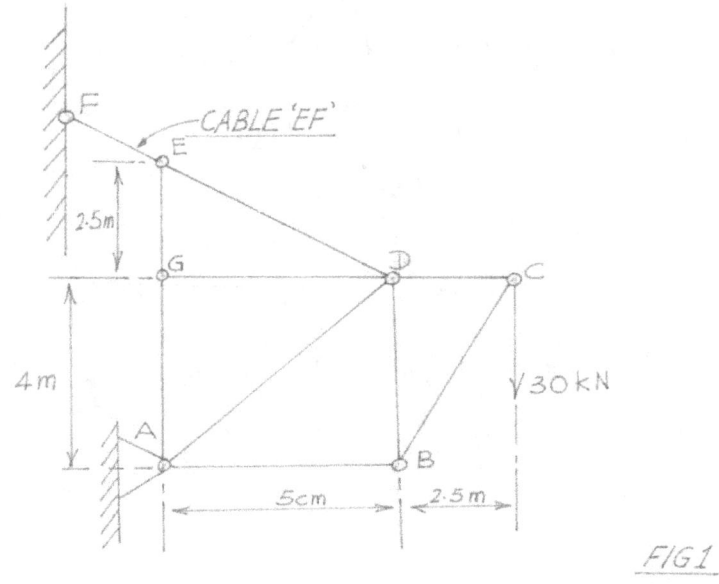

FIG 1

The line of action in the cable must be along EF. This force, let us call it F_{EF}; the load of 30kN at C; and, the reaction at A constitute a system of three non-parallel forces. And for the system to be in equilibrium, the lines of action of the forces must be concurrent.

Therefore, in order to determine the line of action of the reaction at A, line FE is projected to meet the line of action of the external load 30kN acting through C, at Q.

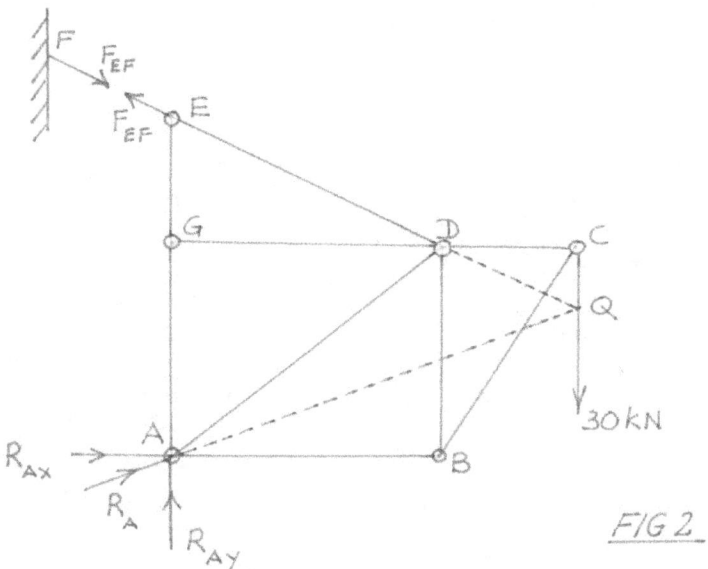

FIG 2

The reaction at A must therefore pass through Q. In Fig. 2, I have shown this construction. In the same diagram the components of the reaction at A are shown, R_A being equal to $\sqrt{R_{AY}^2 + R_{AX}^2}$.

In order to make the force diagram, a drawing of the framework is first prepared using an appropriate scale. Such a drawing, labelled in Bow's notation is provided at Fig. 3.

For the force diagram itself, a vertical line representing the external load of 30kN to some suitable scale is set off. This is line '*l* m' in Fig. 4. Then through '*l*' a line is drawn parallel to LR of Fig. 3; and similarly through 'm' a line is drawn parallel to the one joining A to Q, i.e. the line MR. Where these lines through *l* and m meet, establishes the point 'r'. Evidently vector mr and r*l* represent the reaction at A and tension in the cable respectively. Accordingly, the reaction at A is magnitude 37.5kN and it acts upwards to the right at 20° to the horizontal; the tension in the cable is 39kN and acts in the direction 'r*l*',, the force triangle '*l* mr/s' being in equilibrium with the arrows representing the forces proceeding anti-clockwise from '*l*' to 'm', 'm' to 'r,s', and from 'r,s,' back to '*l*'.

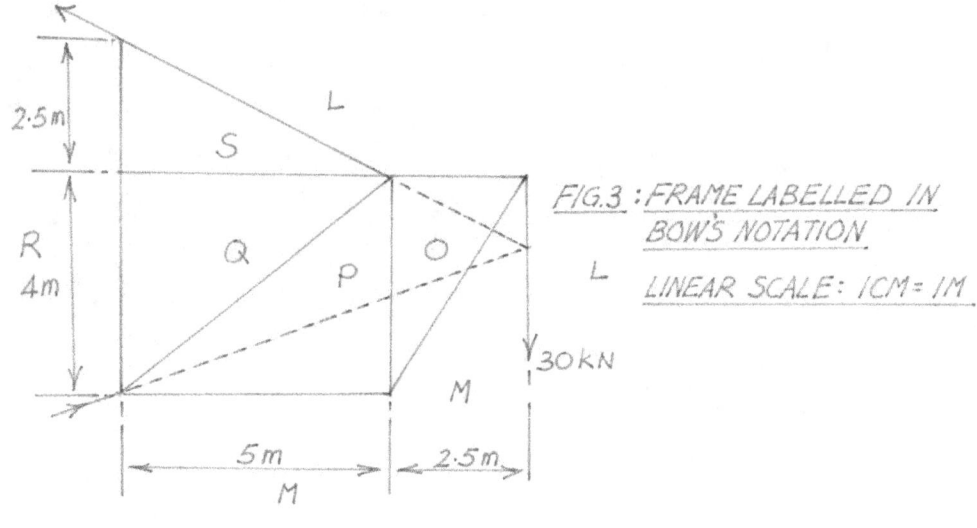

FIG.3 : FRAME LABELLED IN
        BOW'S NOTATION

L   LINEAR SCALE: 1CM = 1M

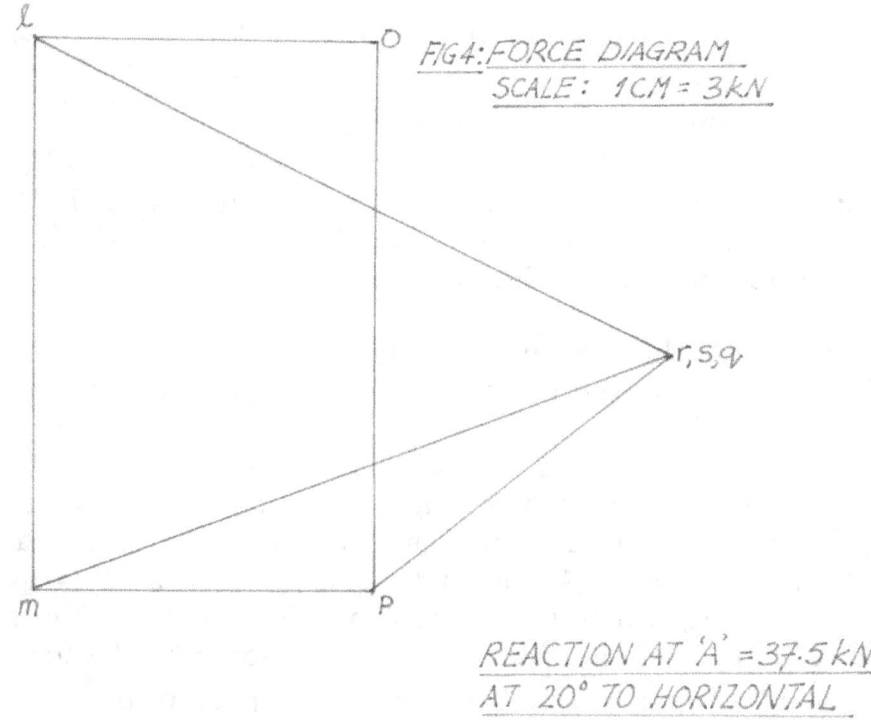

FIG4: FORCE DIAGRAM
      SCALE: 1CM = 3kN

REACTION AT 'A' = 37.5 kN
AT 20° TO HORIZONTAL

The remainder of the force diagram is completed in the usual manner.

The values of the scaled forces taken from Fig. 4 and indications of whether they are struts or ties are given in the table below.

~ 128 ~

TABLE 1

Member as designated in Fig. 1	Member as designated in Force Diagram	Magnitude in kN	Strut(S) or Tie(T)
AB	mp	-18.6	S
BC	om	-35.7	S
CD	ol	18.9	T
DE	ls	39	T
EG	sr	0	-
GD	qs	0	-
AG	qr	0	-
AD	pq	-20.7	S
BD	op	30	T

To identify the struts and ties we consider the joints individually. Commencing with joint 'C' in Fig. 1 where we know that force LM is downwards, the relevant triangle in the force diagram is lmo. Starting at 'l' and moving to m, the only path back to 'l' is from 'm' to 'o' and from 'o' to 'l'. See Fig. 5a.

FIG. 5a

FIG. 5b

When the directions $\overrightarrow{mo}$ and $\overrightarrow{ol}$ are inserted in the members at the joint it is clear that BC is a strut and CD a tie. See Fig. 5(b).

Next we consider joint B. With reference to Fig. 4, the relevant triangle is omp. Because we previously identified the force 'om' to be a 'compression' (i.e. the member OM is a strut) it should be clear that at joint B the direction of the arrow representing om at B must be in the direction shown in Fig. 6b.

FIG. 6a

FIG. 6b

When this direction is inserted in the relevant force triangle at Fig. 6a, the arrows must proceed in an anti-clockwise manner to maintain equilibrium. These directions : $\overrightarrow{mp}$ to the right; and 'po'↑ upwards are then transferred to Fig. 6(b). Thus, AB is a strut and 'BD' a tie.

Finally, we consider joint D.

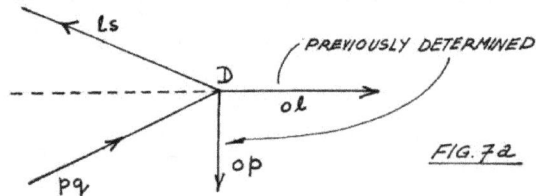

FIG. 7a

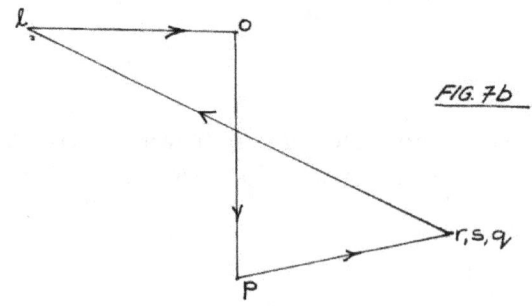

FIG. 7b

From previous analysis 'o*l*' the force in CD was identified as a tie, likewise 'op' the force in BD. Therefore straightaway in Fig. 7a showing the lines of action of the forces at joint 'D', we enter the directions of the forces in CD and BD. The relevant polygon in the force diagram is 'loprsq'. Inserting the directions of 'BD' and 'CD' i.e. from 'l' to 'o' and from 'o' to 'p', equilibrium is maintained by going from 'p' to r,s,q and back to '*l*' from r,s,q. The directions $\overrightarrow{pq}$ and $\overrightarrow{sl}$ are then inserted in Fig. 7(a) on which basis we see that 'pq' is a strut and '*l*s' a tie. Refer to TABLE 1.

Q14. Solve Question 2 by using the following :

method of joints
method of sections
method of tension coefficients

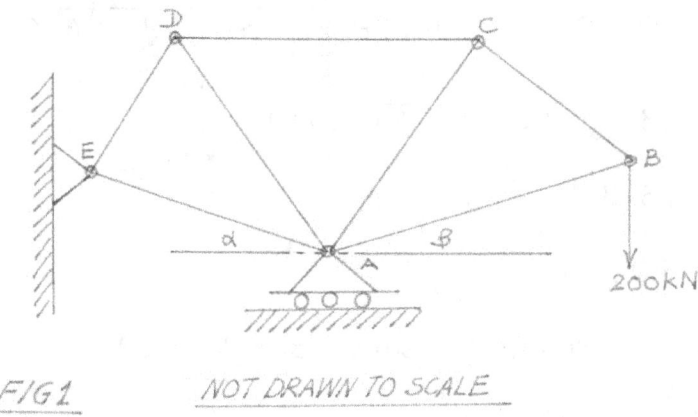

FIG 1   NOT DRAWN TO SCALE

Because the length of members and the angles these make with the horizontal and vertical are necessary for the calculations, let us obtain this information at once:

$$\text{Tan}\,\alpha \ = \ 3/6 \ = \ 0.5$$

$$\therefore \quad \alpha \ = \ 26.56°$$

$$\text{Tan}\,\beta \ = \ 3/7.5 \ = \ 0.4$$

$$\therefore \quad \beta \ = \ 21.8°$$

$$\text{Also} \quad \text{Tan}30° \ = \ \frac{AT}{6}$$

$$\therefore \quad AT \ = \ \frac{6}{\sqrt{3}} = 6\frac{\sqrt{3}}{3}\,m = 2\sqrt{3}\,\text{m}$$

so that $\qquad$ AD = AC = $\sqrt{48}$ m

Further $\qquad$ AE = $\sqrt{45}$ m

$\qquad$ AB = $\sqrt{65.25}$ m

Applying cosine law to triangles ADE and ABC

In triangle ADE : $(DE)^2 = (\sqrt{48})^2 + (\sqrt{45})^2 - 2\sqrt{48}\sqrt{45}\ Cos33.44^o$

$\qquad$ = 48 + 45 - 2 $\quad$ $\sqrt{2160\ (0.8345)}$

i.e. $\quad$ $(DE)^2 = 93 - 77.6 = 15.4$

$\therefore$ $\qquad$ DE = 3.924 m

In triangle ABC : $(BC)^2 = (\sqrt{48})^2 + (\sqrt{65.25})^2 - 2\sqrt{(48)65}\ Cos38.2°$

$\qquad$ = 48 + 65.25 - $2\sqrt{3120\ (0.7859)}$

$\qquad$ = 25.29

$\therefore$ $\qquad$ BC = 5.029m

Applying the Sine rule to the same triangles we have :

$$\frac{3.924}{Sin33.44^o} = \frac{\sqrt{48}}{SinDEA}$$

$\therefore$ $\qquad$ SinDEA = $\dfrac{0.5511 \times 6.328}{3.924}$

$\qquad$ = 0.9730

$\therefore$ $\qquad$ $\angle$DEA = 76.65°

and

$$\frac{5.029}{Sin38.2} = \frac{\sqrt{48}}{SinABC}$$

$$\therefore \quad SinABC = \frac{\sqrt{48}(0.6184)}{5.029}$$

$$= 0.8519$$

$$\therefore \quad \angle ABC = 58.42°$$

Therefore

$$\angle ADE = 69.96° \quad and \quad \angle ACB = 83.38°$$

All these values of length and of angles are entered in Fig. 2.

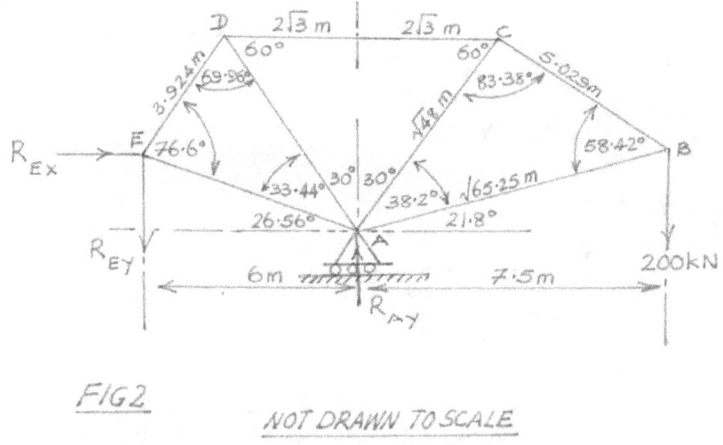

FIG 2.   NOT DRAWN TO SCALE

Because of the rollers at A the reaction there, R_{AY}, is entirely vertical. Evidently R_{EX} is zero and taking moments about A proves that R_{EY} = 250kN. Therefore R_{AY} = 450kN.

Forces by Method of Joints.

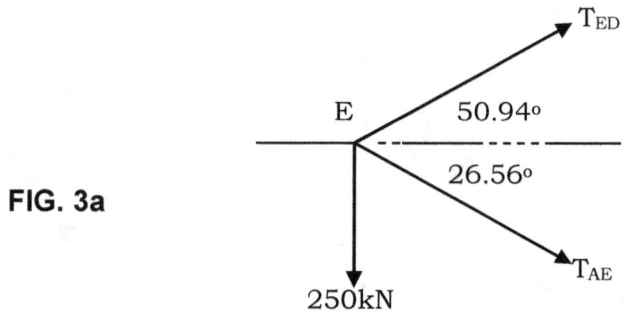

FIG. 3a

Consider Fig. 3a

For $\quad \sum F_X = 0$

$T_{ED}Cos50.04^o + T_{EA}Cos26.56^o = 0$

i.e.

$\qquad T_{ED}(0.6423) = -0.8945T_{AE}$

$\qquad$ or $\quad T_{ED} = -1.393T_{AE}$

For $\quad \sum F_y = 0$

$\qquad T_{ED}Sin50.04^o - T_{AE}Sin36.56^o - 250 = 0$

from which

$\qquad 1.5148T_{AE} = -250$

$\qquad$ or $\quad T_{AE} = -165kN$

and $\qquad T_{ED} = +229.8kN$

Consider joint 'B' next. See Fig. 3b

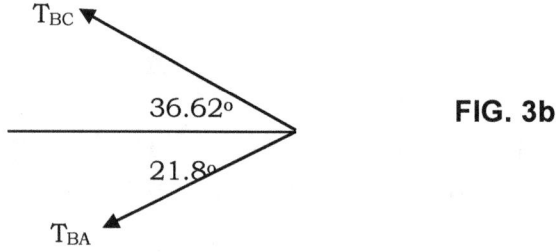

FIG. 3b

For $\quad \sum F_X = 0$

$\qquad T_{BC}Cos36.62 + T_{BA}Cos21.8^o = 0$

$\qquad$ i.e. $\quad T_{BC} = -1.1569T_{BA}$

and

For $\quad \sum F_Y = 0$

$\qquad T_{BC}Sin36.62^o - T_{BA}Sin21.8^o - 200 = 0$

from which

T_{BA} = -188kN

T_{BC} = +218kN

Because the magnitude and nature of the forces in AE and AB are unknown, there are only 2 unknowns at joint A at this juncture. Therefore you may now consider equilibrium at joint 'A'. See Fig. 3c.

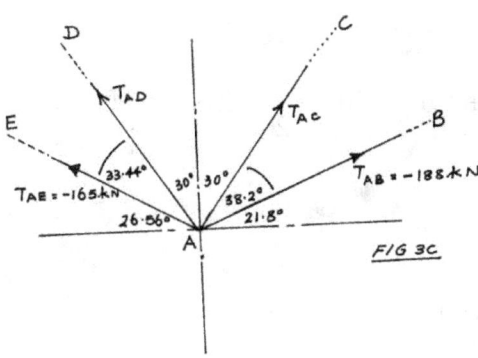

For $\sum F_X = 0$

$T_{AE}Cos26.56 + T_{AD}Cos60° - T_{AC}Cos60° - T_{AB}Cos21.8° = 0$

i.e. $-165Cos26.56 + T_{AD}(0.5) - T_{AC}(0.5) - (-188)Cos21.8° = 0$

i.e. $-147.6 + 0.5T_{AD} - 0.5T_{AC} + 174.6 = 0$

i.e. $T_{AD} - T_{AC} = -54$

For $\sum F_y = 0$

$T_{AE}Sin26.56 + T_{AD}Sin69° + T_{AC}Sin60° + T_{AB}Sin21-8° = 0$

∴ $T_{AD} + T_{AC} = -363.8$

so that T_{AD} = -204kN

T_{AC} = -150kN

Thus, the members AE, AD, AC and AB are all struts.

Now, take joint 'D'. See Fig. 3d.

~ 135 ~

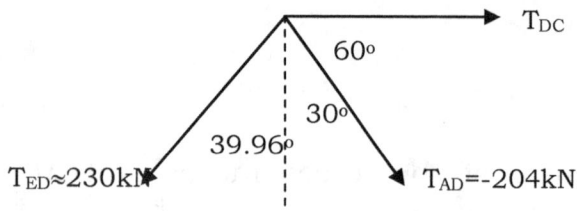

FIG. 3d

For $\sum F_X = 0$

$$T_{ED}\text{Sin}39.96° - T_{AD}\text{Cos}60° - T_{DC} = 0$$

i.e. $230(0.6423) - (204)\cdot\dfrac{1}{2} - T_{DC} = 0$

$+ 102 - T_{DC} = 0$

$\therefore \quad T_{DC} = 250\text{kN}$

Summarising the results, we have :

AE = -165kN, Strut   ;   AC = -150kN, Strut

ED = 230kN, Tie   ;   AB = -188kN, Strut

BC = 218kN, Tie

CD = 250kN, Tie

AD = -204kN, Strut

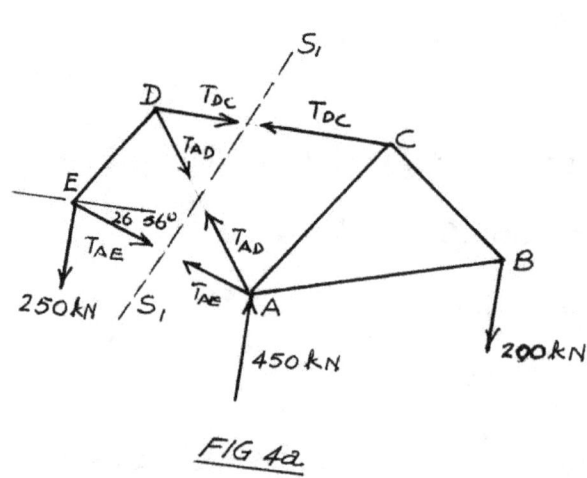

FIG 4a

Forces by Method of Sections

Cut the truss by section S_1S_1. The internal forces exposed are in members AD, AE and CD. For each of the 2 sections of the truss to remain in equilibrium the forces exposed must be balanced by any applied load and external reaction acting on each section. Thus for the left-hand part internal forces in AE, AD and DC must be balanced by the reaction at E. Similarly, for the right-hand section, the internal forces in AE, AD and DC must balance the external reaction at A and the load at B.

Considering equilibrium of the structure to left of section S_1S_1, we have :

For $\sum F_X = 0$

$$T_{DC} + T_{AD}Cos60° + T_{AE}Cos26.56 = 0$$

and

For $\sum F_y = 0$

$$T_{AD}Sin60° + T_{AE}Sin26.56° + 250 = 0$$

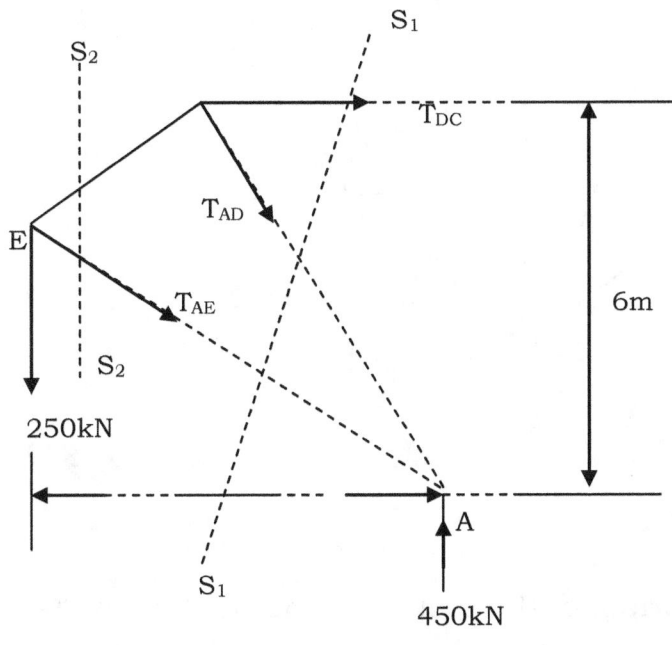

FIG. 4b

The lines of action of T_{AE} and T_{AD} pass through 'A'; thus these forces have no moments about this point; only the unknown force T_{DX} and the 250kN reaction at E have moments about A. See Fig. 4b.

so for $\sum M_A = 0$

$$250(6) - T_{DC}(6) = 0$$

Substitution of this result in the equation for $\sum F_X = 0$

$$250 + 0.5T_{AD} + T_{AE}(0.8945) = 0$$

As before $\sum F_Y = 0$, gives

$$T_{AD} \cdot \left(\frac{\sqrt{3}}{2} \right) + T_{AE}(0.4471) + 250 = 0$$

Solving these 2 equations gives

$$T_{AE} = -165.5\text{kN}$$
$$T_{AD} = -204\text{kN}$$

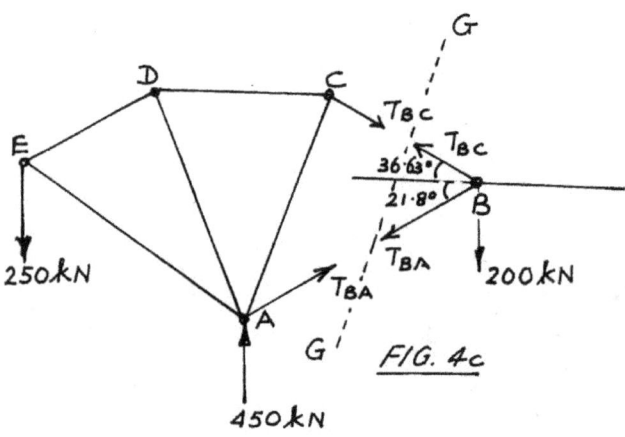

FIG. 4c

Consider next equilibrium of the section to the right of cut 'GG'

For $\sum F_X = 0$

$$T_{BC}Cos36.62 + T_{BA}Cos21.8° = 0$$

and

For $\sum F_Y = 0$

$$T_{BC}Sin36.62° - T_{BA}Sin21.8° - 200 = 0$$

from which $T_{BC} = +218kN$ and $T_{BA} = -188kN$

With reference to Fig. 4b, consider equilibrium of section to left of cut 'S₂S₂'.

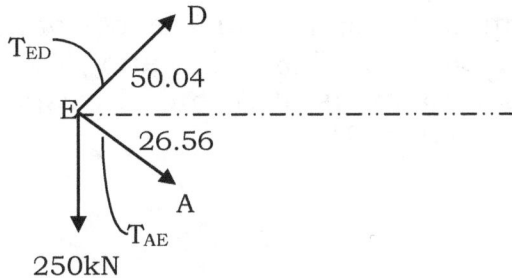

For $\sum F_X = 0$

$$T_{ED}Cos50.04° + T_{EA}Cos26.56° = 0$$

For $\sum F_Y = 0$

$$T_{ED}Sin50.04° - T_{EA}Sin26.56° - 250 = 0$$

It is already known that $T_{AE} = -165.5kN$

∴ $T_{ED} = 230kN$

There is only one force remaining to be determined i.e. T_{AC}. Instead of taking a section that would simplify its determination let us by way of illustrating the method more expansively, consider the section KK in Fig. 4d.

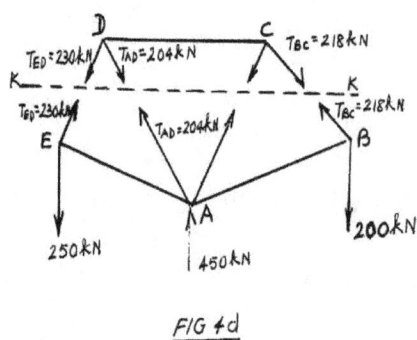

FIG 4d

There are only internal forces exposed in the upper section of the frame; no external reactions; no imposed loads. In the bottom section these same exposed internal forces T_{ED}, T_{AD}, T_{AC} and T_{BC} must be in equilibrium with the external reactions at A and E, respectively 450kN and 250kN; and the load of 200kN at B.

Taking the top section first :

For $\quad \sum F_Y = 0$

$\quad\quad T_{AD}\cos30° + T_{ED}\cos39.96° + T_{AC}\cos30° + T_{BC}\cos53.38° = 0$

i.e. $\quad -204(0.866) + 230(0.7665) + T_{AC}(0.866) + 218(0.5965) = 0$

or $\quad -176.7 + 176.3 + 0.866T_{AC} + 130 = 0$

$\quad\quad$ i.e. $\quad 0.866T_{AC} = -129.6$

$\quad\quad\quad\quad T_{AC} = -150kN$

Now, the bottom section

$450 - 250 - 200 + T_{ED}\cos39.96° + T_{AD}\cos30° + T_{AC}\cos30° + T_{BC}\cos43.38 = 0$

$230(0.7665) + (-204)(0.866) + 218(0.5965) = 0$

$176.3 - 176.7 + T_{AC}(0.866) + 130 = 0$

$\quad\quad \therefore \quad T_{AC} = -150kN$

Let us summarise the results :

AE = -165.5kN, Strut

ED = 230kN, Tie

BC = 218kN, Tie

CD = 250kN, Tie

AD = -204kN, Strut

AC = -150kN, Strut

AB = -188kN, Strut

Forces by the method of Tension Coefficients

Recall the equilibrium relationships for plane frames, i.e. say X- and Y-axes, viz :

For X-direction

$\sum$ tension coefficient (projected length of member in positive direction of X- axis) + X = 0; and

For Y-direction

$\sum$ tension coefficient (projected length of member in positive direction of Y-axis) + Y = 0

Let us redraw the truss and label it fully; also define the positive directions of the co-ordinate-axis system. See Fig. 5

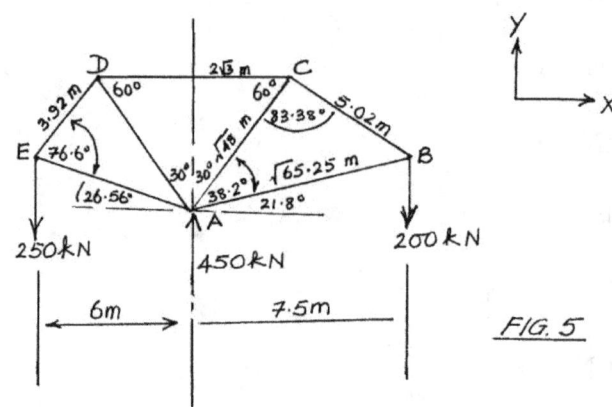

FIG. 5

Considering joint E first, you have for X-direction.

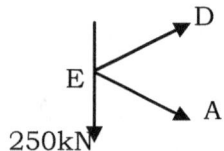

250kN

ED (projection of ED   +   EA(projection of   = 0
      on X-axis)   EA on X-axis)

Note that instead of writing the coefficient of ED as t_{ED} we merely used the line designation; the same was done for t_{EA}.

∴   ED(EL) + EA(MA) = 0
or   ED(3.994Cos50.04) + EA(6) = 0

Observe carefully that ED and EA considered as ties have their projections in the positive direction of the X-axis. The latter equation reduces to

2.52ED + 6EA = 0   . . . . . (i)

For the same joint but considering the Y-direction

ED(Projection of ED   + EA(Projection of = -250= 0
        on Y-axis)   EA on Y-axis)

i.e.   ED(EL) – EA(EM) – 250 = 0

~ 142 ~

the minus signs because the projection of EA on the Y-axis, EA being considered a tie, is in the negative direction of this axis; and the external reaction of 250kN at E is also in the negative direction of Y-axis.

$$\therefore \quad 3ED - 3EA = 250 \qquad \qquad \cdots \cdots \text{(ii)}$$

Solving equations (i) and (ii) for Ed and EA gives

$$8.52ED = 500$$

$$\text{or} \quad ED = 58.69$$

$$\text{and } 3EA = -73.94$$

$$\text{i.e.} \quad EA = -24.65$$

i.e. tension coefficient of Ed = 58.69 and tension coefficient of EA = -24.65.

Consider next joint 'D'

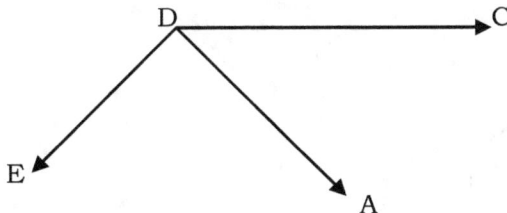

Remember always in tension-coefficient analysis that all members are assumed to be ties. Generally speaking this applies to the other methods of analysis as well.

For X-direction

DE(projection of DE + DA(projection of DA + on X-axis) on X-axis)

DC(projection of DC = 0 on x-axis)

i.e. DE(-LE) + DA(NA) + DC(NU) = 0

$$\text{or} \quad -DE(3.924 \cos 50.04°) + DA(\sqrt{48} \cos 60°)$$
$$+ DC(4\sqrt{3}) = 0$$

or $-2.52DE + 3.46DA + 6.92DC = 0$ . . . . . (iii)

Expanding the equilibrium equation in the Y-direction

DE(Projection of   +   DA(Projection of +
    DE on Y-axis)   DA on Y-axis)

DC(Projection of DC   =   0
    on Y-axis)

i.e. $DE(-DL) + DA(-DN)   =   0$

∴ $-3DE - 6DA = 0$ . . . . . (iv)

Solving (iii) and (iv)

$DA = -29.35$

$DA = 36.05$

Now, joint 'B'

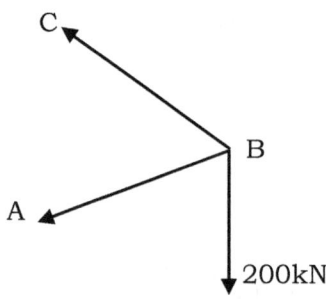

For X-direction

BA(projection of BA   + BC(projection of BC = 0
    in X-direction)   in X-direction)

i.e. $BA(-WA) + BC(-BG)   =   0$

or $-BA(7.5) - BC(5.029 Cos 36.62°) = 0$

or $-7.5BA - 4BC = 0$ . . . . . (v)

<u>For Y-direction</u>

BC(projection of BC + BA(projection of +Y_B = 0
    on Y-axis)   BA on Y-axis)

i.e.   BC(3) – BA(3) – 200 = 0

or   3BC – 3BA = 200   . . . . . (vi)

Solving (v) and (vi) for BA and BC you should get

    -34.5BA = 800

    i.e. BA = -23.2

so that
3BC – 3(-23.2)   = 200

    ∴   BC = 43.5

Finally, let us consider joint 'C'

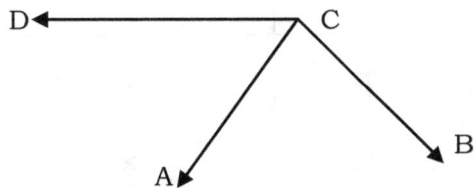

<u>For X-direction</u>

CB(projection of + CA(projection of +
    CB on X-axis)   CA on X-axis)

CD(projection of CD = 0
    on X-axis

i.e.   CB(GB) + CA(-AU) + CD(-4$\sqrt{3}$) = 0

or   CB(5.029Cos36.62°) – CA(3.46) - 6.92CD = 0

or   4CB – 3.46CA – 6.92CD = 0   . . . . . (vii)

Expanding the equilibrium equation in the y-direction

CB(projection of   + CA(projection of CA) +
    CB on Y-axis)   on Y-axis)

CD(projection of   = 0
    CD on Y-axis)

i.e.   CB(-CG) + CA(-CU) + 0 = 0

or   -3CB – 6CA = 0   . . . . . (viii)

But tension coefficient of BC = tension coefficient of CB

$\therefore$ -3(43.5) – 6CA = 0

$\therefore$ CA = -21.75

The values of tension coefficients and lengths of members of the framework being known, it merely remains to compute the magnitude of the force in each member by multiplying its tension coefficient by its length, as was done in table following.

Member	Tension Coefficient (t) kN/m	Length	Force = t x ℓ kN	Strut(S) or Tie(T)
AE	-24.65	$\sqrt{45}$	-165.4	S
ED	+58.69	3.924	+230.2	T
BC	+43.5	5.029	+218.8	T
CD	+36.05	6.92	+249.5	T
AD	-29.35	$\sqrt{48}$	-203.3	S
AC	-21.75	$\sqrt{48}$	-150.7	S
AB	-23.2	$\sqrt{65.25}$	-187.4	S

A comparison table is now drawn up showing the results for the 3 methods.

Member	Method of Joints kN	Method of Sections kN	Method of Tension Coefficients kN
AE	-165,S	-165.5,S	-165.4, S
ED	+230,T	+230,T	+230.2, S
BC	+218,T	+218,T	+218.8, T
CD	+250,T	+250,T	+249.5, T
AD	-204,S	-204,S	-203.3, S
AC	-150,S	-150,S	-150.7, S
AB	-188,S	-188,S	-197.4, S

Pretty good correspondence you would agree!

Q15. You are asked to design an umbrella truss for a promenade at a shopping plaza. The configuration together with estimated loading due to cladding and wind forces is shown in Fig. 1. Determine graphically the forces in the members of the principal rafter AJ and in the other bars which make up the left-hand half of the truss. The stanchion MJ is $(3+\sqrt{3})$ metres long.

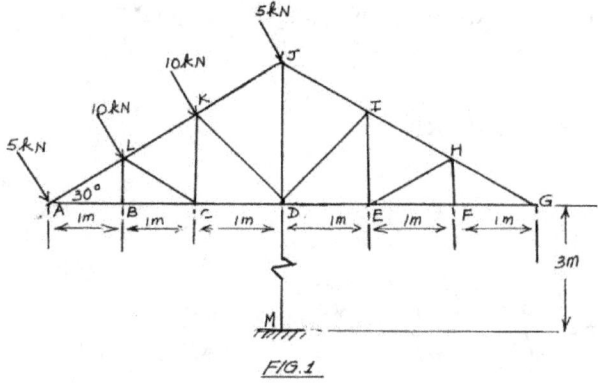

FIG.1

The reason why you are told the stanchion is $(3+\sqrt{3})m$ in length is to let you know that the truss is cantilevered on both sides from a single stanchion, JM, JD being $\sqrt{3}$ metres in length. It should be clear that for dead load due to cladding, the only reaction would be a vertical one at D. If however there are wind forces, the effects of which in combination with dead loads give rise to the inclined forces as indicated in Fig. 1, then we would expect a horizontal reaction at D and another at J.

In Fig. 2, I have shown horizontal and vertical reactions at D, viz R_{DH} and R_{DV} respectively and a horizontal reaction, R_{JH} at J. The vertical component of the reaction at D supports the dead load of the truss and the vertical components of the wind load. In Fig. 2, I have assumed the directions of R_{JH} and R_{DH}.

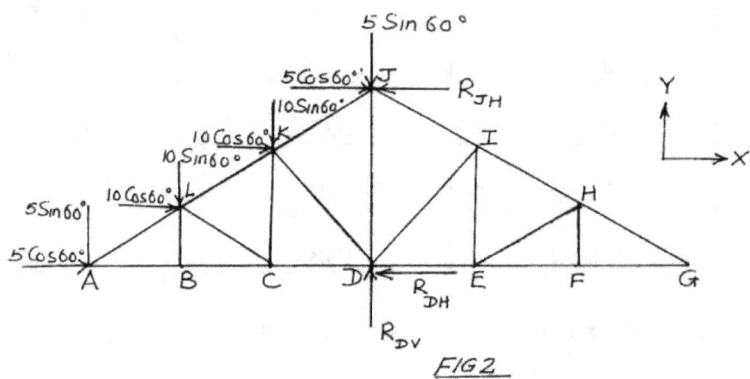

FIG 2

Also, as indicated, the wind loads at the joints A, L, K and J, have been resolved into their horizontal and vertical components

For $\sum F_X = 0$: $R_{DH} + R_{JH} = 30\cos60° = 15\text{kN}$

For $\sum F_Y = 0$: $R_{DV} = 30\sin60° = 30\dfrac{\sqrt{3}}{2}$

$\therefore$ $R_{DV} = 15\sqrt{3}\text{ kN}$

For $\sum M_J = 0$

$$R_{DH}\left(\frac{3}{\sqrt{3}}\right) - 5\left(\frac{3}{\sqrt{3}} - \frac{2}{\sqrt{3}}\right) - 10\frac{\sqrt{3}}{2}(1) - 5\left(\frac{3}{\sqrt{3}} - \frac{1}{\sqrt{3}}\right)$$

$$-10\frac{\sqrt{3}}{2}(2) - \frac{5}{2}\left(\frac{3}{\sqrt{3}}\right) - 5\frac{\sqrt{3}}{2}(3) = 0$$

i.e. $R_{DH}\left(\dfrac{3}{\sqrt{3}}\right) - \dfrac{5}{\sqrt{3}} - 5\sqrt{3} - \dfrac{10}{\sqrt{3}} - 10\sqrt{3} - \dfrac{15}{2\sqrt{3}} - \dfrac{15}{2}\sqrt{3} = 0$

or $3R_{DH} - 5 - 5(3) - 10 - 10(3) - \dfrac{15}{2} - \dfrac{15(3)}{2} = 0$

from which $\qquad 3R_{DH} = 90 \quad$ or $\quad R_{DH} = 30kN$

and because $R_{DH} + R_{JH} = 15kN$, it follows that $R_{JH} = -15kN$

Therefore the correct direction of R_{JH} is from left to right as shown in Fig. 3a.

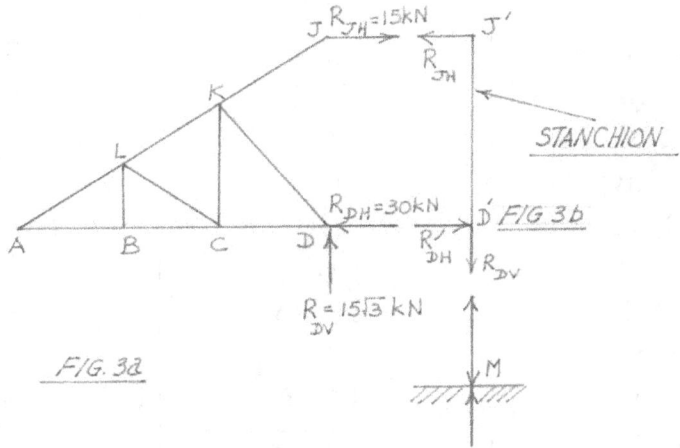

The reactions of the truss on the stanchion are shown in Fig. 3b.

What you have to remember is that JD is part of the stanchion and also a component part of the truss, providing the support reactions indicated in Fig. 3c. The truss is in fact supported at J and at D. In Fig. 3c you will observe that the right-hand portion of the truss has been removed. For loading of the right-hand portion, separate reactions at J and D would have to be computed for that particular circumstance and a separate force diagram constructed. In such a case the final load on the stanchion due to loading on the left-hand side of the truss and that on the right-hand side of the truss would be obtained by superposition.

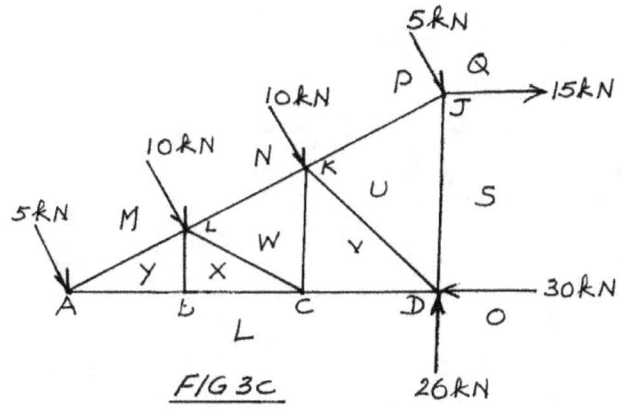

The force diagram for the loading configuration of Fig. 3c is presented as Fig 5. The magnitude and nature of each force is given in the table following.

Member	Magnitude kN	Strut(S) or Tie(T)
AB/ly	-10	S
BC/lx	-10	S
AL/my	-8.75	S
BL/xy	0	-
LK/nw	14.75	T
LC/wx	-11.5	S
CK/vw	6	T
CD/lv	-20.25	S
JK/pu	20	T
JD/su	-14.25	S
KD/uv	-15.25	S

As an additional note of interest, the bending moment at point 'D' on the stanchion is $R_{JH}(\sqrt{3})$ kNm anticlockwise; and at 'M' $R_{JH}\{(3+\sqrt{3})\}-30(3)$ kN i.e. 26kNm, anticlockwise and approximately 19kNm, clockwise, respectively. See Fig. 6. Evidently at some point between 'J' and 'M', the bending moment is zero. Can you locate it? Based on these findings, the bending stress in the stanchion can be determined. [Note : bending moment is zero at a distance of 1.27m from M.] You may wish to come back to this problem after reading the chapter on Bending Moment.

FIG. 4 SCALED DIAGRAM
OF UMBRELLA

TRUSS
SCALE : 1cm=0.4m

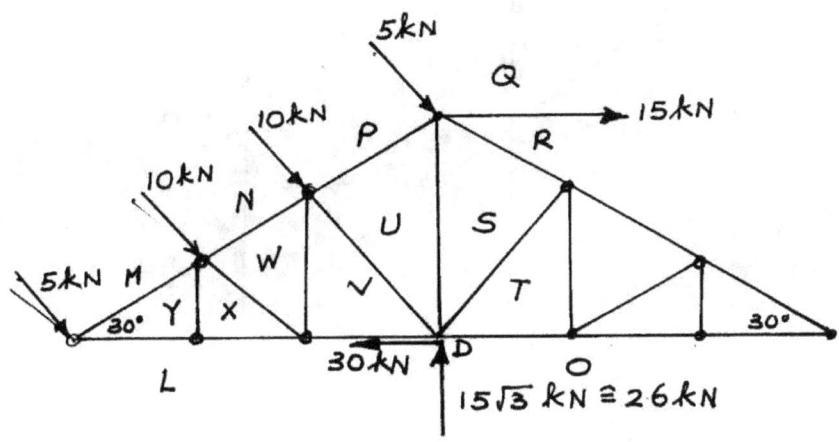

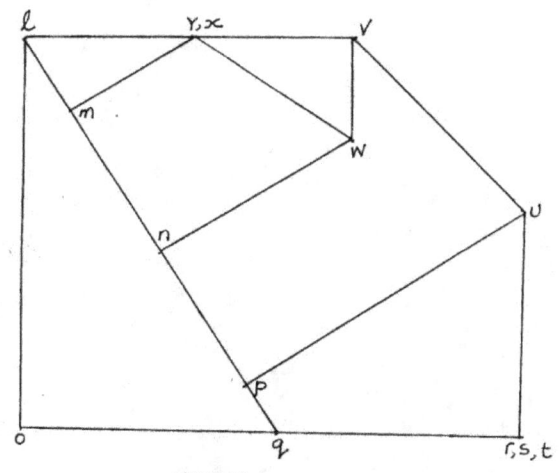

FIG. 5
FORCE DIAGRAM (ORIGINAL DIAGRAM
WAS DRAWN TO SCALE OF 1cm=2.5kN

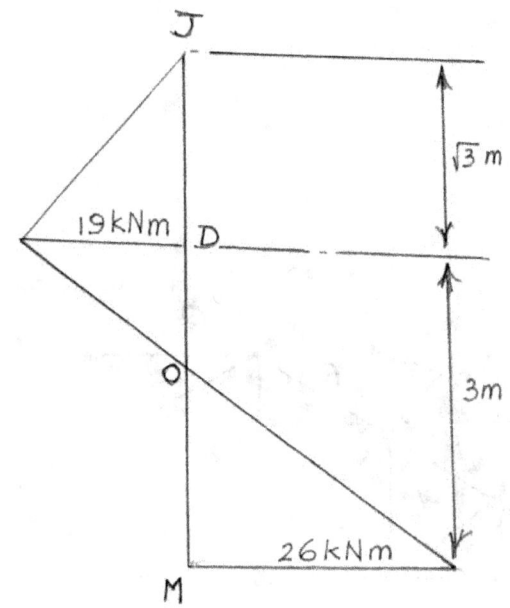

FIG 6: BENDING MOMENT DIAGRAM FOR
STANCHION JD

CHAPTER 3

CENTROIDS AND CENTRES OF MASS

The <u>centroid of an area</u> is a point generally located within the area but which also could be outside it, and where the total area may be thought of as being concentrated. Similarly, the centroid of a volume or a line may be thought of as being the point where, respectively, the total volume or total length is concentrated; all abstractions, these statements are really!

Now, the <u>centre of mass of a solid object</u> is that point generally in the object where its total mass may, as in the cases of centroids of areas, of volumes, and of lengths of line, be thought of as being concentrated. But is a centre of a mass of an object always its centroid? Let us answer this question by considering the solid steel ball depicted in Fig. 1a. The ball is assumed to be made of homogeneous material throughout. The ball's centroid coincides with its geometric centre; and its centroid is also its centre of mass. Look now at the drawing labelled Fig. 1b which is a three-dimensional representation of the steel ball in every particular except of course that it has no mass. Evidently, the geometric centre of the drawing is coincident with the centroid and centre of mass of the solid steel ball shown in Fig. 1a. Why is this so? It is because the steel ball is made of homogeneous material. That is to say the density of the steel ball being uniform, the location of its centroid does not differ from the position of the centroid in the drawing at Fig. 1b – the wholly geometric shape which was assumed to be without mass. We may therefore conclude that provided geometric shapes are equal, centroids of objects are coincident. Having said that it should be pointed out that the position of the centroid may not necessarily coincide with the position of the centre of mass for the same object, albeit they are of identical geometric shapes. A difference occurs when variable density is a factor. Consider a fabricated cylindrical bar of uniform section throughout, one half of which is made of, say, carbon fibre and the other of steel. Clearly, on the basis of the foregoing, the centroid of the bar is at its geometric centre – at a point on the junction of the carbon fibre and steel – whereas its centre of mass is evidently somewhere on the steel-half side of the bar. Thus, a centroid of an object coincides with its centre of mass only when the object is homogeneous, i.e. of constant density.

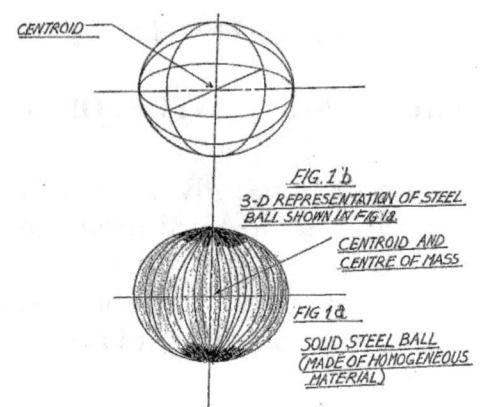

Determination of the centroids of structural sections is an important prerequisite in the calculation of their bending stresses when under load. Cross-sections of some of the more popular structural components are shown in <u>Fig. 2</u>. I use the term centroid in this text in reference to lines, areas, volumes and homogeneous solids. On the other hand centre of mass is used in reference to solid bodies, generally but not exclusively confined to those with varying density.

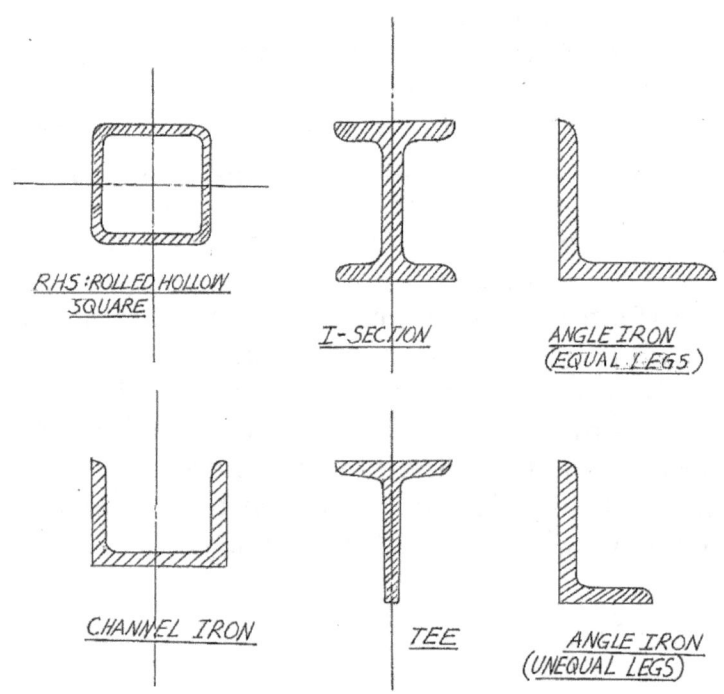

R.H.S :ROLLED HOLLOW
SQUARE

I-SECTION

ANGLE IRON
(EQUAL LEGS)

CHANNEL IRON

TEE

ANGLE IRON
(UNEQUAL LEGS)

FIG.2   SOME POPULAR STRUCTURAL SECTIONS

Centroids of Areas and Volumes

(i)   Centroids of Area

Consider a plane area made up of 'n' component areas a_1, a_2, a_3.... a_n of which their centres of areas are at distances x_1, x_2, x_3... x_n respectively from the Y-axis. Also, let the corresponding distances of their centres of areas from the X-axis be y_1, y_2, y_3, ... y_n respectively. The coordinates $\bar{x}$, $\bar{y}$ of the centroid of the composite area are given by :

Distance from Y-axis : $\bar{x} = \dfrac{a_1 x_1 + a_2 x_2 + a_3 x_3 + ...a_n x_n}{a_1 + a_2 + a_3 + ...a_n}$

or more concisely,

$$\bar{x} = \frac{\sum ax}{\sum a}$$

Similarly,

Distance from X-axis : $\bar{y} = \dfrac{\sum ay}{\sum a}$

(ii)   Centroids of Volumes

Assume that each of the same 'n' component areas in the preceding article has a uniform but unique thickness, say, 't'. Evidently each area 'a' multiplied by its thickness 't' is its volume. Therefore, the set of 'n' component volumes, respectively V_1, V_2, V_3,.... V_n may be conveniently represented in a 3-dimensional region, say, R^3 : the X- and Y-axes as before, and a third axis, Z for the thickness dimension. Now if the distances of each centre of volume from the YZ-, XY- and XY-planes are x,y, and z respectively then the coordinates $\bar{x}$, $\bar{y}$ and $\bar{z}$ of the centroid of the composite volume are :

$$\bar{x} = \frac{V_1 x_1 + V_2 x_2 + V_3 x_3 + ... + V_n x_n}{V_1 + V_2 + V_3 + ... + V_n}$$

$$\bar{y} = \frac{V_1 y_1 + V_2 y_2 + V_3 y_3 + ... + V_n y_n}{V_1 + V_2 + V_3 + ... + V_n} \quad \text{and}$$

$$\bar{z} = \frac{V_1 z_1 + V_2 z_2 + V_3 z_3 + ... + V_n z_n}{V_1 + V_2 + V_3 + ... + V_n}$$

It should be noted that because the areas 'a' remain the same and the thickness 't' of each area is uniform, the xs in this analysis are the same as in the previous. The foregoing result may be expressed more concisely as:

$$\bar{x} = \frac{\sum Vx}{\sum V} \quad ; \quad \bar{y} = \frac{\sum Vy}{\sum V} \quad \text{and} \quad ; \quad \bar{z} = \frac{\sum Vz}{\sum V}$$

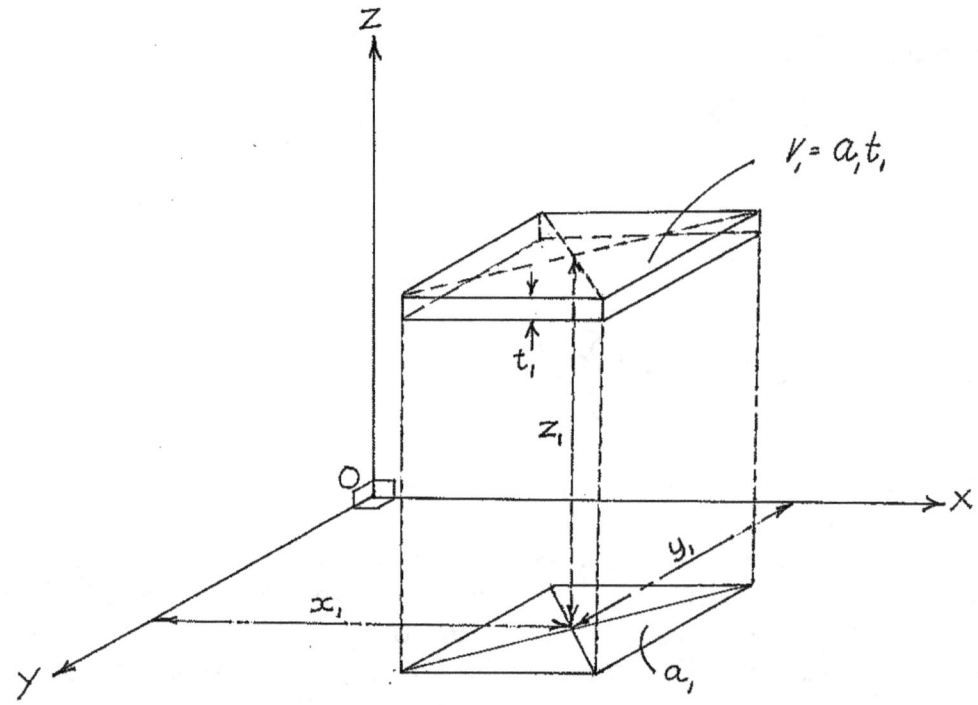

FIG3   A TYPICAL IDEALISED COMPONENT AREA AND COMPONENT VOLUME

Typical components each of area of volume are shown in Fig. 3

Let us now illustrate the determination of centroid of an area and of a volume by means of the following examples.

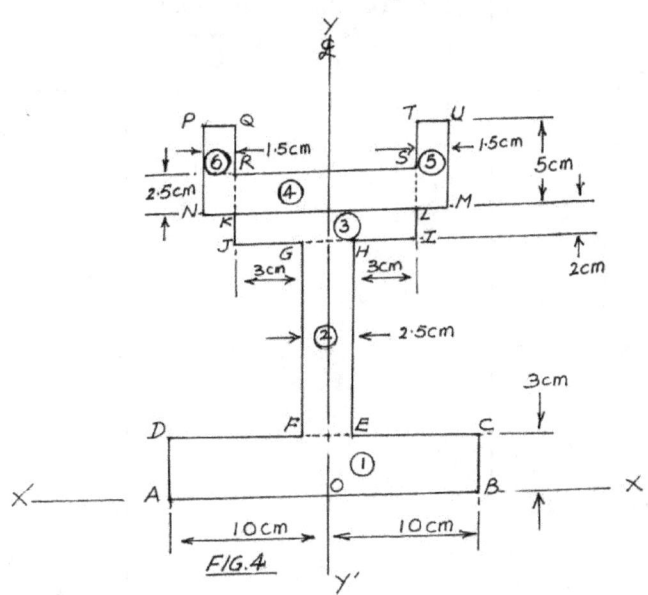

FIG.4

Illustrative Example 1

To determine the centroid of the cross-sectional area shown in Fig. 4:

The position of a centroid has to be expressed with reference to datum axes. For this illustration it is observed that the area is symmetrical about its centre line. Therefore the Y-axis is chosen to coincide with this line; the X-axis runs along the base of the area; and, the origin of the axes is at 'O'

Evidently $\bar{x} = 0$ because the centroid must be on the Y-axis.

The cross-sectional area is divided into a number of common geometrical shapes and the distances of the centre of these shapes from the Y-axis recorded. For example, for the flange ABCD we have, say, $a_1 = 20 \times 3 = 60 cm^2$; and its centre of area $y_1 = 3/2cm$ i.e. 1.5cm from the X-axis.

I have shown the details of the calculations in the following tabulation.

Area Designation 'a' (cm)^2	Distance 'y' of geometric centre of area from X-axis (cm)	Product 'ay' (cm)^3
ABCD : 20 x 3 = 60	1.5	90
EFGH : 12 x 2.5 = 30	9	270
IJKL : 8.5 x 2 = 17	16	272
KRSL : 8.5 x 2.5 = 21.25	18.25	387.8
NPQK + *TLUM* $= 2\{1.5 \times 5\} = 15$	19.5	292.5
$\sum a = 143.25$ (cm)^2		$\sum ay = 1312.3$ (cm)^3
so that	$\bar{y} = \dfrac{1312.3}{143.25} = 9.2 cm$	

Therefore, the coordinates of the centroid of the cross-sectional area shown in Fig. 4 are (0, 9.2)

Illustrative Example 2

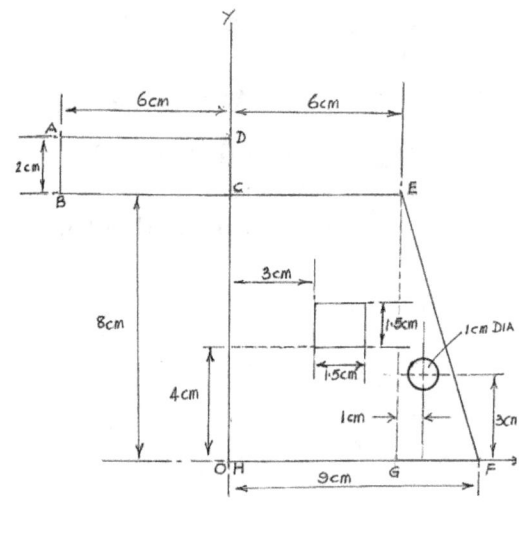

FIG. 5

~ 158 ~

To determine the coordinates of the centroid of the area shown in Fig. 5 with reference to the axes OX and OY as indicated.

Observing that the area ABCD is to the left of the OY axis, the distance of its centroid must be assigned a negative sign. Without much ado therefore we may write down the expressions for $\bar{x}$ and $\bar{y}$, viz.

$$\bar{x} = \frac{(6\times2)(-3) + (8\times6)(3) - (1.5\times1.5)(3.75) + \left(\dfrac{8\times3}{2}\right)(7) - \pi(1)^2(7)}{(6\times2) + 8(6) - (1.5\times1.5) + \left(\dfrac{8\times3}{2}\right) - \pi(1)^2}$$

$$= \frac{161.6}{66.6}$$

$$\bar{x} \approx 2.4cm$$

and

$$\bar{y} = \frac{(6\times2)(9) + (8\times6)(4) - (1.5\times1.5)(4.75) + \dfrac{8(3)}{2}\left(\dfrac{8}{3}\right) - \pi(1)^2(3)}{(6\times2) + 8(6) - (1.5\times1.5) + \left(\dfrac{8\times3}{2}\right) - \pi(1)^2}$$

$$= \frac{311.3}{66.6}$$

i.e. $\bar{y} \approx 4.7cm$

Therefore the coordinates of the centroid of the area are (2.4, 4.7)

Illustrative Example 3

To determine the coordinates of the centroid of the volume of the homogenous body the elevation of which in the XY-plane is shown in Fig. 6. Each of the three sections whose faces are ABCD, BEFG and FHIK are uniform thickness of 3cm, 5cm and 4cm respectively.

As in the previous illustration, details of the required workings are shown in tabular form for ease of illustration.

The X-, Y-, and Z- reference axes are shown in the accompanying diagrams.

Volume Designation (cm)³	'y' from X-Z plane (cm)	'x' from YZ-plane (cm)	'z' from XY-plane (cm)
'ABCD', 3cm thick = 8 x 2 x 3 = 48	6	1	1.5
'BEFG', 5cm thick = 2 x 10 x 5 = 100	1.5	4	2.5
'FHIK' 4cm thick = 2 x 5 x 4 = 40	2.5	9	2

Based on these data,

$$\bar{x} = \frac{48(1) + 100(4) + 40(9)}{48 + 100 + 40} = \frac{808}{188} = 4.3cm$$

$$\bar{y} = \frac{48(6) + 100(1.5) + 40(2.5)}{48 + 100 + 40} = \frac{538}{188} = 2.9cm$$

$$\bar{z} = \frac{48(1.5) + 100(2.5) + 40(2)}{48 + 100 + 40} = \frac{402}{188} = 2.1cm$$

Therefore the coordinates of the centroid of the volume of the object shown in Fig. 6 with reference to the X-, Y-, and Z- axes as shown in the diagram are (4.3, 2.9, 2.1).

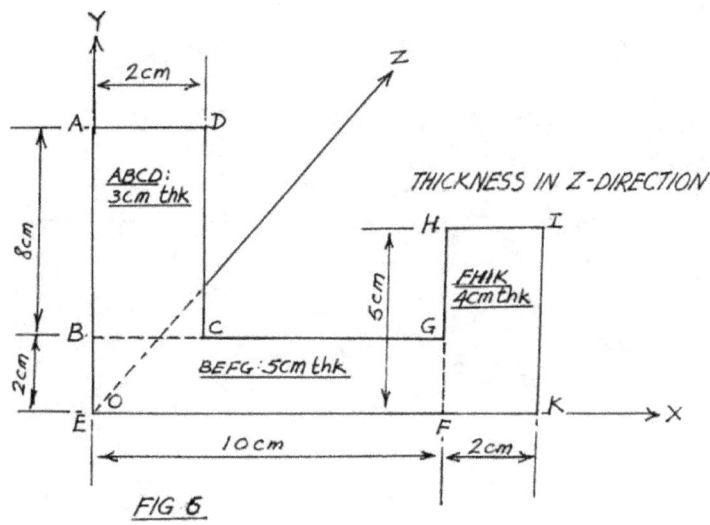

FIG 6

(iii)   Centre of Mass of Heterogeneous Solids

I commence this section with the concept of the mathematical fiction of a point considered as a particle with mass 'm' whose position is fixed in the conventional X-, Y-, Z-system of orthogonal axes and with the definition of first moments of said mass 'm' with reference to the YZ-, XZ-, and XY-planes respectively, viz M_{YZ}, M_{XZ} and M_{XY} and given by the following:

$$M_{YZ} = mx; \quad M_{XZ} = my; \quad M_{XY} = mz$$

where x, y and z are distances of 'm' from the ZY-, XZ- and XY-planes respectively as shown in Fig. 7.   Imagine now a set of points or particles of masses m_1, m_2, m_3,..., m_n in the same X-, Y-, and Z-space with 'x' coordinates = x_1, x_2, x_3,..., x_n; 'y' coordinates : y_1, y_2, y_3, ... y_n; and 'z' coordinates : z_1, z_2, z_3...., z_n, respectively.

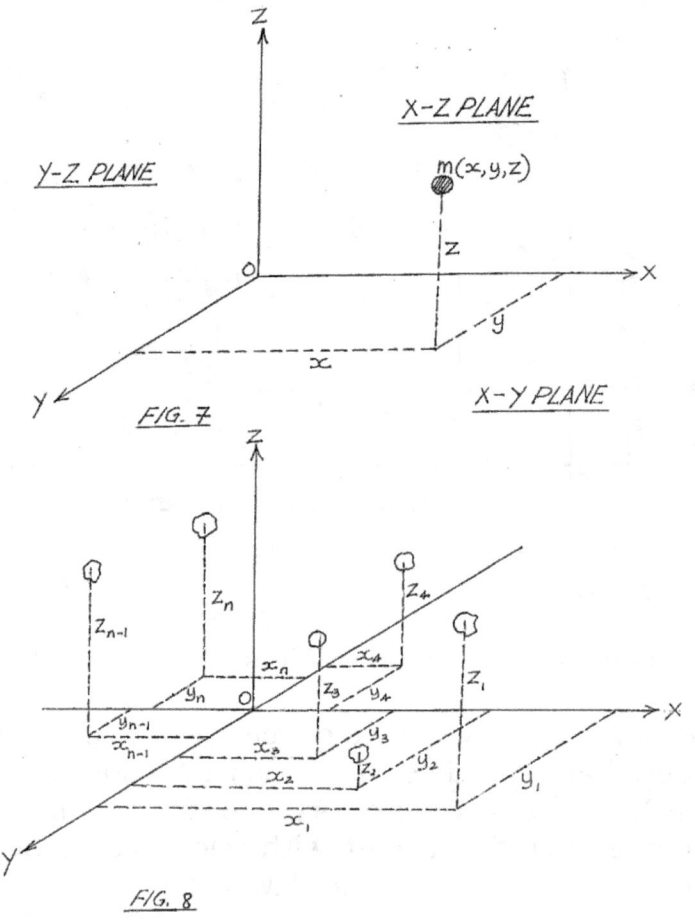

FIG. 7

FIG. 8

Refer to Fig. 8. Based on the definition of first moment of area :

$$M_{ZY} = m_1x_1 + m_2x_2 + m_3x_3 \text{ " } ..... + m_nx_n = \sum_1^n m_1 x_1$$

$$M_{XZ} = m_1y_1 + m_2y_2 + m_3y_3 + ..., m_ny_n = \sum_1^n m_1 y_1$$

$$M_{XY} = m_1z_1 + m_2z_2 + m_3z_3 + ... + m_nz_n = \sum_1^n m_1 z_1$$

from which we may write

$$\overline{x} = \frac{\sum_1^n m_1 x_1}{\sum_1^n m_1}$$

~ 162 ~

$$\bar{y} = \frac{\sum\limits_{1}^{n} m_1 y_1}{\sum\limits_{1}^{n} m_1}$$

$$\bar{z} = \frac{\sum\limits_{1}^{n} m_1 z_1}{\sum\limits_{1}^{n} m_1}$$

where $\bar{x}, \bar{y}, \bar{z}$ are the coordinates of all the centre of masses of all the particles.

Illustrative Example 4

Suppose particles of 10, 5 and 6 units of mass are located at points with X, Y and Z coordinates respectively (2, -5, 8), (-3, 1, 7) and (5, -2, -4), find the centre of mass of the system.

$$\bar{x} = \frac{10(2) + 5(-3) + 6(5)}{10 + 5 + 6} = \frac{35}{21} = \frac{5}{3}$$

$$\bar{y} = \frac{10(-5) + 5(1) + 6(-2)}{10 + 5 + 6} = \frac{57}{21} = -\frac{19}{7}$$

$$\bar{z} = \frac{10(8) + 5(7) + 6(-4)}{10 + 5 + 6} = \frac{91}{21} = \frac{13}{3}$$

$\therefore$ Coordinates of the centre of mass of the system are $\frac{5}{3}, -\frac{19}{7}, \frac{13}{3}$.

Thus the system of masses considered in this example is equivalent to a particle of mass of 21 units located at a point with coordinates $\frac{5}{3}, -\frac{19}{7}, \frac{13}{3}$

Consider next an infinitesimal volume $\delta V = \delta x\, \delta y\, \delta z$ in a region in conventional X, Y, Z space. If this volume has a density 'ρ' which is independent of its position in the region, that is to say density is constant, then its mass δM at position with coordinates x, y, z is given by:

$$\delta M = \rho\, \delta x\, \delta y\, \delta z$$

or $\qquad M = \rho \iiint dx\ dy\ dz$

and proceeding along similar lines as in the case of the particles, the first moments of mass are :

$$M_{YZ} = \rho \iiint x\ dx\ dy\ dz$$

$$M_{XZ} = \rho \iiint yd\ x\ dy\ dz \text{, and}$$

$$M_{XY} = \rho \iiint z\ dx\ dy\ dz$$

from which

$$\bar{x} = \frac{\rho \iiint x\ dx\ dy\ dz}{\rho \iiint dx\ dy\ dz} = \frac{\iiint x\ dx\ dy\ dz}{\iiint dx\ dy\ dz}$$

Similarly,

$$\bar{y} = \frac{\iiint y\ dx\ dy\ dz}{\iiint dx\ dy\ dz}$$

$$\bar{z} = \frac{\iiint z\ dx\ dy\ dz}{\iiint dx\ dy\ dz}$$

As you can see, these coordinates do not depend on density at all, the mass being homogeneous; centroid and centre of mass coincide.

Suppose now that density 'ρ' is varying throughout the region in X-, Y-, and Z-space that is to say the body is heterogeneous. This condition may be conveniently expressed by writing density as a function of x, y and z viz. $\rho(x, y, z)$. Accordingly

$$\delta M = \rho(x, y, z) \delta x\ \delta y\ \delta z$$

i.e. $\qquad M = \iiint \rho(x, y, z) dx\ dy\ dz$

from which, as before

$$\bar{x} = \frac{\iiint \rho(x,y,z)x \; dx \; dy \; dz}{\iiint \rho(x,y,z)dx \; dy \; dz}$$

$$\bar{y} = \frac{\iiint \rho(x,y,z)y \; dx \; dy \; dz}{\iiint \rho(x,y,z)dx \; dy \; dz}$$

and

$$\bar{z} = \frac{\iiint \rho(x,y,z)z \; dx \; dy \; dz}{\iiint \rho(x,y,z)dx \; dy \; dz}$$

These coordinates are functions of the density of the heterogeneous body and define its centre of mass as opposed to its centroid; centroid and centre of mass do not here coincide as I shall illustrate shortly. Observe that the terms $\rho(x,y,z)$ in the latter expressions do not cancel out because they are themselves functions of the variables x, y and z.

Illustrative Example 5

To determine the centroid of a solid homogeneous steel hemisphere of radius 'r'. See Fig. 9.

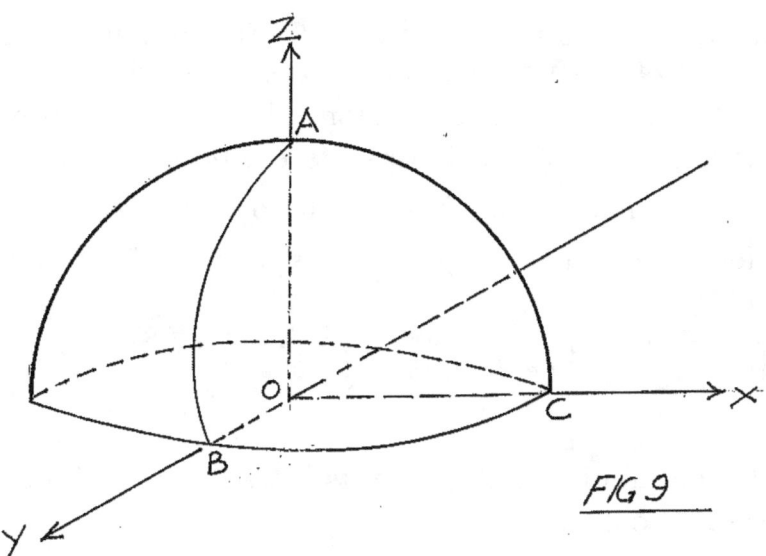

FIG 9

Before solving this problem which I propose to tackle using triple integration, let me provide the foundation as it were for the superstructure. Mark you, there are many other ways – most of them easier – by which this problem and others similar to it could be solved. However, the method chosen is meant to illustrate the one used in the preceding theoretical analysis. I should also add that later in this chapter other results will be derived for determination of centroids and centres of masses based on polar, cylindrical and spherical coordinates. Now to the solution.

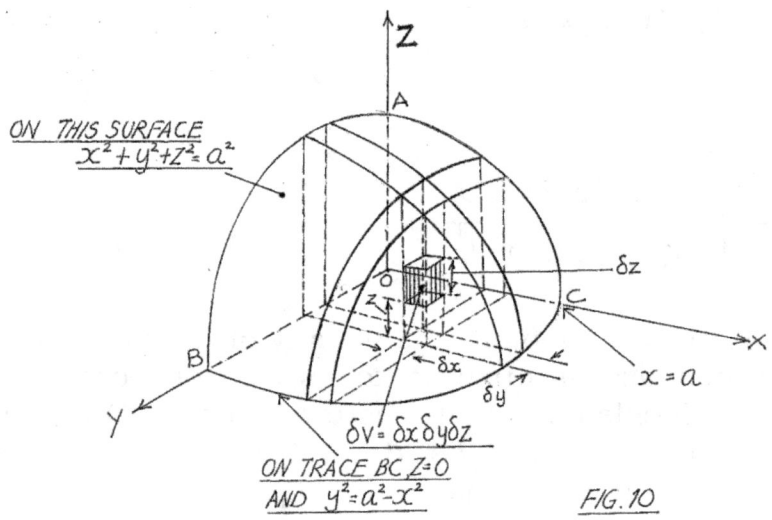

FIG. 10

In Fig. 10, one quadrant of the solid hemisphere is shown. The infinitesimal volume $\delta V = \delta x\,\delta y\,\delta z$ is shown. It is what the mathematicians call a parallelepiped: "a solid body of which each face is a parallelogram."

First the infinitesimal volumes are summed up starting from the XY-plane where z = 0 and up to the surface of the hemisphere where $z = f(x,y)$, viz. z² = a² − x² − y². Next the column is moved along the slice from the XZ-plane where y = 0 to the perimeter BC of the hemisphere where $y = f(x)$, viz. $y^2 = a^2 - x^2$; and finally the entire slice moves from $x = 0$ to $x = a$. Hence $\iiint dx\,dy\,dz$ may now be written for the quadrant of the hemisphere with the appropriate limits, as follows :

$$\int_0^a \int_{y=0}^{y=f(x)} \int_{z=0}^{z=f(x,y)} dz\,dy\,dx = \int_0^a \int_0^{y=\sqrt{a^2-x^2}} \int_0^{z=\sqrt{a^2-x^2-y^2}} dz\,dy\,dx$$

Knowing that the equation of a sphere is $x^2 + y^2 + z^2 = a^2$ in which 'a' is the radius, this integral becomes

$$\int_0^a \int_{y=0}^{y=\sqrt{a^2-x_2}} \left[z \right]_0^{z=(a^2-x^2-y^2)^{1/2}} dy\,dx$$

$$= \int_0^a \int_0^{\sqrt{a^2-x^2}} (a^2 - x^2 - y^2)^{1/2} \, dy \, dx$$

Substituting $(a^x - x^2) = u^2$

$$(a^2 - x^2 - y^2)^{1/2} = (u^2 - y^2)^{1/2}, \text{ and}$$

Putting $y = u Sin\theta$; with $dy = u Cos\theta \, d\theta, (u^2 - y^2)^{1/2} = u Cos\theta$

$$\therefore \quad \int_0^{\sqrt{a^2-x^2}} u^2 Cos^2\theta \, d\theta$$

$$\int Cos^2\theta \, d\theta = \frac{Cos\theta \, Sin\theta}{2} + \frac{1}{2}\theta$$

We can change the limits of integration. When $y = \sqrt{a^2 - x^2}$ and $u = a^2 - x^2$; $Sin\theta = 1$, i.e. $\theta = \frac{\pi}{2}$.

$$\therefore \quad \int_0^{\pi/2} Cos^2 \, d\theta = 0 + \frac{1}{2}\left(\frac{\pi}{2}\right) = \frac{\pi}{4}$$

$$\therefore \quad \int_0^{\sqrt{a^2-x^2}} u^2 Cos^2\theta \, d\theta = u^2 \frac{\pi}{4}$$

$$= \frac{\pi}{4}(a^2 - x^2)$$

We are now left with the evaluation of $\int_0^a \frac{\pi}{4}(a^2 - x^2) dx$. Integrating we get

$$\frac{\pi}{4}\left[a^2 x - \frac{x^3}{3}\right]_0^a = \frac{\pi}{4} \cdot \frac{2}{3} a^3$$

Since we have dealt with only one quadrant, the result for the entire hemisphere i.e. four quadrants is $4\left(\dfrac{\pi}{4}\cdot\dfrac{2}{3}a^3\right)$. Accordingly, the volume of the hemisphere is $\dfrac{2}{3}\pi\,a^3$; and its mass $\dfrac{2}{3}\rho\pi\,a^3$ where ρ = density.

Now to determine $\bar{z}$. We know that $\bar{z}=\dfrac{M_{XY}}{M}$.

Accordingly,

$$M_{XY}=\int_0^a\ \int_0^{\sqrt{a^2-x^2}}\int_0^{\sqrt{a^2-x^2-y^2}} z\,dz\ dy\ dx.$$

$\therefore$ $$M_{XY}=\int_0^a\ \int_0^{\sqrt{a^2-x^2}}\left[\frac{z^2}{2}\right]dy\ dx$$

Putting $u^2=a^2-x^2,$ and

$y = u\mathrm{Sin}\,\theta$

$\left[\dfrac{z^2}{2}\right]dy$ becomes

$\dfrac{1}{2}u^3\int \mathrm{Cos}^3\theta\ d\theta$

$=\dfrac{u^3}{2}\left\{\dfrac{1}{3}\mathrm{Cos}^3\theta\ \mathrm{Sin}\theta+\dfrac{2}{3}\mathrm{Sin}\theta\right\}$

which reduces to $\dfrac{u^2}{3}\sqrt{a^2-x^2}$, so that the final evaluation is :

$$\frac{1}{3}\int_0^a (a^2-x^2)^{3/2}dx$$

Writing $x=a\mathrm{Sin}\theta$, this integral reduces to

$$\frac{1}{3}\int a^4\mathrm{Cos}^4\theta\ d\theta.$$

Employing Wallis' Reduction formula, we get

$$M_{XY} = \frac{a^4}{3}\left[\frac{1}{4}Cos^3\theta\, Sin\theta + \frac{3}{8}Cos\theta\, Sin\theta + \frac{3}{8}\theta\right]_0^{\frac{\pi}{2}}$$

$$= \frac{a^4}{3}\left(0 + 0 + \frac{3}{8}\cdot\frac{\pi}{2}\right)$$

$$= \frac{a^4\pi}{16}$$

For the entire hemisphere $M_{XY} = 4\left(\frac{a^4\pi}{16}\right)$, i.e. $M_{XY} = \frac{a^4\pi}{4}$. Including density

'ρ' $M_{XY} = \rho\frac{a^4\pi}{4}$, from which $\bar{z} = \frac{M_{XY}}{M}$; so that $\bar{z} = \rho\frac{a^4\pi}{4} / \frac{2}{3}\rho\pi\, a^3 = \frac{3a}{8}$.

Thus, the centroid of a solid hemisphere (it being understood that it is homogeneous) is at a distance of three-eights its radius from its flat surface.

Let us see how incorporation of the effect of variable density i.e.

$\rho = f(x, y, z)$ changes this result.

Illustrative Example 6

Determine the centre of mass of a solid steel hemisphere of radius 'a' above the XY-plane assuming that the density of every point within it varies in direct proportion to its distance from the XY-plane as in Figs. 9 and 10. I have chosen a simple case:

Given density 'ρ' = kz

$$\therefore \quad \bar{z} = \frac{\int_0^a \int_{y=0}^{y=f(x)} \int_{z=0}^{z=\sqrt{a^2-x^2-y^2}} \rho z\; dz\; dy\; dx}{\int_0^a \int_0^{y=f(x)} \int_0^{\sqrt{a^2x^2y^2}} \rho\; dz\; dy\; dx}$$

$$= \frac{\iiint kz\cdot z\; dz\; dy\; dx}{\iiint kz\; dz\; dy\; dx} \quad \cdots\cdots\cdots\cdots \text{(i)}$$

Based on a result obtained in Illustrated Example 5 it is known that

$$\therefore \quad \int_0^a \int_0^{\sqrt{a^2-x^2}} \int_0^{\sqrt{a^2-x^2-y^2}} kz\; dz\; dy\; dx \text{ for the entire hemisphere } M = \frac{ka^4\pi}{4}.$$

Therefore we need only find the result of the numerator in (i) above.

Integrating once with respect to 'z', we get

$$\int_o^a \int_0^{\sqrt{a^2-x^2}} \left[\frac{kz^2}{3}\right]^{\sqrt{a^2-x^2-y^2}} dy$$

$$\left[\frac{kz^3}{3}\right]_0^{\sqrt{a^2-x^2-y^2}} dy = \int_0^{\sqrt{a^2-x^2}} \frac{k}{3}(a^2-x^2-y^2)^{3/2} dy$$

Putting $u^2 = a^2 - x^2$ and

$$y = uSin\theta$$

$$\frac{k}{3} \int (a^2-x^2-y^2)^{3/2} dy \text{ reduces to}$$

$$\frac{k}{3} \cdot u^4 Cos^4\theta \, d\theta = \frac{ku}{3}\left[\frac{1}{4}Cos^3\theta \, Sin\theta + \frac{3}{8}Sin\theta + \frac{3}{8}\theta\right]_0^{\frac{\pi}{2}}$$

$$= \frac{ku^4\pi}{16} = \frac{k\pi}{16}(a^2-x^2)^2$$

Finally for $\dfrac{k\pi}{16}\displaystyle\int_0^a (a^2-x^2)^2 \, dx$

Substitute $x = aSin\theta$

$$dx = aCos\theta \, d\theta$$

to obtain $\dfrac{k\pi \, a}{16}\displaystyle\int_0^{\pi/2} Cos^5\theta \, d\theta$

Using Wallis' Reduction Formula

$$\int Cos^5\theta \, d\theta = \left[\frac{1}{5}Cos^4\theta \, Sin\theta + \frac{4}{15}Cos^2\theta \, Sin\theta + \frac{8}{15}Sin\theta\right]_0^{\frac{\pi}{2}}$$

$$= \frac{8}{15}$$

$$\therefore \qquad \frac{k\pi a}{16} \int_0^{\pi/2} Cos^5\theta \, d\theta = \frac{ka^5\pi}{30}$$

Accordingly, for the entire hemisphere, $M_{XY} = \dfrac{4ka^5\pi}{30}$

i.e. $\qquad\qquad M_{XY} = \dfrac{2}{15}ka^5\pi$

from which $\qquad\qquad \bar{z} = \dfrac{M_{xy}}{M}$

and because $\qquad\qquad M = \dfrac{ka^4\pi}{4}$

$$\bar{z} = \frac{8a}{15}$$

So you see, the centre of mass of the heterogeneous hemisphere is $\dfrac{19a}{120}$ i.e.

$\left(\dfrac{8a}{15} - \dfrac{3a}{8}\right)$ above the position of the centroid of its homogeneous counterpart.

Let us leave the world of calculus for a while and look now at some simple practical technical determinations.

Illustrative Example 7

An automotive technician wishes to determine the distance from either of its end-supports of the centre of mass of the crankshaft shown in Fig. 11, assuming the material of the crank web and crank pins to have different densities. Take $g = 9.81 \text{m/s}^2$

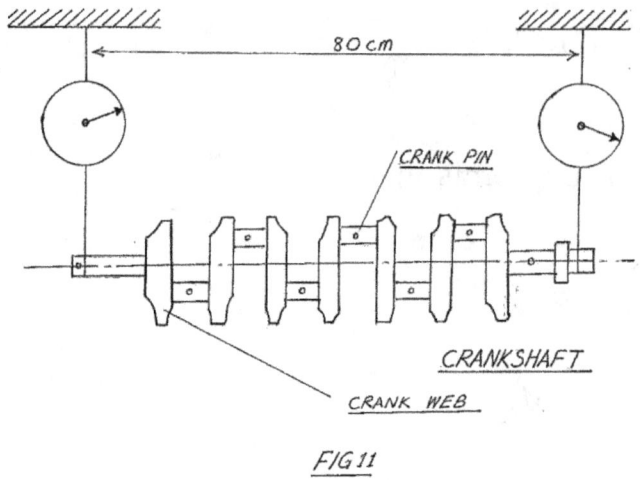

FIG 11

An arrangement is set up as shown in the diagram; the scales are calibrated in kilograms. The scale on the left records 7.5kg; that on the right is 12.5kg. The distance between the spring scales is 75cm.
Evidently, the mass of the crankshaft is 20kg, i.e. (7.5 + 12.5)kg.

Taking moments from the left-hand end support, assuming the centre of mass to be x *cm* from that end; and working in centimetres

$$20g(x) = 12.5(g)(75)$$

i.e. $x = 46.875$ cm from the left-hand support.

Illustrative Example 8

A designer of carnival costumes proposes to incorporate in a headpiece a part consisting of a very thin uniform, copper sheet in the shape shown in Fig. 12. In order to distribute the weight of the headpiece evenly on the head of the masquerader he seeks to determine the centre of mass of the copper sheet, the density of which is assumed heterogeneous. How might he set about the task? Any advice?

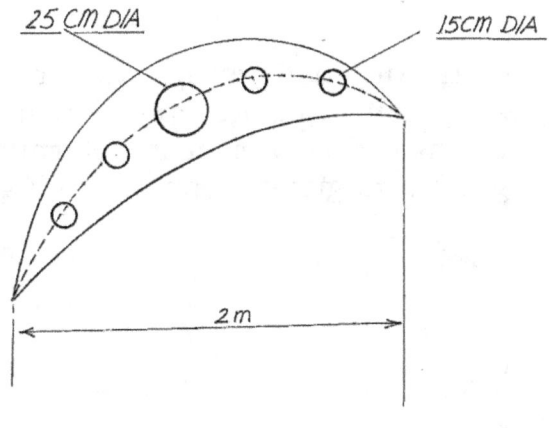

25 CM DIA   15cm DIA

2 m

FIG.12

Advice: Make a template of the headpiece to exact shape, thickness and mass. Suspend template by a light string from any two points from a fixed support, one suspension at a time. Since the line of action of the weight of the template must pass through the point of suspension, scribe the vertical lines on the template from each of the two suspensions. You may need to place a back-up board for this purpose. Where the two lines intersect is the centroid of the headpiece. Note that it is possible that the lines may intersect at a point outside the headpiece.

Illustrative Example 9

A Field Engineer has to distribute loads evenly on a raised platform on which several pieces of heavy equipment are to be positioned. One of the problems she has to resolve in relation to this project is determination of the centre of mass of a machine having the configuration shown in Figs. 13a and 13b.

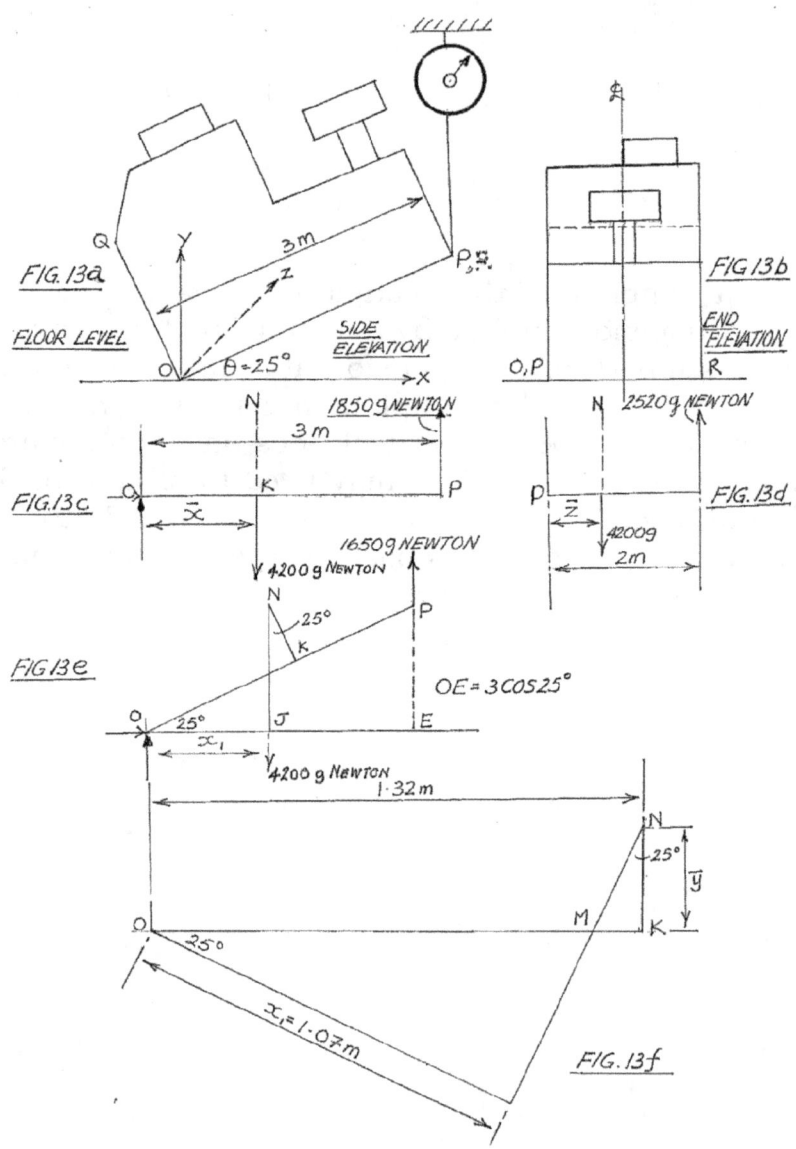

A special rig is brought in to lift the machine; first at the edge along P and next at the edge along R. When the lifting device just takes the strain before tilting at the edge PR, i.e. which $\theta = 0^\circ$ the scale which is calibrated in kg reads 1850kg; and when $\theta = 25^\circ$ the reading is 1650kg. When the lifting device just takes the strain before tilting along side OP, the scale reading is 2520kg.

The manufacturer of the machine mass quotes its total mass as 4200kg. Show how the Field Engineer can use these data to determine the location of the centre of mass of the machine. Take g = 9.81m/s².

First of all the bottom of the machine is designated the XZ-plane; the XY- and YZ-planes are mutually perpendicular to this plane; and the origin is at 'θ'.
With $\theta = 0^o$ and with point 'N' as the centre of mass, take moments about 'O'. See Fig. 13c.

$$4200(g)(\bar{x}) = 1850g(3)$$

$$\therefore \quad \bar{x} = 1.32m.$$

Refer now to Fig. 13e.

Taking moments about 'O'

$$4200(g)(x_1) = 1650g(OE)$$

$$= 1650g(3Cos25^o)$$

$$= 1650(g)(0.9063)$$

from which $\quad x_1 = 1.07m$

Observe that distances $\bar{x}$ and x_1 are shown in relation to the distance of the centre of mass above P as in Fig. 13f.

Referring to this diagram

$$\frac{OJ}{OM} = Cos25^o$$

$$\therefore \quad OM = OJ/0.9063.= 1.07/0.9063$$

$$\text{i.e.} \quad OM = 1.2m$$

so that $\quad MK = 0.12m$ (i.e. 1.32m – 1.2m)

Also $\quad \dfrac{\bar{y}}{MK} = Tan65^o$

or $\quad \bar{y} = 0.12(0.4663)$

$$\therefore \quad \bar{y} = 0.056m$$

When the machine is just about to tilt along edge OP, and we take moments about 'P'

$$4200g(\bar{z}) = 2520g(2)$$

$$\therefore \qquad \bar{z} = 1.2m$$

Accordingly, the centre of mass of the machine is 1.32m ahead of point 'O' and 0.56m (5.6cm) above it on the side elevation, and 1.2m from the edge OP.

Here is one for you to solve.

The front axle of a large Sports Utility Vehicle with a wheel base of 3m is tipped upwards on a weighbridge at the Licensing Authority at an angle of 15°. The rear axle of the vehicle is on level terrain but not on the weighbridge at this weighing. The weighbridge's scale which is calibrated in kilograms registers 1800kg. When the front axle is still on the weighbridge but level with the rear axle which is on level ground and still off the weighbridge, the scale registers 900kg. Locate the centre of mass of the car with reference to its rear axle. The mass of the car is 2000kg. Assume g = 9.81 m/s². See Figs 14 and 15 where I have given some pointers to aid your solution of the problem.

Ans: Centre of mass = 1.35m in front of the rear axle and 0.56m above it.

How would you find the other coordinate of the centre of mass?

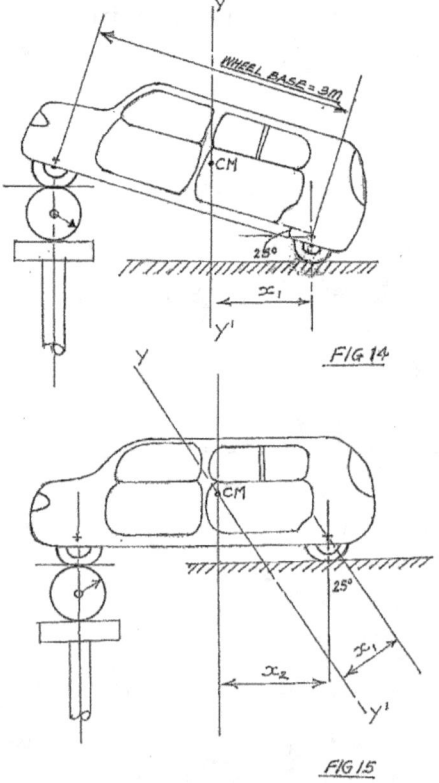

FIG 14

FIG 15

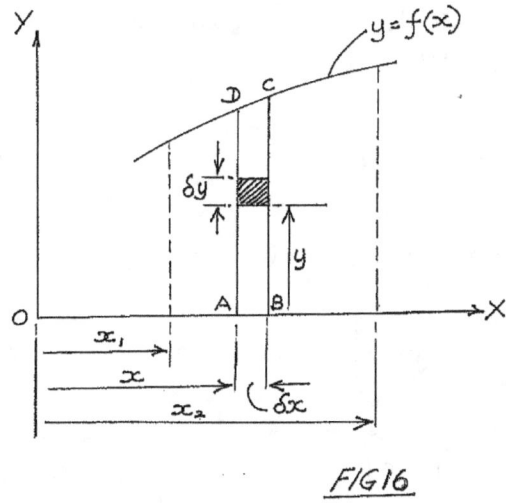

FIG 16

Referring to Fig. 16, the area of the infinitesimal element shown shaded is $\delta y \delta x$ and its moments of area about the X- and Y-axes are respectively:

$$\delta M_x = y \, \delta y \, \delta x \, ; \quad \delta M_y = x \, \delta y \, \delta x$$

[Note that 'M' here refers to moment of area and not to mass].

Accordingly, the coordinates of the centroid of the area with respect to the X- and Y-axes respectively are obtained by dividing the moment of the area by the area, thus:

$$\bar{x} = \frac{\iint y \, dy \, dx}{\iint dy \, dx} \, ; \quad \bar{y} = \frac{\iint x \, dy \, dx}{\iint dy \, dx}$$

For those students who are meeting multiple integrals for the first time a brief note on the method is in order. In solving Illustrative Example 5, I explained how the limits of the triple integral being evaluated were arrived at. But I shall now illustrate the general method by explaining the steps required; a double-integral problem should suffice.

Suppose a double integral 'A' is given by :

$$A = \int_{x_1}^{x_2} \int_{0}^{y=f(x)} dy \, dx$$

The inner integral sign $\int_{0}^{y=f(x)}$, belongs to dy and $\int_{x_1}^{x_2}$ belongs to dx. In Fig. 16 look at the element $\delta y \, \delta x$ and you will see that the ordinate 'y' goes from the X-axis i.e. where $y = 0$ up to meet the curve where $y = f(x)$; hence the limits attached to the inner integral. Similarly, 'x' goes from $x = x_1$ to $x = x_2$; hence

~ 177 ~

the limits for the outer integral sign. The method is perhaps best explained by working on example. Suppose:

$$A = \int_0^2 \int_0^{a^2-x^2} (x+y)dy \; dx$$

Working on the inner integral first, we write:

i.e.
$$A = \int_0^2 \left[\int_0^{a^2-x^2} (x+y)dy \right] dx$$

$\therefore$
$$A = \int_0^2 \left\{ \left[xy + \frac{y^2}{2} \right]_0^{a^2-x^2} \right\} dx$$

$$= \int_0^2 \left[x(a^2 - x^2) + \frac{(a^2 - x^2)^2}{2} \right] dx$$

Evaluating the outer integral next, we have

$$A = \frac{1}{2} \int_0^2 (xa^2 - x^3 + a^4 - 2a^2x^2 + x^4)dx$$

$$= \frac{1}{2} \left[x^2a^2 - x^4 + a^4x - 2a^2\frac{x^3}{3} + \frac{x^5}{5} \right]_0^2$$

$$= 2a^2 - 4 + 2a^4 - \frac{16a^2}{3} + \frac{32}{5}$$

$\therefore A$
$$= 2a^4 - \frac{10a^2}{3} + \frac{12}{5}$$

or A
$$= \frac{30a^4 - 50a^3 + 36}{15}$$

Considering triple integrals, suppose a volume is given by

$$V = \int_0^c \int_0^{y=f(x)} \int_0^{z=f(x,y)} dz \; dy \; dx$$

the procedure is similar to that followed for double integrals. The innermost integral sign $\int_0^{z=f(x,y)}$ refers to dz; the next $\int_0^{y(f(x))}$ refers to dy and the outermost $\int_0^c$ refers to dx.

Take a look again at Fig. 10 which was drawn for Illustrative Example 5. The volume of the hemisphere was obtained by considering one quadrant of it. Accordingly,

$$V = \int_{x=0}^{x=a} \int_0^{y=\sqrt{a^2-x^2}} \int_0^{z=\sqrt{a^2-x^2-y^2}} dz\ dy\ dx$$

You should return to the solution of the Illustrative Example 5 and observe that when the innermost integral was being evaluated x and y were treated as 'constants'. Likewise for the integral with dy, 'x' was treated as a constant.

Illustrative Example 10

Let us illustrate the use of the foregoing by determining the centroid of the area or region enclosed by the parabola $y^2 = 2x$, the X-axis, and the ordinates at $x = 2$ and $x = 4$.

The first thing to do is to make an appropriate sketch delimiting the region defined in the problem, as for example Fig. 17.

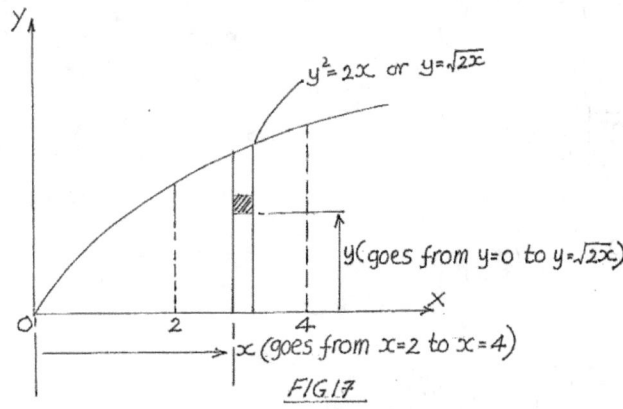

FIG 17

Note for any value of x, the corresponding value of $y = \sqrt{2x}$

$$\therefore \quad A = \int_2^4 \int_0^{\sqrt{2x}} dy\ dx$$

$$= \int_2^4 \left[y \right]_0^{\sqrt{2x}} dx$$

$$= \int_2^4 (2x)^{1/2}\, dx$$

$$A = \sqrt{2} \left[\frac{2}{3} x^{3/2} \right]_2^4$$

$$A = \frac{2}{3}\sqrt{2} \left\{ 2^3 - \left(\sqrt{2}\right)^3 \right\} \text{ square units}$$

Now, with 'M' referring to Moment

$$M_Y = \int_2^4 \int_0^{\sqrt{2x}} x\, dx\, dy$$

$$= \int_2^4 \left[y \right]_0^{\sqrt{2x}} x\, dx$$

$$\therefore \quad M_Y = \int_2^4 \left(\sqrt{2x}\right) x\, dx$$

$$= \sqrt{2} \int_2^4 x^{3/2}\, dx$$

$$= \sqrt{2} \left[\frac{2}{5} (x)^{5/2} \right]_2^4$$

i.e.
$$M_y = \frac{2\sqrt{2}}{5} \left[2^5 - \left(\sqrt{2}\right)^5 \right]$$

$$\therefore \quad \bar{x} = \frac{M_Y}{A} = \frac{\dfrac{2\sqrt{2}}{5}\left[2^5 - \left(\sqrt{2}\right)^5 \right]}{\dfrac{2\sqrt{2}}{3}\left[2^3 - \left(\sqrt{2}\right)^3 \right]}$$

i.e.
$$\bar{x} = \frac{3}{5} \frac{\left[2^5 - \left(\sqrt{2}\right)^5 \right]}{2^3 - \left(\sqrt{2}\right)^3}$$

or $\qquad \bar{x} = \dfrac{3}{5} \dfrac{(2^5 - 4\sqrt{2})}{(2^3 - 2\sqrt{2})}$

or $\qquad \bar{x} \approx 3$

Continuing,

$$M_X = \int_2^4 \int_0^{\sqrt{2x}} y \, dy \, dx$$

$$= \int_2^4 \left[\dfrac{y^2}{2} \right]_0^{\sqrt{2x}} dx$$

$$= \dfrac{1}{2} \int_2^4 2x \, dx$$

$$= \dfrac{1}{2} \left[x^2 \right]_2^4$$

i.e. $\qquad M_x = 6$

$\therefore \qquad \bar{y} = \dfrac{M_X}{A}$

$$= \dfrac{6}{\dfrac{2\sqrt{2}}{3}\left(2^3 - 2\sqrt{2}\right)} \qquad\qquad = \dfrac{6}{\dfrac{2\sqrt{2}}{3}}(2^3 - 2\sqrt{2})$$

$$\bar{y} = \dfrac{9}{(8\sqrt{2} - 4)} \approx 1.2$$

$\therefore$ Coordinates of the centroid of the demarcated area in Fig. 17 are

(3, 1.2).

Special Note

The reader is requested to note the following very carefully. When double integrals are used to obtain moments about the orthogonal axes, the elemental areas '$\delta y \, \delta x$' are merely multiplied by the moment arms 'x' and 'y' as the case may be, to obtain the moments of these elements about the relevant axis; 'x' for moments about the Y-axis and 'y' for moments about the X-axis.

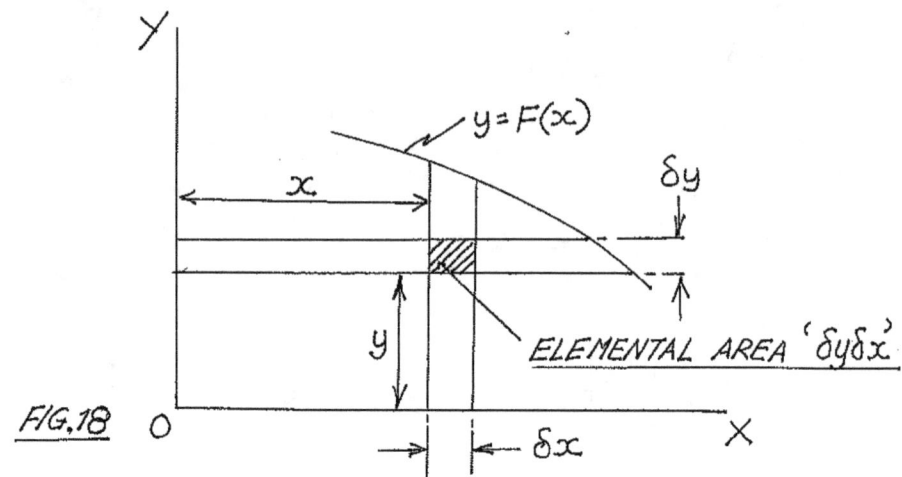

FIG.18

Accordingly, with reference to Fig. 18

$$\delta M_Y = \delta y \; \delta x(x)$$

and

$$\delta M_X = \delta y \; \delta x(y)$$

from which

$$\bar{y} = \frac{\iint x \; dy \; dx}{\iint dy \; dx}$$

Similarly,

$$\bar{x} = \frac{\iint y \; dy \; dx}{\iint dy \; dx}$$

With the use of single integrals, when strips are employed to determine areas as in Fig. 19, the situation is slightly different.

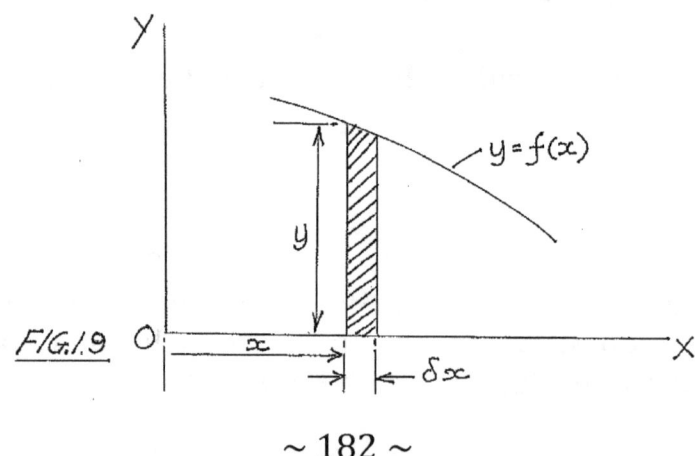

FIG.19

The elemental area here is $y\, \delta x$ and its moment about the Y-axis is :

$$\delta M_Y = y\, \delta x (x)$$

$$= xy\, \delta x$$

or $\quad M_Y = \int xy\ dx$

But its moment about the X-axis is

$$\delta M_X = y\, \delta x \left(\frac{y}{2}\right)$$

Why $\frac{y}{2}$? Because that is the distance of the centroid of the elemental strip from the X-axis.

There is also another possibility which I will now illustrate in Fig. 20.

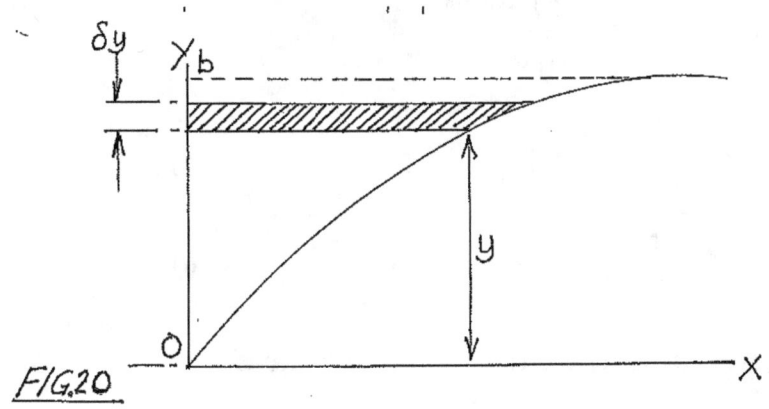

FIG.20

Let us suppose that we wanted to locate the coordinates of the centroid of the area delimited by the portion of the curve $y = f(x)$, the origin '0' and $y = b$. If we chose to use the horizontal strip shown shaded in Fig. 13, then

$$\delta A = x\, \delta x$$

$$\delta M_y = x\, dy \left(\frac{x}{2}\right)$$

$$\delta M_X = x\, \delta y (y)$$

I will now demonstrate these applications.

Illustrative Example 11

Let us find the coordinates of the centroid of the quadrant of a circle $x^2 + y^2 = r^2$.
See Fig. 21

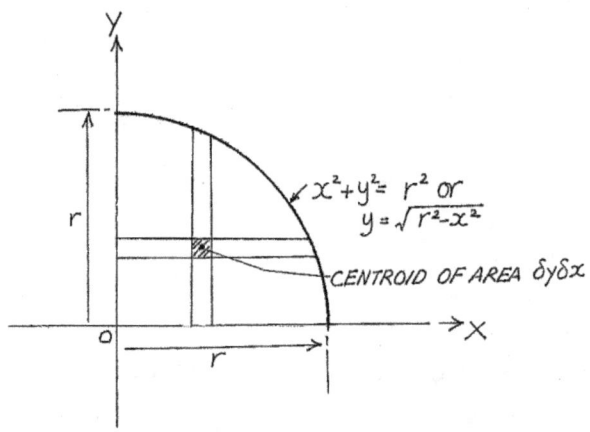

$$\underline{FIG.21}$$

By double integrals
$$\bar{y} = \frac{\iint y \, dy \, dx}{\iint dy \, dx}$$

i.e.
$$\bar{y} = \frac{\int_0^r \int_0^{\sqrt{r^2-x^2}^2} y \, dy \, dx}{\int_0^r \int_0^{\sqrt{r^2-x^2}} dx \, dy} = \frac{\int_0^r \int_0^{\sqrt{r^2-x^2}} y \, dy \, dx}{\int_0^r \int_0^{\sqrt{r^2-x^2}^2} dx \, dy}$$

Evaluating the integral in the numerator, we have

$$\int_0^r \int_0^{\sqrt{r^2-x^2}} y \, dy \, dx = \int_0^r \left[\frac{y^2}{2}\right]_0^{\sqrt{r^2-x^2}} dx$$

$$= \frac{1}{2}\int_0^r (r^2 - x^2) \, dx$$

$$= \frac{1}{2}\left[r^2 x - \frac{x^3}{3}\right]_0^r$$

$$= \frac{r^3}{3}$$

We need not work out the denominator integral. It has to be $\dfrac{\pi r^2}{4}$ i.e. one-fourth the area of the circle of radius 'r'.

$$\therefore \quad \bar{y} = \frac{r^3 \cdot 4}{3.\pi r^2}$$

i.e. $\quad \bar{y} = \dfrac{4r}{3\pi}$

For the $\bar{x}$-coordinate

$$\bar{x} = \frac{\iint x\, dy\, dx}{\pi r^2 / 4}$$

Let us evaluate $\iint x\, dy\, dx$

$$\int_{o}^{r} \int_{0}^{\sqrt{r^2-x^2}} dy\, x\, dx = \int_{o}^{r} x\, dx[y]_{0}^{\sqrt{r^2-x^2}}$$

$$= \int_{o}^{r} x\left(\sqrt{r^2-x^2}\right) dx$$

$$= \left[(r^2-x^2)^{3/2}\right]_{r}^{0}$$

$$= \frac{r^2}{3}$$

$$\therefore \quad \bar{x} = \frac{4r}{3\pi}$$

When as shown in Fig. 22 we use the strip in the single integral form:

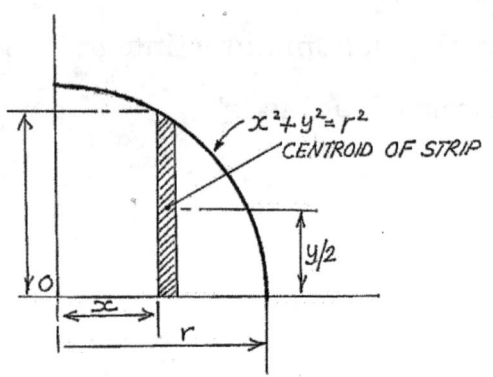

FIG.22

$$\bar{x} = \frac{\int_0^r yx \ dx}{\pi r^2 / 4}$$

$$= \frac{\int_o^r x\sqrt{(r^2 - x^2)}dx}{\pi r^2 / 4}$$

$$= \frac{r^2 \cdot 4}{3 \cdot \pi r^2}$$

i.e. $\quad \bar{x} = \dfrac{4r}{3\pi}$

Also,

$$\bar{y} = \frac{\int_o^r y \ dx \cdot \frac{y}{2}}{\pi r^2 / 4}$$

$$\int_o^r y \ dx \cdot \frac{y}{2} = \frac{1}{2} \int_o^r y^2 dx$$

$$= \frac{1}{2} \int_o^r (r^2 - x^2) \ dx$$

$$= \frac{1}{2} \left[r^2 x - \frac{x^3}{3} \right]_0^r$$

$$= \frac{r^3}{3}$$

and $\therefore$ $\bar{y} = \frac{r^3}{3} \Big/ \frac{\pi r^2}{4}$

i.e. $\bar{y} = \frac{4r}{3\pi}$ as before.

Centroids of Area by Double Integration : Polar Coordinates

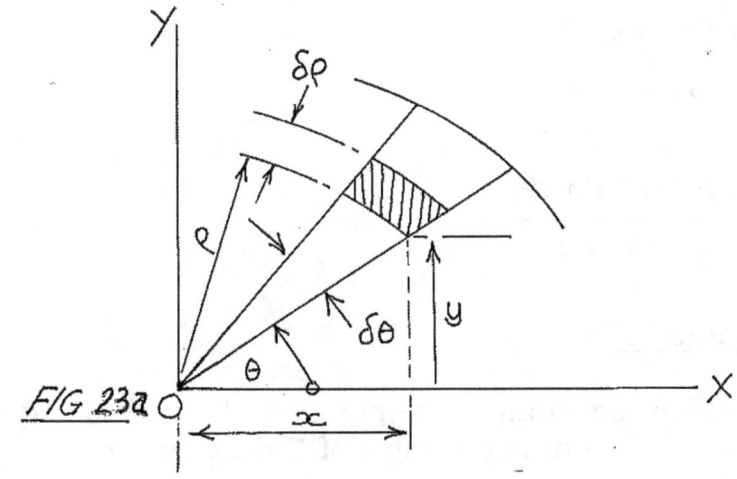

FIG. 23a

The elemental area shown shaded in Fig. 23a is $\rho \, \delta\theta \, \delta\rho$

Referring to Fig. 23a

$$x = \rho \, Cos\theta$$

$$y = \rho \, Sin\theta$$

$\therefore$ Moment of the elemental area about the Y-axis is $= \rho \, \delta\theta \, \delta\rho(x)$

$\therefore$ $\bar{y} = \dfrac{\iint x\rho \, \delta\theta \, \delta\rho}{\iint \rho \, \delta\theta \, \delta\rho}$

$\delta\theta$ and $\delta\rho$ having approached the limit we may write $d\theta$ and $d\rho$ instead. In most of what was done before we assumed, without saying so that, that as infinitesimals approached their limits $\delta x, \delta y$ became dx and dy respectively. Substituting $x = \rho\,Cos\theta$ in the last equation produces

$$\bar{y} = \frac{\iint \rho\,Cos\theta\,\rho\,d\rho\,d\theta}{\iint \rho\,d\theta\,d\rho}$$

i.e.
$$\bar{y} = \frac{\iint P^2 Cos\theta \cdot d\rho \cdot d\theta}{\iint \rho\,d\theta\,d\rho}$$

By a similar token

$$\bar{x} = \frac{\iint y\rho\,d\theta\,d\rho}{\iint \rho\,d\theta\,d\rho}$$

i.e.
$$\bar{x} = \frac{\iint \rho^2 Sin\theta \cdot d\rho \cdot d\theta}{\iint \rho\,d\theta\,d\rho}$$

Illustrative Example 12

Determine using polar coordinates the position of the centroid of a circular arc of radius 'r', the arc subtending an angle of '2α' at its centre.

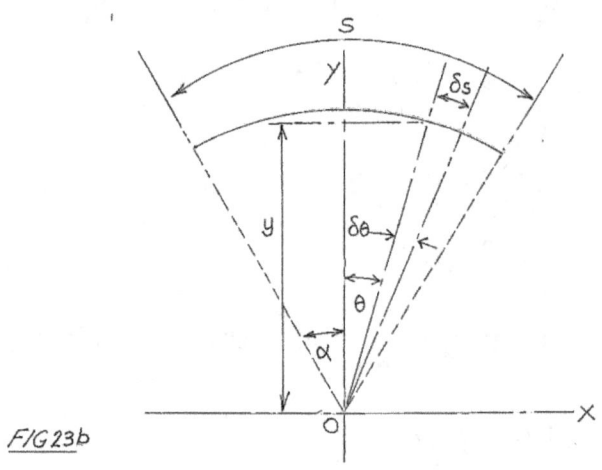

FIG 23b

We note that $\bar{x} = 0$, because the centroid will lie on the Y-axis. Referring to Fig. 23b, the moment arm of the elemental length of arc δs

$$y = r\,Cos\theta$$

But, $\qquad ds = rd\theta$

and $\quad s = 2r\alpha$

$$\therefore \quad \bar{y} = \int \frac{yds}{s}$$

$$= \frac{2\int_{0}^{\alpha} rCos\theta \cdot r\,d\theta}{2r\alpha}$$

$$= \frac{2\int_{0}^{\alpha} r^{2}Cos\theta\,d\theta}{2r\alpha} = \frac{rSin\alpha}{\alpha}$$

Centroids of Solid by Triple Integration:

(i) **Rectangular Coordinates**

The basic relationships to be used here were developed when we defined centroids and centres of mass of continuous bodies. Here, a further illustrative example is provided to reinforce the method.

Illustrative Example 13

Determine the coordinates of the centroid of the homogeneous body whose base in the X-Y plane is delimited by the lines $x = 0$, $x = 3$; $y = 0$, $y = 5$ and whose sides are parallel to the X-axis. The top surface of the body is bounded by the plane $5z = 20 - (x + y)$.

Begin by making a sketch of the body. It is important to do so as it will help you to visualise the problem.

At $\quad x = 0, \; y = 0 \; : \; z = 4$

" $\quad x = 0, \; y = 5 : z = 3$

" $\quad y = 0, \; x = 3 : \; z = 3.4$

" $\quad y = 5, \; x = 3 : \; z = 2.4$

Use these data to prepare the sketch labelled Fig. 23c.

Referring to Fig. 23c, it is easy to see that an infinitesimal volume, say, δV is given by:

$$\delta V = \delta x \; \delta y \; \delta z$$

from which

$$V = \iiint dx \; dy \; dz$$

Looking again at the diagram it is observed that the limits of integration are as

follows:

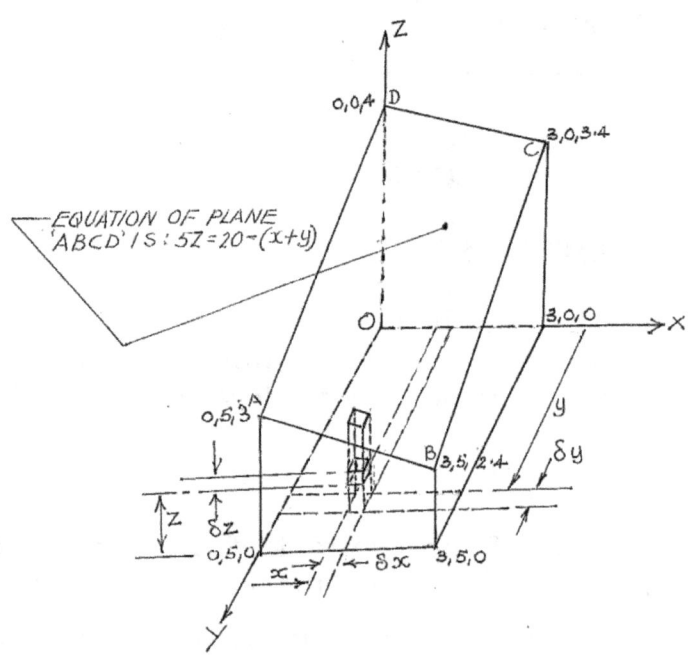

FIG 23c

For z: from $z = 0$, to $z = 4 - \dfrac{1}{5}(x + y)$

" x: " $x = 0$, to $x = 3$

" y: " $y = 0$, to $y = 5$

$\therefore$ $V = \displaystyle\int_0^3 \int_0^5 \int_0^{z = 4 - \frac{1}{5}(x+y)} dz \; dy \; dx$

$$= \int_0^3 \int_0^5 \left\{ 4 - \frac{1}{5}(x+y) \right\} dy \; dx$$

$$= \int_0^3 \left[4y - \frac{1}{5}xy - \frac{y^2}{10} \right]_0^5$$

$$= \int_0^3 (17.5 - x) dx$$

$$= \left[17.5x - \frac{x^2}{2} \right]_0^3$$

i.e. $V = 48$ cubic units.

To determine the coordinates of the centroid, $\bar{x}, \bar{y}, \bar{z}$ of the centroid, the following must be evaluated

$$\bar{x} = \frac{\iiint y \; dz \; dy \; dx}{\iiint dz \; dy \; dx} ;$$

$$\bar{y} = \frac{\iiint x \; dz \; dy \; dx}{\iiint dz \; dy \; dx}, \text{ and, } \quad \bar{z} = \frac{\iiint z \; dz \; dy \; dx}{\iiint dz \; dy \; dx}$$

Starting with

$$\iiint dz \; ydy \; dx = \int_0^3 \int_0^5 [z]_0^{z=4-\frac{1}{5}(x+y)} y \; dy \; dx$$

$$= \int_0^3 \int_0^5 \left(4y \; dy - \frac{1}{5}xy \; dy - \frac{y^2}{5} dy \right) dx$$

$$= \int_0^3 \left[2y^2 - \frac{1}{10}xy^2 - \frac{y^2}{15} \right]_0^5 dx$$

$$= \int_0^3 \left(50 - 2.5x - \frac{25}{3} dx \right.$$

$$= \left[\frac{125x}{3} - \frac{2.5x^2}{2}\right]_0^3$$

$$\therefore \quad \overline{x} = \frac{113.7.5}{48} = 2.37 \text{ units, say, } \overline{x} = 2.4 \text{ units}$$

In the previous working some of the intermediate lines were omitted in the interest of space; the same shall be done in the following for $\overline{y}$ and $\overline{z}$. Continuing in order to find $\overline{y}$ we evaluate

$$\iiint dz\,dy\,xdx = \int_0^3 \int_0^5 \int_0^{z=4-\frac{1}{5}(x+y)} dz\,dy\,x\,dx$$

$$= \int_0^3 \int_0^5 \left\{\left(4 - \frac{1}{5}x - \frac{1}{5}y\right)dy\right\}x\,dx$$

$$= \int_0^3 \left[4y - \frac{1}{5}xy - \frac{1}{10}y^2\right]_0^5 x\,dx$$

$$= \int_0^3 \left(20 - x - \frac{25}{10}\right)x\,dx$$

$$= \left[\frac{17.5x^2}{2} - \frac{x^3}{3}\right]_0^3$$

$$= 78.75 - 9 = 69.75 \text{ (units)}^4$$

$$\therefore \qquad \overline{y} = \frac{69.75}{48} = 1.45 \text{ units, say, } 1.5 \text{ units.}$$

Finally,

$$\iiint z\,dz\,dy\,dx = \int_0^3 \int_0^5 \int_0^{z=4-\frac{1}{5}(x+y)} z\,dz\,dy\,dx$$

Writing $z = 4 - \frac{1}{5}(x+y)$ as $z = 4 - a$, with $a = \frac{1}{5}(x+y)$

$$z^2 = 16 - \frac{8x}{5} - \frac{8y}{5} + \frac{x^2}{25} + \frac{2}{5}xy + y^2$$

$$\therefore \quad \iiint z\, dz\, dy\, dx = \int_0^3 \int_0^5 \left[\frac{z^2}{2}\right]_0^{4-\frac{1}{5}(x+y)} dy\, dx$$

$$= \frac{1}{2} \int_0^3 \left[16y + \frac{x^2 y}{25} - \frac{8}{5}xy - \frac{8y^2}{10} + \frac{xy^2}{5} + \frac{y^3}{3}\right]_0^5$$

$$= \frac{1}{2} \int_0^3 \left(60 + \frac{x^2}{5} - 3x + \frac{125}{3}\right) dx$$

$$= \frac{1}{2} \left[60x + \frac{x^3}{15} - \frac{3x^2}{2} + \frac{125x}{3}\right]_0^3$$

$$= \frac{1}{2} \left[305 - \frac{351}{30}\right] = 146.65$$

$$\therefore \quad \bar{z} = \frac{146.65}{48} = 3.05 \text{ units, say, 3 units}$$

The coordinates of the centroid of the body are (2.4, 1.5, 3).

(ii) Triple Integration: Cylindrical Coordinates

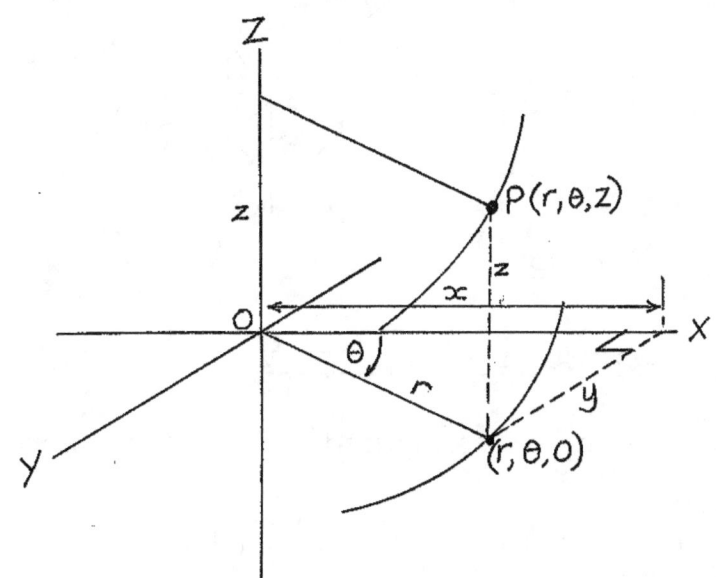

FIG 24a

$$\frac{y}{r} = Sin\theta \qquad \text{or} \qquad y = rSin\theta$$

$$\frac{x}{r} = Cos\theta \qquad \text{or} \qquad x = rCos\theta$$

$$Z = z$$

Readers coming across specification of the position of a point in cylindrical coordinates for the first time may wonder why the coordinates are described as such. Well, very simply because the point is on a surface which is generated by a constant radius 'r' and a fixed value of 'z' as axis. Such a surface is that of a circular cylinder. After all, the equation of a right circular cylinder is $r = R$, in which R = radius of cylinder. See Fig. 24a and 24b. The surfaces of constant θ are planes through the Z-axis making an angle 'θ' with the XZ-plane as shown in the same diagram. The Z-coordinate defines a surface parallel to the XY-plane and perpendicular to the Z-axis. Accordingly the cylindrical coordinates of point 'P' in Fig. 24a (r,θ,z); i.e. $x = rCos\theta$, $y = rSin\theta$, z=z.

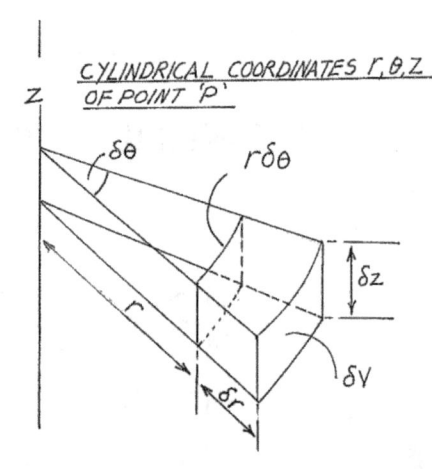

FIG 24.b

INFINITESIMAL VOLUME $\delta V = \delta r. r\delta\theta.\delta z$

The elemental volume δV configured at Fig. 17b is expressed as:

$$\delta V = \left(r + \frac{\delta r}{2}\right)\delta r\ \delta\theta\ \delta z) = r\ \delta r\ \delta\theta\ \delta z + \frac{(\delta r)}{2}\ \delta\theta\ \delta z$$

Which upon ignoring the second order quantity $(\delta r)^2/2$ may be expressed by:

$$\delta V = (\delta r)(r\delta\theta)(\delta z)$$

As these infinitesimals approach zero,

$$dV = dr\, r\, d\theta\, dz \qquad \text{or} \qquad dV = r\, dr\, d\theta\, dz$$

Illustrative Example 14

Determine, using cylindrical co-ordinates the centroid of a solid homogeneous right circular cone of base radius 'r' and height 'h', assuming the density of the material which is constant is 'ρ'.

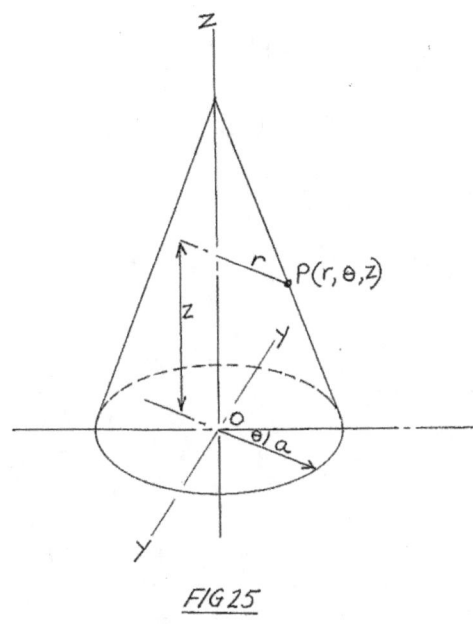

FIG 25

In Fig. 25, point P is on the surface of the cone whose height is 'h'. The cylindrical coordinates of P are: (r,θ,z). We know that:

$$V = \iiint r\, dr\, d\theta\, dz$$

Noting that the radius of the base of the cone is 'a', we obtain from geometrical considerations, having designated 'r' as radius 'z' from the base

$$\frac{r}{a} = \frac{h-z}{h}$$

or $\qquad r = \dfrac{a}{h}(h-z)$

The limits of integration are: for 'r', from 0 to $\dfrac{a}{h}(h-z)$; for 'θ', from 0 to 2π; and, for z from 0 to h.

~ 195 ~

$$\therefore \qquad V = \int_0^h \int_0^{2\pi} \int_o^{\frac{a}{h}(h-z)} r\, dr\, d\theta\, dz$$

$$= \int_0^{2\pi} \int_0^{2\pi} \left[\frac{1}{2} r^2 \right]_0^{\frac{a}{h}(h-z)} dz\, d\theta$$

$$= \int_0^h \int_0^{2\pi} \frac{1}{2}\left[\frac{a^2}{h^2}(h^2 - 2hz + z^2) \right] dz\, d\theta$$

$$= \frac{1}{2} \int_0^{2\pi} \left[\frac{a^2}{h^2}(h^2 z - hz^2 + \frac{z^3}{3}) \right]_o^h d\theta$$

$$= \frac{1}{2} \int_0^{2\pi} \frac{a^2}{h^2} (h^3 - h^3 + \frac{h^3}{3}) d\theta$$

$$\therefore \qquad V = \int_0^{2\pi} \frac{a^2 h}{6} d\theta$$

$$= \frac{a^2 h 2\pi}{6}$$

$$V = \frac{\pi\, a^2 h}{3}$$

Because density ρ is constant

$$M = \frac{\rho\, \pi\, a^2 h}{3}$$

Because of symmetry $\bar{x} = \bar{y} = 0$, since the centroid must lie on the Z-axis.

Now, the moment about the XY-plane is given by:

$$M_{XY} = \iiint \rho z\, r\, dr\, d\theta\, dz$$

$$= \rho \int_0^h \int_0^{2\pi} \left[\frac{1}{2} r^2 \right]_o^{\frac{a}{h}(h-z)} z\, r\, dr\, d\theta\, dz$$

$$= \rho \int_{o}^{2\pi} \int_{o}^{h} \frac{1}{2} \left[\frac{a^2}{h^2} (h^2 - 2hz + z^2)z \; dz \; d\theta \right)$$

$$\rho \int_{0}^{2\pi} \int_{0}^{K} \frac{1}{2} \left[\frac{a^2}{h^2} \left(\frac{h^2 2}{2} - \frac{2hz^2}{3} + \frac{z^4}{4} \right) d\theta \right.$$

$$= \rho \int_{0}^{2\pi} \frac{1}{2} \frac{a^2}{h^2} \left(\frac{h^4}{2} - \frac{2h^4}{3} + \frac{h^4}{4} \right) d\theta$$

$$= \frac{1}{2} \rho \int_{0}^{2\pi} \left[\frac{a^2}{h^2} \left(\frac{3}{4}h^4 - \frac{2}{3}h^4 \right) \right] d\theta$$

$$\frac{1}{2} \rho \int_{0}^{2\pi} \left(\frac{a^2}{h^2} \right) \left(\frac{h^2}{12} \right) d\theta$$

$$= \frac{1}{2} \rho \int_{0}^{2\pi} \left(\frac{a^2 h^2}{12} \right) d\theta$$

$$= \frac{1}{2} \rho \cdot \frac{a^2 h^2}{12} \cdot 2\pi$$

$$\therefore \quad M_{XY} = \frac{\rho \pi a^2 h}{12}$$

and, $\bar{z} = \dfrac{M_{XY}}{M}$, $\quad \therefore \quad \bar{z} = \dfrac{\rho \pi a^2 h^2}{12} \cdot \dfrac{3}{\rho \pi a^2 h} = \dfrac{h}{4}$

Illustrative Example 15

Determine the centre of mass of a right circular conical object having density 'ρ' varying directly with the distance of point 'P' on its surface from the axis of the cone. Fig. 25 also refers.

We are given: $\rho(x, y, z) = kr$, k being the proportionality constant. Let us therefore deal with the problem using cylindrical coordinates. Accordingly, mass 'M' is given by:

$$M = \iiint \rho \; r \; dr \; d\theta \; dz$$

Referring to Fig. 1 $r = \dfrac{a}{h}(h-z)$ and putting $\rho = Kr$

$$\therefore \qquad M = k \int_0^h \int_0^{2\pi} \int_0^{\frac{a}{r}(h-z)} r^2 \, dr \, d\theta \, dz$$

$$= k \int_0^h \int_0^{2\pi} \left[\frac{r^3}{3}\right]_0^{\frac{a}{h}(h-z)} d\theta \, dz$$

$$= k \int_0^h \int_0^{2\pi} \frac{1}{3} \frac{a^2}{h^3}(h-z)^3 \, d\theta \, dz$$

$$= \frac{2\pi \, k}{3} \int_0^h \frac{a^3}{h^3}(h-z)^3 \, dz$$

$$M = \pi \frac{ka^3 h}{6}$$

and moment about the Z-Y plane, M_{XY} is to be determined from

$$M_{XY} = k \iiint z \, r^2 \, dr \, d\theta \, dz$$

$$= k \int_0^h \int_0^{2\pi} \int_0^{\frac{a}{h}(h-z)} z \, r^2 \, dr \, d\theta \, dz$$

$$= k \int_0^h \int_0^{2\pi} \frac{1}{3} \frac{a^3}{h^3}(h-z)^3 \, z \, d\theta \, dz$$

from which

$$M_{XY} = \frac{\pi \, ka^3 h^4}{30}$$

$$\therefore \qquad \bar{z} = \frac{\pi ka^3 h^2}{30} \bigg/ \frac{\pi ka^3 h}{6}$$

$$\text{i.e.} \qquad \bar{z} = \frac{h}{5}$$

The centre of mass of the cone has coordinates $\bar{x} = 0$; $\bar{y} = 0$; $\bar{z} = \dfrac{h}{5}$.

(iii) Triple Integration: Spherical Coordinates

Spherical co-ordinates of a point 'E' are shown in Figs 26 and 26a. The distance of E from the origin 0 is 'r'; the angle 'ϕ' is that between 'r' and the Z-axis; and, 'θ' is the angle that the plane through OE and the Z-axis makes with the YZ-plane. Rectangular coordinates x, y and z may thus be converted to spherical coordinates by substituting:

$$y = r Sin\phi \; Sin\theta$$
$$x = r Sin\phi \; Sin\theta$$
$$z = r Cos\phi$$

as shown in Fig. 26.

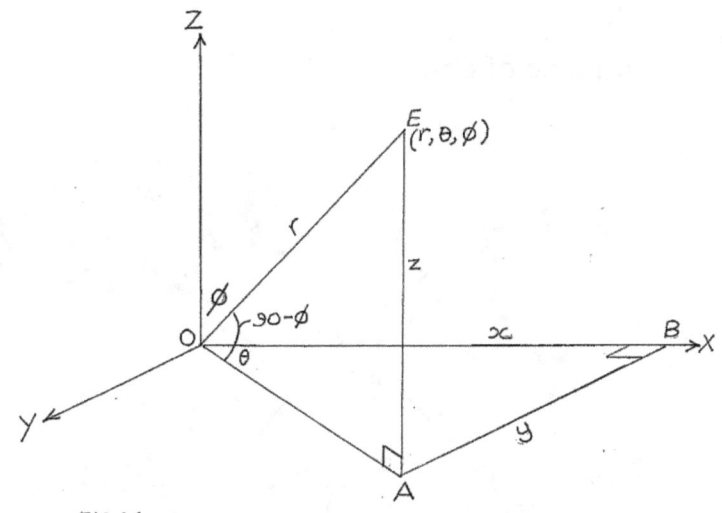

FIG.26
RELATIONSHIP BETWEEN X, Y, r, θ AND Φ IN SPHERICAL COORDINATES

IN △ OAE : AE = z = r Sin(90°-φ) = r Cosφ;
ALSO, OA = r Cos(90°-φ) = r Sinφ
IN △ OAB : x = OB = OA Cosθ = r Sinφ Cosθ;
ALSO, y = AB = OA Sinθ = r Sinφ Sinθ

Referring to Fig. 26b it is seen that the elemental volume δV, EJVMQPST may be expressed by

$$\delta V = \delta r (r\, \delta\phi)(r Sin\phi \; \delta\theta)$$

$$\delta V = r^2 Sin\phi \; \delta r \; \delta\theta \; \delta\phi$$

Illustrative Example 16

Q. Using spherical coordinates, determine the volume of a sphere of radius 'r'. Referring to the elemental volume shown in Fig. 26b and considering the eight octants of the sphere

$$V = 8 \int_0^{\pi/2} \int_0^{\pi/2} \int_0^r r^2 \, dr \, d\theta \, Sin\phi \, d\phi$$

$$= 8 \int_0^{\pi/2} \int_0^{\pi/2} \frac{r^2}{3} d\theta \, Sin\phi \, d\phi$$

$$= \frac{8r^3}{3} \cdot \frac{\pi}{2} \int_0^{\pi/2} Sin\phi \, d\phi$$

$$= \frac{4\pi r^3}{3} \left[-Cos\phi \right]_0^{\pi/2}$$

i.e. $V = \frac{4}{3}\pi r^3$, a piece of cake.

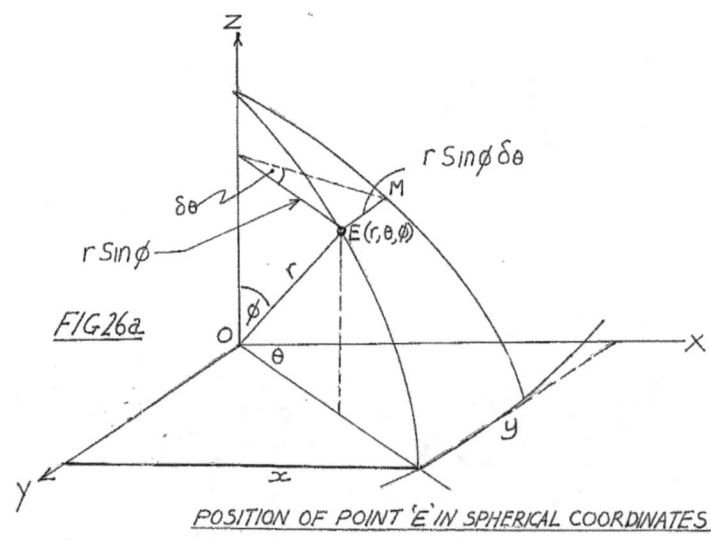

FIG 26a

POSITION OF POINT 'E' IN SPHERICAL COORDINATES

~ 200 ~

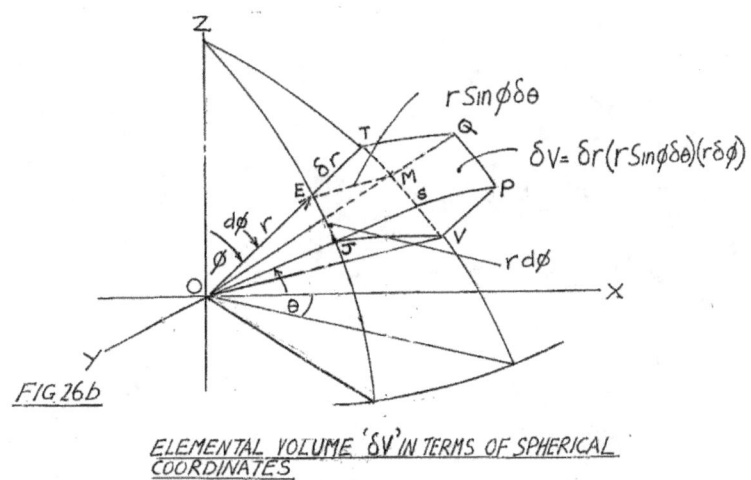

FIG 26b

ELEMENTAL VOLUME 'δV' IN TERMS OF SPHERICAL COORDINATES

Illustrative Example 17

Q   Using spherical coordinates, find the centre of mass of a hemispherical body of radius 'r', assuming that its density 'ρ' at every point is directly proportional to the square of the distance of the point from the axis of symmetry.

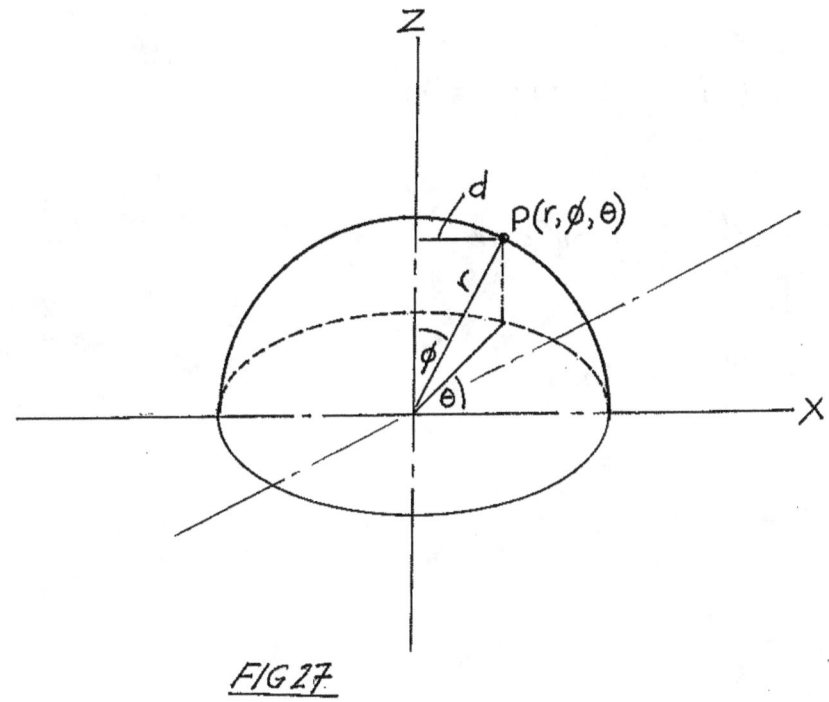

FIG 27

Given: Density '$\rho' = kd^2$, K being the proportionality constants.

Referring to Fig. 27

$$d = r^2 - z^2$$

Putting $\quad z = r\,Cos\phi$

$$d^2 = r^2 - r^2 Cos^2\phi$$

$$= r^2(1 - Cos^2\theta)$$

$$= r^2 Sin^2\theta$$

$$\therefore \quad \rho = kr^2 Sin^2\phi$$

Now $\quad M = 4 \int\limits_0^{\pi/2} \int\limits_0^{\pi/2} \int\limits_0^{r} \rho r^2 Sin\phi \; d\theta \; d\phi \; dr$

so that

$$M = 4 \int\limits_0^{\pi/2} \int\limits_0^{\pi/2} \int\limits_0^{r} kr^4 Sin^3\phi \; d\theta \; d\phi \; dr$$

$$= 4k \int\limits_0^{\pi/2} \int\limits_0^{\pi/2} \frac{r^2}{5} d\theta \; Sin^3\phi \; d\phi$$

$$= 4k \frac{r^5}{5} \cdot \frac{\pi}{2} \int\limits_0^{\pi/2} Sin^3\phi \; d\phi$$

$$= \frac{2\pi kr^5}{5} \int\limits_0^{\pi/2} Sin\phi \; d\phi$$

$$\int\limits_0^{\pi/2} Sin^3\phi \; d\phi = \left[-Cos\frac{\pi}{2} + \frac{1}{3}Cos^3\frac{\pi}{2} \right] - \left[-Cos0 + \frac{1}{3}Cos0 \right] = 1 - \frac{1}{3} = \frac{2}{3}$$

$$\therefore \quad M = \frac{2\pi kr^5}{5} \cdot \frac{2}{5}$$

$$M = \frac{4\pi}{15} kr^5$$

Continuing,

$$\int dM_{XY} = \int\int\int z\rho\, r^2 Sin\phi\, d\theta\, d\phi\, dr$$

Putting $\qquad z = rCos\phi$ and $\rho = kr^2 Sin^2\phi$

$$\therefore \qquad M_{XY} = 4 \int_0^{\frac{\pi}{2}} \int_0^{\frac{\pi}{2}} \int_0^{r} rCos\phi\, kr^4 Sin^3\phi\, d\theta\, d\phi\, dr$$

$$= 4k \int_0^{\frac{\pi}{2}} \int_0^{\frac{\pi}{2}} \int_0^{r} r^5 Cos\phi\, Sin^3\phi\, d\theta\, d\phi\, dr$$

$$= 4k \int_0^{\frac{\pi}{2}} \int_0^{\frac{\pi}{2}} \frac{r^6}{6} Cos\phi\, Sin^3\phi\, d\theta\, d\phi$$

$$= 4k \cdot \frac{r^6}{6}\frac{\pi}{2} \int_0^{\frac{\pi}{2}} Cos\phi\, Sin^3\phi\, d\phi$$

$$= \frac{\pi k r^6}{3} \int_0^{\frac{\pi}{2}} Cos\phi\, Sin^3\phi\, d\phi$$

$$= \frac{\pi k r^6}{3} \int_0^{\frac{\pi}{2}} Sin^3\theta(Sin\phi)$$

$$= \frac{\pi}{3} kr^6 \left[\frac{Sin^4\phi}{4}\right]_0^{\pi/2}$$

$$= \frac{\pi}{3} kr^6 \left(\frac{1}{4}\right)$$

i.e. $\qquad M_{XY} = \dfrac{\pi k r^6}{12}$

so that $\qquad \bar{z} = M_{XY}/M$

$$= \frac{\pi k k r^6}{12}$$

$$\bar{z} = \frac{5r}{16}$$

Therefore the coordinates of the centre of mass of the hemisphere are:

$$\bar{x} = o; \quad \bar{y} = o; \quad \bar{z} = \frac{5r}{16}$$

Here is one for you to solve.

Using spherical as opposed to rectangular and cylindrical coordinates show that the centroid of a hemispherical body of constant density and radius 'r' is distant $\frac{3r}{8}$ from its base.

THE THEOREMS OF PAPPUS

These theorems, of which there are two, establish relationships between surface areas and volumes of revolution and the centroids of the lines and rotated areas respectively which are associated with these surface areas and swept volumes.

When you ponder the theorems, their sheer simplicity engenders the feeling that had not Pappus invented them you might have arrived yourself at the same result, heuristically. This is not to disparage or deny the genius of Pappus. But let us digress for a moment.

The theorems of Pappus are named after the Greek geometer Pappus of Alexandria who flourished around the end of the third century of the Christian era. He was unquestionably a mathematician of the first rank. I have deliberately refrained from using the epithet 'excellent' in describing his many-faceted contribution to mathematics, because nowadays the word excellent is used to describe so many widely varying standards that it has ceased to have any meaning at all; Absolutely!

Before getting caught up in the mathematical intricacies of Pappus' Theorems, I would like to proffer a few words on insight which I contend anyone would find useful in the study of any subject. Pappus' work has provided me with an opportunity to do so. Pappus wrote extensively. He is perhaps best known for his eight-volume "Collection." This work is incomplete in that the first book was lost and in many of the other volumes, portions are missing. But in all his writings in the set of volumes Pappus did something which I have come to believe is an essential element in the mastery of any art or science; any subject it seems to me. Pappus studied extensively the works of the great mathematicians who went before him; and, it is from this deep acquaintanceship with these that his own original contributions were spawned. Centuries later, hear what the renowned artist Sir Joshua Reynolds said on this very theme. Mark you, Sir Josh is here referring to the art of painting: "The more extensive, therefore, your acquaintance is with the works of those who have excelled, the more extensive will be your powers of invention – and what may appear still more like a paradox, the more original will be your conceptions." Enter Sir Arthur Quiller-Couch in his delightful *'vade-mecum'* "On the art of Writing", who said of these words of Sir Joshua: "And will anyone (....) tell me that what Reynolds said of his painting is not today, for us, applicable to writing?" Sir Arthur might well have included: scientific and mathematical inquiry; in short, the entire academic epitome!

Thus my brothers and sisters I exhort you to seek out the works of the best "spiritual begetters of practice" in whatever branch of study you are pursuing; develop an extensive acquaintanceship then create your own inventions, and thereby push the frontiers of knowledge forward and upwards. Amen!

The two theorems of Pappus which it is claimed were rediscovered more than twelve centuries after Pappus' time by a Swiss mathematician Paul Gueldin (sometimes 'Guldinus') may be stated as follows.

FIRST THEOREM OF PAPPUS

The volume of a solid of revolution swept out by rotating an area bounded by a plane curve or curves on one side of an axis in its plane, is equal to the magnitude of the bounded area times the circumference of the circle in which the radius is equal to the distance of the centroid from the axis of rotation. This definition is illustrated by way of Fig 28. Notice that the region of the bounded area 'A' has to lie on one side, or it may just touch the side, of the axis of revolution. It must not however cut the axis of revolution.

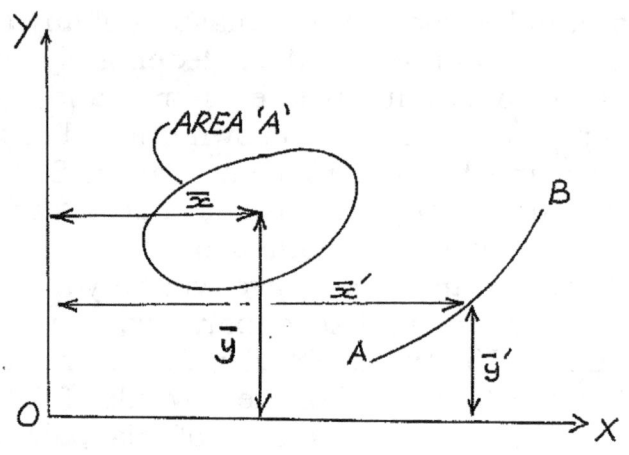

FIG 28

Accordingly the volume of solid of revolution, V_x resulting from the rotation of an area A in Fig. 28 about the X-axis is given by:

$$V_X = A(2\pi\overline{y})$$

Similarly, the corresponding volume of revolution about the Y-axis is given by:

$$V_Y = A(2\pi\overline{x})$$

where $\overline{x}, \overline{y}$ are the centroidal distances of area 'A; from the Y- and X-axis, respectively.

In the literature you may sometimes come across Pappus' Theorem under the hyphenated name Pappus-Guldin; Guldin meaning Gueldin.

SECOND THEOREM OF PAPPUS

The area of the surface of revolution about an axis, resulting from the rotation of an arc of a plane curve, which lies entirely on one side or only touches but does not cut said axis, equals the length of the arc multiplied by the length of the path of its centroid.

Referring again to Fig. 28, the plane curve AB has its centroid at a distance $\overline{y'}$ from the X-axis. According to Pappus' second theorem:

Swept surface are due to =   Length of arc 'AB' × Distance of
rotation about X-axis   centroid of arc 'AB' from X-axis

Similarly,

Swept surface area due to =   Length of arc 'AB' × Distance of
rotation about Y-axis   centroid of arc 'AB' from Y-axis

Illustrative Example 19

Q.   A triangle is formed by lines $3y = 4x$, $y = 0$, and the perpendicular on to the first line from point (5,0). Use Pappus' Theorem to show that the volumes obtained by revolving this triangle about OX and OY are $48\pi/5$ and $136\pi/5$ respectively. Verify these results by another method. (Geary, Lowry and Hayden : their question; my solution)

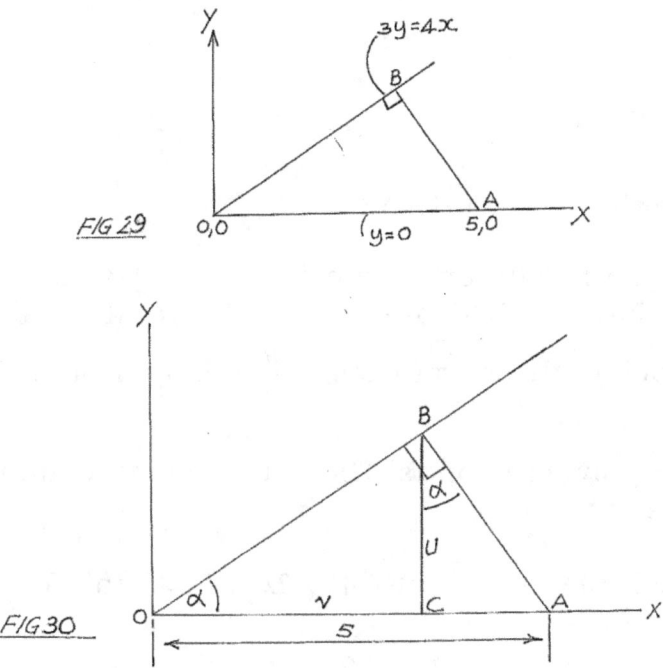

FIG 29

FIG 30

Be careful to note that the perpendicular AB is from the first line $3y = 4x$ as shown in Fig. 29. In Fig. 30, I have drawn a larger diagram in the interest of clarity. There

$$\text{Tan}\,\alpha = \frac{AB}{OB} = \frac{4}{5}$$

Thus, we may treat OAB as a 3, 4, 5 right-angled triangle. Now in triangle OCB, let OC = v and BC = u. In this triangle

$$\text{Tan}\,\alpha = \frac{u}{v}$$

i.e. $\quad \dfrac{4}{3} = \dfrac{u}{v}$

Also $\quad u^2 + v^2 = (3)^2$

$\therefore \quad u^2 + \dfrac{9u^2}{16} = 9$

or $\quad u = \dfrac{12}{5}$

$\therefore \quad u = 2.4$

and $\quad v = 1.8$

which means that $\quad CA = 3.2$

Let us treat the revolution of the triangle OAB as a revolution of the 2 component triangles OCB and CAB. For rotation about the X-axis the centroid of each of these triangles is $\dfrac{u}{3}$ i.e. 0.8 from this axis.

Therefore in applying Pappus' Theorem we may write for rotation about t he X-axis:

Area of $OCB \times 2\pi(0.8) +$ Area of $CAB \times 2\pi(0.8) =$ Total Vol. required.

$\therefore \quad \left(\dfrac{2.4 \times 1.8}{2}\right) \times 2\pi(0.8) + \left(\dfrac{2.4 \times 3.2}{2}\right) \times 2\pi(08) = V_X$

i.e. $\quad V_Y = (4.32\pi + 7.68\pi)0.8$

$= 9.6\pi \quad \text{or} \quad 9\,^3\!/_5\,\pi$

or $\quad V_X = \dfrac{48}{5}\pi \qquad\qquad$ QED

For rotation about the Y-axis

Area of $OCB \times 2\pi\left(\dfrac{2}{3}v\right) +$ Area of $CAB \times 2\pi\left(\dfrac{AC}{3} + v\right) = V_Y$

$$\therefore \quad \left(\frac{2.4 \times 1.8}{2}\right) \times 2\pi(1.2) + \left(\frac{2.4 \times 3.2}{2}\right) \times 2\pi\left(\frac{3.2}{3}1.8\right) = V_Y$$

$$5.184\pi + 22.016\pi = V_Y$$

$$\therefore \quad V_Y = 27.2\pi$$

$$\text{or} \quad V_Y = 27\frac{1}{5}\pi$$

$$\text{i.e} \quad V_Y = \frac{136}{5}\pi \qquad \text{Q.E.D}$$

In order to verify these results, reference is made to Fig. 31. Projecting line AB until it meets the Y-axis in 'E', we may write:

$$\frac{u}{OE} = \frac{AC}{OA}$$

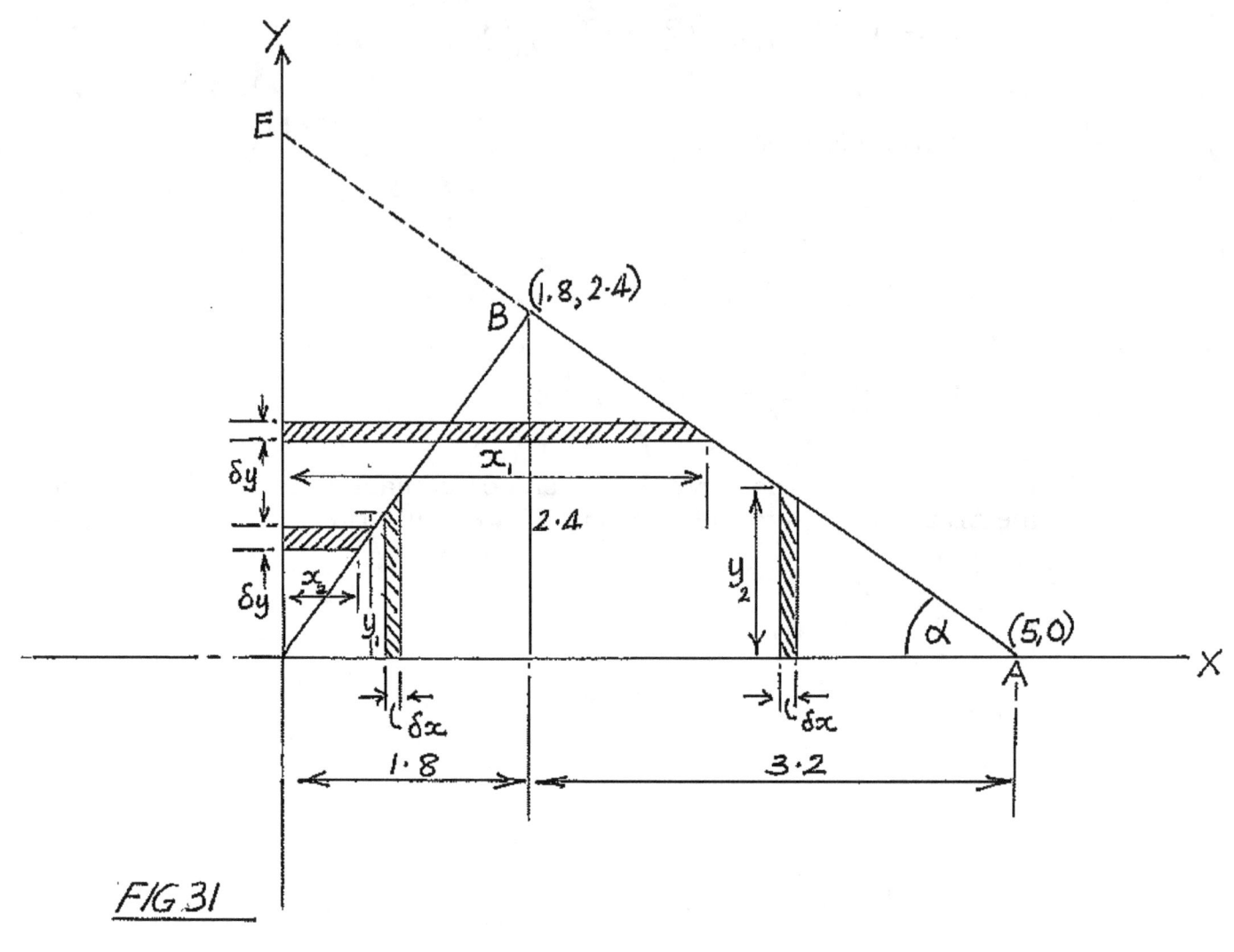

FIG 31

i.e. $\quad \dfrac{2.4}{OE} = \dfrac{3.2}{5}$

or $\quad OE = \dfrac{12}{3.2}$

$\quad\quad\quad\ = 3.75$

Also, the slope of line AB = $-\dfrac{2.4}{3.2} = -\dfrac{3}{4}$. Therefore the equation of AB can now be expressed in intercept form by the following:

$$y = -\frac{3}{4}x + 3.75$$

or $\quad 4y = -3x + 15$

or $\quad x = -\dfrac{4y}{3} + 5$

Considering the volume 'V_Y' of area OAB for rotation about the Y-axis it is seen that

$$V_Y = \pi \int_0^{2.4} x_1^2 \, dy - \pi \int_0^{2.4} x_2^2 \, dy$$

i.e. $\quad V_Y = \pi \int_0^{2.4} \left(-\dfrac{4y}{3} + 5\right)^2 dy - \pi \int_0^{2.4} \left(\dfrac{3y}{4}\right)^2 dy$

$$= \pi \int_0^{2.4} \left(\dfrac{16y^2}{9} - \dfrac{40y}{3} + 25 - \dfrac{9y^2}{16}\right) dy$$

$$= \pi \int_0^{2.4} \left(\dfrac{175}{144} y^2 - \dfrac{40y}{3} + 25\right) dy$$

$$= \pi \left|\dfrac{175y^3}{432} - \dfrac{20y^2}{3} + 25y\right|_o^{2.4}$$

$$= \pi(5.6 - 38.4 + 60)$$

$$= \pi(27.2)$$

$$V_Y = \dfrac{136\pi}{5} \qquad \text{Check}$$

For rotation of the area about the X-axis we consider the 2 component areas OCB and CAB. Thus volume V_x is given by:

$\therefore \quad V_X = \pi \int_0^{1.8} y_1^2 \, dx + \pi \int_{1.8}^{5} y_2^2 \, dx$

$$= \pi \int_0^{1.8} \left(\dfrac{4x}{3}\right)^2 dx + \pi \int_{1.8}^{5} \left(-\dfrac{3x}{4} + 3.75\right)^2 dx$$

$$= \pi \int_0^{1.8} \dfrac{16x^2}{9} \, dx + \pi \int_{1.8}^{5} \left(\dfrac{9x^2}{16} - \dfrac{90x}{16} + \dfrac{225}{16}\right) dx$$

i.e. $\quad V_X = \left|\dfrac{16x^3}{27}\right|_0^{1.8} + \pi\left|\dfrac{9x^3}{48} - \dfrac{45x^2}{16} + \dfrac{225x}{16}\right|_{1.8}^5$

$$= \pi(3.456) + \pi\{23.4375 - 70.3125 + 70.3125$$

$$- (1.0935 - 9.1125 + 25.3125)\}$$

$$= 3.456\pi + \pi(23.4375 - 17.2935)$$

$$= 3.456\pi + 6.144\pi$$

$$= 9.6\pi$$

or $\quad V_X = \dfrac{48\pi}{5}$, $\qquad$ Check.

The results obtained by use of Pappus' Theorem in the first part are therefore confirmed.

TABLE 1

Centroids of some common geometrical areas and bodies of constant density

Area/Volume	Figure	Centroidal Position $\bar{x}, \bar{y}$
Rectangle		$\bar{x} = \dfrac{b}{2}; \quad \bar{y} = \dfrac{a}{2}$
Triangle		$\bar{x} = \dfrac{b}{3}; \quad \bar{y} = \dfrac{h}{3}$
Circle or sphere		$\bar{x} = \bar{y} = \bar{z} = 0$
Semi-Circle		$\bar{y} = \dfrac{4r}{3\pi}$
Quadrant of a Circular Arc		$\bar{x} = \dfrac{2r}{\pi}; \bar{y} = \dfrac{2r}{\pi}$

Area/Volume	Figure	Centroidal Position $\bar{x}, \bar{y}$
Quadrant of a Circle		$\bar{x} = \dfrac{4r}{3\pi} ; \bar{y} = \dfrac{4r}{3\pi}$
Conic Volume (with material of constant density)		$\bar{y} = \dfrac{h}{4}$
Hemispherical Volume (with material of constant density)		$\bar{y} = \dfrac{3r}{8}$
Circular Sector		$\bar{x} = \dfrac{2r Sin\alpha}{3}$
Parabola		$\bar{x} = \dfrac{3b}{5} ; \bar{y} = \dfrac{3h}{8}$

SOLVED AND OTHER PROBLEMS

Q.1   Find the centroid of the unequal-flanged I-Section shown in Fig. 1.

[Note : If a beam with such a cross-section were to be subjected to a bending moment along its longitudinal axis, it would be necessary to find the position of the centroid of the section as a pre-requisite to determination of the magnitude of the bending and shearing stresses in the beam due to the bending moment].

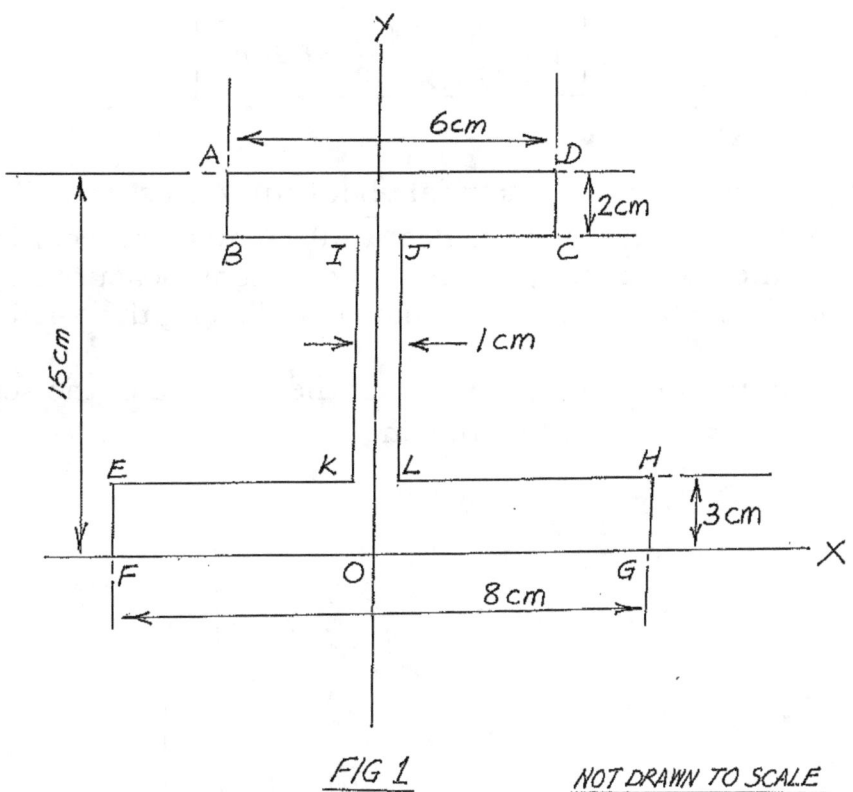

FIG 1   NOT DRAWN TO SCALE

The first step is to locate reference axes in respect of which the position of the centroid is to be reported.   Because the cross-section is symmetrical about a vertical line through it, this line is chosen as a Y-axis; the X-axis is coincident with the base of the bottom flange.

For the sake of convenience, the cross-section is divided into 3 parts : two flanges viz. *ABIJCD''* and *'EFGHLK'* and one web : *'IKLJ'*.

The necessary calculations are given in the following tabulation.

Area (cm)^2 A	Distance of its centroid 'y' from X-axis (cm)	Ay (cm)^3
ABIJCD=6 x 2 = 12	14	168
EFGHLK = 8 x 3 x 24	1.5	36
IKLJ = 10 x 1 = 10	8	80
$\therefore$	$\bar{y} = \dfrac{168+36+80}{12+24+10} = 6.2cm$	

Because the section is symmetrical about the Y-axis, the 'X' coordinate of the centroid is on this axis; hence $\bar{x} = 0$. Thus, the coordinates of the centroid of the section are (0, 6.2cm), i.e. 6.2cm measured upwards from the bottom face *FG* of the bottom flange and along the Y-axis.

Q2. You are required to find the position of the centroid of the segment of the parabola $y^2 = 4x$ truncated by the line $x = 3$. See Fig. 1.

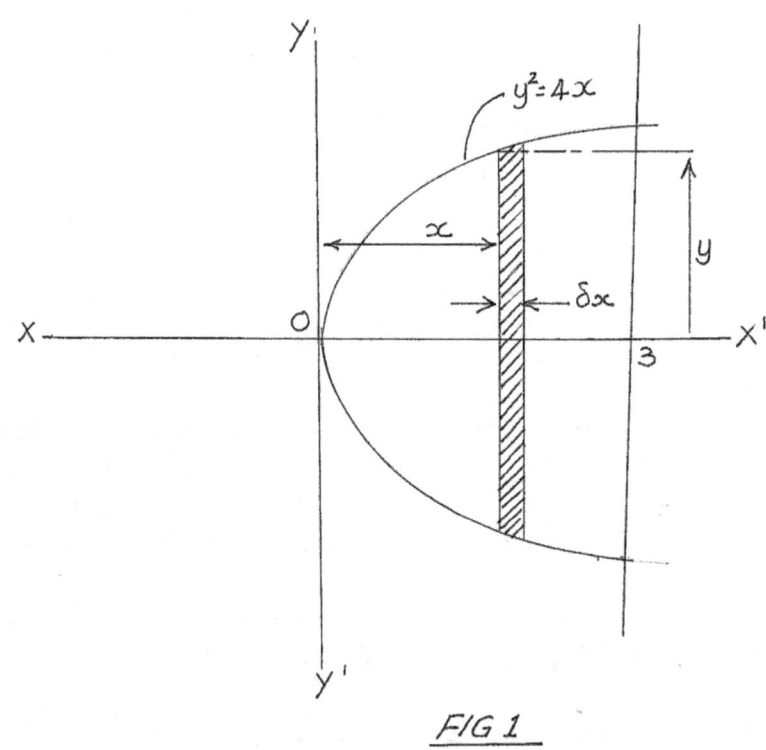

FIG 1

~ 216 ~

The first thing to record is the evident fact that because of symmetry, the 'Y' coordinate of the centroid is zero, i.e. $\bar{y} = 0$. Now the moment about the Y-axis of the elemental area δA shown shaded in the diagram is:
$y \, \delta x \cdot x$

$\therefore$ Moment of top segment about the Y-axis, $= \int_0^3 y \, dx \cdot x$

So that for both top and bottom segments, the moment about the Y-axis $= 2\int_0^3 y \, dx \cdot x$

From $y^2 = 4x$, $y = 2\sqrt{x}$,

$\therefore \quad 2\int_0^3 y \, dx.x = 2\int_0^3 2\sqrt{x}.(x)dx$

Evaluating this integral produces $\quad 4\left[\frac{2}{5}(x)^{\frac{5}{2}}\right]_0^3$

$\therefore \quad M_y = 4\left[\frac{2}{5} \cdot \left(\sqrt{3}\right)^5\right]$

$= 4 \cdot \frac{2}{5} \cdot 9\sqrt{3}$

$= \frac{72\sqrt{3}}{5}$

Next, we determine the area of the truncated portion of the segment. Designating this area, A, then

$A = 2\int_0^3 y \, dx$

$= 2\int_0^3 2\sqrt{x} \, dx$

$= 4\left[\frac{2}{3}\left(\sqrt{x}\right)^3\right]$

$= 4\left(\frac{2}{3} \cdot 3\sqrt{3}\right)$

i.e. $A = 8\sqrt{3}$

∴ $\bar{x} = \dfrac{72\sqrt{3}}{5.8\sqrt{3}} = 1.8$

The coordinates of the centroid are therefore (1.8, 0).

Q3.   Determine the coordinates of centroid of the area in the region between the parabola $y^2 = 4x$ and the lines $y = 4$ and $x = 5$. See Fig. 1.

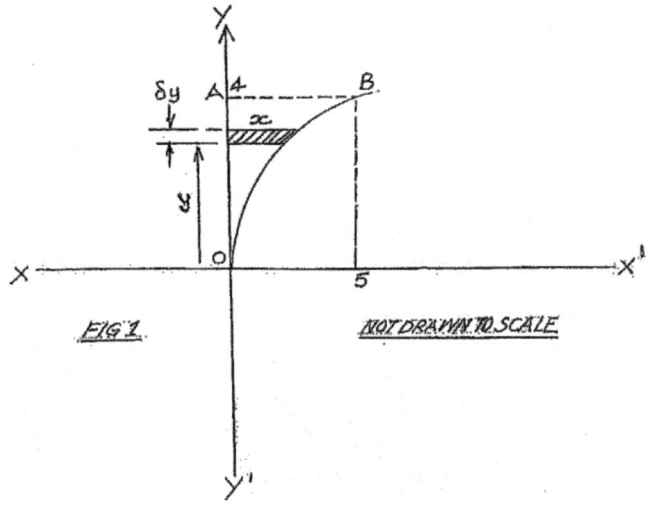

FIG 1.   NOT DRAWN TO SCALE

The area in which we are interested in OAB in Fig. 1

Shaded elemental area $= \delta A = x \, \delta y$; but $x = \dfrac{y^2}{4}$

∴ $A = \int x \, dy = \int_0^4 \dfrac{y^2}{4} \cdot dy$

or $A = \left[\dfrac{y^3}{12} \right]_0^4 = \dfrac{16}{3}$ sq. units

For moments of the element $\delta A = x \, \delta y$ about the X-axis

$\delta M_X = \delta A \cdot y$

and about the Y-axis

$$\delta M_Y = \delta A \cdot \frac{x}{2}$$

Evaluating these integrals

$$M_X = \int x \, dy \cdot y \;\; = \;\; \int_0^4 \frac{y^2}{4} \cdot y \, dy$$

$$M_x = \int_0^4 \frac{y^3}{4} \, dy \;\; = \;\; \left[\frac{y^4}{16}\right]_0^4 \;\; = \;\; \frac{256}{16} = 16$$

$$\therefore \qquad \bar{y} = \frac{M_X}{A} \;\; = \;\; \frac{16.3}{16} = 3$$

Continuing

$$M_Y = \int x \, dy \cdot \frac{x}{2} = \frac{1}{2} \int x^2 dy$$

But $\quad x^2 = \dfrac{y^4}{16}$

$$\therefore \qquad M_Y = \frac{1}{2} \int_0^4 \frac{y^4}{16} \, dy = \frac{1}{32} \left[\frac{y^5}{5}\right]_0^4 = \frac{1}{32}\left(\frac{1024}{5}\right)$$

$$M_Y = \frac{32}{5}$$

from which

$$\bar{x} = \frac{M_Y}{A} = \frac{32}{5} \cdot \frac{3}{16}$$

$$\bar{x} = \frac{6}{5}$$

$$\therefore \qquad \text{Coordinates of centroid} = \left(\frac{6}{5},\, 3\right)$$

Q4. Sketch the portions in the 1st and 2nd quadrant of the curve $\rho = 4Sin^2\theta$ and determine the coordinates of the centroid of the area in the first quadrant.

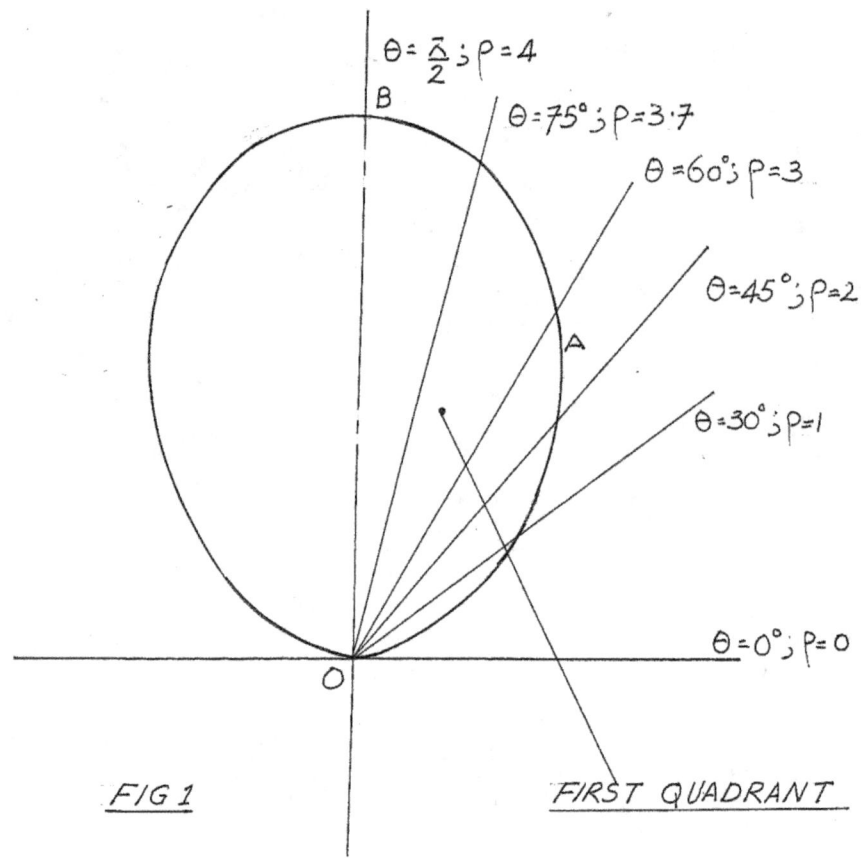

FIG 1

FIRST QUADRANT

A rough sketch of $\rho = 4Sin^2\theta$ is shown in Fig. 1; two quadrants are shown. For the area in the first quadrant 'OAB' :

$$\overline{x} = \frac{\int_0^{\pi/2} \int_0^{4Sin^2\theta} \rho^2 d\rho \, Cos\theta \, d\theta}{\int_0^{\pi/2} \int_0^{4Sin^2\theta} \rho \, d\rho \, d\theta}$$

Evaluating the numerator integral first

$$\int_0^{\pi/2} \int_0^{4Sin^2\theta} \rho^2 d\rho \, Cos\theta \, d\theta$$

$$= \int_0^{\pi/2} \left[\frac{\rho^3}{3} \right]_0^{4Sin^2\theta} Cos\theta \, d\theta$$

~ 220 ~

$$\int_0^{\pi/2} \frac{64}{3} Sin^6\theta \, Cosd\theta = \frac{64}{3}\int_0^{\pi/2} Sin^6\theta \, d \, (Sin\theta)$$

$$= \frac{64}{3}\left[\frac{Sin^7\theta}{7}\right]_0^{\frac{\pi}{2}} = \frac{64}{3}\cdot\frac{1}{7}$$

$$= \frac{64}{21}$$

Now to the evaluation of the denominator integral, viz

$$\int_0^{\pi/2}\int_0^{4Sin^2\theta} \rho \, d\rho \, d\theta$$

$$\int_0^{\pi/2}\left[\frac{\rho^2}{2}\right]_0^{4Sin^2\theta} d\theta$$

$$\int_0^{\pi/2} \frac{16}{2} Sin^4\theta \, d\theta = 8\int_0^{\pi/2} Sin^4\theta \, d\theta$$

Using Wallis's reduction formula for $I_n = \int_0^{\pi/2} Sin^n x$, when n is even, viz

$$I_n = \frac{(n-1)(n-3)....3.1}{n(n-2)....4.2}\cdot\frac{\pi}{2}$$

We get,

$$I_4 = \frac{3.1}{4(2)}\cdot\frac{\pi}{2} = \frac{3\pi}{16}$$

$$\therefore \quad 8\int_0^{\pi/2} Sin^4\theta \, d\theta = 8\left(\frac{3\pi}{16}\right)$$

$$= \frac{3\pi}{2}$$

Just to break the monotony of all these vapid integrations, I cannot resist telling you about John Wallis, the eminently distinguished author of the reduction formula used earlier in the main text and in a number of the previous problems. He was by all accounts an arithmetic prodigy. Born in 1606 at Ashford in Kent, England, he was what some of his detractors called a late developer. This so-called late developer became Savilian Professor of Geometry at Oxford University in 1649. That status was accorded him in consequence of his brilliant display of talent "in deciphering intercepted Royalist documents for (...) Parliamentarians." At least he did something of significance to earn the rank of 'Professor'. He is perhaps best known for the astounding mathematical feat of extracting mentally... that is without recourse to anything written.... the square root of a number comprising fifty-three digits. He was 54 years old at the time, and it took him a month to accomplish the task. He died in his 86th year. Some late developer! Back to our problem. Based on what has been done so far:

$$\bar{x} = \frac{64}{21} \Big/ \frac{3\pi}{2}$$

i.e. $$\bar{x} = \frac{128}{63\pi}$$

Continuing,

$$\bar{y} = \int_0^{\pi/2} \int_0^{4Sin^2\theta} \rho^2 dp \, Sin\theta \, d\theta$$

$$= \int_0^{\pi/2} \left[\frac{\rho^3}{3} \right]_0^{4Sin^2\theta} Sin\theta \, d\theta$$

$$= \int_0^{\pi/2} \frac{64}{3} Sin^6\theta \cdot Sin\theta \, d\theta$$

$$= \frac{64}{3} \int_0^{\pi/2} Sin^7\theta \, d\theta$$

Professor Wallis's Reduction for $\int_0^{\pi/2} Sin^n d\theta$ when n is odd is

$$I_n = \frac{(n-1)(n-3)....4.2}{n(n-2)....5...3}$$

$$\therefore \quad \int_0^{\pi/2} Sin^7\theta \, d\theta = \frac{6\cdot4\cdot2}{7\cdot5\cdot3} = \frac{48}{105}$$

~ 222 ~

$$\therefore \qquad \frac{64}{3} \int_0^{\pi/2} Sin^7\theta \, d\theta = \frac{64}{3} \cdot \frac{48}{105} = \frac{1024}{105}$$

$$\therefore \qquad \overline{y} = \frac{1024}{105} \Big/ \frac{3\pi}{2} = 2048/989.5$$

i.e. $\qquad \overline{y} = 2.1$

Here is a slightly more difficult assignment.

Q5. What are the coordinates of the centroid of the area enclosed by one arch of the cycloid : $x = a(\theta - Sin\theta)$, $y = a(1 - Cos\theta)$, shown in Fig. 1.

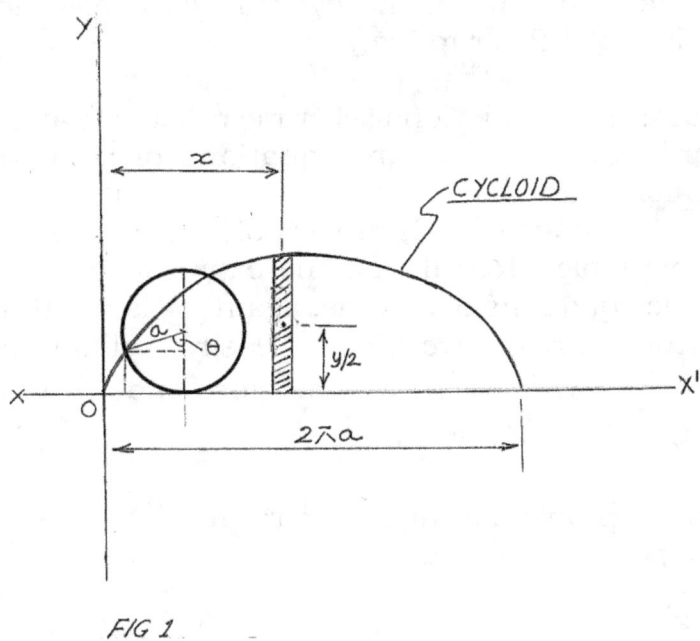

FIG 1

You may be interested to know that the cycloid was at one time called the 'Helen of Geometers'. This came about because investigating its properties was fraught with as much controversy at the time as did the description of that most famous figure of Greek mythology "Helen of Troy" – said to be indirectly the cause of the Trojan War. She was claimed by some to be an epic heroine and by others a goddess.

Did you know the cycloid, also called the "tautochrone of Huygens" is known as the brachistochrone? A three-in-one curve! The curve of the cycloid upside-down (the brachistochrone alias tautochrone) is the line of quickest descent down which any heavy body would slide. And if you thought Sir Christopher Wren was only famous for being the architect of St. Paul's Cathedral, think again, for it was he who rectified the cycloid (i.e. a straight line was found whose length equalled the curved arch of the cycloid). Is the curve of the skateboarder's rink a brachistochrone? Investigate.

I recall my youthful days when fun was had by placing a short stick in the groove of the rim of a discarded bicycle wheel, holding it standing aright on the ground and by manipulating the stick, accelerating the wheel to roll and to attain quite some speed I might add, in more or less a straight line, as I ran behind it, urging it along with the stick.

It was only long afterwards when I had embarked on engineering studies that I discovered that as a wheel rolls on a fixed straight line, every point on its circumference traces a curve called a cycloid, relative to the ground, assuming no slipping occurs. Have you ever rolled a roller?

Now, the expressions:

$$x = a(\theta - Sin\theta) \text{ and } y = a(1 - Cos\theta)$$

are parametric equations of the cycloid, with reference to the circle or wheel of radius 'a'.. Refer to Fig. 1

What are parametric equations? Let me answer that with reference to 'x' and 'y' coordinates. These are equations of a curve with 'x' and 'y' coordinates for every point on the curve. And these xs and ys are expressed as functions of a third variable. In the case of our cycloid, 'θ' is the third variable. Recall that the expressions : $x = rCos\theta$ and $rSin\theta$ are parametric equations of a circle; again 'θ' is the third variable. In the case of the circle we can easily eliminate 'θ', thus : $x^2 + y^2 = r^2Cos^2 + r^2Sin^2\theta = r^2$. We therefore end up with the rectangular equation of the circle $x^2 + y^2 = r^2$.

To return to the problem at hand, we recall

$$\bar{x} = \frac{\int x \, dA}{dA}$$

and

$$\bar{y} = \frac{\int y \, dA}{dA}$$

Now, since $\quad x = F(\theta)$ and $y = Q(\theta)$

$$dx = F'(\theta)d\theta; \quad dy = Q'(\theta)d\theta$$

$$\therefore \quad dA = y \, dx$$

$$= Q(\theta)F'(\theta)d\theta \text{ so that}$$

$$\bar{x} = \frac{\int xQ(\theta)F'(\theta)d\theta}{Q(\theta)F'(\theta)d\theta}$$

and

$$\bar{y} = \frac{\int \frac{y}{2}Q(\theta)F'(\theta)d\theta}{Q(\theta)F'(\theta)d\theta}$$

Let us proceed to evaluate $\bar{x}$. Here $Q(\theta) = a(1 - Cos\theta); \quad F(\theta) = a(\theta - Sin\theta)$

$\therefore \qquad F'(\theta)d\theta = a(1 - Cos\theta)d\theta$

$\therefore \qquad \bar{x} = \dfrac{\int a(\theta - Sin\theta)a(1 - Cos\theta)a(1 - Cos\theta)d\theta}{\int a(1 - Cos\theta)a(1 - Cos\theta)d\theta}$

$$= \frac{a^3 \int (\theta - Sin\theta)(1 - Cos\theta)^2\, d\theta}{a^2 \int (1 - Cos\theta)^2\, d\theta}$$

Evaluating the numerator

$$a^3 \int (\theta - Sin\theta)(1 - 2Cos\theta + Cos^2\theta)d\theta$$

$$= a^3 \int \theta - 2\theta\, Cos\theta + \theta\, Cos^2\theta - Sin\theta + 2Sin\theta\, Cos\theta - Cos^2 Sin\theta)d\theta$$

$$= a^3 \int \theta\, d\theta - 2\int \theta Cos\theta\, d\theta + \theta\int Cos^2\theta\, d\theta - \int Sin\theta\, d\theta$$

$$+ \int Sin2\theta\, d\theta - \int (1 - Sin^2\theta)Sin\theta\, d\theta$$

$$= a^3 \int \theta\, d\theta - 2\int \theta Cos\theta\, d\theta + \theta\int \theta\left(\frac{Cos2\theta}{2} + \frac{1}{2}\right)d\theta - \int Sin\theta\, d\theta$$

$$+ \int Sin2\theta\, d\theta - \int Sin\theta\, d\theta + \int Sin^3\theta\, d\theta$$

$$= a^3 \left\{ \int \theta \, d\theta - 2\int \theta \, Cos\theta \, d\theta + \frac{1}{2}\int \theta \, Cos2\theta + \int \frac{\theta}{2} \, d\theta \right.$$

$$\left. - 2\int Sin\theta \, d\theta + \int Sin2\theta \, d\theta + \int Sin^3\theta \, d\theta \right.$$

$$= a^3 \left\{ \frac{3}{2}\int \theta \, d\theta - 2\int \theta \, Cos\theta \, d\theta + \frac{1}{2}\int \theta \, Cos2\theta + 2\int Sin\theta \, d\theta \right.$$

$$\left. + \int Sin2\theta \, d\theta + \int Sin^3\theta \, d\theta \right.$$

Working out each integral separately, the following results are obtained. Note that I have not bothered to show arbitrary constants.

$$\frac{3}{2}\int \theta \, d\theta = \frac{3\theta^2}{4}$$

$$2\int \theta \, Cos\theta \, d\theta = 2(\theta \, Sin\theta + Cos\theta)$$

$$\frac{1}{2}\int \theta \, Cos2\theta \, d\theta = \frac{1}{4}\theta \, Sin2\theta + \frac{1}{8}Cos2\theta$$

$$2\int Sin\theta \, d\theta \qquad = -2Cos2\theta$$

$$\int Sin2\theta \, d\theta \qquad = -\frac{1}{2}Cos2\theta$$

$$\int Sin^3\theta \, d\theta = -\frac{1}{3}Sin^2\theta \, Cos\theta - \frac{2}{3}Cos\theta$$

One arch of the cycloid embraces one rotation of the rolling wheel or circle. Accordingly in evaluating each interval the limits are : from $\theta = 0$, to $\theta = 2\pi$.

Summarising,

$$a^3 \int_0^{2\pi} (\theta - Sin\theta)(1 - Cos\theta)^2 \, d\theta$$

$$= a^3 \left[\frac{3\theta^2}{4} + 2\theta \, Sin\theta + 2Cos\theta + \frac{\theta \, Sin2\theta}{4} + \frac{1}{8}Cos2\theta \right.$$

$$-2Cos\theta - \frac{1}{2}Cos2\theta - \frac{1}{3}Sin^2\theta \, Cos\theta - \frac{2}{3}Cos\theta \Big]_0^{2\pi}$$

$$= a^3 \left(3\pi^2 + 0 + 2 + 0 + \frac{1}{8} - 2 - \frac{1}{2} - 0 - \frac{2}{3}\right) - \left(0 + 0 + 2\right.$$

$$\left. + 0 + \frac{1}{8} - 2 - \frac{1}{2} - 0 - \frac{2}{3}\right)$$

$$= a^3 \left\{\left(3\pi^2 - \frac{25}{24}\right) - \left(\frac{-25}{24}\right)\right\}$$

$$= 3a^3\pi^2$$

Let us now evaluate the denominator :

$$\int dA = \int Q(\theta)F'(\theta)d\theta$$

$$Q(\theta) = a(1 - Cos\theta); \quad F'(\theta)d\theta = a(1 - Cos\theta)d\theta$$

$$\therefore \quad \int_0^{2\pi} dA = \int_0^{2\pi} a^2 (1 - Cos\theta)^2 \, d\theta$$

$$= a^2 \int_0^{2\pi} (1 - 2Cos\theta + Cos^2\theta)d\theta$$

$$= a^2 \left[\left(\theta - 2Sin\theta + \frac{1}{2}Cosa\theta \, Sin\theta + \frac{1}{2}\theta\right)\right]_0^{2\pi}$$

$$= a^2 \left[\left(\frac{3\theta}{2} - 2Sin\theta + \frac{1}{4}Sin2\theta\right)\right]_0^{2\pi}$$

$$= a^2 \left[(3\pi - 0 + 0) + (0 - 0 + 0)\right]$$

$$= 3a^2\pi$$

$$\therefore \quad \bar{x} = \frac{3a^3\pi^2}{3a^2\pi}$$

i.e. $\quad \bar{x} = \pi \, a$

Without further ado we can get right into this, the concluding part of the assignment. Noting that the centroid of the strip is $y/2$ from the X-axis, the moment about the X-axis is : $\frac{y}{2}A(\theta)F'(\theta)d\theta$

$$\bar{y} = \frac{\int \frac{y}{2} Q(\theta)F'(\theta)d\theta}{\int Q(\theta)F'(\theta)d\theta}$$

$$\bar{y} = \frac{\frac{1}{2}\int a(1-Cos\,\theta)a(1-Cos\,\theta)a(1-Cos\,\theta)d\theta}{3a^2\pi}$$

Inserting our limits

i.e.
$$\bar{y} = \frac{\frac{a^2}{2}\int_0^{2\pi}(1-Cos\,\theta)^3\,d\theta}{3a^2\pi}$$

$$\frac{\frac{a^3}{2}\int_0^{2\pi}(1-3Cos\,\theta+3Cos^2\theta-Cos^3\theta)d\theta}{3a^2\pi}$$

Working on the numerator integral

$$3\int Cos^2\theta\,d\theta = 3\left(\frac{1}{2}Cos\,\theta\,Sin\,\theta+\frac{1}{2}\theta\right),$$

$$\int Cos^3\theta\,d\theta = \frac{1}{3}Cos^2\theta\,Sin\,\theta+\frac{2}{3}\int Cos\,\theta\,d\theta$$

$$= \frac{1}{3}Cos^2\theta\,Sin\,\theta+\frac{2}{3}Sin\,\theta$$

$\therefore$
$$\bar{y} = \frac{a^3}{2}\left[\theta-3Sin\,\theta+\frac{3}{2}Cos\,\theta+\frac{3}{2}\theta-\frac{1}{3}Cos^2\theta\,Sin\,\theta\right.$$

$$\left.-\frac{2}{3}Sin\,\theta\right]_0^{2\pi}/3a^2\pi$$

~ 228 ~

i.e. $\quad \overline{y} = \dfrac{a^3}{2}\{(2\pi - 0 + 0 + 3\pi - 0 - 0) -$

$$(0 - 0 + 0 + 0 - 0 - 0)\}/3a^2\pi$$

or $\quad = \dfrac{5\pi a^3}{6a^2\pi}$

$$= \dfrac{5a}{6}$$

Therefore the coordinates of the centroid of one arch of the given cycloid are $\left(\pi a,\ \dfrac{5a}{6}\right)$.

Q6. Find the coordinates of the centroid of the area of the cardioid[2] $\rho = a(1 + Cos\theta)$, shown in Fig. 1.

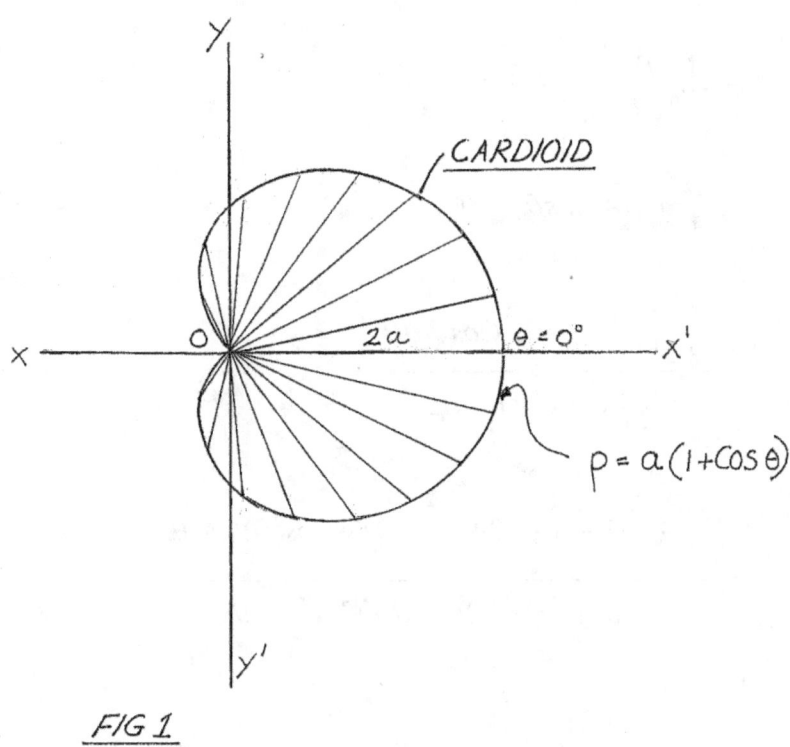

FIG 1

---

[2] Like the cycloid, the cardioid is a complex curve. It is the curve that resembles a heart, and was first studied by the Dutchman, Koersma in 1689. It is in fact the path traced by a point on the circumference of a circle rolling without slipping on another stationary circle of the same radius. By the time the rolling circle completes its traverse returning to its original starting point at the end of the trace it would have made two revolutions completely.

By virtue of symmetry about the X-axis $\bar{y} = 0$.

Now $M_Y = \iint \rho^2 d\rho \, Cos\theta \, d\theta$

$$= 2\int_0^\pi \int_0^{a(1+Cos\theta)} \rho^2 d\rho \, Cos\theta \, d\theta$$

$$= 2\int_0^\pi \left[\frac{\rho}{3}\right]_0^{a(1+Cos\theta)} Cos\theta \, d\theta$$

$$= \frac{2}{3}\int_0^\pi \left\{ a^3(1+Cos)^3 \right\} Cos\theta \, d\theta$$

$$= \frac{2a^3}{3}\int (1+Cos\theta)^3 Cos\theta \, d\theta$$

Also   Area $= 2\int_0^\pi \rho \, d\rho \, d\theta$

$$= 2\int_0^\pi \frac{\rho^2}{2}\Big|_0^{a(1+Cos\theta)} d\theta$$

$$= \int_0^\pi a^2(1+Cos\theta)^2 d\theta$$

$\therefore$ $$\bar{x} = \frac{\dfrac{2a^3}{3}\int(1+Cos\theta)^3 Cos\theta \, d\theta}{a^2\int_0^\pi (1+Cos\theta)^2 d\theta}$$

$$= \frac{\dfrac{2a^3}{3}\int_0^\pi (1+3Cos\theta+3Cos^2\theta+Cos^3\theta)Cos\theta d\theta}{a^2\int_0^\pi (1+2Cos\theta+Cos^2\theta)d\theta}$$

~ 230 ~

i.e. $\bar{x} = \dfrac{2a^3}{3}\left[\left\{Sin\,\theta + \dfrac{3}{2}Cos\,\theta\,Sin\,\theta + \dfrac{3}{2}\theta + (Cos^2\theta\,Sin\,\theta\,Sin\,\theta)\right.\right.$

$$\left.\left. + \dfrac{1}{4}Cos^3\theta\,Sin\,\theta + \dfrac{3}{8}Cos\,\theta\,Sin\,\theta + \dfrac{3}{8}\theta\right\}\right]_0^\pi$$

$$\overline{\hspace{13cm}}$$

$$a^2\left[\theta + 2Sin\,\theta + \dfrac{1}{2}Cos\,\theta\,Sin\,\theta + \dfrac{1}{2}\theta\right]_0^\pi$$

$\therefore$ $\bar{x} = \dfrac{\dfrac{2a^3}{3}\left(\dfrac{3\pi}{2} + \dfrac{3\pi}{8}\right)}{a^2\left(\pi + \dfrac{\pi}{2}\right)}$

$$= \dfrac{2a^3}{3} \cdot \dfrac{15\pi}{8} \cdot \dfrac{2}{3a^2\pi}$$

$$\bar{x} = \dfrac{5a}{6}$$

Therefore the position of the coordinates of the centroid of the cardioid are $\left(\dfrac{5a}{6},\ 0\right)$

I did not show the details of the integration of the numerator in the expression for $\bar{x}$. The reader might wish to test his or her skills by checking the results.

Q7. What is the length of one arch of the cycloid $x = a(\theta - Sin\,\theta)$, $y = a(1 - Cos\,\theta)$?

Determine the coordinates of the centroid of a uniform piece of wire with the shape of this curve.

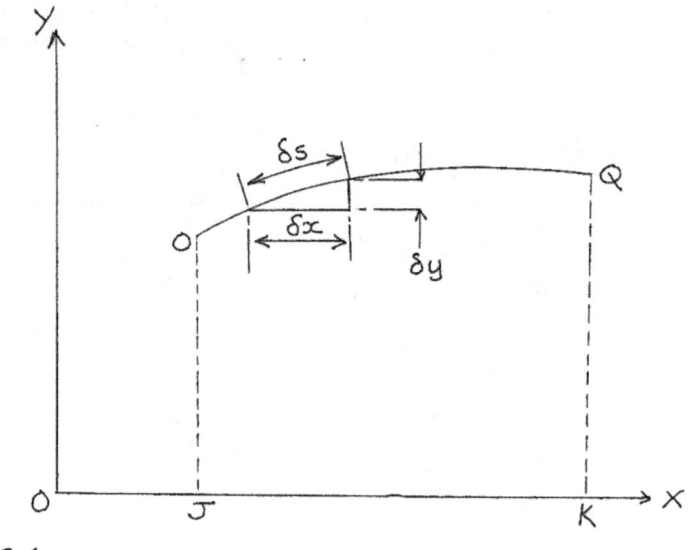

FIG 1

With reference to the curve OQ shown in Fig. 1, the length ds may be expressed as

$$(\delta s)^2 = (\delta x)^2 + (\delta y)^2$$

$\therefore$ Length of $OQ = \int_{j}^{K} \sqrt{1 + \left(\dfrac{dy}{dx}\right)^2} \, dx$

Here 'x' and 'y' are expressed in terms of the parameter 'θ',

i.e. $x = a(\theta - Sin\theta)$, $y = a(1 - Cos\theta)$.

$\therefore$ $\left(\dfrac{ds}{d\theta}\right)^2 = \left(\dfrac{dx}{d\theta}\right)^2 + \left(\dfrac{dy}{d\theta}\right)^2$

$\therefore$ $s = \int \sqrt{\left(\dfrac{dx}{d\theta}\right)^2 + \left(\dfrac{dy}{d\theta}\right)^2} \cdot d\theta$

For the cycloid with 'a' being the radius of the generating circle,

$$x = a(\theta - Sin\theta)$$

$$y = a(1 - Cos\theta)$$

$$\frac{dx}{d\theta} = a(1 - Cos\theta)$$

~ 232 ~

$$\frac{dy}{d\theta} = aSin\,\theta$$

The limits of integration for one arch of the cycloid are $\theta = 0$; $\theta = 2\pi$.

Accordingly

$$s = \int_0^{2\pi} \left\{ \sqrt{a(1-Cos\,\theta)^2 + (aSin\,\theta)^2} \right\} d\theta$$

$$= a\int_0^{2\pi} \sqrt{(1-2Cos\,\theta + Cos^2\theta + Sin^2\theta)} \cdot d\theta$$

$$= a\int_0^{2\pi} \sqrt{2(1-Cos\,\theta)} d\theta$$

But $Cos\,\theta$ may be expressed in the form :

$$Cos\,\theta = Cos^2\frac{\theta}{2} - Sin^2\frac{\theta}{2} = 1 - 2Sin^2\frac{\theta}{2}$$

$$\therefore \qquad s = a\int_0^{2\pi} \sqrt{2\left(1-1+2Sin^2\frac{\theta}{2}\right)} \cdot d\theta$$

$$\text{or} \qquad s = a\int_0^{2\pi} \sqrt{4Sin^2\frac{\theta}{2}} \cdot d\theta$$

$$= 2a\int_0^{2\pi} Sin\frac{\theta}{2} d\theta$$

$$= 2a\left[-2Cos\frac{\theta}{2} \right]_0^{2\pi}$$

$$= -4a\left[Cos\,\pi - Cos\,0 \right]$$

$$= -4a(-1-1) = -4a(-2)$$

$$\therefore \qquad s = 8a$$

i.e. the length of one arch of the cycloid = 8a. Continuing, let $\bar{x}$ and $\bar{y}$ be the coordinates of the centroid of the uniform wire; also let 'λ', be the density of the material of the wire, assumed constant throughout its length and 'a' its cross sectional area, also assumed constant.

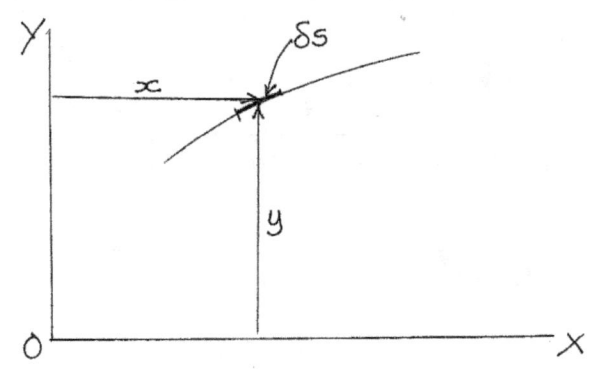

With reference to Fig. 2

$$\overline{x} = \frac{\lambda \int_{0}^{2\pi} x \, ds}{\lambda \int_{0}^{2\pi} ds}$$

and

$$\overline{y} = \frac{\int_{0}^{2\pi} y \, ds}{\int_{0}^{2\pi} ds}$$

Let us evaluate $\int_{0}^{2\pi} x \, ds$

From previous working it was found that

$$\frac{ds}{d\theta} = \sqrt{\left(\frac{dx}{d\theta}\right)^2 + \left(\frac{dy}{d\theta}\right)^2} = \sqrt{\{a(1 - Cos\,\theta)\}^2 + (aSin\,\theta)^2}$$

i.e.

~ 234 ~

$$ds = 2a \, Sin\frac{\theta}{2}\,d\theta$$

$$\therefore \quad \int_0^{2\pi} x \, ds = \int_0^{2\pi} a(\theta - Sin\theta)2a \, Sin\frac{\theta}{2}\,d\theta$$

$$= 2a^2 \int_0^{2\pi}(\theta - Sin\theta)Sin\frac{\theta}{2}\,d\theta$$

$$= 2a^2\left[\int \theta \, Sin\frac{\theta}{2} - \int Sin\theta \, Sin\frac{\theta}{2}\,d\theta\right]$$

Now
$$\int \theta \, aSin\frac{\theta}{2} = -2\theta \, Cos\frac{\theta}{2} + 4Sin\frac{\theta}{2}$$

$$\therefore \quad 2a^2\int_0^{2\pi} \theta \, Sin\frac{\theta}{2}\,d\theta = 2a^2\left[-2\theta \, Cos\frac{\theta}{2} + 4Sin\frac{\theta}{2}\right]_0^{2\pi}$$

$$= 2a^2\left[-2.2\pi(Cos\pi) + 4(0) - (-0 + 0)\right]$$

$$= 2a^2\left\{(-2)(2\pi)(-1)\right\} = 8\pi a^2$$

And
$$2a^2\int -Sin\theta \, Sin\frac{\theta}{2}\,d\theta = -2a^2\int 2Sin\frac{\theta}{2}Cos\frac{\theta}{2}Sin\frac{\theta}{2}\,d\theta$$

$$= -4a^2\int Sin^2\frac{\theta}{2}Cos\frac{\theta}{2}\,d\theta$$

$$= -4a^2\int Sin^2\frac{\theta}{2}\,d\left(2Sin\frac{\theta}{2}\right)$$

$$= -8a^2\left[\frac{1}{3}Sin^3\frac{\theta}{2}\right]_0^{2\pi}$$

$$= 0$$

$$\therefore \quad \int_0^{2\pi} x \, ds = 8\pi a^2$$

From previous working it was determined that $\int_0^{2\pi} ds = 8a$

$$\therefore \quad \bar{x} = \frac{\int_0^{2\pi} x \, ds}{\int_0^{2\pi} ds}$$

$$= \frac{8\pi a^2}{8a}$$

$$\therefore \quad \bar{x} = \pi a$$

Now

$$\bar{y} = \frac{\int_0^{2\pi} y \, ds}{\int_0^{2\pi} ds}$$

Evaluating the numerator integral

$$\int_0^{2\pi} y \, ds = \int_0^{2\pi} a(1 - Cos\theta) \cdot 2a Sin \frac{\theta}{2} d\theta$$

$$= 2a^2 \int_0^{2\pi} (1 - Cos) Sin \frac{\theta}{2} d\theta$$

$$= 2a^2 \left[\int_0^{2\pi} Sin \frac{\theta}{2} \cdot d\theta - \int_0^{2\pi} Cos\theta \, Sin \frac{\theta}{2} d\theta \right]$$

$$\int_0^{2\pi} Cos\theta \, Sin \frac{\theta}{2} = \int_0^{2\pi} \left(Cos^2 \frac{\theta}{2} - Sin^2 \frac{\theta}{2} \right) Sin \frac{\theta}{2} d\theta$$

$$= \int_0^{2\pi} \left(1 - 2 Sin^2 \frac{\theta}{2} \right) Sin \frac{\theta}{2} d\theta$$

$$= \int_0^{2\pi} \left(Sin \frac{\theta}{2} d\theta - 2 Sin^3 \frac{\theta}{2} \right) d\theta$$

$$\therefore \quad \int_0^{2\pi} y \, ds = 2a^2 \left[\int_0^{2\pi} Sin \frac{\theta}{2} d\theta - \int_0^{2\pi} Sin \frac{\theta}{2} d\theta + 2 \int_0^{2\pi} Sin^3 \frac{\theta}{2} d\theta \right]$$

$$= 4a^2 \int_0^{2\pi} Sin^3 \frac{\theta}{2} d\theta$$

$$= 4a^2 \left[-\frac{1}{3} Sin^2 \frac{\theta}{2} Cos \frac{\theta}{2} - \frac{4}{3} Cos \frac{\theta}{2} \right]_0^{2\pi}$$

$$= 4a^2 \left[-0 - \frac{4}{3}(-1) - \left\{ -\frac{4}{3}(1) \right\} \right]$$

$$= 4a^2 \left(\frac{4}{3} + \frac{4}{3} \right)$$

$$= \frac{32a^2}{3}$$

and as $\qquad \int_0^{2\pi} ds = 8a$

$$\bar{y} = \frac{32a}{3(8a)}$$

$$\bar{y} = \frac{4a}{3}$$

Therefore the coordinates of the centroid of a uniform wire, in the form of one arch of the cycloid : $x = a(\theta - Sin\theta); \ y = a(1 - Cos\theta)$ are $\left(\pi a, \ \frac{4a}{3} \right)$.

Q8. Determine the coordinates of the centroid of the hatched area which is the region within the circle with equation $r = Cos\theta$ and outside the cardioid $r = 1 + Cos\theta$, as shown in Fig. 1 : "the horned moon" [i.e. the crescent moon] of Samuel Taylor Coleridge's "Rime of the Ancient Mariner".

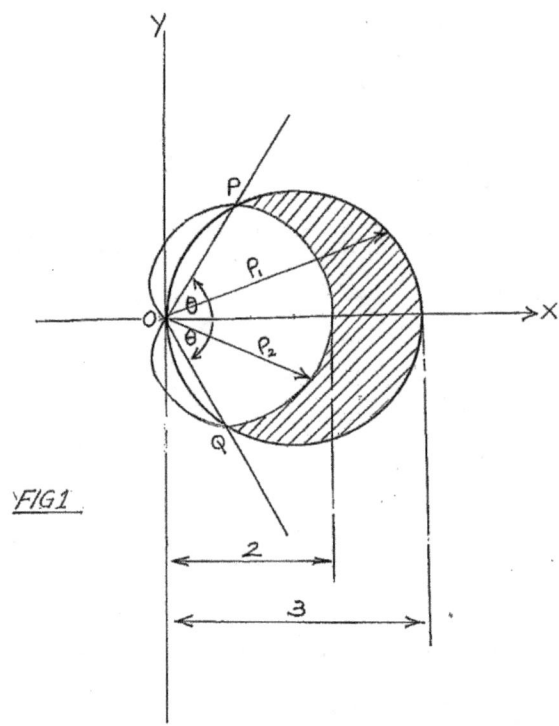

FIG1

Referring to Fig. 1, we can straightway write :

$$\bar{x} = \frac{\iint \rho_1^2 Cos\theta \, d\rho \, d\theta}{\iint (\rho_1 - \rho_2) d\rho \, d\theta} - \frac{\iint \rho_2^2 Cos\theta \, d\rho \, d\theta}{\iint (\rho_1 - \rho_2) d\rho \, d\theta}$$

The hatched area is symmetrical about the X-axis. Therefore $\bar{y} = 0$.

The cardioid and the circle intersect at P and Q. At these two points

$$3Cos\theta = 1 + Cos\theta$$

$\therefore$ $2Cos\theta = 1$ or $Cos\theta = \dfrac{1}{2}$

~ 238 ~

So that P is at 60° or $\dfrac{\pi}{3}$; and Q is at -60° or $-\dfrac{\pi}{3}$

Accordingly, we may now write in the limits of integration thus :

$$\bar{x} = \dfrac{\displaystyle\int_{-\pi/3}^{\pi/3}\int_{0}^{3Cos\theta}\rho_1^2 d\rho\, Cos\theta\, d\theta - \int_{-\pi/3}^{\pi/3}\int_{0}^{3Cos\theta}\rho_2^2 d\rho\, Cos\theta\, d\theta}{\displaystyle\int_{-\pi/3}^{\pi/3}\int_{0}^{3Cos\theta}\rho_1 d\rho\, d\theta - \int_{-\pi/3}^{\pi/3}\int_{0}^{Cos3\theta}\rho_2\, d\rho\, d\theta}$$

$$\bar{x} = \dfrac{\displaystyle\int_{-\pi/3}^{\pi/3}\left[\dfrac{\rho_1^3}{3}\right]_0^{3Cos\theta} Cos\theta\, d\theta - \int_{\pi/3}^{\pi/3}\left[\dfrac{\rho_2^3}{3}\right]_0^{1+Cos\theta} Cos\theta\, d\theta}{\displaystyle\int_{-\pi/3}^{\pi/3}\left[\dfrac{\rho_1^2}{2}\right]_0^{3Cos\theta} d\theta - \int_{-\pi/3}^{\pi/3}\left[\dfrac{\rho_2^2}{2}\right]_0^{1+Cos\theta} d\theta}$$

Let us evaluate the numerator integral separately, the limits being $-\dfrac{\pi}{3}\, to +\dfrac{\pi}{3}$.

We may simplify this integral by writing :

$$\bar{x} = \dfrac{1}{3}\int_{-\pi/3}^{\pi/3}(3Cos\theta)^3 Cos\theta\, d\theta - \dfrac{1}{3}\int_{-\pi/3}^{\pi/3}(1+Cos\theta)^3 Cos\theta\, d\theta$$

$$= \dfrac{1}{3}\int_{-\pi/3}^{\pi/3}\{(3Cos\theta)^3 Cos\,\theta - (1+Cos\theta)^3 Cos\theta\}d\theta$$

$$= \dfrac{1}{3}\int_{-\pi/3}^{\pi/3} 27Cos^4\theta - Cos\{1+Cos^3\theta + 3Cos\theta(1+Cos\theta)\}d\theta$$

$$= \dfrac{1}{3}\left[\int_{-\pi/3}^{\pi/3} 27Cos^4\theta - Cos\theta(1+Cos^3\theta + 3Cos\theta + 3Cos^2\theta)\right]d\theta$$

$$= \dfrac{1}{3}\left[\int_{-\pi/3}^{\pi/3} (27Cos^4\theta - Cos\theta - Cos^4\theta - 3Cos^2\theta - 3Cos^3\theta)d\theta\right]$$

$$= \dfrac{1}{3}\left[\int_{-\pi/3}^{\pi/3} (26Cos^4\theta - 3Cos^3\theta - 3Cos^2\theta - Cos\theta)d\theta\right]$$

I assume by now that you are quite familiar with the reduction formulae for $\int Sin^n\theta\, d\theta$ and $\int Cos^n\theta\, d\theta$. No? Then I shall state them again

$$\int Sin^n\theta\, d\theta = -\frac{1}{n}Sin^{n-1}\theta\, Cos\theta + \frac{n-1}{n}\, I_{n-2}, \text{ with } I_1 = \int Sin\theta\, d\theta \text{ and } I_0 = \int d\theta.$$

$$\int Cos^n\theta\, d\theta = +\frac{1}{n}Cos^{n-1}\theta\, Sin\theta + \frac{n-1}{n}\, I_{n-2}, \text{ with } I_1 = \int Cos\theta\, d\theta \text{ and } I_0 = \int d\theta$$

Use them often enough and they will stick in your head!

Accordingly. Let us evaluate integrals separately.

$$\int Cos^4\theta\, d\theta = \frac{1}{4}Cos^3\theta\, Sin\theta + \frac{3}{4}I_2$$

$$I_2 = \int Cos^2\theta\, d\theta = \frac{1}{2}Cos\theta\, Sin\theta + \frac{1}{2}I_0$$

$$= \frac{1}{2}Cos\theta\, Sin\theta + \frac{\theta}{2}$$

$$\therefore \quad \int Cos^4\theta\, d\theta = \frac{1}{4}Cos^3\theta\, Sin\theta + \frac{3}{4}\left(\frac{1}{2}Cos\theta\, Sin\theta + \frac{\theta}{2}\right)$$

$$= \frac{1}{4}Cos^3\theta\, Sin\theta + \frac{3}{8}Cos\theta\, Sin\theta + \frac{3}{8}\theta$$

Also

$$\int Cos^3\theta\, d\theta = \frac{1}{3}Cos^2\theta\, Sin\theta + \frac{2}{3}I_1$$

$$= \frac{1}{3}Cos^2\theta\, Sin\theta + \frac{2}{3}\int Cos\theta\, d\theta$$

$$= \frac{1}{3}Cos^2\theta\, Sin\theta + \frac{2}{3}Sin\theta$$

And

$$\int Cos^2\theta\, d\theta = \frac{1}{2}Cos\theta\, Sin\theta + \frac{1}{2}I_0 = \frac{1}{2}Cos\theta\, Sin\theta + \frac{1}{2}\int d\theta$$

$$= \frac{1}{2} Cos\theta\ Sin\theta + \frac{1}{2}\theta$$

Finally

$$\int Cos\theta\ d\theta = Sin\theta$$

You would have observed that constants of integration were omitted in the statement of each result. Strictly speaking a constant could have been added to each result but inasmuch as we knew that our limits are from $\theta = -\frac{\pi}{3}\ to +\frac{\pi}{3}$, it the constant that was left out. We shall evaluate each result separately in order to avoid confusion.

$$\int_{-\pi/3}^{\pi/3} Cos^4\theta = \frac{1}{4}\left(\frac{1}{2}\right)^3\left(\frac{\sqrt{3}}{2}\right) + \frac{3}{8}\left(\frac{1}{2}\right)\left(\frac{\sqrt{3}}{2}\right) + \frac{3}{8}\cdot\frac{\pi}{3}$$

$$-\left\{\frac{1}{4}\left(\frac{1}{2}\right)^3\left(-\frac{\sqrt{3}}{2}\right) + \frac{3}{8}\left(\frac{1}{2}\right)\left(-\frac{\sqrt{3}}{2}\right) + \frac{3}{8}\left(-\frac{\pi}{3}\right)\right\}$$

$$= \frac{1}{4}\cdot\frac{1}{8}\cdot\frac{\sqrt{3}}{3} + \frac{3\sqrt{3}}{32} + \frac{\pi}{8} - \left\{\frac{1}{4}\left(\frac{1}{8}\right)\left(-\frac{\sqrt{3}}{2}\right) - \frac{3\sqrt{3}}{32} - \frac{\pi}{8}\right\}$$

$$= \frac{\sqrt{3}}{64} + \frac{3\sqrt{3}}{32} + \frac{\pi}{8} + \frac{\sqrt{3}}{64} + \frac{3\sqrt{3}}{32} + \frac{\pi}{8}$$

$$= \frac{7\sqrt{3}}{64} + \frac{\pi}{8} + \frac{7\sqrt{3}}{64} + \frac{\pi}{8}$$

$$= \frac{14\sqrt{3}}{64} + \frac{\pi}{4} = \frac{7\sqrt{3}}{32} + \frac{\pi}{4}$$

$$\int Cos^3\theta\ d\theta = \frac{1}{3}\left(\frac{1}{2}\right)^2\left(\frac{\sqrt{3}}{2}\right) + \frac{2}{3}\ \frac{\sqrt{3}}{2} - \left\{\frac{1}{3}\left(\frac{1}{2}\right)^2\left(-\frac{\sqrt{3}}{2}\right) + \frac{2}{3}\left(-\frac{\sqrt{3}}{2}\right)\right\}$$

$$= \frac{1}{3}\left(\frac{1}{4}\right)\frac{\sqrt{3}}{2} + \frac{\sqrt{3}}{3} - \left\{\frac{1}{3}\left(\frac{1}{4}\right)\left(\frac{1}{4}\right)\left(\frac{-\sqrt{3}}{2}\right) - \frac{\sqrt{3}}{3}\right\}$$

$$= \frac{\sqrt{3}}{24} + \frac{\sqrt{3}}{3} + \frac{3}{24} + \frac{\sqrt{3}}{3}$$

$$= \frac{2\sqrt{3}}{24} + \frac{2\sqrt{3}}{3}$$

$$= \frac{\sqrt{3}}{12} + \frac{2\sqrt{3}}{3} = \frac{9\sqrt{3}}{12} = \frac{3\sqrt{3}}{4}$$

$$\int_{-\pi/3}^{\pi/3} Cos^2\theta \, d\theta = \frac{1}{2}\left(\frac{1}{2}\right)\left(\frac{\sqrt{3}}{2}\right) + \frac{1}{2}\left(\frac{\pi}{3} - \right) - \left\{\frac{1}{2}\left(\frac{1}{2}\right)\left(-\frac{\sqrt{3}}{2}\right) + \frac{1}{2}\left(-\frac{\pi}{3}\right)\right\}$$

$$= \frac{\sqrt{3}}{8} + \frac{\sqrt{\pi}}{6} + \frac{\sqrt{3}}{8} + \frac{\pi}{6} = \frac{2\sqrt{3}}{8} + 2\left(\frac{\pi}{6}\right)$$

$$= \frac{\sqrt{3}}{4} + \frac{\pi}{3}$$

$$\int_{-\pi/3}^{\pi/3} Cos\theta \, d\theta = \frac{\sqrt{3}}{2} - \left(-\frac{\sqrt{3}}{2}\right)$$

$$= \sqrt{3}$$

$$\therefore \quad \frac{1}{3} \int_{-\pi/3}^{\pi/3} (26Cos^4\theta - 3Cos^3\theta - 3Cos^2\theta - Cos\theta)d\theta$$

$$= \frac{1}{3}\left[26\left(\frac{7\sqrt{3}}{32} + \frac{\pi}{4}\right) - 3\left(\frac{3\sqrt{3}}{4}\right) - 3\left(\frac{\sqrt{3}}{4} + \frac{\pi}{3}\right) - \sqrt{3}\right]$$

$$= \frac{1}{3}\left[26(0.38 + 0.78) - 3.9 - 3(0.43 + 1.05) - 1.732\right]$$

$$= \frac{1}{3}\left[30.16 - 39 - 4.44 - 1.73\right] = \frac{1}{3}(30.16 - 10.07)$$

$$= \frac{20.09}{3} = 6.7$$

Continuing, the denominator which is the area of the hatched region

$$\frac{1}{2}\iint (\rho_1 - \rho_2)\, d\rho\, d\theta$$

$$= \frac{1}{2}\left[\int_{-\pi/3}^{\pi/3} \left[\rho_1^2\right]^{3Cos\theta}\, d\theta - \int_{-\pi/3}^{\pi/3} \left[\rho_2^2\right]_0^{1+Cos\theta}\, d\theta \right]$$

$$= \frac{1}{2}\int_{-\pi/3}^{\pi/3} \left\{9Cos^2\theta - (1 + 2Cos\theta + Cos^2\theta)\right\}\, d\theta - \frac{\pi}{3}$$

$$= \frac{1}{2}\int_{-\frac{\pi}{3}}^{\frac{\pi}{3}} \left(8Cos^2\theta - 2Cos - 1\right)\, d\theta$$

$$= \frac{1}{2}\left[8\left(\frac{1}{2}Cos\theta\ Sin\theta + \frac{1}{2}\theta\right) - 2Sin\theta - \theta \right]_{-\frac{\pi}{3}}^{\frac{3\pi}{}}$$

$$= \frac{1}{2}\left[\left\{8\left(\frac{\sqrt{3}}{8} + \frac{\pi}{6}\right) - 2\left(\frac{\sqrt{3}}{2}\right) - \frac{\pi}{3}\right\} \right.$$

$$\left. -8\left(-\frac{\sqrt{3}}{8} - \frac{\pi}{6}\right) - 2\left(-\frac{\sqrt{3}}{2}\right) - \left(-\frac{\pi}{3}\right) \right]$$

$$= \frac{1}{2}\left(\sqrt{3} + \frac{\pi}{3} - \sqrt{3} - \frac{\pi}{3} + \sqrt{3} + \frac{4\pi}{3} - \sqrt{3} - \frac{\pi}{3} \right)$$

$$= \frac{1}{2}\left(\frac{8\pi}{3} - \frac{2\pi}{3} \right) = \frac{1}{2}\left(\frac{6\pi}{3} \right)$$

$$= \pi$$

So that $\overline{x} = 6.7/\pi$

$$= 2.1$$

∴   Co-ordinates of the centroid are (2.1, 0)

Q9. What are the coordinates of the centroid of the area of one loop of the curvilinear form whose polar equation is given by $\rho = aSin2\theta$?

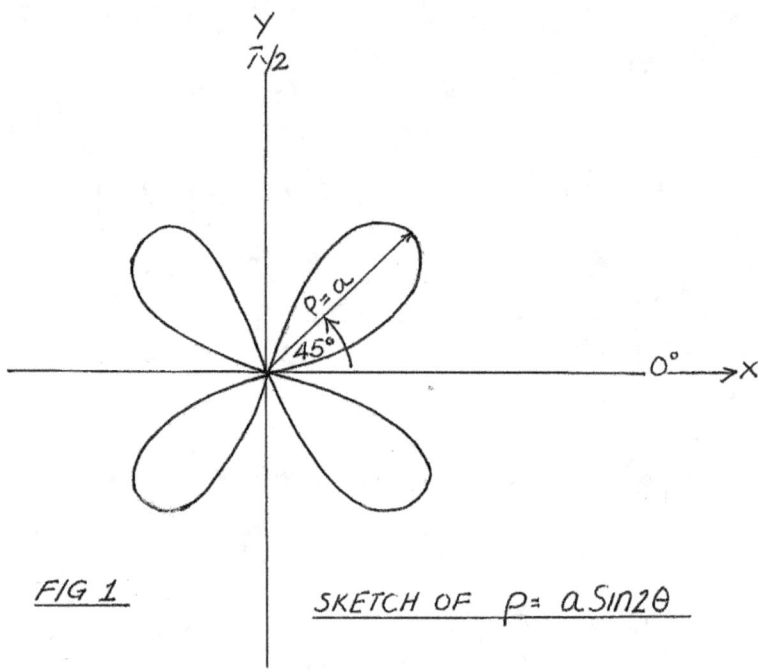

FIG 1.   SKETCH OF $\rho = aSin2\theta$

The 4 loops of the curve $\rho = aSin2\theta$ are shown in Fig. 1.

$$\bar{x} = \frac{\iint \rho^2 d\rho \; Cos\theta \; d\theta}{\iint \rho \; d\rho \; d\theta}$$

Evaluating the numerator integral first,

$$\int_0^{\pi/2} \int_0^{aSin2\theta} \rho^2 d\rho \; Cos\theta \; d\theta = \int_0^{\pi/2} \left[\frac{\rho^3}{3}\right]_0^{aSin2\theta} Cos\theta \; d\theta$$

$$= \frac{1}{3}\int_0^{\pi/2} a^3 Sin^3 2\theta \; Cos\theta \; d\theta \qquad \cdots\cdots\cdots\cdots \text{(i)}$$

Now, $Sin2\theta = 2Sin\theta \; Cos\theta$

$\therefore \qquad Sin^3 2\theta = 8Sin^3\theta \; Cos^3\theta$

$\therefore \qquad$ Expression (i) now becomes $\dfrac{8a^3}{3}\displaystyle\int_0^{\pi/2} Sin^3\theta \; Cos^4\theta \; d\theta$

~ 244 ~

But $Cos^2\theta = (1 - Sin^2\theta)$

$\therefore \qquad Cos^4\theta = (1 - Sin^2\theta)^2 = (1 - 2Sin^2\theta + Sin^4\theta)$

$\therefore \qquad \dfrac{8a^3}{3}\displaystyle\int_0^{\pi/2} Sin^3\theta\, Cos^4\theta\, d\theta = \dfrac{8a^3}{3}\int_0^{\pi/2} Sin^3\theta(1 - 2Sin^2\theta + Sin^4\theta)d\theta$

$\dfrac{8a^3}{3}\displaystyle\int_0^{\pi/2}(Sin^3\theta - 2Sin^5\theta + Sin^7\theta)\, d\theta$

Let us use a reduction formula for $\int Sin^n\theta\, d\theta$ viz.

$I_n = -\dfrac{1}{n}Sin^{n-1}\theta\, Cos\theta + \dfrac{n-1}{n}I_{n-2},$ noting that $I_1 = \int Sin\theta\, d\theta,$ and $I_0 = \int d\theta$

Hence,

$\displaystyle\int Sin^3\theta\, d\theta = -\dfrac{1}{3}Sin^2\theta\, Cos\theta + \dfrac{2}{3}I_1$

$= -\dfrac{1}{3}Sin^2\theta\, Cos\theta + \dfrac{2}{3}(-Cos\theta) = -\dfrac{1}{3}Sin^2\theta\, Cos\theta - \dfrac{2}{3}Cos\theta$

$\displaystyle\int Sin^5\theta\, d\theta = -\dfrac{1}{5}Sin^4\theta\, Cos\theta + \dfrac{4}{5}I_3$

$= -\dfrac{1}{5}Sin^4\theta\, Cos\theta + \dfrac{4}{5}\left\{-\dfrac{1}{3}Sin^2\theta\, Cos\theta - \dfrac{2}{3}Cos\right\}$

$= -\dfrac{1}{5}Sin^4\theta\, Cos\theta - \dfrac{4}{15}Sin^2Cos\theta - \dfrac{8}{15}Cos\theta$

$\displaystyle\int Sin^7\theta\, d\theta = -\dfrac{1}{7}Sin^6\theta\, Cos\theta + \dfrac{6}{7}I_5$

$= -\dfrac{1}{7}Sin^6\theta\, Cos\theta + \dfrac{6}{7}\left(-\dfrac{1}{3}Sin^4\theta\, Cos\theta\right.$

$\left. -\dfrac{4}{15}Sin^2Cos - \dfrac{8}{15}Cos\theta\right)$

$$\therefore \quad \int Sin^7\theta \, d\theta = -\frac{1}{7}Sin^6\theta \, Cos\theta - \frac{2}{7}Sin^4\theta \, Cos\theta$$

$$= \frac{8}{35}Sin^2\theta \, Cos\theta - \frac{16}{35}Cos\theta$$

Applying our limits $\frac{\pi}{2}$ and 0

$$\int_0^{\pi/2} Sin^3\theta \, d\theta = \left[\frac{1}{3}Sin^3\theta \, Cos\theta - Cos\theta\right]_0^{\frac{\pi}{2}} = (0-0)\left(0-\frac{2}{3}\right) = \frac{2}{3}$$

$$-\int_0^{\pi/2} Sin^5\theta \, d\theta = \left[+\frac{1}{5}Sin^4Cos\theta + \frac{4}{15}Sin^2Cos\theta + \frac{8}{15}Cos\theta\right]_0^{\frac{\pi}{2}}$$

$$= (0+0+0)-\left(\frac{8}{15}\right) = -\frac{8}{15}$$

$$\int_0^{\pi/2} Sin^7\theta \, d\theta = \left[-\frac{1}{7}Sin^6\theta \, Cos\theta - \frac{2}{7}Sin^4\theta \, Cos - \frac{8}{35}Sin\theta^2Cos\theta\right.$$

$$\left.-\frac{16}{35}Cos\right]_0^{\frac{\pi}{2}}$$

$$(0-0-+0)-\left(0-0-0-\frac{16}{35}\right) = \frac{16}{35}$$

$$\therefore \quad \frac{8a^3}{3}\int_0^{\pi/2}\left(Sin^3\theta - 2Sin^5\theta + Sin^6\theta\right) d\theta = \frac{8a^2}{3}\left\{\frac{2}{3} - \frac{2(8)}{15} + \frac{16}{35}\right\}$$

$$= \frac{8a^3}{3}\left(\frac{2}{35}\right)$$

$$\therefore \quad \int_0^{\pi/2}\int_0^{aSin\rho} \rho^2 \, d\rho Cos\theta \, d\theta = \frac{16a^3}{105}$$

In order to obtain the area of one loop we must evaluate

$\int_0^{\pi/2} \int_0^{aSin2\theta} \rho \, d\rho \, d\theta$, the denominator integral.

But, $\int_0^{\pi/2} \int_0^{aSin2\theta} \rho \, d\rho \, d\theta = \int_0^{\pi/2} \frac{1}{2} \left[\rho^2 \right]_0^{aSin2\theta} d\theta$

$$= \frac{1}{2} \int_0^{\pi/2} a^2 Sin2\theta \, d\theta$$

$$= \frac{a^2}{2} \int_0^{\pi/2} \frac{1}{2} d\theta - \frac{Cos4\theta}{2} d\theta$$

$$= \frac{a^2}{2} \left[\frac{\theta}{2} - \frac{1}{2} Sin4\theta \right]_0^{\frac{\pi}{2}}$$

$$= \frac{a^2}{2} \left\{ \left(\frac{\pi}{3} - 0 \right) - (0 - 0) \right\}$$

i.e.　Area $= \dfrac{\pi a^2}{8}$

so that　$\bar{x} = \dfrac{\int_0^{\pi/2} \int_0^{aSin2\theta} \rho^2 d\rho \, Cos\theta \, d\theta}{\int_0^{\pi/2} \int_0^{aSin2\theta} \rho \, d\rho \, d\theta}$

reduces　$\bar{x} = \dfrac{16a^2}{105} \cdot \dfrac{8}{\pi a^2}$

i.e.　$\bar{x} = \dfrac{128a}{105\pi}$

Continuing　$\bar{y} = \dfrac{\iint \rho^2 d\rho \, Sin\theta \, d\theta}{\iint \rho \, d\rho \, d\theta}$

The limits of integration are : for 'ρ'= 0 to $a\sin2\theta$ and for 'θ': from 0 to $\dfrac{\pi}{2}$.

Accordingly we write :

$$\overline{y} = \frac{\int_0^{\pi/2} \int_0^{aSin2\theta} \rho^2 d\rho \; Sin\theta \; d\theta}{\pi \; a^2 / 8}$$

Let us evaluate the numerator integral of the latter expression.

$$\int_0^{\pi/2} \int_0^{aSin2\theta} \rho^2 d\rho \; Sin\theta \; d\theta$$

$$= \int_0^{\pi/2} \left[\frac{\rho^3}{3} \right]_0^{aSin\theta} Sin\theta \; d\theta$$

$$= \int_0^{\pi/2} \frac{a^3}{3} Sin^3 2\theta \cdot Sin\theta \; d\theta$$

$$= \frac{a^3}{a} \int Sin^3 2\theta \cdot Sin\theta \; d\theta$$

$$= \frac{a^3}{3} \int (2Sin\theta \; Cos\theta)^3 \cdot Sin\theta \; d\theta$$

$$= \frac{a^3}{3} 8 \int Sin^3\theta \; Cos^3\theta \; Sin\theta \; d\theta$$

$$I' = \frac{8a^3}{3} \int_0^{\pi/2} Sin^4\theta \; Cos^3\theta \; d\theta$$

$$Sin^2\theta = 1 - Cos^2\theta$$

$$\therefore \quad Sin^4\theta = \left(1 - Cos^2\theta^2\right)^2 = 1 - 2Cos^2\theta + Cos^4\theta$$

$$I' = \frac{8a^3}{3} \int \left(Cos^3\theta - 2Cos^5\theta + Cos^7\theta\right) d\theta$$

$$\int Cos^3 = \frac{1}{3} Cos^2\theta \; Sin\theta + \frac{2}{3} I_1 = \frac{1}{3} Cos^2\theta \; Sin\theta + \frac{2}{3} Sin\theta$$

With limits θ and $\frac{\pi}{2}$: $\frac{1}{2} Cos^2\theta \; Sin\theta + \frac{2}{3} Sin\theta = \frac{3}{3} - (0-0) = \frac{2}{3}$

$$\int Cos^5\theta = \frac{1}{5}Cos^4\theta\ Sin\theta + \frac{4}{5}I_3$$

$$= \frac{1}{5}Cos^4\theta\ Sin\theta + \frac{4}{5}\left(\frac{1}{3}Cos^2\theta\ Sin\theta + \frac{2}{3}Sin\theta\right)$$

$$= \frac{1}{5}Cos^5\theta\ Sin\theta + \frac{4}{15}Cos^2\theta\ Sin\theta + \frac{8}{15}Sin\theta$$

With the same limits

$$\int Cos^5 = 0 + 0 + \frac{8}{15} - (0+0+0\_ = \frac{8}{15}$$

$$\therefore \qquad -2Cos^5\theta = -\frac{16}{15}$$

$$\int Cos^7\theta = \frac{1}{7}Cos^6\theta\ Sin\theta + \frac{6}{7}I_5$$

$$= \frac{1}{7}Cos^6\theta\ Sin\theta + \frac{1}{7}\left(\frac{1}{5}Cos^4\theta\ Sin\theta + \frac{4}{15}Cos^2\theta\ Sin\theta + \frac{8}{15}Sin\theta\right)$$

$$= \frac{1}{7}Cos^6\theta\ Sin\theta + \frac{6}{35}Cos^4\theta\ Sin\theta + \frac{8}{35}Cos^2\theta\ Sin\theta$$

$$= +\frac{16}{35}$$

$$\therefore \qquad I' = \frac{8a^3}{3}\left(\frac{2}{3} - \frac{16}{15} + \frac{16}{35}\right)$$

$$I' = \frac{8a^2}{3} \times \frac{2}{35} = \frac{16a^3}{105}$$

$$\therefore \qquad \overline{y} = \frac{16a^3 \cdot 8}{105 \cdot \pi\ a^2}$$

$$\overline{y} = \frac{128a}{105\pi}$$

Collecting our results : $\bar{x} = \bar{y} = \dfrac{128a}{105\pi}$

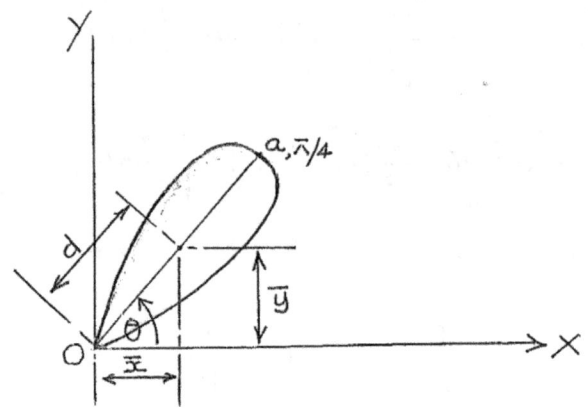

FIG 2

$Tan\theta = \dfrac{128a}{105\pi} / \dfrac{128a}{105\pi} = 1$ i.e. $\theta = 45°$

Therefore the centroid lies on the axis of the loop which is what we should have expected based on considerations of symmetry. Distance 'd' in Fig. 2 given by:

$$d = \sqrt{\left(\dfrac{128a}{105\pi}\right)^2 + \left(\dfrac{128a}{105\pi}\right)^2} = 0.55a.$$

Q.10 Revise what you know of Pappus' Theorem, then determine the centroid of the right-angled triangular area shown in Fig. 1 by calculating the volumes of the cones obtained by rotating the triangle about the X- and Y-axes.

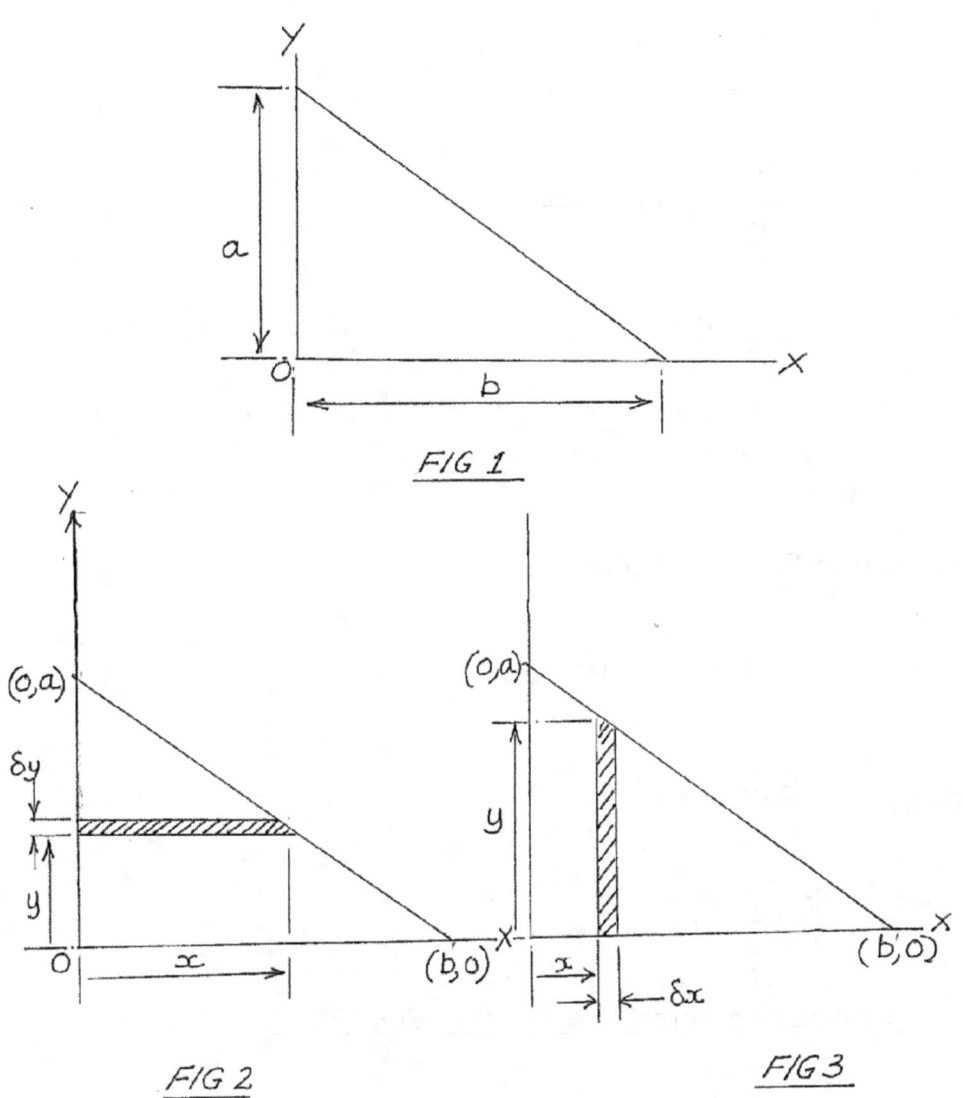

FIG 1

FIG 2

FIG 3

(a) <u>For rotation about the Y-axis. See Fig. 2</u>

$$dV = \int_0^a \pi\, x^2\, dy$$

But $\dfrac{x}{b} = \dfrac{a-y}{a}$

i.e. $x = \dfrac{b}{a}(a-y)$

$\therefore$ $V = \pi \int_0^a \left\{ \dfrac{b}{a}(a-y) \right\}^2 dy$

$$= \frac{\pi b^2}{a^2} \int_0^a \left(a^2 - 2ay + y^2\right) dy$$

$$= \frac{\pi b^2}{a^2} \left| a^2 y - ay^2 + \frac{y^3}{3} \right|_0^a$$

$$= \frac{\pi b^2}{a^2} \cdot \frac{a^3}{3}$$

i.e. $\quad V = \dfrac{\pi b^2 a}{3}$

According to Pappus' Theorem

(Area of the triangle) $(2\pi \bar{x}) = \dfrac{\pi b^2 a}{3}$

so that, $\quad \dfrac{ab}{2} \cdot 2\pi \bar{x} = \dfrac{\pi \, b^2}{3}$

$\therefore \quad \bar{x} = \dfrac{b}{3}$

(b) For rotation about the X-axis. See Fig. 3

$$V = \int_0^b \pi \, y^2 \, dx$$

But $\quad \dfrac{y}{a} = \dfrac{b-x}{b}$

or $\quad y = \dfrac{a}{b}(b-x)$

$\therefore \quad V = \dfrac{\pi \, a^2}{b^2} \int_0^b (b-x)\, dx$

i.e. $\quad V = \dfrac{\pi a^2}{b^2} \left| b^2 x - bx^2 + \dfrac{x^3}{3} \right|_0^b$

$$\therefore \qquad V = \frac{\pi\, a^2 b}{3}$$

Again by Pappus' Theorem

$$(\text{Area of the triangle})\ (2\pi\bar{y}) = \frac{\pi\, a^2 b}{3}$$

so that $\qquad \dfrac{ab}{2}\cdot 2\pi\bar{y} \qquad = \dfrac{\pi\, a^2 b}{3}$

i.e. $\quad \bar{y} = \dfrac{a}{3}$

Q11. The form of a washer is obtained by revolving a trapezium with parallel sides of length 'a' and '$2a$' at a distance 'a' apart about an axis which is parallel to these sides and at a distance $\dfrac{3}{2}a$ from the shorter side and a distance $\dfrac{5}{2}a$ from the longer side. Find the volume and total surface area of the washer. (Problem set by Geary, Lowry and Hayden: the solution is mine).

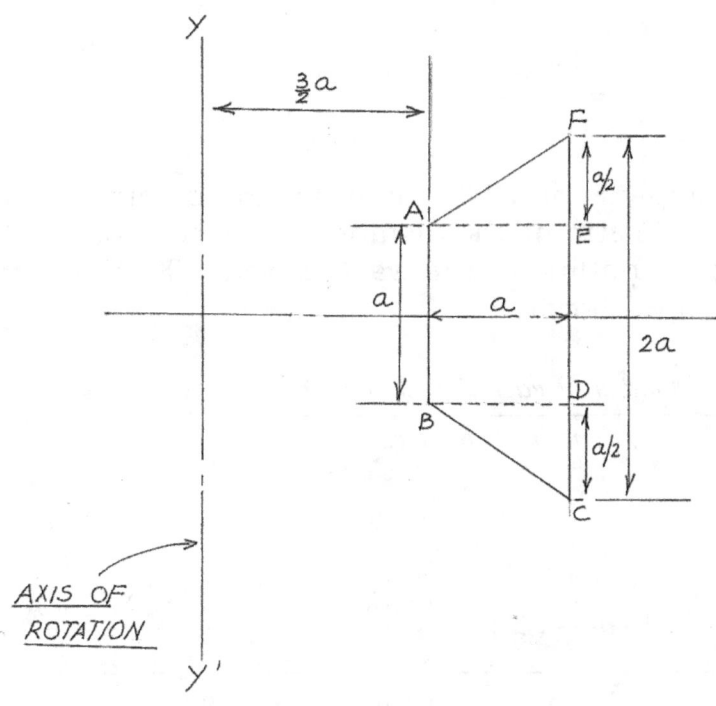

FIG 1

~ 253 ~

Diagram labelled Fig. 1 reflects the description and data provided. The form is divided into three parts : square '*ABDE*'; and two triangles '*BCD*' and *AEF* of identical configuration.

Accordingly, the centroid of *ABDE* is '2a' from the axis of rotation *YY'*; so that of triangle *AEF* , $\left(\dfrac{3}{2}a+\dfrac{2a}{3}\right)$ i.e. $\dfrac{13a}{6}$ from *YY'*; and that of triangle also $\dfrac{13a}{6}$ from *YY'*. The area of each triangle is $\dfrac{a^2}{4}$.

Applying Pappus' Theorem for volumes of revolution :

$$V_{YY'} = \text{(Area of ABDE) } 2\pi\,(2a) +$$

$$2\text{(Area of triangle AEF)} \cdot 2\pi \cdot \dfrac{13a}{6}$$

$$= a^2 \cdot 4\pi a + 2\left(\dfrac{a^2}{4}\right)\cdot \dfrac{13\pi a}{3}$$

$$= 4\pi a^3 + 13\dfrac{\pi a^3}{6}$$

i.e. $\qquad V_{YY'} = \dfrac{37\pi a^3}{6}$

In order to apply Pappus' Theorem for the determination of surface area we must first determine the distance of the centroid of the system of lines comprising the outline of the washer from *YY'*. We obtain this distance by applying the following:

$$\bar{x} = \dfrac{\sum Moment\ of\ each\ line\ about\ YY'}{\sum length\ of\ all\ lines}$$

Therefore

$$\bar{x} = \dfrac{a\left(\dfrac{3}{2}a\right) + a\dfrac{\sqrt{5}}{2}(2a) + \dfrac{a}{2}\left(\dfrac{5a}{2}\right) + a\left(\dfrac{5a}{2}\right) + \dfrac{a}{2}\left(\dfrac{5a}{2}\right) + \dfrac{a\sqrt{5}}{2}(2a)}{a + \dfrac{a\sqrt{5}}{2} + \dfrac{a}{2} + a + \dfrac{a}{2} + \dfrac{a\sqrt{5}}{2}}$$

Note that $AF = BC = \sqrt{\left(\dfrac{a}{2}\right)^2 + a^2} = \dfrac{a\sqrt{5}}{2}$

$\therefore$

$$\bar{x} = \dfrac{\dfrac{3a^2}{2} + a^2\sqrt{5} + \dfrac{5a^2}{4} + \dfrac{5a^2}{2} + \dfrac{5a^2}{4} + a^2\sqrt{5}}{3a + a\sqrt{5}}$$

$$= \dfrac{\dfrac{3a^2}{2} + 5a^2 + 2a^2\sqrt{5}}{a\left(3 + \sqrt{5}\right)}$$

$$= \dfrac{\dfrac{13a^2}{2} + 2a^2\sqrt{5}}{a\left(3 + \sqrt{5}\right)}$$

$$\bar{x} = \dfrac{13a^2 + a^2 4\sqrt{5}}{2a\left(3 + \sqrt{5}\right)}$$

As the surface area swept = Length of centroidal path $\times$ length of lines,

we have Swept surface area $= 2\pi\left\{\dfrac{a^2(13 + 4\sqrt{5})}{2a\left(3 + \sqrt{5}\right)}\right\} \cdot a\left(3 + \sqrt{5}\right)$

$\therefore$ Swept surface area $= \pi a^2\left(13 + 4\sqrt{5}\right)$

Q.12 The tetrahedron shown in Fig. 1 is bounded by the coordinates axes and the plane 'ABC' passing through the points A, B and C. Find the volume of the tetrahedron and determine the coordinates of its centroid assuming constant density.

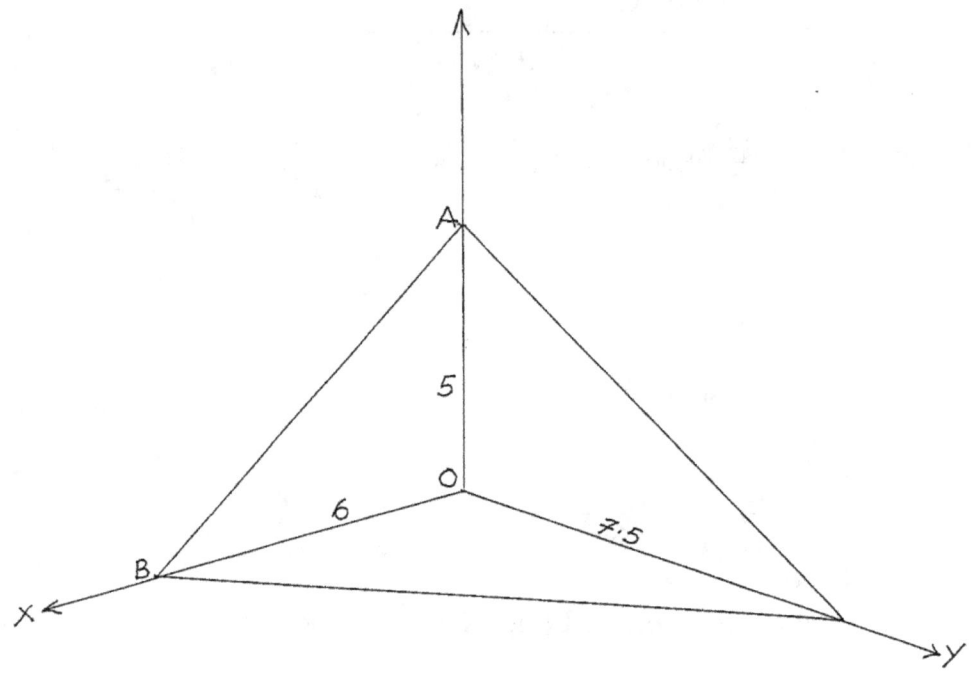

Fig. 1

The equation of the plane ABC is obtained from:

$$\frac{x}{6} + \frac{y}{7.5} + \frac{z}{5} = 1.$$

i.e. $5x + 4y + 6z = 30$

or $z = 5 - \frac{5x}{6} - \frac{2y}{3}$

Also, the equation of the line BC is, by inspection

$$y = -\frac{7.5x}{6} + 7.5$$

or $y = 7.5 - 1.25x$

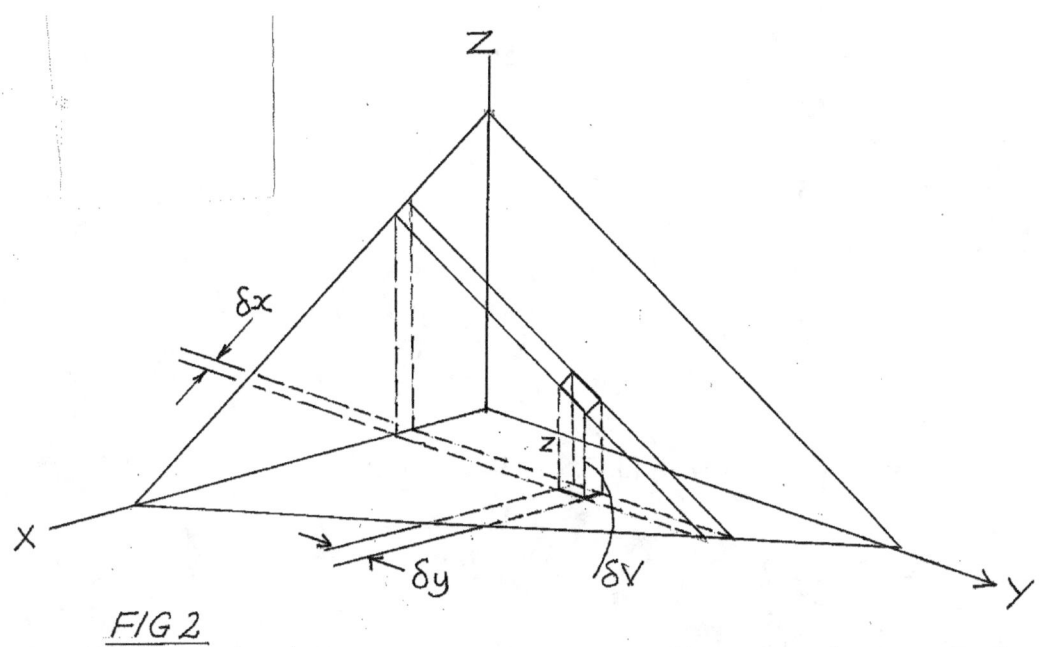

FIG 2

Referring to Fig. 2

$$\delta V = z \ \delta y \ \delta x$$

$$= \left(5 - \frac{5x}{6} - \frac{2y}{3}\right) dy \ dx$$

from which

$$V = \int_{x=0}^{x=6} \int_{y=0}^{y=7.5-1.25x} \left(5 - \frac{5x}{6} - \frac{2y}{3}\right) dy \ dx$$

$$= \int_{0}^{6} \int_{0}^{7.5-1.25x} \left\{5\left(\frac{1-x}{6}\right) dy - \frac{2y}{3}\right\} dy \ dx$$

$$= \int_{0}^{6} \left| 5\left(\frac{1-x}{6}\right) y - \frac{y^2}{3} \right|_{0}^{(7.5-1.25x)} dx$$

$$= \int_{0}^{6} \left\{5\left(1 - \frac{x}{6}\right)(7.5 - 1.25) - \frac{1}{3}(7.5 - 1.25x)^2\right\} dx$$

$$= \int_{0}^{6} 5\left(7.5 - 1.25x - 1.25x + 1.25\frac{x^2}{6}\right) dx$$

~ 257 ~

$$-\frac{1}{3}\int_{0}^{6} 56.25 - 18.75x + 1.5625x^2\,)dx$$

$$=\left[5\left(7.5x - 2.50\frac{x^2}{2} + \frac{1.25x^3}{18}\right)\right]_{0}^{6}$$

$$-\frac{1}{3}\left[56.25x - 18.75\frac{x^2}{2} + 1.5625\frac{x^3}{3}\right]_{0}^{6}$$

$$= 5(45 - 45 + 15) - \frac{1}{3}(337.5 - 337.5 + 112.5)$$

$$V = 37.5(cm)^3$$

This result can be checked, based on the knowledge that for intercepts *a,b* and *c* on the x-, y- and z- coordinate axes, the volume, V, of such a tetrahedron $= \frac{1}{6}abc$. Accordingly $V = \frac{1}{6}(6)(7.5)(15) = 37.5(cm)^3$.

Now, to a determination of the centroids.

For this purpose, refer to Fig. 3

The volume δV of the triangular element *EJP* of thickness 'δx'
$$=\frac{1}{2}(JP)(EJ)\cdot\delta x = \frac{1}{2}\alpha\beta\,\delta x$$

The distance of this element's centroid from the YZ-plane is '*x*'.

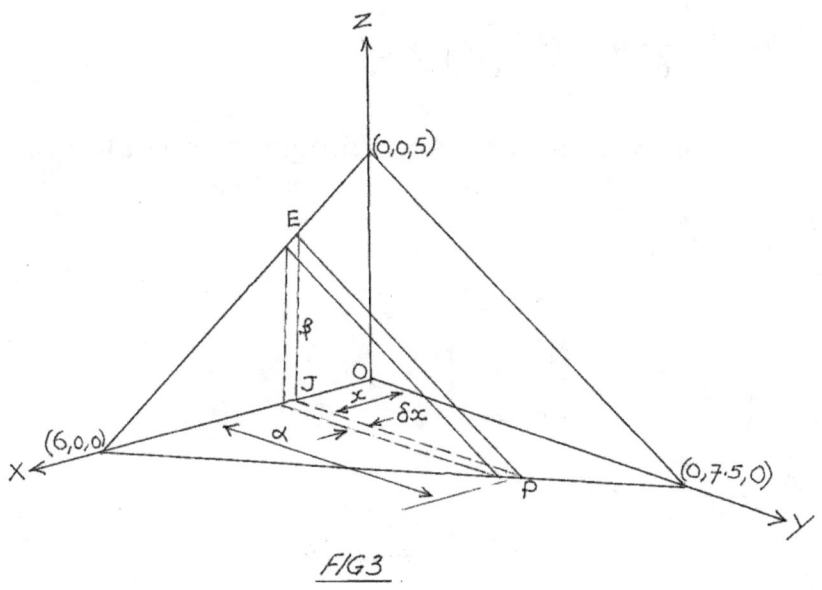

FIG 3

i.e. $\qquad \delta V = \dfrac{1}{2}\alpha\beta\ \delta x$

$$\bar{x} = \dfrac{\int dV \cdot x}{V}$$

i.e. $\qquad \bar{x} = \dfrac{\int \dfrac{1}{2}\alpha\beta(x)dx}{V}$

Inspection of Fig. 3 reveals the following geometrical relationship :

$$\dfrac{\alpha}{7.5} = \dfrac{6-x}{6}$$

$\therefore \qquad \alpha = \dfrac{7.5}{6}(6-x)$

Also $\qquad \dfrac{\beta}{5} = \dfrac{6-x}{6}$

or $\qquad \beta = \dfrac{5(6-x)}{6}$

$\therefore \qquad M_x = \dfrac{1}{2}\int \alpha\ \beta\ x\ dx$

~ 259 ~

$$= \frac{1}{2} \int \frac{7.5}{6} (6-x) 5 \left(\frac{6-x}{6} \right) x \; dx$$

The limits of x are zero and 6. Accordingly we need to evaluate.

$$\frac{37.5}{72} \int_0^6 x(6-x)^2 \, dx$$

$$= \frac{37.5}{72} \left| 36x - 6x^2 + \frac{x^3}{3} \right|_0^6$$

$$= \frac{37.5}{72} (72)$$

$$= 37.5$$

which means that

$$\bar{x} = \frac{M_X}{V}$$

$$= \frac{37.5}{37.5}$$

$$\therefore \qquad \bar{x} = 1 \; cm$$

Let us now consider the distance of the centroid from the XY-plane. Referring once more to Fig. 3, it is seen that the distance of the centroid of the element EJP of thickness δx from the XY-plane is $\dfrac{\beta}{3}$

$$\therefore \qquad \delta M_y = \frac{\beta}{3} \cdot \frac{1}{2} \alpha \; \beta \; \delta x$$

$$= \frac{1}{6} \alpha \; \beta^2 \delta x$$

As before $\alpha = 7.5 \left(\dfrac{6-x}{6} \right)$

$$\beta = \frac{5(6-x)}{6}$$

$$\therefore \qquad M_Y = \frac{1}{6} \int_0^6 7.5\left(\frac{6-x}{6}\right) \cdot \left\{\frac{5}{6}(6-x)\right\}^2 dx$$

$$= \frac{187.5}{1296} \int_0^6 (6-x)^3 \, dx$$

$$= \frac{187.5}{1296} \left| \left(216x - \frac{x^4}{4} - 54x^2 + 6x^3\right) \right|_0^6$$

$$= \frac{187.5}{1296}(1296 - 324 - 1944 + 1296)$$

$$= \frac{187.5}{1296}(324)$$

$$= 46.875$$

$$\therefore \qquad \overline{z} = 46.875 / 37.5$$

$$\overline{z} = 1.25cm$$

Finally we determine the distance of the centroid from the XZ-plane, i.e. $\overline{y}$. As was done previously, the element of area $\frac{1}{2}\alpha\beta$ has its own centroid at a distance of $\frac{1}{3}\alpha$ from the XZ-plane. Accordingly we may write without further ado

$$M = \int_0^6 \left(\frac{1}{2}\alpha\beta\right)\left(\frac{1}{3}\alpha\right) dx$$

$$= \frac{1}{6} \int_0^6 \left\{\frac{7.5}{6}(6-x)\right\}^2 \cdot 5\frac{(6-x)}{6} dx$$

$$= \frac{281.25}{1296} \int_0^6 (6-x)^3 \, dx$$

Using the previous outcome

$$M = \frac{281.25}{1296}(324)$$

$$\bar{y} = \frac{281.25(324)}{1296(37.5)}$$

$$\bar{y} = 1.875cm$$

In summary, the coordinates of the centroid of the tetrahedron shown in Fig. 1 are : 1cm, 1.87cm, 1.25cm.

The discerning reader would realize that the volume of the tetrahedron could just as easily have been obtained by writing.

$$\delta V = \frac{1}{2}\alpha\,\beta\,\delta x$$

i.e. $$V = \frac{1}{2}\int_0^6 \frac{7.5}{6}(6-x)5\left(\frac{6-x}{6}\right)dx$$

$$= \frac{37.5}{72}\int_0^6 (36-12x+x^2)dx$$

$$= \frac{37.5}{72}\left|36x - 6x^2 + \frac{x^3}{3}\right|_0^6$$

$$= \frac{37.5}{72}(216 - 216 + 72)$$

i.e. $$V = 37.5(cm)^3$$

Q13   The volume show in Fig. 1 is bounded on the YZ-plane by an elliptical area with major and minor axes of 15cm and 10cm respectively, on the XY-plane by the quadrant of a circle of radius 10cm and on the XY-plane by the plane $x = 10 - \frac{2z}{3}$. Calculate the volume of the solid assuming its density is constant and determine the coordinates of its centroid.

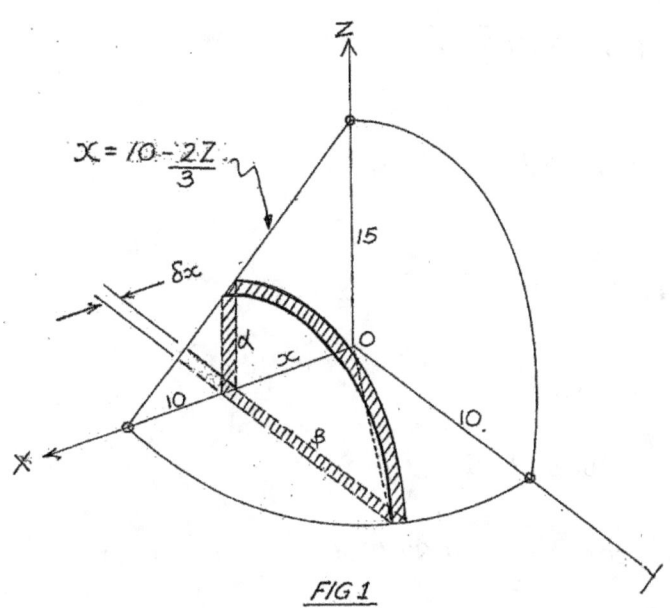

$$x = \frac{10-2Z}{3}$$

FIG 1

Area of elemental strip, equivalent to ¼ area of ellipse = $\dfrac{\pi\alpha\beta}{4}$

Volume of   "   "   "   "   "   " = $\dfrac{\pi}{4}\alpha\beta \cdot \delta x$

On the basis of geometrical considerations.

$$\frac{\alpha}{15} = \frac{10-x}{10}$$

$\therefore \qquad \alpha = \dfrac{15}{10}(10-x)$

or $\qquad = \dfrac{3}{2}(10-x)$

Likewise

$$10^2 = \beta^2 + x^2$$

or

$$\beta = \sqrt{10^2 - x^2}$$

Accordingly

$$\delta V = \frac{\pi}{4} \frac{3}{2} (10 - x)\sqrt{10^2 - x^2} \cdot \delta x$$

$$= \frac{3\pi}{8} (10 - x)\sqrt{10^2 - x^2} \cdot \delta x$$

$$\therefore \qquad V = \frac{3\pi}{8} \int (10 - x)\sqrt{(10^2 - x^2)} \cdot dx$$

Putting $\qquad x = 10 Sin\theta$

$$dx = 10 Cos\theta \; d\theta$$

The limits are : for $x = 0$; $\;\; 10 Sin\theta = 0$; $\;\; \therefore \;\; \theta = 0$; and,

for $x = 10$; $\;\; 10 Sin\theta = 10$; $\;\; \therefore \;\; \theta = \dfrac{\pi}{2}$

$$\therefore \qquad V = \frac{3\pi}{8} \int_{0}^{\pi/2} 10(1 - Sin\theta) 10 Cos\theta (10 Cos\theta \; d\theta)$$

$$= \frac{3000\pi}{8} \int^{\pi/2} (1 - Sin\theta)(Cos^2\theta) d\theta$$

$$= \frac{3000\pi}{8} \int^{\pi/2} \{(1 - Sin\theta)(1 - Sin^2\theta) + Sin^3\theta\} d\theta$$

$$= \frac{3000\pi}{8} \int^{\pi/2} (1 - Sin\theta - Sin^2\theta) d\theta + \frac{3000\pi}{8} \int_{0}^{\frac{\pi}{2}} Sin^3\theta \; d\theta$$

$$= \frac{3000\pi}{8} \left| \{\theta + Cos\theta - \left(\frac{-Sin\theta \; Cos\theta}{2} \frac{\theta}{2}\right) \right|_{0}^{\frac{\pi}{2}}$$

$$+ \int_{0}^{\pi/2} Sin^3\theta \; d\theta$$

$$= \frac{3000\pi}{8} \left[\left\{\frac{\pi}{2} - \frac{\pi}{4} - 1 - 0\right\} + \frac{3000\pi}{8} \int_{0}^{\pi/2} Sin^3\theta \; d\theta \right.$$

Now $\displaystyle\int_{0}^{\frac{\pi}{2}} Sin^3\theta \; d\theta = -\frac{1}{3} Sin^2\theta \; Cos\theta + \frac{2}{3} \int Sin\theta \; d\theta$

~ 264 ~

$$= -\frac{1}{3} Sin^2\theta\, Cos - \frac{2}{3} Cos\theta$$

Substituting for θ

$$\int_0^{\pi/2} Sin^3\theta\, d\theta = 0 - 0 - 0 + \frac{2}{3}$$

$$\therefore \quad V = \frac{3000\pi}{8}\left(\frac{\pi}{2} - \frac{\pi}{4} - 1 + \frac{2}{3}\right)$$

$$V = \frac{3000\pi}{8}\left(\frac{\pi}{4} - \frac{1}{3}\right)(cm)^3$$

or $\quad V = 169.5\pi \quad$ or $\quad 532(cm)^3$

Let us first of all find the distance of the centroid from the YZ-plane, i.e. $\overline{x}$. Referring to Fig. 1, it is seen that the distance of the centroid of the elemental volume from the YZ-plane is 'x'.

$$\therefore \quad \delta M_{YZ} = \delta V(x)$$

$$= \frac{\pi}{4}\alpha\beta\, dx \cdot x$$

$$= \frac{\pi}{4}\cdot\frac{3}{2}(10-x)\sqrt{10^2 - x}\cdot x\, dx$$

Employing the same substitution as before, viz $x = 10 Sin\theta$

$$dM_{YZ} = \frac{3\pi}{8}\cdot 10(1 - Sin\theta)10 Cos\theta \cdot 10 Sin\theta \cdot 10 Cos\theta\, d\theta$$

i.e. Total moments $\quad M_{YZ} = \dfrac{30{,}000\pi}{8}(1 - Sin\theta)Sin\theta\, Cos^2\theta\, d\theta$

$$= \frac{30{,}000\pi}{8}(1 - Sin\theta)Sin\theta(1 - Sin^2\theta)d\theta$$

$$= \frac{30{,}000\pi}{8}\int_0^{\pi/2}(Sin\theta - Sin^2\theta - Sin^3\theta + Sin^4\theta)d\theta$$

Let us write down the results of these integrals separately :

$$\int Sin\theta\, d\theta = -Cos\theta \;\; ; \;\; \int Sin^2\theta\, d\theta = -\frac{1}{2}Sin\theta\, Cos\theta + \frac{\theta}{2} \;\; ;$$

$$\int Sin^2 d\theta = -\frac{1}{2}Sin\theta Cos\theta + \frac{\theta}{2}$$

$$\int Sin^3\theta\, d\theta = -\frac{1}{3}Sin^2\theta\, Cos\theta - \frac{2}{3}Cos\theta \;\; ;$$

$$\int Sin^4\theta = -\frac{1}{4}Sin^3\theta\, Cos\theta + \int Sin^2\theta\, d\theta = -\frac{1}{4}Sin^3\theta\, Cos\theta - \frac{3}{8}Sin\theta\, Cos\theta + \frac{3\theta}{8}$$

$$\therefore \;\; M_{YZ} = \frac{30000\pi}{8}\left| -Cos\theta + \frac{1}{2}Sin\theta\, Cos - \frac{\theta}{2} + \frac{1}{3}Sin^2\theta\, Cos\theta + \frac{2}{3}Cos\theta + \right.$$

$$= \frac{1}{4}Sin^3\theta\, Cos\theta - \frac{3}{8}Sin\theta\, Cos\theta + \frac{3\theta}{8}\left.\right|_0^{\frac{\pi}{2}}$$

$$= \frac{30\,000\pi}{8}\left(1 - \frac{\pi}{4} - \frac{2}{3} + \frac{3\pi}{16} \right)$$

$$= \frac{30\,000\pi}{8}(1.589 - 1.455)$$

$$= 3750\pi(0.134)$$

$$= 502.5\pi(cm)^4$$

And because $V = 169.5\pi(cm)^3$

$$\overline{x} = \frac{502.5\pi}{169.5\pi} = 2.96cm$$

$$x \approx 3cm$$

Continuing, the distance of the centroid of the solid from the XZ-plane is $\overline{y}$. But the distance of the centroid of the elemental volume from this plane is $\frac{4\beta}{3\pi}$. Accordingly

$$\delta M_{XZ} = \frac{1}{4}\pi\alpha\beta \cdot \delta x \cdot \frac{4\beta}{3\pi}$$

$$= \frac{1}{3}\alpha\beta^2 \ \delta x$$

$$= \frac{1}{3} \cdot \frac{3}{2}(10-x)\left(\sqrt{10^2-x^2}\right)^2 \delta x$$

$$= \frac{1}{2}(10-x)(100-x^2)\delta x$$

As before let us put $x = 10 Sin\theta$, noting as before that the limits of x are : 0 and 10.

$$\therefore \qquad \delta M_X = \frac{1}{2}(10 - 10 Sin\theta)100(1 - Sin^2\theta)10Cos\theta \ d\theta$$

$$= 5000(1 - Sin\theta)(1 - Sin^2\theta)Cos\theta \ d\theta$$

$$dM_{XZ} = 5000(1 - Sin\theta)(1 - Sin^2\theta)Cos\theta \ d\theta$$

$$= 5000(1 - Sin\theta^2\theta - Sin\theta + Sin^3\theta)Cos\theta \ d\theta$$

$$= 5000(1 - Sin^2\theta - Sin\theta + Sin^3\theta)d(Sin\theta)$$

so that we may write at once

$$M_{XZ} = 5000\left[\left(Sin\theta - \frac{Sin^3\theta}{3} - \frac{Sin^2\theta}{2} + \frac{Sin^4\theta}{4}\right)\right]_0^{\frac{\pi}{2}}$$

$$= 5000\left(1 - \frac{1}{3} - \frac{1}{2} + \frac{1}{4}\right)$$

$$= 5000(5/12)$$

$$\overline{y} = \frac{25000}{12(169.5\pi)}$$

so that $\overline{y} = 3.9cm$

Finally, let $\bar{z}$ be the distance of the z-component of the centroid from the XY-plane. The two important things you needed to know in order to determine $\bar{y}$, and now, $\bar{z}$ is that the distances of the centroid of an elliptical area from the axes of the ellipse are respectively $\dfrac{4a}{3\pi}$ and $\dfrac{4\beta}{3\pi}$. In this case therefore the distance of the centroid of the elemental volume from the XY-plane is $\dfrac{4\alpha}{3\pi}$.

$$\therefore \quad \delta M_{XY} = \left(\frac{1}{4}\pi\alpha\beta \; \delta x\right)\left(\frac{4\alpha}{3\pi}\right)$$

$$= \frac{1}{3}\alpha^2 \beta \; \delta x$$

$$= \frac{1}{3}\left\{\frac{3}{2}(10-x)\right\}^2 \cdot \sqrt{10^2 - x^2} \cdot \delta x$$

$$= \frac{1}{3}\left(\frac{9}{4}\right)(10-x)^2 \sqrt{10^2 - x^2} \cdot \delta x$$

Putting $x = Sin\theta$ as before

$$\delta M_{XY} = \frac{3}{4\pi}(10 - 10Sin\theta)^2 10Cos\theta \cdot 10Cos\theta \; d\theta$$

$$= \frac{3}{4}\cdot 100(1 - Sin\theta)^2 100Cos^2\theta \; d\theta$$

$$= \frac{30,000}{4}(1 - 2Sin\theta + Sin^2\theta)Cos^2\theta \; d\theta$$

$$= 7500(1 - 2Sin\theta + Sin^2\theta)(1 - Sin^2\theta)d\theta$$

$$= 7500(1 - 2Sin\theta + 2Sin^3\theta - Sin^4\theta)d\theta$$

$$\therefore \quad M_{XY} = 7500\int_0^{\pi/2}(1 - 2Sin\theta + 2Sin^3\theta - Sin^4\theta)d\theta$$

$$\int Sin^3\theta \; d\theta = -\frac{1}{3}Sin^2\theta \; Cos + \frac{2}{3}\int Sin\theta \; d\theta = -\frac{1}{3}Sin^2\theta \; Cos\theta - \frac{2}{3}Cos\theta$$

$$\int Sin^4\theta = -\frac{1}{4} Sin^3\theta\, Cos\theta + \frac{3}{4}\int Sin^2\theta$$

$$\therefore \int Sin^4\theta\, d\theta = -\frac{1}{4} Sin^3\theta\, Cos\theta + \frac{3}{4}\left(-\frac{1}{2} Sin\theta\, Cos\theta + \frac{1}{2}\theta\right)$$

$$= -\frac{1}{4} Sin^3\theta\, Cos\theta - \frac{3}{8} Sin\theta\, Cos\theta + \frac{3}{8}\theta$$

Accordingly

$$M_{XY} = 7500\left\{\theta + 2Cos\theta + 2\left(-\frac{1}{3} Sin^2\theta\, Cos\theta - \frac{2}{3}Cos\theta\right)\right.$$

$$\left. + \frac{1}{4} Sin^3\theta\, Cos\theta + \frac{3}{8} Sin\theta\, Cos\theta - \frac{3}{8}\theta\right\}$$

$$= 7500\left(\theta + 2Cos\theta - \frac{2}{3} Sin^2\theta\, Cos\theta - \frac{4}{3}Cos\theta\right.$$

$$\left. + \frac{1}{4} Sin^3\theta\, Cos\theta + \frac{3}{8} Sin\theta\, Cos\theta - \frac{3}{8}\theta\right)$$

$$= 7500\left|\frac{5\theta}{8} + 2Cos\theta - \frac{2}{3} Sin^2\theta\, Cos\theta - \frac{4}{3}Cos\theta\right.$$

$$M_{XY} = \frac{5\pi}{16} - 2 + \frac{4}{3}$$

$$\therefore \qquad M_{XY} = 7500(0.3116)$$

$$= 2337(cm)^4$$

so that $\qquad \bar{z} = \frac{2337}{169.5\pi}\, cm$

or $\qquad \bar{z} = 4.4cm$

Therefore, the volume of the solid is $169.5\pi(cm)^3$ or $532(cm)^3$ and the coordinates of the centroid are 3cm, 3.9cm, and 4.4cm.

Q14. The solid of constant density shown in Fig 1 is bounded by a square in the YZ-plane; and quadrants of a circle of the same radius in the XY- and XZ-planes. Show that the volume of the solid is $\dfrac{2000(cm)}{3}$, and locate the coordinates of its centroid.

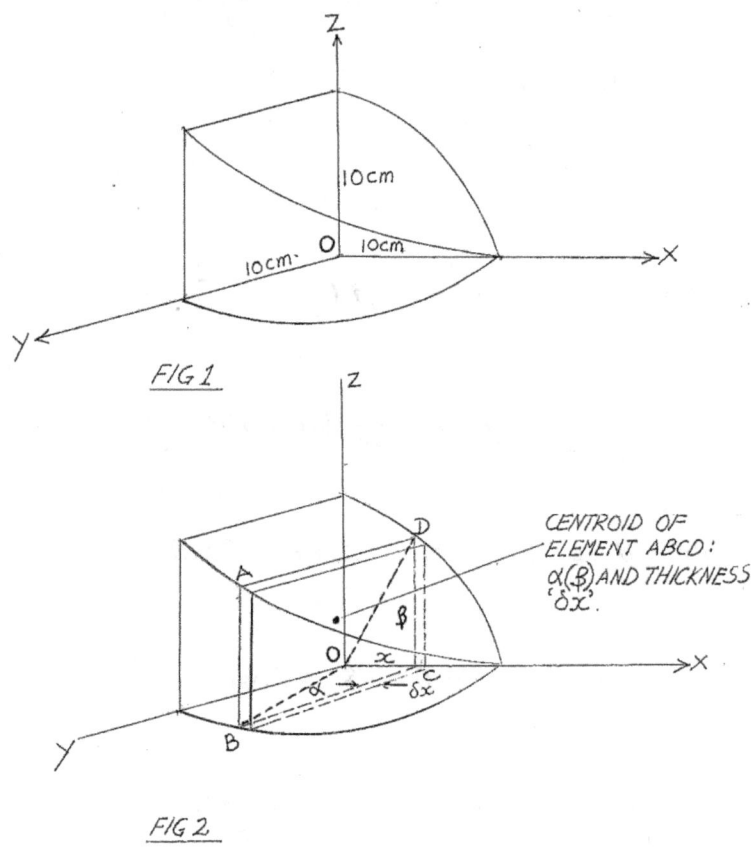

FIG 1

FIG 2

In Fig. 2, I have chosen an element '$ABCD$' parallel to the YZ-plane. Its dimensions are 'α' by 'β' and its thickness is δx. Accordingly its volume is $\alpha\beta\,dx$. Symbolically, we may write

$$\delta V = \alpha\beta\,\delta x$$

Referring again to the diagram, it is deduced that

$$10^2 = \alpha^2 + x^2$$

also, $10^2 = \beta^2 + x^2$

Therefore

$$\delta V = \left(\sqrt{100 - x^2}\right)\left(\sqrt{100 - x^2}\right)\delta x$$

$$= (100 - x^2) \delta x$$

The limits of x are $x = 0$; $x = 10$

$$\therefore \qquad V = \int_0^{10} (100 - x^2) dx$$

$$\left[100x = \frac{x^3}{3} \right]_0^{10}$$

$$V = \frac{2000}{3} (cm)^3 \qquad\qquad \text{Q.E.D.}$$

Consider next, the moment δM of the elemental volume about OY. It is given by the product of this volume and the distance of the volume's centroid from YZ-plane. This distance is 'x'.

$$\therefore \qquad \delta M = (\alpha \beta \ \delta x) x$$
$$= \alpha \beta x \ \delta x$$

Substituting the expressions for α, β as previously obtained

$$\delta M = \left(\sqrt{100 - x^2} \right) \sqrt{100 - x^2} \, x \ \delta x$$

$$= (100 - x^2) x \ \delta x$$

$$\therefore \qquad M = \int_0^{10} (100x - x^3) dx$$

$$= \left| 50x^2 - \frac{x^4}{x} \right|_0^{10}$$

$$= 2500$$

$$\therefore \qquad \overline{x} = \frac{2500}{V}$$

$$= \frac{2500(3)}{2000}$$

$$\overline{x} = 3.75 cm$$

Continuing, it is observed that the centroid of the elemental strip is distant $\frac{\alpha}{2}$ from the XZ-plane. Therefore, here we have for moment about the X-axis.

$$\delta M = (\alpha \beta \; \delta x) \cdot \frac{\alpha}{2}$$

$$= \frac{1}{2} \alpha^2 \beta \; \delta x$$

The geometrical relationships between α, β, x and the radius of 10cm still hold.

$$\therefore \qquad \delta M = \frac{1}{2}(100 - x^2)(\sqrt{100 - x^2} \cdot \delta x)$$

$$= \frac{1}{2}(100 - x^2)^{\frac{3}{2}} \delta x$$

Substituting as before $x = 10 Sin\theta$

$$M = \frac{1}{2} \int_0^{\pi/2} 1000 Cos^3\theta \cdot 10 Cos\theta \; d\theta$$

$$= \frac{10000}{2} \int_0^{\pi/2} Cos^4\theta \; d\theta$$

But $\int Cos^4 d\theta = \frac{1}{4} Cos^3\theta \; Sin\theta + \frac{3}{4} \int Cos^2\theta \; d\theta$

$$= \frac{1}{4} Cos^3 Sin\theta + \frac{3}{4}\left(\frac{1}{2} Cos\theta \; Sin\theta + \frac{\theta}{2}\right)$$

i.e. $\int Cos^4\theta \; d\theta = \frac{1}{4} Cos^3 Sin\theta + \frac{3}{8} Sin\theta \; Cos\theta + \frac{3}{8}\theta$

so that

$$M = 5000 \left| \frac{1}{4} Cos^3 Sin\theta + \frac{3}{8} Sin\theta \; Cos\theta + \frac{3}{8}\theta \right|$$

$$M = 5000\left(\frac{3\pi}{16}\right)$$

from which

$$\overline{y} = \frac{5000}{V}\left(\frac{3\pi}{16}\right)$$

$$= \frac{5000}{2000}\left(\frac{3\pi}{16}\right)\cdot 3$$

i.e. $\overline{y} = \dfrac{45\pi}{32}\,cm$

Finally, the distance of the centroid of the element is distance $\dfrac{\beta}{2}$ from the XY-plane. The moment here is

$$\delta M = \left(\alpha\beta\ \delta x\right)\frac{\beta}{2}$$

$$= \frac{1}{2}\alpha\beta^2\,\delta x$$

$$= \frac{1}{2}\sqrt{10^2 - x^2}\,(10^2 - x^2)\delta x$$

$$= \frac{1}{2}(10^2 - x)^{3/2}\,\delta x$$

from which

$$M = \frac{1}{2}\int(10^2 - x^2)^{3/2}\,dx$$

With the substitution $x = 10Sin\theta$, we obtain the same result as for $\overline{y}$.

Therefore

$$M = 5000\left(\frac{3\pi}{16}\right)$$

from which $\bar{z} = \dfrac{45\pi}{32}\, cm$

The co-ordinates of the centroid are $\left(3.75,\ \dfrac{45\pi}{32},\ \dfrac{45\pi}{32}\right)$ in centimeters.

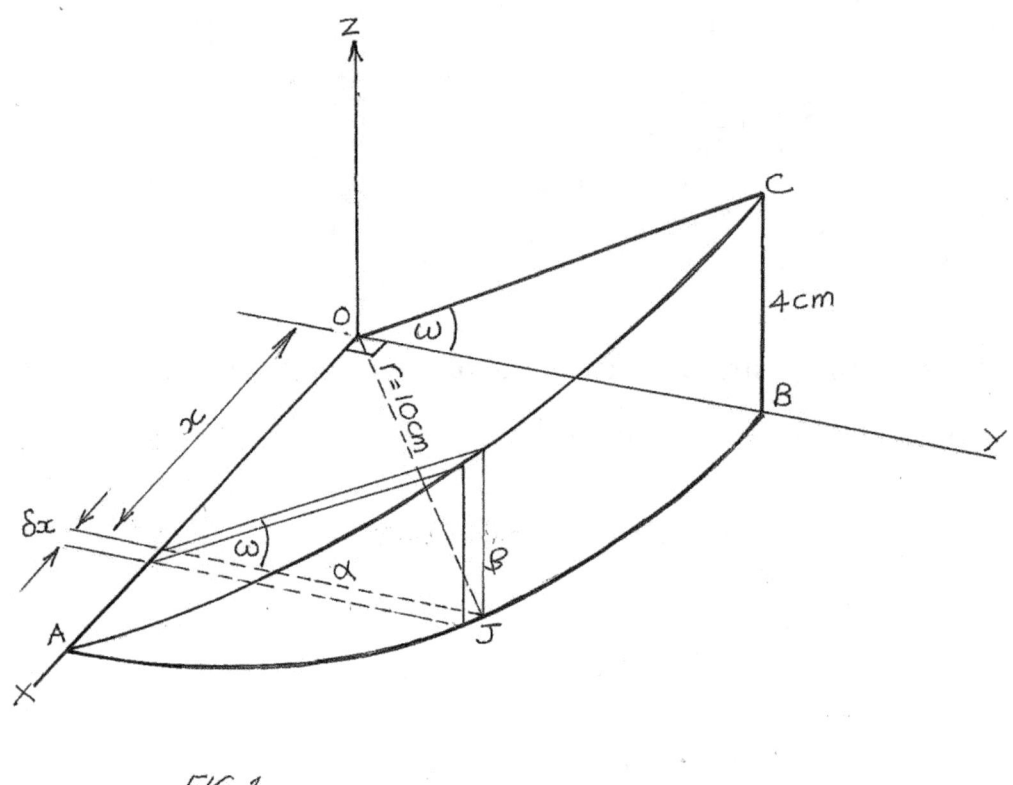

FIG 1

Q15   Determine the coordinates of the centroid of the wedge 'OABCO' shown in Fig. 1. OBA is a quadrant of a circle of radius 10cm. Assume constant density.

A triangular strip 'EJP'' of thickness 'δx' is demarcated as indicated in the diagram. It is perpendicular to the XZ plane.

$\therefore \qquad \delta V = \dfrac{1}{2}\alpha\beta \cdot \delta x$

$\alpha^2 + x^2 = (OJ)^2$

i.e. $\alpha^2 + x^2 = 10^2$

$\therefore \qquad \alpha = \sqrt{100 - x^2}$

Also

$$\frac{\beta}{\alpha} = Tan\alpha = \frac{4}{10} = 0.4$$

or $\qquad \beta = 0.4\alpha$

$$= 0.4\sqrt{100 - x^2}$$

$\therefore \qquad \delta V = \frac{1}{2}\left(\sqrt{100 - x^2}\right)0.4\sqrt{100 - x^2} \cdot \delta x$

$$= 0.2(100 - x^2)\delta x$$

so that

$$V = 0.2\int_0^{10}(100 - x^2)dx$$

$$= 0.2\left|100x - \frac{x^3}{3}\right|_0^{10}$$

$$= 0.2\left(\frac{2000}{3}\right)(cm)^3$$

i.e. $\qquad V = \frac{400}{3} cm^3$

The centroid of the element is distant 'x' from OY. Therefore for moments about this axis – the Y-axis:

$$\delta M_Y = \left(\frac{1}{2}\alpha\beta\right)x \; \delta x$$

$$= \frac{1}{2}\left(\sqrt{100 - x^2}\right)(0.4)\left(\sqrt{100 - x^2}\right) \cdot x \; \delta x$$

i.e. $\qquad M_Y = 0.2\int_0^{10}(100 - x^2)x \; dx$

$$= 0.2 \left[\frac{100x^2}{2} - \frac{x^4}{4} \right]_0^{10}$$

$$= 0.2(5000 - 2500)$$

$$= 500$$

so that $\quad \bar{x} = \dfrac{500}{400} \cdot 3$

$$\bar{x} = \frac{15}{4} \ cm.$$

The centroid of the triangular element EJP is distant $\dfrac{2}{3}\alpha$ from the OX axis. Therefore the moment δM of this element about OX is

$$\delta M_x = \frac{1}{2}\alpha\beta \cdot \frac{2}{3}\alpha\beta \ \delta x$$

$$= \frac{1}{3}\alpha^2 \beta \ \delta x$$

$$= \frac{1}{3}\left(100 - x^2\right)(0.4)\sqrt{100 - x^2} \ \delta x$$

$$M_x = \frac{0.4}{3}\int_0^{10} (100 - x^2)^{3/2} dx$$

Employing the substitution $x = 10 Sin\theta$ as was done in a previous problem, it is found that

$$M_X = \frac{0.4}{3} \cdot 10000 \cdot \frac{3\pi}{16} = 250\pi$$

so that $\quad \bar{y} = \dfrac{250\pi}{400} \cdot 3 = \dfrac{75\pi}{40}$

$$\bar{y} = \frac{15\pi}{8}$$

Now to $\bar{z}$. The centroid of the elemental triangle is at a distance of $\frac{1}{3}\beta$ from the XY-plane. The moment is

$$\delta M_{XY} = \left(\frac{1}{2}\alpha\beta\right)\cdot\frac{1}{3}\beta\ \delta x$$

$$= \frac{1}{6}\alpha\beta^2 \delta x$$

$$= \frac{1}{6}\left(\sqrt{100-x^2}\cdot\right)\beta^2 \cdot \delta x$$

Putting $\beta = 0.4\alpha$

$$= 0.4\sqrt{100-x^2}$$

$$\delta M_{XY} = \frac{1}{6}\sqrt{100-x^2}\,(0.16)(100-x^2)\delta x$$

$$= \frac{0.16}{6}(100-x^2)^{3/2}\delta x$$

Or $M_{XY} = \frac{0.16}{6}\int_0^{10}(100-x^2)^{3/2}\,dx$

Using a result obtained earlier

$$M_{XY} = 0\cdot\frac{16}{6}\left[\frac{(1000)}{}\right]\frac{3}{16}\pi$$

So that

$$\bar{z} = \frac{0.16}{6}\cdot\frac{30,000\pi\cdot3}{16\ \ 400}$$

$$= \frac{16\times300\pi\times3}{6\times16\times400}$$

$\therefore\qquad \bar{z} = \frac{3\pi}{8}\,cm$

The coordinates of the centroid are (in centimeters) : $\left(\dfrac{15}{4}, \dfrac{15\pi}{8}, \dfrac{3\pi}{8}\right)$

Q16. Show that the volume of the soild shown in Fig. 1 is 40(cm)³ and find the coordinates of its centroid. Assume constant density.

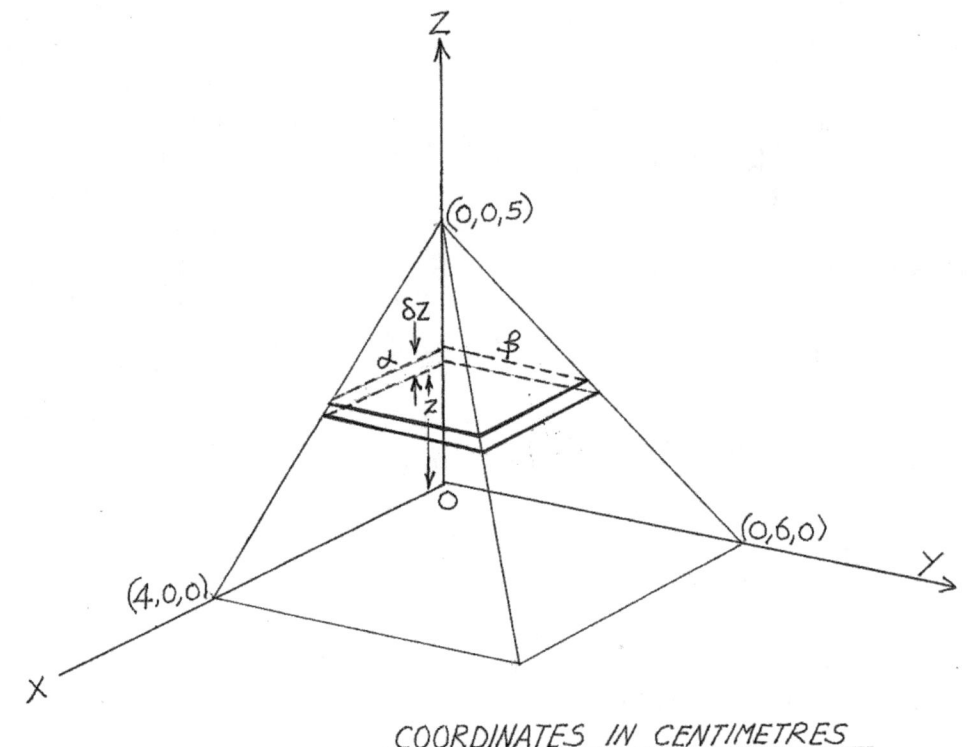

COORDINATES IN CENTIMETRES

FIG 1

Note : Coordinates are in centimetres

The elemental volume δV shown in Fig. 1 may be expressed as:

$$\therefore \quad \delta V = Area \times thickness$$
$$= \alpha\beta\,\delta z$$

By geometrical considerations
$$\frac{\alpha}{4} = \frac{5-z}{5}$$

Also,

$$\frac{\beta}{6} = \frac{5-z}{5}$$

Therefore

$$\alpha = \frac{4(5-z)}{5}$$

and

$$\beta = \frac{6(5-z)}{5}$$

from which it follows that :

$$\delta V = \frac{4(5-z)}{5} \cdot \frac{6(5-z)}{5} \cdot \delta z$$

$$= \frac{24}{25}(5-z)^2 \, \delta z$$

and $\quad V = \dfrac{24}{25} \displaystyle\int_0^5 (5-z)^2 \, dz$

$$= \frac{24}{25} \int_0^6 (25 - 10z + z^2) \, dz$$

$$= \frac{24}{25} \left| 25z - \frac{10z^2}{2} + \frac{z^3}{3} \right|_0^5$$

$$= \frac{24}{25} \left(125 - 125 + \frac{125}{3} \right)$$

i.e $\quad V = 40(cm)^3,$ $\qquad$ Q.E.D

Next, let us determine $\bar{z}$. We observe that the elemental section is distant 'z' from the XY-plane. Therefore the requisite moment 'M_{XY} is given by: $\delta V(z)$, i.e.

$$\delta M_{XY} = (\alpha \beta \, \delta z) z$$

$$= \frac{24}{25} z(5-z)^2 \, dz$$

$$= \frac{24}{25} z(25 - 10z + z^2) \delta z$$

Or $\quad M_{XY} = \frac{24}{25} \int_0^5 (25z - 10z^2 + z^3) dz$

$$= \frac{24}{25} \left| \frac{25z^2}{2} - \frac{10z^3}{3} + \frac{z^4}{4} \right|_0^5$$

$$= \frac{24}{25} \left(\frac{625}{2} - \frac{10.125}{3} + \frac{625}{4} \right)$$

$$= \frac{24}{25} \left(\frac{625}{12} \right)$$

i.e. $\quad M_{XY} = 50$

so that

$$\bar{z} = \frac{M_{XY}}{V} = \frac{50}{40}$$

$$\bar{z} = 1.25 cm$$

Now to $\bar{x}$

The elemental volume has its own centroid at a distance of $\frac{'\alpha'}{2}$ from the YZ-plane. Therefore its moment about OZ, dM is given by

$$\delta M_{YZ} = (\alpha \beta \, \delta z) \frac{\alpha}{2}$$

i.e. $\quad \delta M_{YZ} = \frac{\alpha^2 \beta}{2} \delta z$

$$= \left\{ \frac{4}{4}(5-z) \right\}^2 6 \left(\frac{5-z}{5} \right) \cdot \frac{\delta z}{2}$$

$$= \frac{16}{25} \cdot \frac{6}{10} (5-z)^3 \, \delta z$$

from which

$$M_{YZ} = \frac{48}{125} \int_0^5 (5-z)^3 \, dz$$

$$= \frac{48}{125} \int_0^5 \left\{ 125 - z^3 - 15z(5-z) \right\} dz$$

$$= \frac{48}{125} \int_0^5 (125 + 15z^2 - 75z - z^3) \, dz$$

$$= \frac{48}{125} \left[125z + \frac{15z^3}{3} - \frac{75z^2}{2} - \frac{z^4}{4} \right]_0^5$$

$$= \frac{48}{125} (156.25)$$

$$= 48(1.25)$$

$$M_{YZ} = 60 \, (cm)^4$$

$$\therefore \qquad \bar{x} = \frac{60}{40} \, cm$$

or $\qquad x = 1.5 cm$

Finally,

$$\bar{y} = \frac{\int (\alpha\beta \, dz) \frac{\beta}{2}}{V}$$

Now $\quad \frac{1}{2} \int_0^5 \alpha\beta^2 \, dz = \frac{1}{2} \int_0^5 \frac{4}{5}(5-z) \left\{ \frac{6}{5}(5-z) \right\}^2 dz$

$$= \frac{1}{2} \cdot \frac{4}{5} \cdot \frac{36}{25} \int_0^5 (5-z)^3 \, dz$$

$$= \frac{72}{125}(156.25) = 90(cm)^4$$

which means that

$$\bar{y} = \frac{90}{40}$$

or $\quad \bar{y} = 2.25cm$

In summary therefore the coordinates of the centroid are, in centimetres : (1.5, 2.25, 1.25).

Q17   The areas of the base and top of a regular frustum of a cone *hcm* in height are respectively, in sq.cm, A_1 and A_2 where $A_1 > A_2$. Show that the centroid is on the axis of the frustum at a distance from the base equal to :

$$\frac{h}{4}\frac{(A_1 + 2\sqrt{A_1 A_2} + 3A_2)}{(A_1 + \sqrt{A_1 A_2} + A_2)}cm$$

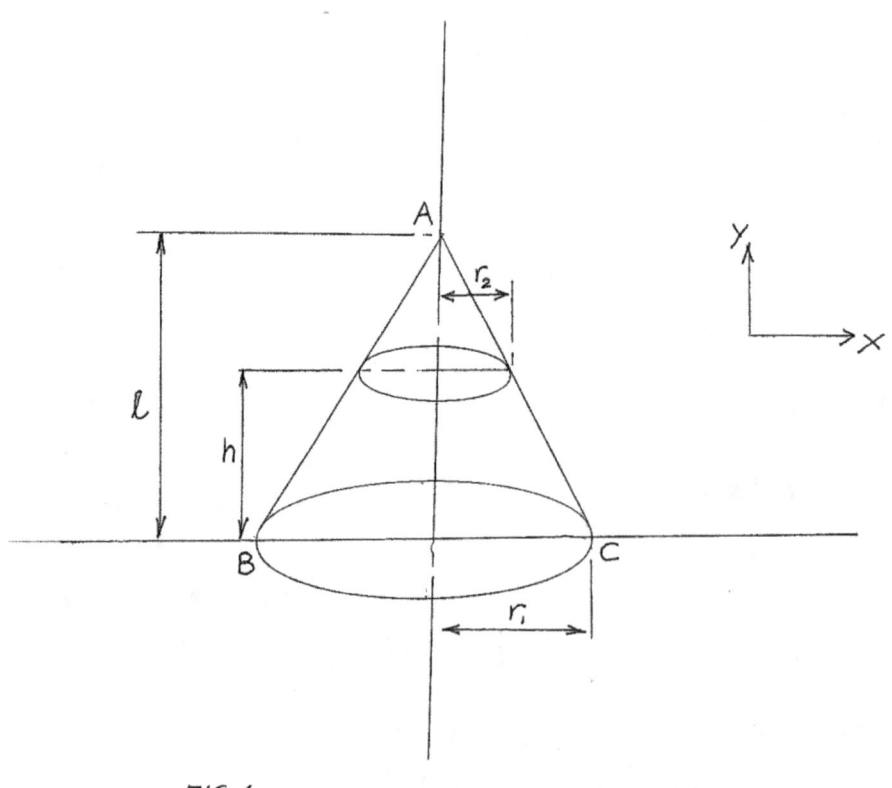

FIG 1

Let $\bar{y}$ = centroid of frustum

Assuming constant density,

$$\bar{y} = \frac{\left\{\begin{array}{l}Volume\ of\ l\arg er\ cone\ ABC\,(dis\tan ce\ of\ its\ centroid\ from\ BC)\\ -Volume\ of\ smaller\ cone\ ADE\,(dis\tan ce\ of\ its\ centroid\ from\ BC\end{array}\right\}}{Volume\ of\ l\arg er\ cone - Volume\ of\ smaller\ cone}$$

Considering the geometry of Fig. 1

$$\frac{r_2}{r_1} = \frac{\ell}{\ell+h}$$

where r_2 = radius of base of smaller cone, cm

r_1 =   "   "   "   " larger   "   , cm

ℓ = total height of larger cone,

h = height of frustum, cm

$$\therefore \quad \bar{y} = \frac{\frac{1}{3}\pi r_1^2(\ell+h)\cdot\left(\frac{\ell+h}{4}\right) - \frac{1}{3}\pi r_2^2 \ell\left(\frac{\ell+h}{4}\right)}{\frac{1}{3}\pi r_1^2(\ell+h) - \frac{1}{3}\pi r_2^2 \ell}$$

Eliminating π and simplifying

$$\bar{y} = \frac{r_1^2(\ell+h)^2 - r_2^2 \ell(\ell+4h)}{4r_1^2(\ell+h) - 4r_2^2 \ell}\,cm$$

Substituting $\ell = \dfrac{r_2 h}{(r_1 - r_2)}$

$$(\ell+h)^2 = \left\{\frac{r_2 h}{(r_1 - r_2)} + h\right\}^2 = h^2 \frac{(r_2 + r_1 - r_2)}{(r_1 - r_2)} = \frac{h^2 r_1^2}{(r_1 - r_2)^2}$$

Also

$$r_2^2 \ell(\ell + 4h) = r_2^2 \left[\frac{r_2 h}{r_1 - r_2} \left\{ \frac{r_2 h}{r_1 - r_2} + 4h \right\} \right]$$

$$= r_2^3 h^2 \frac{(4r_1 - 3r_2)}{(r_1 - r_2)^2}$$

$$\therefore \quad \overline{y} = \frac{\dfrac{r_1^2 h^2 r_1^2}{(r_1 - r_2)^2} - \dfrac{r_2^3 h^2 r_1}{(r_1 - r_2)^2} + \dfrac{3r_2^2 h^2}{(r_1 - r_2)^2}}{4r_1^2 (\ell + h) - 4r_2^2 \ell} \, cm$$

Let us reduce occasion for error by evaluating the terms in the denominator separately.

$$4r_1^2 (\ell + h) = 4r_1^2 \left\{ \frac{r_2 h}{(r_1 - r_2)} + h \right\}$$

$$= \frac{4r_1^2 \, hr_1}{(r_1 - r_{2)}}$$

$$= \frac{4r_1^3 \, h}{r_1 - r_2}$$

$$\text{i.e.} \quad 4r_2^2 \ell \quad = \frac{4r_2^2 \, r_2 h}{(r_1 - r_2)} = \frac{4r_2^3 h}{(r_1 - r_2)}$$

$$\therefore \quad \overline{y} = \frac{\dfrac{h^2}{(r_1 - r_2)^2} \left(r_1^4 - 4r_1 \, r_2^3 + 3r_2^4 \right)}{\dfrac{4h}{(r_1 - r_2)} \left(r_1^3 - r_2^3 \right)} \, cm$$

$$= \frac{h}{4(r_1 - r_2)} \frac{\left(r_1^4 - 4r_1 \, r_2^3 + 3r_2^4 \right)}{\left(r_1^3 - r_2^3 \right)}$$

$$= \frac{h}{4(r_1 - r_2)^2} \frac{\left(r_1^4 - 4r_1 \, r_2^3 + 3r_2^4 \right)}{\left(r_1^2 + r_1 \, r_2 + r_2^2 \right)}$$

i.e.
$$\bar{y} = \frac{h}{4} \frac{(r_1^4 - 4r_1 r_2^3 + 3r_2^4)}{(r_1^2 - 2r_1 r_2 + r_2^2)(r_1^2 + r_1 r_2 + r_2^2)} cm$$

$$= \frac{h}{4} \frac{(r_1^2 + 2r_1 r_2 + 3r_2^2)}{(r_1^2 + r_1 r_2 + r_2^2)} cm$$

Noting that $r_1 = \sqrt{\frac{A_1}{\pi}}$; $r_2 = \sqrt{\frac{A_2}{\pi}}$

$$\bar{y} = \frac{\frac{h}{4}\left(\frac{A_1}{\pi} + 2\frac{\sqrt{A_1 A_2}}{\pi} + \frac{3A_2}{\pi}\right)}{\left(\frac{A_1}{\pi} + \frac{\sqrt{A_1 A_2}}{\pi} + \frac{A_2}{\pi}\right)} cm$$

i.e.
$$\bar{y} = \frac{h}{4}\left(\frac{A_1 + 2\sqrt{A_1 A_2} + 3A_2}{A_1 + \sqrt{A_1 A_2} + A_2}\right) cm \qquad \text{Q.E.D}$$

Q18. Find the volume of the solid shown in Fig. 1, below by considering an elemental section parallel to the XY-plane such as that sketched in the diagram, and confirm your result by means of Pappus' Theorem. Determine also the coordinates of the centroid of the volume. Assume constant density.

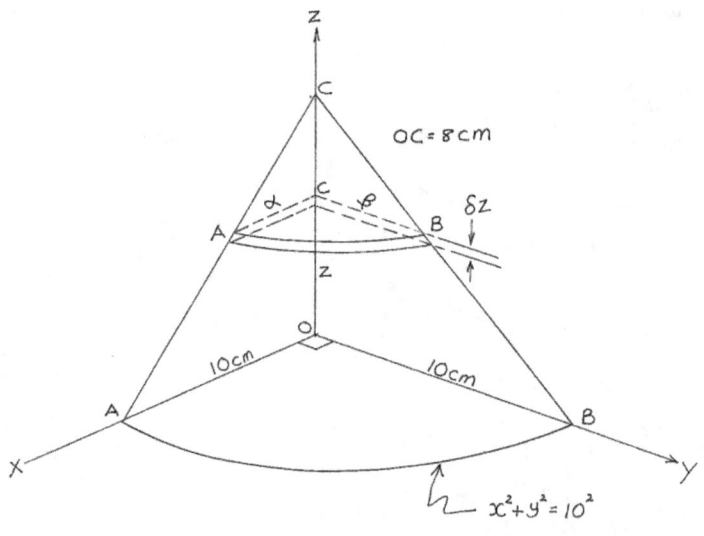

FIG 1

Referring to the diagram it is observed that the element is the quadrant of a circle; 'α' and 'β' are identical. Let the thickness of the quadrant be 'δz'. Therefore, the volume of the element is :

$$\delta V = \left(\frac{1}{4}\pi\alpha^2\right)\delta z \quad \text{or} \quad \frac{1}{4}\pi\beta^2)\delta z$$

Let us work with 'α'

$$\therefore \qquad V = \frac{\pi}{4}\int \alpha^2 dz$$

But $\quad \dfrac{\alpha}{10} = \dfrac{8-z}{8}$

or $\qquad \alpha = \dfrac{5}{4}(8-z)$

$$\therefore \qquad V = \frac{\pi}{4}\int_0^8 \left\{\frac{5}{4}(8-z)\right\}^2 dz$$

$$= \frac{25\pi}{64}\int_0^8 (64-16z+z^2)dz$$

$$= \frac{25\pi}{64}\left|64z-8z^2+\frac{z^3}{8}\right|_0^8$$

$$= \frac{25\pi}{64}\left(\frac{512}{3}\right)$$

i.e. $\qquad V = \dfrac{200\pi}{3}(cm)^3$

For $\bar{z}$

$$\delta M = \delta V \cdot z$$

$$= \left(\frac{1}{4}\pi\alpha^2 \cdot \delta z\right)\cdot z$$

$$= \frac{1}{4}\pi\left\{\frac{5}{4}(8-z)\right\}^2 \cdot z \; \delta z$$

$$\therefore \; M = \frac{25\pi}{64}\int_0^8 (64-16z+z^2)z \; dz$$

$$= \frac{25\pi}{64}\int_0^8 (64z-16z^2+z^3)dz$$

$$= \frac{25\pi}{64}\left|\frac{64z^2}{2} - \frac{16z^3}{3} + \frac{z^4}{4}\right|_0^8$$

$$= \frac{25\pi}{64}\left(2048 - \frac{8192}{3}1024\right)$$

$$= \frac{25\pi}{64}\cdot\frac{1024}{3}$$

i.e. $\dfrac{400\pi}{3}$ and

$$\bar{z} = \frac{400\pi \times 3}{3} \frac{}{200\pi}$$

$$\bar{z} = 2cm.$$

For $\quad \bar{x}$

$$\delta M = \left(\frac{1}{4}\pi\alpha^2\delta z\right)\cdot\frac{4\beta}{3\pi}$$

Why $\dfrac{4\beta}{3\pi}$? Because that is the distance of the centroid of the element from the X-axis, so

$$\delta M = \frac{\alpha^2\beta}{3}\delta z$$

But $\quad \alpha = \beta$

$$\therefore \qquad \delta M = \frac{1}{3}\alpha^3 \delta z$$

$$\therefore \qquad M = \frac{1}{3}\int_0^8 \left\{\frac{5}{4}(8-z)\right\}^3 dz$$

$$= \frac{125}{192}\int_0^8 (512 - 192z + 24z^2 - z^3)dz$$

$$= \frac{125}{192}\left|512z - 96z^2 + 8z^3 - \frac{z^4}{4}\right|_0^8$$

$$= \frac{125}{192}(1024)$$

$$= \frac{2000}{3}$$

$$\therefore \qquad \bar{x} = \frac{2000}{3}\cdot\frac{3}{200\pi}$$

$$\bar{x} = \frac{10}{\pi} cm.$$

I leave it to the reader t show that $\bar{y}$ is also equal to $\frac{10}{\pi} cm.$

Let us now using one of Pappus' theorems confirm the result obtained for the volume of the solid. Just to refresh your memory : one of Pappus' theorems states that the volume swept out by revolving an area about an axis in its plane which does not cut the boundaries of the area is equal to the area multiplied by the length of the path of the centroid of the area around the axis of revolution. Referring to Fig. 1, it is seen that when the area *AOC* is rotated about axis *OZ* which as you can see is in the same plane as the area, the radius of rotation being the distance of the centroid of triangle *AOC* from the Z-axis, the resulting volume would be 4 times the volume of the solid represented as Fig. 1.

Accordingly,

$$\text{Area of triangle } AOC = \frac{8\times 10}{2} = 40cm^2$$

Distance of centroid of rotation 'z' $= \dfrac{10}{3}\,cm$

$\therefore$ Length of path of centroid $= 2\pi \cdot \dfrac{10}{3} = \dfrac{20\pi}{3}$

And so according to Pappus

Revolving area $= 40cm^2$

Length of path of centroid $= \dfrac{20\pi}{3}$

$\therefore$ 4(Swept Volumes) $= 40\left(\dfrac{20\pi}{3}\right)$

$\therefore$ Volume of Solid $= \dfrac{40}{4}\left(\dfrac{20\pi}{3}\right)$

$= \dfrac{200\pi}{3}$, Check

By working with triangular area OBC you would obtain confirmation of this result.

Q19. Using spherical coordinates, find the centre of mass of a hemispherical body of radius 'r', assuming that its density 'ρ' at every point is directly proportional to the square root of the distance of the point from the axis of symmetry.

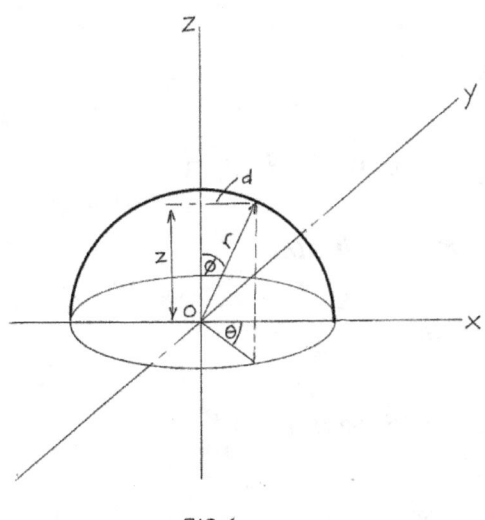

FIG 1

~ 289 ~

Volume of Element, $\delta V = r^2 Sin\phi \; \delta r \; \delta\theta \; \delta\phi$

But $\rho = K\sqrt{d} = K\sqrt{r^2 - z^2}$

$$= K\sqrt{(r^2 - r^2 Cos^2\phi)}$$

$$= Kr \cdot Sin\phi$$

$\therefore$ Mass of element δM

$$= Kr^3 Sin^2\phi \; dr \; d\theta \; d\phi$$

So that total mass, $M = 4K\int_0^{\pi/2} \int_0^{\pi/2} r^3 dr \; d\theta \; Sin^2\phi \; d\phi$

$$= 4K\int_0^{\pi/2} \int_0^{\pi/2} \frac{r^4}{4} d\theta \; Sin^2\phi \; d\phi$$

$$4K\int_0^{\pi/2} \cdot \frac{r^4}{4} \cdot \frac{\pi}{2} Sin^2\phi \; d\phi$$

$$= \frac{\pi K r^4}{2}\int_0^{\pi/2} Sin^2\phi \; d\phi$$

$$= \frac{\pi K r^4}{2}\left[\frac{\phi}{2} - \frac{1}{4}Sin2\phi\right]_0^{\frac{\pi}{2}}$$

$$M = \frac{\pi^2 K r^4}{8}$$

Taking moments about the X-Y plane

$$M_{XY} = \iiint \rho \; zr^2 Sin\phi \; dr \; d\theta \; d\phi$$

Substituting $\rho = KrSin\phi$

$$= \iiint z \; Kr^3 Sin^2\phi \; dr \; d\theta \; d\phi$$

Substituting $z = rCos\phi$

$$M_{XY} = 4K \int_0^{\pi/2} \int_0^{\pi/2} \int_0^r r^4 Cos\phi \, Sin^2\phi \, dr \, d\theta \, d\phi$$

$$= 4K \int_0^{\pi/2} \int_0^{\pi/2} \frac{r^5}{5} Sin^2\phi \, Cos\phi \, d\theta \, d\phi$$

$$= \frac{2\pi K r^5}{5} \cdot \int_0^{\pi/2} Sin^2\phi \, Cos\phi \, d\phi$$

$$= 2\pi \frac{K r^5}{5} \int_0^{\pi/2} Sin^2\phi \, d(Sin\phi)$$

$$= 2\pi \frac{K r^5}{5} \left[\frac{Sin^3\phi}{3} \right]_0^{\pi/2} = \frac{2\pi K r^5}{5} \left[\frac{1}{3} - 0 \right]$$

i.e. $\quad M_{XY} = \dfrac{2\pi K r^5}{15}$

From which $\quad \bar{z} = \dfrac{2\pi K r^5}{15} \cdot \dfrac{8}{\pi^2 K r^4}$

i.e. $\quad \bar{z} = \dfrac{16r}{15\pi}$

Q20. By means of spherical coordinates, determine the mass of a spherical object of radius '*a*' when the density at each point in the object is directly proportional to the distance of each point from its centre. See Fig. 1.

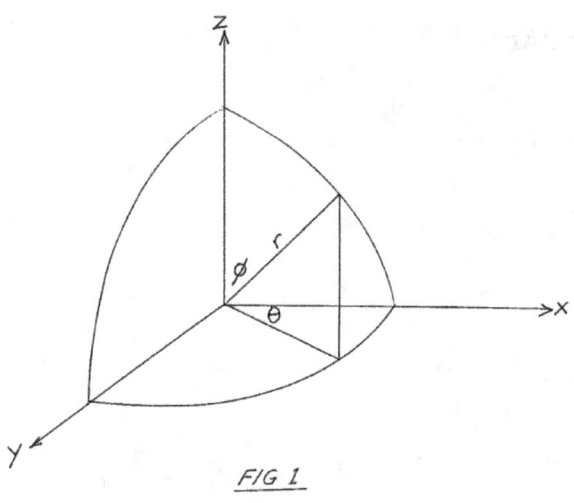

FIG 1

In Fig. 1, I have shown one-eight of the spherical object. We know that:
$$\delta V = r^2 \delta r \, Sin\phi \, \delta\phi \, \delta\theta$$

We are told that density at each point is directly proportional to distance from centre which means that $\rho = Kr$ where $K = $ constant of proportionality.

$$\therefore \quad \delta M = \rho r \, dr \, Sin\phi \, \delta\phi \, \delta\theta \quad \text{becomes}$$

$$\therefore \quad \delta M = (Kr)r \, \delta r \, Sin\phi \, \delta\phi \, \delta\theta$$

so that $$M = K \iiint r^3 dr \, Sin\phi \, d\phi \, d\theta$$

To obtain the entire mass of the sphere we must account for eight component parts such as the one shown in Fig. 1

$$\therefore \quad M = 8K \int_{\theta=0}^{\theta=\frac{\pi}{2}} \int_{\phi=0}^{\phi=\frac{\pi}{2}} \int_{0}^{r} r^3 dr \, Sin\phi \, d\phi \, d\theta$$

Let us now begin to integrate

$$M = 8K \int_{0}^{\pi/2} \int_{0}^{\pi/2} \left[\frac{r_4}{4} \right] Sin\phi \, d\phi \, d\theta$$

$$= 2Kr^4 \int_{0}^{\pi/2} \left[-Cos\phi \right]_{0}^{\pi/2} d\theta$$

$$= 2Kr^4 \int_{0}^{\pi/2} -[0-1] \, d\theta$$

$$= 2Kr^4 \cdot \theta$$

$$= 2Kr^4 \cdot \frac{\pi}{2}$$

$$\therefore \quad M = \pi \, Kr^4$$

Q21. Show that the centroid of a solid cone is located at a distance of $\frac{1}{4}$ of its altitude above the base. Assume density 'ρ' is constant throughout. Use rectangular coordinates.

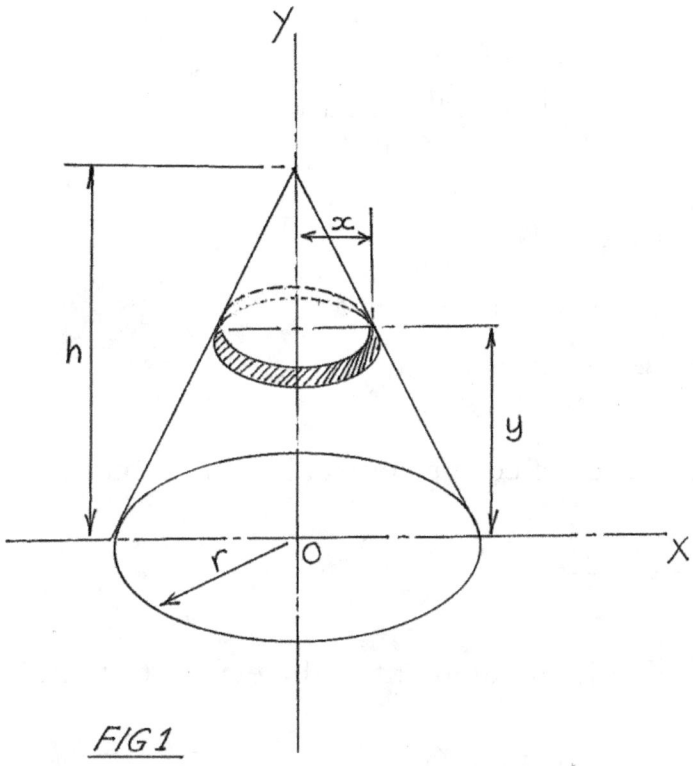

<u>FIG 1</u>

Fig. 1 is the diagram of our cone which is of height 'h' and radius of its base equal to 'r'.

Consider a section of the cone with radius 'x', distant 'y' from the base. Accordingly, if δV be the volume of this elemental section, then

$$\delta V = \pi x^2 \delta y$$

By geometrical considerations:

$$\frac{x}{r} = \frac{h-y}{h}$$

or $\qquad x = \frac{r(h-y)}{h}$

$$\therefore \qquad dV = \pi \left\{ \frac{r}{h}(h-y) \right\}^2 dy$$

$$\therefore \qquad V = \frac{\pi r^2}{h^2} \int_0^h (h^2 - 2hy + y^2)\, dy$$

$$= \frac{\pi r^2}{h^2} \left| h^2 y - \frac{2hy^2}{2} - \frac{y^3}{3} \right|_0^h$$

i.e. $\quad V = \dfrac{1}{3}\pi\, r^2 h$

or   Mass, $M = \dfrac{\rho}{3}\pi r^2 h$

Let $\overline{y} =$ distance of cone's centroid from its base.

i.e. $\quad \overline{y} = \dfrac{M_X}{V}$

Now δM_X which is the moment of the element about the base is given by

$$\delta M_X = \rho\, \pi\, x^2 \delta y(y)$$

$$\delta M_x = \frac{\rho\, \pi r^2}{h^2}(h^2 - 2hy + y^2)y\ \delta y$$

from which

$$M_X = \frac{\rho\, \pi r^2}{h^2} \int_0^h (h^2 y - 2hy^2 + y^3)\, dy$$

$$= \frac{\rho\, \pi r^r}{h^2} \left| \frac{h^2 y^2}{2} - \frac{2hy^3}{3} + \frac{y^4}{4} \right|_0^h$$

i.e. $\quad M_X = \dfrac{\rho\, \pi r^2}{h^2} \left(\dfrac{h^4}{2} - \dfrac{2h^4}{3} + \dfrac{h^4}{4} \right)$

$$= \frac{\rho \pi r^2 h^2}{12}$$

and because $\quad \bar{y} = \dfrac{M_X}{M}$

$$\bar{y} = \frac{\rho \pi^2 h^2}{12\rho} \cdot \frac{3}{\pi r^2 h}$$

or $\quad \bar{y} = \dfrac{h}{4} \qquad\qquad$ Q.E.D

Q22. Determine the coordinates of the centroid of the homogeneous solid configured above the X-Y plane and bounded by it, the plane $z = x$ and the cylinder $x^2 + y^2 = 100$.

Hint : Try $\displaystyle\int_0^{10} \int_0^{\sqrt{1-y^2}} \int_0^x dz\, dx\, dy$ to determine mass of solid and do the rest.

Ans : $\bar{x} = 20/3; \quad \bar{y} = 15/4; \quad \bar{z} = 10/3$

This page is intentionally left blank.

CHAPTER 4

SECOND MOMENT OF AREA AND SECOND MOMENT OF MASS

Inertia of matter, whether animal, vegetable or mineral is that property displayed by things inanimate and animate (including unfortunately among the latter a majority of the species homo sapiens as Nicolo Machiavelli observed many centuries ago), which causes them to want to remain in a state of unchangeableness, a tendency if you will to remain static, mentally and otherwise as the case may be. Restricting ourselves here for the moment to the inanimate, the prestidigitator who performs the spectacular act of pulling with a single jerk a tablecloth or sheet from under a glass of water resting on a table, without the glass spilling its contents, demonstrates the property of inertia. This happens because the frictional force between the glass and the tablecloth or sheet is insufficient to transmit to the glass enough force to cause it to move as the tablecloth or sheet does; and the glass prefers, to put it anthropomorphically, to stay at rest, and it does so, much to the amazement of the audience and to deafening applause and prodigious bows of acknowledgement.

The magician's act which you may try for yourself at home, (use inexpensive glassware), illustrates well a body's tendency to resist motion. But let us digress for a moment. Consider the well-known Newtonian equation $F = ma$ for the linear motion of a body of constant mass, say, 'm'. When we multiply both sides of this equation by 'r' the radial distance from the axis around which the body of mass 'm' rotates, this equation can be converted to its well known rotational counterpart 'T'$= mr^2\alpha$, angular acceleration 'α' being the linear acceleration 'a' divided by 'r' the radial arm, i.e. $\alpha = \dfrac{a}{r}$ or a = αr.

Clearly, if mr^2 is increased then the angular acceleration 'α' will not be maintained at its initial value unless the torque which is 'Fr', designated 'T', is correspondingly increased. As with the glass of water, where the magnitude of the frictional force between the tablecloth and the glass was insufficient to overcome the glass' resistance to move from rest, our body of mass 'm' as oriented has a tendency to resist rotation. In short, it is exhibiting its property of inertia. It is therefore incorrect to assert that the property displayed by the glass has no connection with the moment of inertia of, say, the cross-sectional area of a beam subjected to bending, the plane of such bending being perpendicular to the cross-sectional area. The cross-sectional area of the beam will, like the glass, tend to resist any rotation about is centroidal and neutral axis, about which more later.

The quantity mr^2 is the moment of inertia of the mass 'm' with reference to the rotational axis distant 'r' away from the centroid of the body. In engineering science, moments of inertia are properties not only of masses but also of areas. They are also called respectively, in reference to these, second moment of masses and second moment of area.

The engineering property 'Moment of Inertia' denoted 'I' is always expressed in reference to either a point or line or plane.

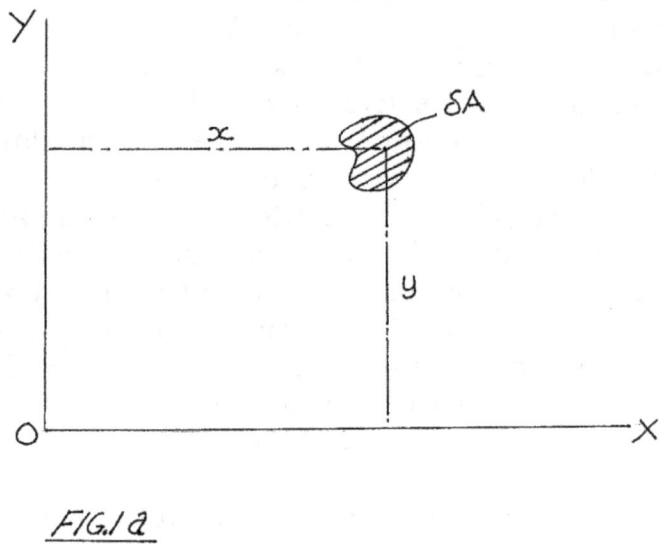

FIG.1a

In Fig. 1a the infinitesimal area 'δA' is situated in the X-Y plane. By definition, the infinitesimal moments of inertia 'δI_X' and 'δI_Y' of this area about the X-, and Y-axes are, respectively, given by:

$$\delta I_X = \delta A(y)^2 \quad \text{or} \quad I_X = \int y^2 dA, \quad \text{and}$$
$$\delta I_Y = \delta A(x)^2 \quad \text{or} \quad I_Y = \int x^2 dA,$$

That is, the moment of inertia of these infinitesimal areas are the products of the areas times their perpendicular distances from the relevant axis, squared.

Consider next Fig. 1b which represents a number of bodies made of different substances, but each of mass $m_1, m_2, m_3 ..... \; m_n$ and at perpendicular distances $r_1, r_2, r_3 ..... r_n$ respectively from a straight line $OX..$ The moments of inertia of these bodies about OX are $m_1 r_1^2, m_2 r_2^3, m_3 r_3^2, ... m_n r_n^2$ respectively from OX.

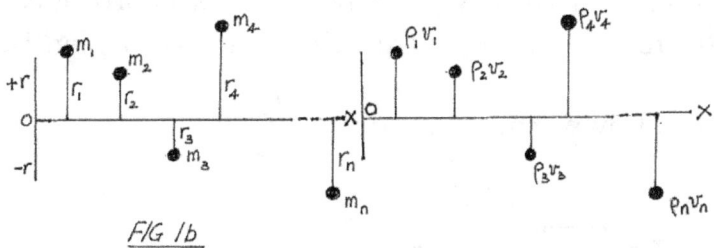

FIG 1b

Imagine all these bodies clustered together to form a single whole but with each part retaining its original distance from OX. Then the moment of inertia of the whole body I_{OX} with reference to axis o-x would be given by

$$I_{OX} = \sum_{i=1}^{n} m_i r_i^2$$

Mass implies volume. Therefore for the composite whole having constituent elements of volume $v_1, v_2, v_3, v_4 \ldots v_n$ and respective densities $\rho_1, \rho_2, \rho_3, \rho_4 \ldots \rho_n$, we may write

$$I_{OX} = \sum_{i=1}^{n} \rho_i \, v_i, \ r_1^2$$

Because some of the distances 'r' in Fig. 1b are negative is of no consequence because in the relationship for I_{OX} distances are squared anyway.

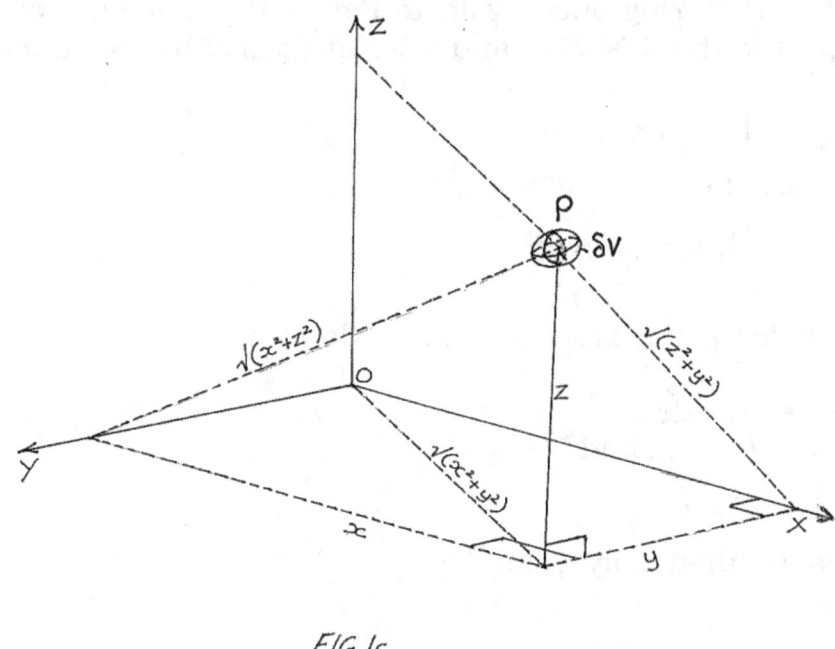

FIG 1c

Imagine now a continuous body having volume 'V' and density varying throughout the space it occupies, i.e. $\rho = f(x, y, z)$. Thus we may write for density at point, say, P within 'V', $\rho = \rho(x, y, x)$. Accordingly 'δM' the mass of an infinitesimal volume 'δV' at P within the space occupied by V is $\rho(x, y, z)\delta V$.

In Fig. 1c I have shown the distances of P from the three rectangular coordinate axes:

$$\text{From } OZ : \sqrt{z^2 + y^2}; \quad \text{From } OY : \sqrt{z^2 + x^2} \text{ and}$$

$$\text{From } OZ : \sqrt{x^2 + y^2}. \quad \text{Accordingly,}$$

$$I_{OX} = \iiint \rho(x, y, z)dV \left\{ \sqrt{(z^2 + y^2)} \right\}^2$$

$$= \iiint (z^2 + y^2)\rho(x, y, z)dV$$

And by similar token,

$$I_{OY} = \iiint (z^2 + x^2)\rho(x, y, z)dV$$

$$I_{OZ} = \iiint (x^2 + y^2)\rho(x, y, z)dV$$

Note that I_{OX}, I_{OY} and I_{OZ} are moments of inertia with reference to the coordinate axes. Referring once again to Fig 1c the moments of inertia of the body with respect to the XY-, XZ-, and YZ-planes may be stated as

$$I_{XY} = \iiint z^2 \rho(x, y, z)dV$$
$$I_{XZ} = \iiint y^2 \rho(x, y, z)dV, \text{ and}$$
$$I_{YZ} = \iiint x^2 \rho(x, y, z)dV$$

Inspection of the foregoing expressions reveals that:

$$I_{OX} = I_{XY} + I_{XZ}$$
$$I_{OY} = I_{XY} + I_{YZ}, \text{ and}$$
$$I_{OZ} = I_{XZ} + I_{YZ}$$

represented diagrammatically in Fig. 1d

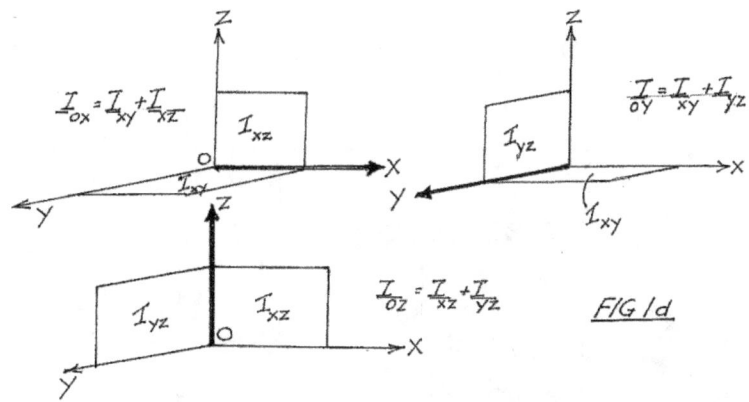

$I_{ox} = I_{xy} + I_{xz}$ I_{xz}

I_{xy}

$\dfrac{I}{oy} = \dfrac{I}{xy} + I_{yz}$

I_{yz}

I_{xy}

I_{yz} I_{xz}

$\dfrac{I}{oz} = \dfrac{I}{xz} + \dfrac{I}{yz}$

FIG ld

Please do not be fazed by these equations involving multiple integrals. In the preceding Chapter on Centroids you were shown how such integrals are easily solved.

Units

Second moment of area is generally expressed in terms of metre or centimetre to the fourth power, i.e. m^4 or $(cm)^4$ respectively; not uncommonly in $(mm)^4$. For second moment of mass, the units are $kg(m)^2$ or $kg(cm)^2$ or $kg(mm)^2$.

ILLUSTRATIVE EXAMPLES

Let us demonstrate applications of the foregoing for the purpose of illustration.

Illustrative Example 1

Determine I_{XX} and I_{YY} for the rectangular area shown in Fig. 2. O is the centroid of the rectangle and the origin of the axes.

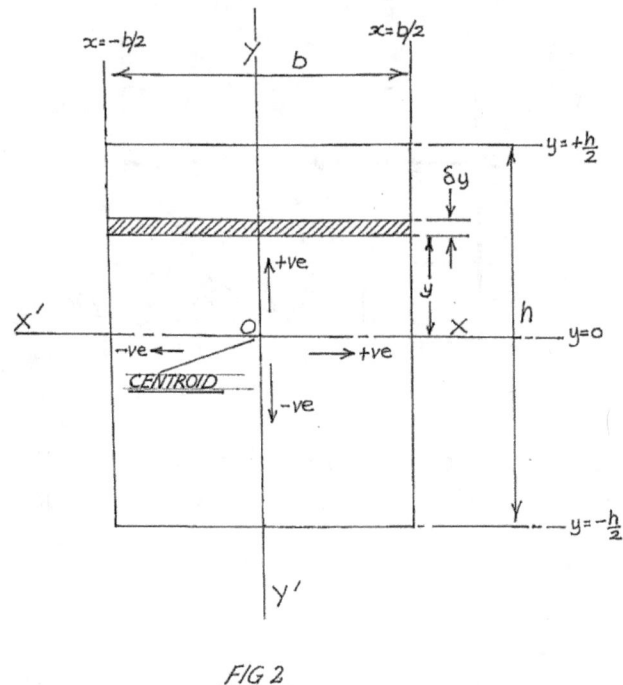

FIG 2

For the horizontal-strip element shown hatched in the diagram, let δA be its area.

$$\therefore \qquad \delta A = b \; \delta y$$

The distance between the strip and the XOX'-axis is $'y'$

Therefore $\quad \int dI_{XX} = \int dA y^2$

becomes $\quad I_{XX} = \int b \, dy(y^2)$

The limits of integration are: $\quad y = -\dfrac{h}{2}; \quad y = +\dfrac{h}{2}$

$$\therefore \qquad I_{XX} = b \int_{-h/2}^{h/2} y^2 dy = b \left[\frac{y^3}{3} \right]_{-h/2}^{h/2}$$

$$= b \left\{ \frac{h^3}{24} - \left(-\frac{h^3}{24} \right) \right\}$$

i.e. $\quad I_{XX} = \dfrac{bh^3}{12}$

Clearly if we considered a vertical strip of thickness dx, then for its area, say, δA_h

$$\delta A_h = h \; \delta x$$

Therefore $\int \delta I_{YY} = \int dA x^2$ becomes the moment of inertia of the rectangular area about the Y-axis.

$$I_{YY} = \int h dx \cdot x^2$$

'x' moves from $-\dfrac{b}{2}$ to $+\dfrac{b}{2}$

$$\therefore \quad I_{YY} = h \int_{-b/2}^{b/2} x^2 dx \qquad I_{YY} = h\int_{-b/2}^{b/2} x^2 dx$$

$$= \left[\frac{x^3}{3}\right]_{-b/2}^{b/2}$$

$$I_{YY} = \frac{hb^3}{12}$$

Illustrative Example 2

Let us consider next the triangular area shown in Fig. 3 and determine its moment of inertia about the *XOX'* -axis through its centroid.

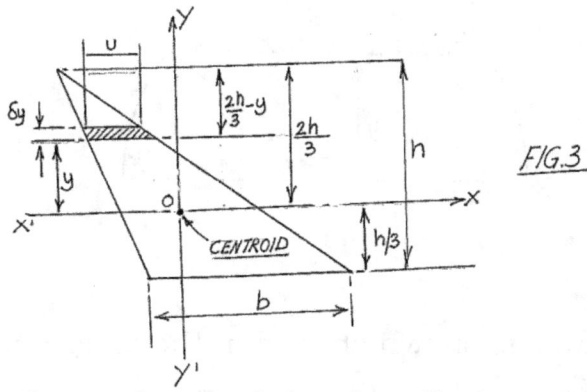

FIG.3

The elemental strip shown hatched in the diagram has a width *'u'* and thickness δy.

Accordingly, $\delta A = u \, \delta y$

and

$$I_{XX} = \int u \, dy \cdot y^2$$

But, $\qquad \dfrac{u}{b} = \left(\dfrac{2h}{3} - y\right)/h$

or $\qquad u = \dfrac{b}{h}\left(\dfrac{2h}{3} - y\right)$

$$\therefore \quad I_{XX} = \frac{b}{h}\int \left(\frac{2h}{3} - y\right) y^2 dy$$

The radius arm *'y'* moves from $y = -\dfrac{h}{3}$ to $\dfrac{2h}{3}$

$$\therefore \quad I_{XX} = \frac{b}{h}\int_{-h/3}^{\frac{2h}{3}}\left(\frac{2h}{3}-y\right)y^2\,dy$$

$$= \frac{b}{h}\left[\frac{2hy^3}{9}-\frac{y^4}{4}\right]_{-h/3}^{2h/3}$$

i.e. $\quad I_{XX'} = \dfrac{bh^3}{36}$

Illustrative Example 3

We now set about to determine the moment of inertia of a semi-circular area of radius 'r' about a diameter sitting on the XOX' axis, as shown in Fig. 4

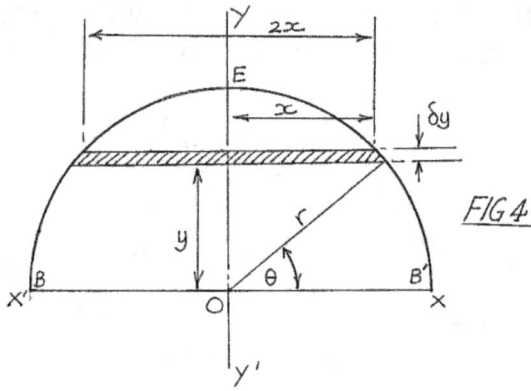

FIG 4

The elemental area $2x\,\delta y$ is shown shaded in the diagram.

By definition, $I_{XOX'} = \int 2x\,dy\cdot y^2$

This integral may be evaluated in at least three ways:

First way: By writing $\dfrac{y}{r} = Sin\theta$ and $\dfrac{x}{r} = Cos\theta$, I_{XX} becomes

$$= 2\int rCos\theta\cdot r^2 Sin^2\theta\,dy$$

Also, because $\quad y = rSin\theta,\ dy = rCos\theta\,d\theta$

$$\therefore \quad I_{XOX'} = 2r^4\int Cos\theta\cdot Sin^2\theta\,Cos\theta\,d\theta$$

$$= 2r^4\int Sin^2\theta\,Cos^2\theta\,d\theta$$

Putting $Sin^2\theta = 1-Cos^2\theta$

$$I_{XOX'} = 2r^4\int(1-Cos^2\theta)Cos^2\theta\,d\theta$$

$$= 2r^4\int Cos^2\theta - Cos^4\theta)d\theta$$

To sweep the entire area of the semi-circle, the elemental strip has to move from BB' to E; that is 'θ' goes from $\theta = 0$ to $\theta = \dfrac{\pi}{2}$.

Accordingly,

$$I_{XOX'} = 2r^4 \int_0^{\frac{\pi}{2}} (Cos^2\theta - Cos^4\theta)d\theta$$

$$= 2r^4 \left[\int_0^{\frac{\pi}{2}} Cos^2\theta \, d\theta - \int_0^{\pi} Cos^4\theta \, d\theta \right]$$

Now

$$\int Cos^2 \, d\theta = \frac{1}{2} Cos\theta \, Sin\theta + \frac{1}{2}\theta$$

and

$$\int Cos^4 \, d\theta = \frac{1}{4} Cos^3\theta \, Sin\theta + \frac{3}{4} \int Cos^2\theta \, d\theta$$

$$= \frac{1}{4} Cos^3\theta \, Sin\theta + \frac{3}{4}\left(\frac{1}{2} Cos\theta \, Sin\theta + \frac{1}{2}\theta \right)$$

$$\therefore \quad I_{XOX^1} = 2r^4 \left[\frac{1}{2} Cos\theta \, Sin\theta + \frac{\theta}{2} - \frac{1}{4} Cos^3\theta \, Sin\theta \right.$$

$$\left. - \frac{3}{8} Cos\theta \, Sin\theta - \frac{3}{8}\theta \right]_0^{\pi/2}$$

$$= 2r^4 \left(\frac{\pi}{4} - \frac{3\pi}{16} \right)$$

So that,

$$I_{XOX'} = \frac{\pi r^4}{8}$$

Second way: By substituting $x = \sqrt{r^2 - y^2}$

$$I_{XOX'} = 2\int y^2 \sqrt{(r^2 - y^2)}dy$$

The limits of y are o and r

$$\therefore \quad I_{XOX'} = 2\int_o^r y^2 \sqrt{(r^2 - y^2)}dy$$

Employing the standard form for $\int y^2 \sqrt{(r^2 - y^2)}dy$ we have,

$$\int y^2 \sqrt{(r^2 - y^2)}dy = -\frac{y}{4}\sqrt{(r^2 - y^2)^3} + \frac{r^2}{8}\left(y\sqrt{r^2 - y^2} + r^2 Sin^{-1}\frac{y}{r} \right)$$

so that

$$2\int_o^r y^2 \sqrt{r^2 - y^2}dy = 2\left[-\frac{y}{4}\sqrt{(r^2 - y^2)^3} + \frac{r^2}{8}\left(y\sqrt{r^2 - y^2} + r^2 Sin^{-1}\frac{y}{r} \right) \right]$$

$$= 2\frac{r^2}{8} \cdot r^2 Sin^{-1} 1 = 2 \cdot \frac{r^4}{8} \cdot \frac{\pi}{2}$$

$$\therefore \quad I_{XOX'} = \frac{\pi r^4}{8}, \quad \text{check}$$

Third way:

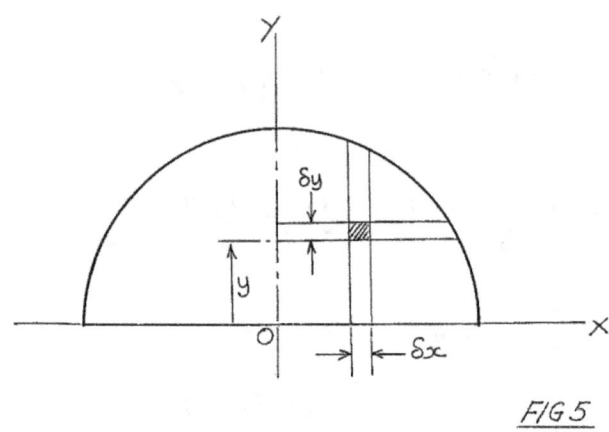

FIG 5

With reference to Fig. 5 we employ double integrals. The area of the shaded element is $\delta y\, \delta x$. Therefore the moment of inertia of this element about the X-axis is:

$$I_{XOX'} = \int\int dy\, dx \cdot y^2$$

The limits are, for y : from $y = 0$ to $y = \sqrt{r^2 - x^2}$ and for x : from $x = -r$ to $x = +r$.

Hence

$$I_{XOX'} = \int_{-r}^{+r} \int_{o}^{\sqrt{r^2-x^2}} [y^2 dy] dx$$

$$= \int_{-r}^{+r} \left[\frac{y^3}{3}\right]_{o}^{\sqrt{r^2-x^2}} dx$$

i.e.

$$I_{XOX'} = \int_{-r}^{+r} \left[\frac{(r^2 - x^2)^{3/2}}{3}\right] dx$$

Substituting

$$x = r Cos\theta \;;\; dx = -r Sin\theta\, d\theta$$

∴

$$I_{XOX'} = -\int_{-r}^{+r} r^3 \left(\frac{Sin^4\theta}{3}\right) r - Sin\theta\, d\theta$$

$$= -\frac{r}{3}\int_{-r}^{+r} Sin^4\theta\, d\theta$$

Let us change the limits of integration.

Using the substitution: $x = r Cos\theta$, we deduce that

when $x = -r$: $Cos\theta = -1$, therefore $\theta = \pi$; and
when $x = +r$: $Cos\theta = 1$, so that $\theta = 0$

Accordingly with the new limits:

$$I_{XOX'} = -\frac{r^2}{3}\int_{\pi}^{0} Sin^4\theta\, d\theta$$

i.e. $\quad I_{xox'} = +\frac{r^4}{3}\int_{0}^{\pi} Sin^4\theta\, d\theta$

Employing the reduction formula

$$I_n = \int Sin^n\theta\, d\theta = -\frac{1}{n} Sin^{n-1}\theta\, Cos\theta + \left(\frac{n-1}{n}\right) I_{m-2}$$

We get

$$\int Sin^4\theta\, d\theta = -\frac{1}{4} Sin^3\theta\, Cos\theta + \frac{3}{4} I_2$$

and $\quad \int Sin^2\theta\, d\theta = -\frac{1}{2} Sin\theta\, Cos\theta + \frac{1}{2} I_0$

$$= -\frac{1}{2} Sin\theta\, Cos\theta + \frac{1}{2}\theta$$

$\therefore \quad \int Sin^4\theta\, d\theta = -\frac{1}{4} Sin^3\theta\, Cos\theta + \frac{3}{4}\left(-\frac{1}{2} Sin\theta\, Cos\theta + \frac{1}{2}\theta\right)$

$$= -\frac{1}{4} Sin^3\theta\, Cos\theta - \frac{3}{8} Sin\theta\, Cos\theta + \frac{3\theta}{8}$$

Accordingly,

$$I_{XOX'} = \frac{r^4}{3}\left[-\frac{1}{4} Sin^3\theta\, Cos\theta - \frac{3}{8} Sin\theta\, Cos\theta + \frac{3}{8}\theta\right]_{0}^{\pi}$$

$$= \frac{r^4}{3}\left(0 - 0 + \frac{3\pi}{8}\right)$$

thus giving $I_{XOX'} = \dfrac{\pi r^4}{8}$, check

RADIUS OF GYRATION

It is just as difficult, perhaps moreso, to attach a physical meaning to the term radius of gyration as it is to the concept of "moment of inertia". One could imagine it to be a distance from a reference axis at which an area or mass is concentrated, with the value of I remaining the same. For example,

$$I_X = Ak_x^2 \quad or \quad k_X = \sqrt{\frac{I_x}{A}},$$

$$I_Y = Ak_y^2 \quad or \quad k_y = \sqrt{\frac{I_Y}{A}}$$

$$I_z = Ak_z^2 \quad or \quad k_z = \sqrt{\frac{I_z}{A}}$$

in which $\quad k_x$ = radius of gyration of area about *X-axis*

$$k_y = \text{radius of gyration of area about } Y\text{-}axis$$
$$k_z = \text{radius of gyration of area about } Z\text{-}axis$$
$$I_X = \text{second moment of area about } X\text{-}axis$$
$$I_Y = \text{second moment of area about } Y\text{-}axis$$
$$I_z = \text{second moment of area about } Z\text{-}axis$$
$$A = \text{area}$$

PARALLEL AXIS THEOREM

This theorem allows determination of the second moment of area about another axis that is parallel to the centroidal axis and at any distance, say, '*d'* from the latter. To demonstrate this theorem, consider the area *ABC* in Fig. 6.

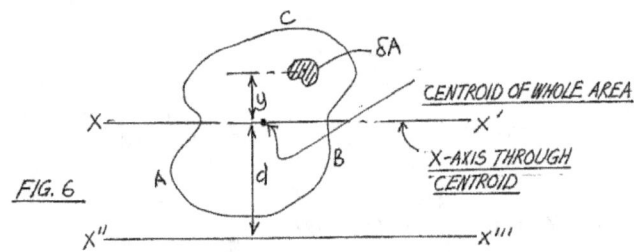

FIG. 6

Axis *XOX'* passes through the centroid of Fig. 6: area *ABC*, of which 'δA' is an elemental area. Accordingly from first principles, we know that $I_{XX'} = \int dA\, y^2$.

In order to determine *I* about the axis *X"X"'* in the same plane as the area *ABC*, parallel to *XOX'* but distant 'd' from it we may write

$$I_{X''X'''} = \int (y+d)^2 dA$$
$$= \int y^2 dA + \int 2yd\, dA + \int d^2 dA$$
$$= I_{XOX'} + 2d \int y\, dA + d^2 A$$

But
$$\bar{y} = \frac{\int y\, dA}{A} \quad \text{or} \quad \int y\, dA = A\bar{y}$$

$$\therefore \quad I_{X''X'''} = I_X + 2d\, A\bar{y} + Ad^2$$

Because $\bar{y}$ locates the distance of the centroid from the centroidal axis which is the reference axis in this case, $\bar{y} = 0$. This enables us to write:

$$I_{X''X'''} = I_{XX'} + Ad^2$$

That is, the second moment of area about an axis parallel to the centroidal axis is equal to the second moment of area about the centroidal axis plus the area times the perpendicular distance between the axes, squared. For example, it was shown that the second moment of area of a triangle about an axis through its centroid is $bh^3/36$. According to the Parallel Axis Theorem, the second moment of area about an axis through its base distant $\dfrac{h}{3}$ away and parallel with the axis through the centroid is:

$$\frac{bh^3}{36} + \frac{bh}{2}\left(\frac{h}{3}\right)^2 = \frac{bh^3}{12}$$

It should be noted that although the two axes involved must be parallel to each other, it is not necessary for them to be in the same plane, as I shall now demonstrate

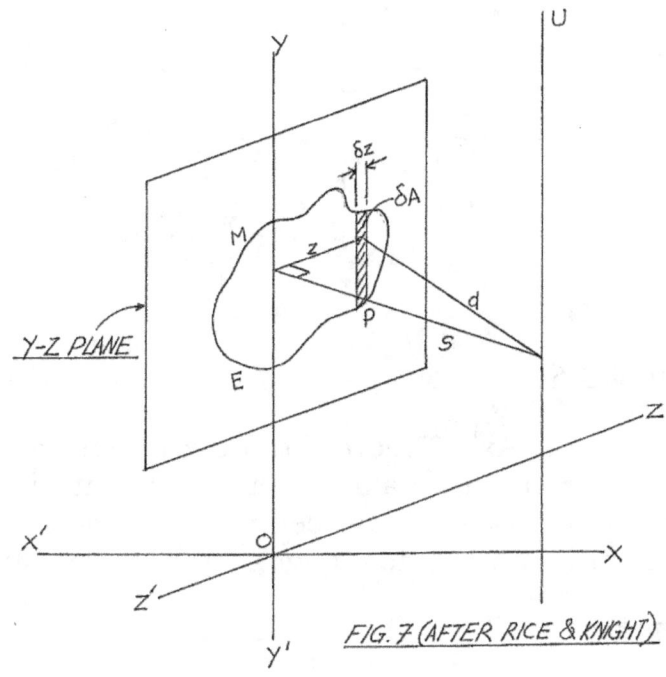

FIG. 7 (AFTER RICE & KNIGHT)

In Fig. 7 the area 'EPM' is in the YZ plane; the axis YOY' its the centroidal axis. By definition therefore, the second moment of area of elemental area δA in the YZ-plane about axis YOY' is $\int z^2\, dA$. Distance 'd' separates element 'δA' from the axis 'U' in the plane distant 's' from the centroidal axis YOY' in the yz plane. From geometrical considerations:

$$d^2 = s^2 + z^2$$

Now, $I_U = \int d^2\, dA$

Therefore $I_U = \int (s^2 + z^2)dA$

i.e. $I_U = \int s^2 dA + \int z^2\, dA$

or $I_U = I_{yoy\prime} + As^2$

~ 309 ~

In words, the second moment of area of an area 'A' about an axis in a plane distant 'S' from the centroidal axis of the area 'A' in another plane, the two axes being parallel to each other, is given by the second moment of area with reference to the centroidal axis plus the plus the product of the area times the distance between the two axes.

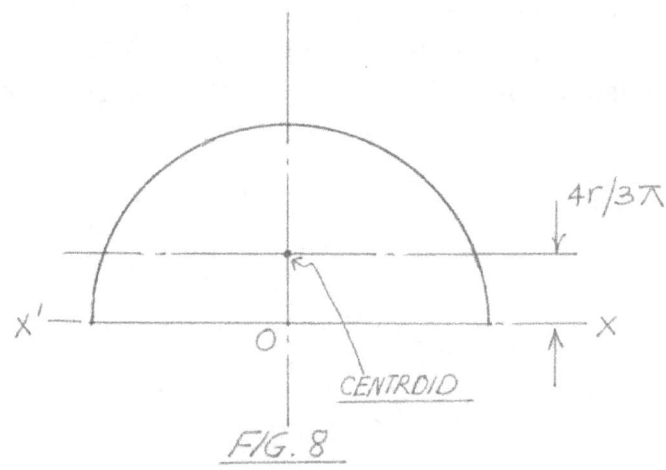

FIG. 8

Illustrative Example 4

Let us apply the Parallel Axis Theorem to determination of the second moment of area of a semi-circle about an axis through its centroid. Refer to Fig. 8. It is known from previous work that the centroid of a semi-circle is at a distance equal to $4r/3\pi$ from its diametrical edge. Also, it was shown earlier that for a semi-circle $I_{XX'} = \dfrac{\pi r^4}{8}$

Therefore

$$\frac{\pi r^4}{8} = I_c + A\left(\frac{4r}{3\pi}\right)^2, \ A \text{ being area.}$$

$$\therefore \quad I_C = \frac{\pi r^4}{8} - \frac{\pi r^2}{2}\left(\frac{16r^2}{9\pi^2}\right)$$

i.e. $\quad I_C = \dfrac{r^4(9\pi^2 - 64)}{72\pi}$

This is the expression for the second moment of area of a semi-circle about its centroid.

For a continuous body of mass 'M', its second moment of mass with reference to its centroid may, as in the case of area just considered, be transferred to an axis parallel to the centroidal axis by employing:

$$I_{U'} = I_{\text{centroidal axis}} + Md^2$$

where $I_{U'}$ = moment of inertia with reference to the parallel axis, and

d = distance between the axes

Illustrative Example 5

Derive relationships for the second moment of area and radius of gyration of a quadrant of a circle of radius 'r' about one edge of the quadrant

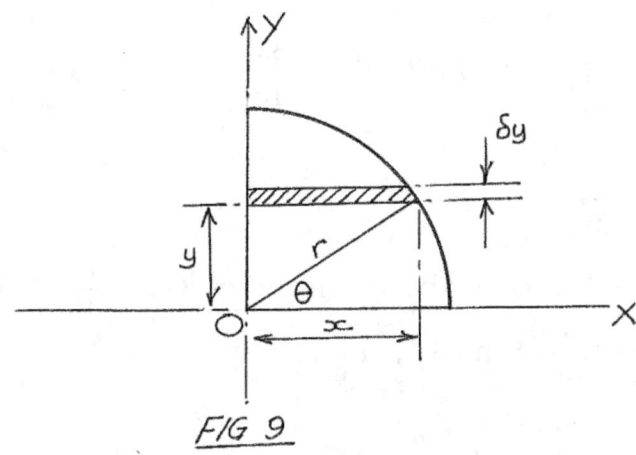

FIG 9

With reference to Fig. 9

$$\delta A = x \, \delta y$$

By definition $I_{OX} = \int y^2 da = \int y^2 x \, dy$ being the second moment of area about OX.

Now $y = rSin\theta$; $x = rCos\theta$

and $dy = rCos\theta \, d\theta$

$$\therefore \quad I_{OX} = \int r^2 Sin^2\theta \cdot rCos\theta . rCos\theta \, d\theta$$

The limits of integration are $\theta = 0$; $\theta = \dfrac{\pi}{2}$

i.e. $$I_{OX} = r^4 \int_0^{\pi/2} Sin^2 Cos^2 \cdot d\theta$$

$$= r^4 \int_0^{\pi/2} Sin^2 (1 - Sin^2\theta) d\theta$$

~ 311 ~

$$= r^4 \left[\int_0^{\pi/2} Sin^2\theta \, d\theta - \int_0^{\pi/2} Sin^4\theta \, d\theta \right]$$

But $\quad \int_0^{\pi/2} Sin^2 d\theta = \left[-\frac{1}{2} Sin\theta \, Cos\theta + \frac{1}{2}\theta \right]_0^{\frac{\pi}{2}} = \frac{\pi}{4}$

also $\quad \int_0^{\pi/2} Sin^4\theta \, d\theta = \left[-\frac{1}{4} Sin^3\theta \, Cos\theta \right]_0^{\frac{\pi}{2}} + \frac{3}{4} \int_0^{\pi/2} Sin^2\theta \, d\theta$

$$= 0 + \frac{3}{4} \left[-\frac{1}{2} Sin\theta \, Cos\theta + \frac{1}{2}\theta \right]_0^{\frac{\pi}{2}}$$

$$= \frac{3\pi}{16}$$

$$r^4 \left[\int_0^{\pi/2} Sin^2 d\theta - \int_0^{\pi/2} Sin^4\theta \, d\theta \right] = r\left(\frac{\pi}{4} - \frac{3\pi}{16} \right)$$

$$= \frac{\pi r^4}{16}$$

that is to say, the value of I about the edge $OX = \dfrac{\pi r^4}{16}$. If we designate the radius of gyration about OX as k_X, then

$$Ak_x^2 = \frac{\pi r^4}{16} \; ; \text{ but } \quad A = \frac{\pi r^2}{4}$$

$\therefore \qquad k_X = r/2$

Illustrative Example 6

Derive relationships for the second moment of area and radius of gyration of the ellipse given by $\dfrac{x^2}{a^2} + \dfrac{y^2}{b^2} = 1$, about the *X-axis* which is the major axis.

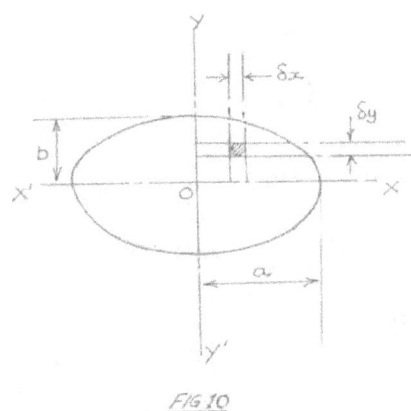

FIG 10

Let us develop these relationships using double integrals to determine the second moment of area about the *X-axis*. Let us deal with one quadrant of the ellipse

In Fig. 10, $\delta A = \delta y\ \delta x$

$\therefore\qquad \delta I = \delta y\ \delta x \cdot y^2$

i.e. $I = \int\int (y^2 dy)dx$

The limits of integration are:

For 'y' : from $y = 0$ to $y = b\sqrt{\left(1 - \dfrac{x^2}{a^2}\right)}$

For 'x' : from $x = 0$ to $x = a$

For the quadrant

$$I_x = \int_0^a \left[\int_0^{b\sqrt{\left(1-\frac{x^2}{a^2}\right)}} y^2 d^2 \right] dx$$

$$= \int_0^a \left[\frac{y^3}{3} \right]_0^{b\sqrt{\left(1-\frac{x^2}{a^2}\right)}} dx$$

$$= \int_0^a \frac{b^3}{3}\left(1 - \frac{x^2}{a^2}\right)^{\frac{3}{2}} dx$$

Putting $x = aCos\theta;\quad dx = -aSin\ dx$

$\therefore\qquad I_x = \dfrac{b^3}{3}\int_0^a Sin^3\theta - aSin\theta\ d\theta$

$$= -\frac{ab^3}{3}\int_0^a Sin^4 d\theta$$

When $x = a$; $Cos\theta = 1$, i.e. $\theta = 0$

When $x = 0$; $Cos\theta = 0$, i.e. $\theta = \dfrac{\pi}{2}$

$$\therefore \quad I_x = -\frac{ab^2}{3} \int_{\frac{\pi}{2}}^{0} Sin^4\theta \; d\theta$$

$$= \frac{ab^3}{3} \int_{0}^{\frac{\pi}{2}} Sin^4\theta \; d\theta$$

Now

$$\int_{0}^{\frac{\pi}{2}} Sin^4\theta \; d\theta - \frac{1}{4} Sin^3\theta \; Cos\theta + \frac{3}{4} \int_{0}^{\frac{\pi}{2}} Sin^2\theta \; d\theta$$

and

$$\int_{0}^{\frac{\pi}{2}} Sin^2\theta \; d\theta = -\frac{1}{2} Sin\theta \; Cos\theta + \frac{1}{2}\theta$$

$$\therefore \quad \int_{0}^{\frac{\pi}{2}} Sin^4\theta \; d\theta = -\frac{1}{4} Sin^3\theta \; Cos\theta + \frac{3}{4}\left(-\frac{1}{2} Sin\theta \; Cos\theta + \frac{1}{2}\theta\right)$$

$$= 0 + \left[\frac{3}{8}\theta\right]_{}^{\frac{\pi}{2}}$$

$$= \frac{3\pi}{16}$$

$$\therefore \quad I_X = \frac{ab^3}{3} \cdot \frac{3\pi}{16}$$

$$I_x = \frac{ab^3\pi}{16} \quad \text{for one quadrant}$$

Accordingly, for the entire ellipse we must multiply this result by *4*. Therefore the second moment of area of the ellipse shown in Fig. 10, about its major axis, is $\pi ab^3/4$. And because the area of the ellipse is πab, its radius of gyration k_x is obtained from

$$(\pi ab)k_x^2 = ab^3 \frac{\pi}{4}, \quad \text{i.e.} \quad k_x = \frac{b}{2}$$

POLAR MOMENT OF INERTIA

The polar moment of inertia of an area or body is the moment of inertia with reference to an axis perpendicular to the plane of the area. This axis, called the polar axis may or may not be at the origin of the orthogonal axes in the plane of the area.

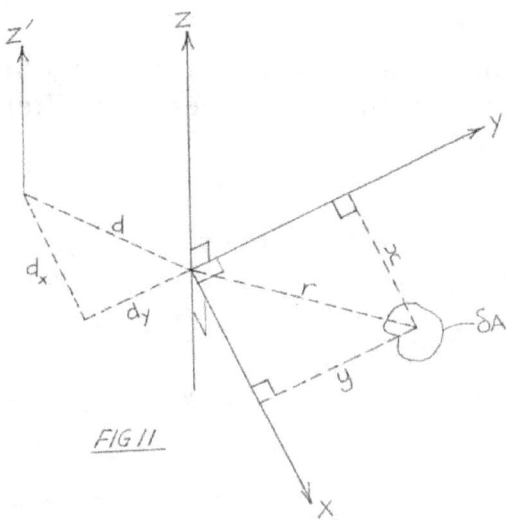

FIG 11

The symbol commonly used to represent polar moment of inertia is 'J'. But you may come across 'J' the symbol also used to designate torsional constant. You shall encounter it in the chapter on Torsion.

With reference to the diagram shown as Fig. 11, and according to definition:

$$J = \int r^2 \, dA$$
$$= \int \left(x^2 + y^2 \right) dA = \int x^2 \, dA + \int y^2 \, dA$$

i.e. $\quad J = I_x + I_y$

If we wish to transfer the polar moment of inertia to another axis parallel to the *Z-axis*, say Z', in Fig. 11 then

$$J_{Z'} = I_x + A\left(d_y\right)^2 + I_Y + A\left(d_x\right)^2$$

in which d_Y and d_X are respectively the distances of the new axis from the *Y-* and *X-axes*, i.e. $d^2 = d_y^2 + d_x^2$

Hence,

$$J_{Z'} = I_X + I_Y + Ad^2$$

or $\quad J_{Z'} = J + Ad^2$, where

$\quad J_{Z'} = $ polar moment of inertia about new axis Z';

$\quad A \;=\;$ area; and

$\quad d \;=\;$ distance between the two axes parallel to each other and perpendicular to the *XY-plane*.

Illustrative Example 7

Determine the polar moment of inertia of a rectangular area of height 'h' and base 'b' about an axis passing through its centroid. Refer to Fig. 12

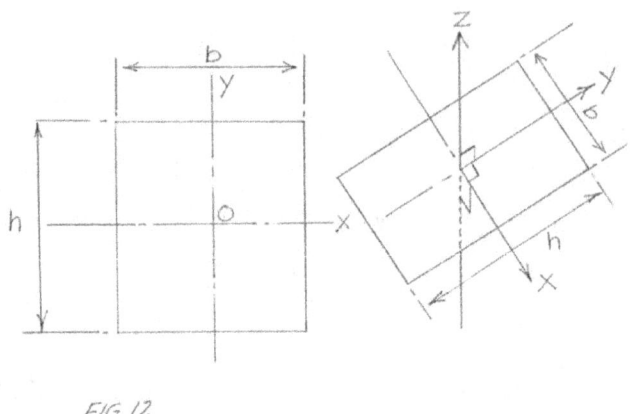

FIG. 12

We know from previous working that $I_{XX} = \dfrac{bh^3}{12}$ and $I_{YY} = \dfrac{hb^3}{12}$, for a rectangular area 'b' by 'h' as indicated.

$$\therefore \qquad J = I_{XX} + I_{YY} = \frac{bh^3}{12} + \frac{hb^3}{12} = \frac{bh}{12}(h^2 + b^2)$$

That's it!

ADDING AND SUBTRACTING COMPONENT SECOND MOMENTS OF AREA IN COMPOSITE AREAS

As the above descriptive heading suggests, a composite area is one which is generally composed of areas of different geometrical shapes; some of the geometrical shapes within the composite area may even be voided. The value of 'I' about the centroidal axis of a composite area is the sum or difference as the case may be of the second moment of area of each component geometrical shape about its own centroidal axis, plus (or minus in the case of a void) the area of each such component times the distance between the component's centroidal axis and the centroidal axis of the entire composite area, squared.

Illustrative Example 8

Let us put these words into action by determining the value of 'I' about the centroidal axis of a composite area comprising a semi-circular area of radius 'r' = 10cm joined to a rectangular area of length = 20cm and depth 'h' = 15cm, as shown in Fig. 13.

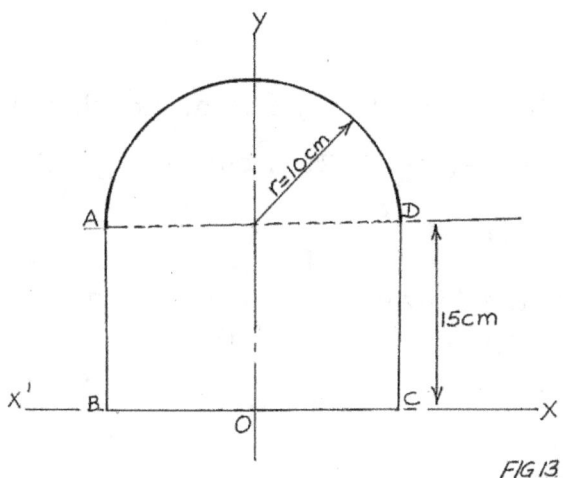

FIG 13

The first requirement is determination of the position of the centroid of the entire composite area; from say the *XX'*-axis first.

Area of semi-circle $\dfrac{\pi r^2}{2} = \dfrac{100\pi}{2} = 157.1 cm^2$

Distance of centroid of semi-circular area from *XX'*- axis $= \dfrac{4r}{3\pi} + 15cm$

$$= \dfrac{40}{3\pi} + 15cm = 19.2cm$$

(c)   Area of rectangle = *300cm²*

Distance of centroid of rectangular area from *XX'*-axis = *7.5cm*

∴   Distance of centroid of composite area from *XX'*

$$= \dfrac{157.1(19.2) + 300(7.5)}{157.1 + 300}$$

$$= 11.52cm$$

The composite area is symmetrical about the axis *YY'*.

It was shown earlier that for a semi-circular area of radius 'r', the value of I_c with reference to its centroid is:

$$I_c = \dfrac{r^4(9\pi^2 - 64)}{72\pi}$$

∴   For $r = 10cm$

$$I_c = \dfrac{10^4(88.85 - 64)}{226.2}$$

$$I_c = 1098.5(cm)^4$$

The distance between the centroid of the semi-circular area and the centroid of the composite area = $\dfrac{4(10)}{3\pi} + (15 - 11.52)$.   Refer to Fig. 10

= *7.68cm*

Therefore by employing the Parallel Axis Theorem, the second moment of area of the semi-circular component $I_{COMP'}$ with reference to the centroid of the composite area, is given by:

$$I_{COMP'} = I_c + Ad^2$$

where I_c = second moment of are of semi-circular component about its
own centroid.

   A = area of semi-circular component
   d = distance between centroid of semi-circular area and centroid of composite area.

Here this sum is given by:

$$1098.5(cm)^4 + \frac{\pi(10)^2}{2}(7.68)^2$$

$$= (1098.5 + 9269)cm^4$$

i.e. $I_{COMP'} = 10,367.5cm^4$

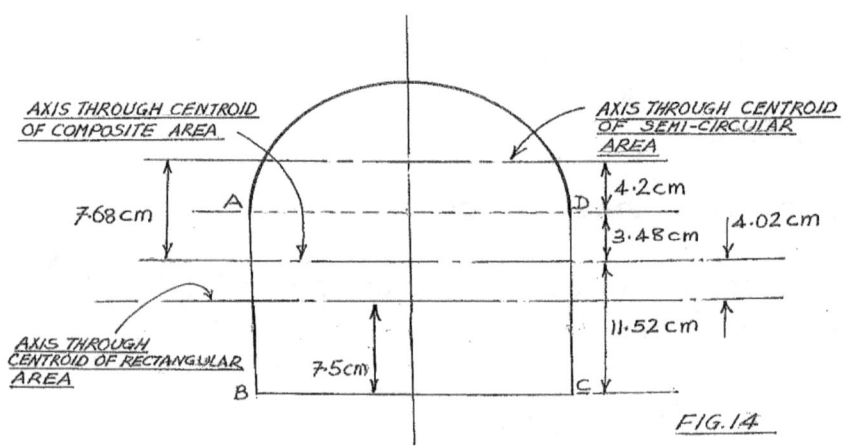

FIG. 14

Repeating the same process for the rectangular area, its second moment of area about its own centroidal axis is $\dfrac{20(15)^3}{12} = 5625cm^4$

Referring to Fig. 14, it is seen that its centroid is at a distance of *4.02cm* from the centroidal axis of the composite area.

Accordingly its second moment of area $I_{COMP'}$ about the composite area's centroid at axis is:

$$= I_c + bh(11.52 - 7.5)^2$$
$$= 5625 + 20(15)(4.02)^2$$
$$= (5625 + 4848)cm^4$$

i.e. $I_{COMP'} = 10473cm^4$

∴ Total second moment of area of the composite area about the composite area's centroidal axis is *(10,367.5 + 10473)cm⁴ = 20,840.5cm⁴.* Evidently, to determine the second moment of the composite area about *XX'* we again employ the Parallel Axis Theorem. In this case the centroid is 11.52cm from the a *XX'*-axis. Therefore

$$I_{XX'} = 20,840.5 + \text{Composite Area } (11.52)^2$$
$$= 28,840.5 + (157.1 + 300)(1.52)^2$$
$$= (20,840.5 + 60662)cm^4$$
$$I_{XX'} = 81,502.5\ cm^4, \text{say, } 81.503cm^4$$

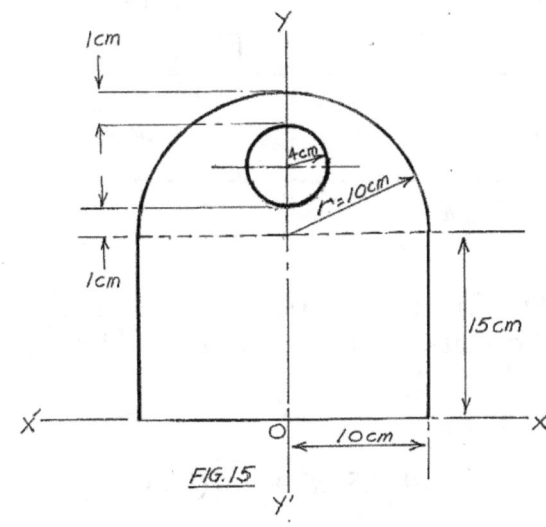

FIG. 15

Now, suppose a circular area of radius = *4cm* is cut out of the semi-circular portion as shown in Fig. 15. What is the second moment of area of the resultant composite area about its centroidal axis?

We proceed as follows:

Area of semi-circular area = *157.1cm² as before*
" " voided circular area = *16π = 50.3cm²*

~ 319 ~

Area of rectangular area $\qquad = 300m^2$

$\therefore$ Distance of centroid of composite area from XX'-axis

$$= \frac{157.1(19.2) - 16\pi(20) + 300(7.5)}{157.1 - 16\pi + 300}$$

$$= \frac{3016 - 1005 + 2250}{407}$$

$$= 4261/407$$
$$= 10.5cm$$

Thus, the centroid of the composite area with the voided circular area is at a distance of 10.5cm from the XX'-axis. See Fig. 16.

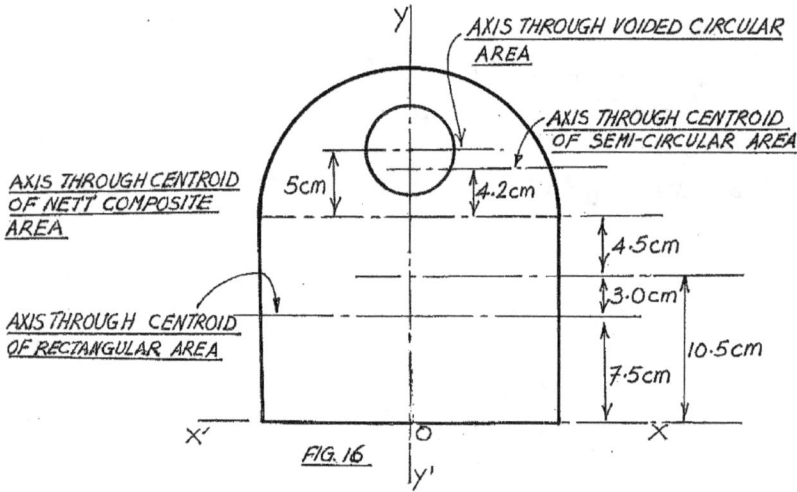

FIG. 16

Applying the Parallel Axis Theorem in order to determine the second moment of area of the composite about its centroidal axis, we have:

For the semi-circular area:

$$1098.5cm^4 + \frac{\pi r^2}{2}(8.7)^2$$

$$(1098.5 + 11891)\ cm^4 = 12989.5cm^4$$

For the circular cut out:

$$\left\{ \frac{\pi(4)}{4} + \pi(4)^2(9.5)^2 \right\}$$

$$= (201 + 4537)\ cm^4 = 4738cm^4$$

For the rectangular area:

$$\{5625 + 20(15)(3)^2\}cm^4$$
$$(5625 + 2700) \ cm^4$$
$$= 8325 \ cm^4$$

$$\therefore \quad I_{centroid} = (12989.5 - 4738 + 8325)cm^4$$
$$= 16,576.5cm^4$$

Therefore the total second moment of area of the composite area about its centroidal axis parallel to XX'- axis is *16,576.5cm⁴.* It should be noted that a composite area may not have any axes of symmetry, in which case it might be useful to establish orthogonal reference axes at the edges of the composite area and to determine $\bar{y}$ and $\bar{x}$ with reference to them. Knowing the location of the centroids of the component areas one can easily determine the distances between these centroids and the centroid of the composite area, thus making it easy to apply the Parallel Axis Theorem. Evidently, in the two examples just worked out involving the semi-circular area atop the rectangular area, it should be clear that the centroid of the composite area remains on the Y-axis because of symmetry. And because the centroid of each geometrical shape is also on the YY' – axis, there is no distance between the two sets of centroid. In that case, the value of I about the YY' – axis for the composite area is merely the sum of the individual moments of area YY'. Hence, for the first example

$I_{YY'}$ = second moment of area of semi-circular area about the YY' – axis plus second moment of area of rectangular area about YY' – axis

$$= \frac{\pi(10)^4}{8} + \frac{15(20)^3}{12}$$
$$= (3928 + 10000) \ cm^4$$
$$\therefore \quad I_{YY'} = 13928cm^4$$

In the example with the circular area voided, the second moment of area of the voided is subtracted

$$I = \frac{\pi(10)^4}{8} + \frac{15(20)^3}{12} - \frac{\pi r^4}{4}$$
$$= 3928 + 10000 - 201$$
$$= 13,727(cm)^4$$

Illustrative Example 9

Determine the second moment of area of the octagonal area about the XX' – axis. See Fig. 17a

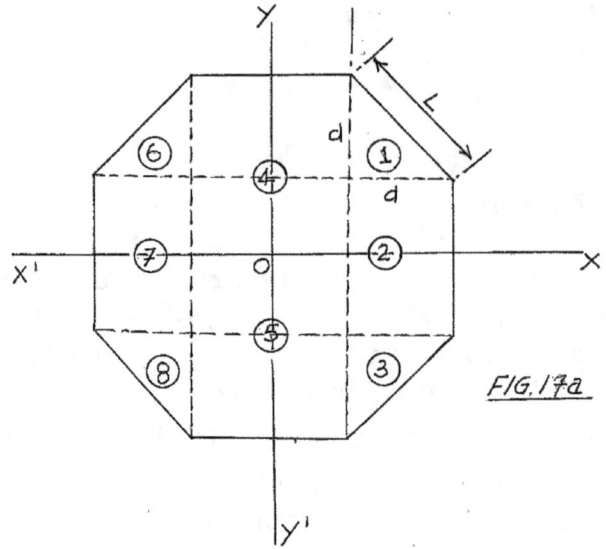

FIG. 17a

Note that the centroid of the octagon is at the origin of the axes

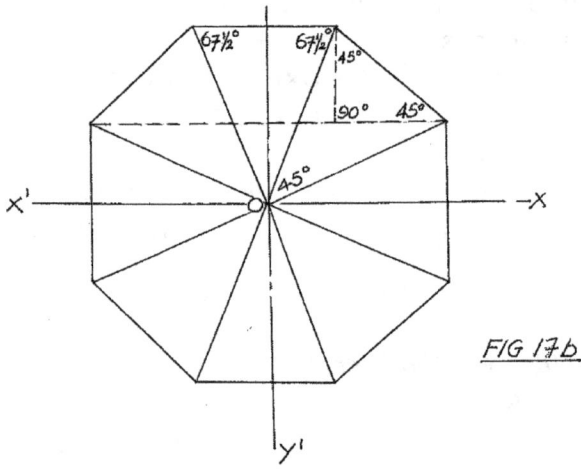

FIG 17b

In Fig. 17b, I have indicated the values of angles required for the calculation. The composite area is divided into 8 components. Areas circled *1, 2* and *3* are identical to those circled *6, 7* and *8* respectively.

Considering triangular component area circled *1*:

$$\frac{d}{L} - Sin45^\circ = \frac{1}{\sqrt{2}} \qquad\qquad \therefore \quad d = \frac{L}{\sqrt{2}}$$

Second moment of this area about its own centroid = $\dfrac{d \cdot d^3}{36} = \dfrac{d^4}{36}$

$\therefore$ Second moment of area about *XX'* is, using the Parallel Axis Theorem:

$$= \frac{d^4}{36} + \frac{d^2}{2} \cdot \left(\frac{d}{3} + \frac{L}{2} \right)^2$$

~ 322 ~

$$= \left(\frac{L}{\sqrt{2}} \right)^4 \cdot \frac{1}{36} + \frac{L^3}{4} \left(\frac{L}{3\sqrt{2}} + \frac{L}{2} \right)^2$$

$$= \frac{L^4}{144} + \frac{L^4}{4} \left(\frac{22}{72} + \frac{1}{3\sqrt{2}} \right)$$

Accordingly, the second moment of areas of the components *1, 3, 6* and *8* about *XX'*.

$$= 4 \left\{ \frac{L^4}{144} + \frac{L^4}{4} \left(\frac{22}{72} + \frac{1}{3\sqrt{2}} \right) \right\}$$

For the component areas marked *2* and *7*:

$$I_X = 2 \left(\frac{dL^3}{12} \right)$$

$$= 2 \frac{L}{\sqrt{2}} \cdot \frac{L^3}{12}$$

$$I_X = \frac{L^4}{6\sqrt{2}}$$

For component areas *4* and *5*:

$$I_{XX} = 2 \left\{ \frac{L \left(d + \frac{L}{2} \right)^3}{3} \right\}$$

$$= \frac{2L}{3} \left(\frac{L}{\sqrt{2}} + \frac{L}{2} \right)^3$$

$$= \frac{L^4}{24\sqrt{2}} (20 + 14\sqrt{2})$$

The student would doubtless have observed that the centroids of the component areas *2* and *7* coincide with the centroid of the entire composite area. Accordingly the distance '*d*' in the Parallel Axis Theorem is zero.

For the rectangular component areas *4* and *5*, the expression (*breadth*)(*depth³*)/3 was used for the determination of I_{XX}.

In order therefore to obtain the second moment of area of the entire octagon about *XX'* we add the separate values obtained thus,

$$I_{XX'} = \left\{ \frac{L^4}{36} + L^4\left(\frac{22}{72} + \frac{1}{3\sqrt{2}} \right) \right\} + \frac{L^4}{6\sqrt{2}} + \frac{L^4}{24\sqrt{2}}(20 + 14\sqrt{2})$$

$$= L^4\left\{ \frac{1}{36} + \frac{22}{72} + \frac{1}{3\sqrt{2}} + \frac{1}{6\sqrt{2}} + \frac{1}{24\sqrt{2}}(20 + 14\sqrt{2}) \right\}$$

$$= L^4(0.027 + 0.31 + 0.24 + 0.118 + 1.2)$$

$$I_{XX'} = 1.895L^4, \text{ say, } 1.9L^4$$

PRODUCT SECOND MOMENT OF AREA (PRODUCT OF INERTIA)

This property of an area is needed in the determination of maximum and minimum values of second moments of area as we shall shortly see and, very importantly in problems of asymmetrical bending. Throughout this text I have used the terms second moment of area and moment of inertia interchangeably; likewise product moment of area and product of inertia.

Product moment of area is defined as $\int xy\, dA$ in which each element of δA is multiplied by the product of its X- and Y- co-ordinates. Refer again to Fig. 1a all the way back to the start of this chapter. The integration extends over the entire area of the relevant plane within the limits, but unlike second moment of area $\int y^2 dA$ or $\int x^2 dA$, product of inertia could be negative or positive depending on whether 'x' and 'y' are positive or negative.

Let us illustrate this distinction at once by considering the case of, say, a rectangular area, like the one shown in Fig. 18. Remember that like second moment of area, product of inertia must be associated with a particular reference axis or axes. For the rectangle, the reference axes are those passing through its centroid, which in this case is where I have placed the origin of the axes $XOX' - YOY'$

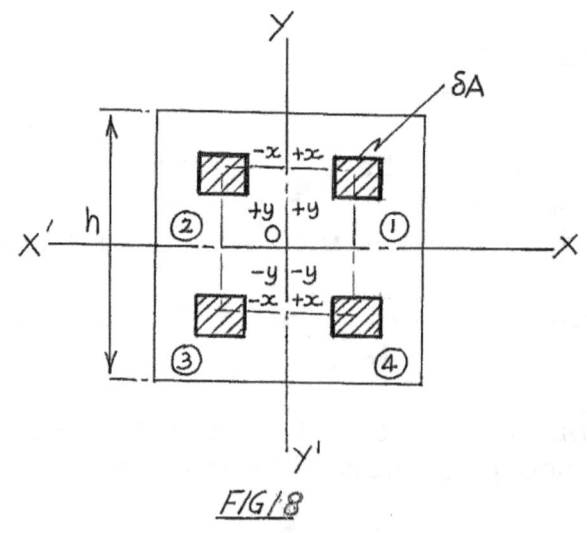

FIG 18

By definition, $\qquad\qquad P_{xy} = \int xy \, dA$

For the 1st quadrant $\qquad\qquad P_{XY} = \int (+x)(+y)dA = +\int xy \, dA$

For the 2nd quadrant $\qquad\qquad P_{XY} = \int (-x)(+y)dA = -\int xy \, dA$

For the 3rd quadrant $\qquad\qquad P_{XY} = \int (-x)(-y)dA = +\int xyd \, A$

and, for the 4th quadrant $\qquad P_{XY} = \int (+x)(-y)dA = -\int xy \, dA$

$\therefore \qquad$ For the entire rectangle $\quad \sum P_{XY} = 0$

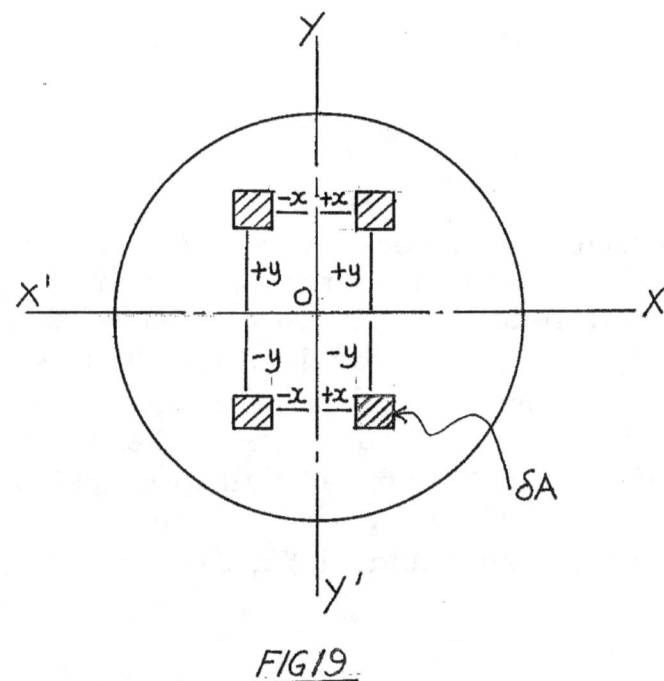

FIG 19.

By similar reasoning it is seen that P_{XY} for a circle about its centroidal axes is also zero. Refer to Fig. 19.

It should be noted that in general, the product of inertia of an area about any set of perpendicular axes, where one or both such axes is or are an axis or axes of symmetry, is always equal to zero.

On this basis therefore, a semi-circle which has one axis of symmetry, has P_{XY} about that axis of symmetry, equal to zero.

What happens when we rotate, say, clockwise the $YOY' - XOX'$ axes through $90°$ as in Fig. 20? Considering the elemental area δA in the first quadrant. Before rotation: $P_{XY} = \int (+x)(+y)dA$

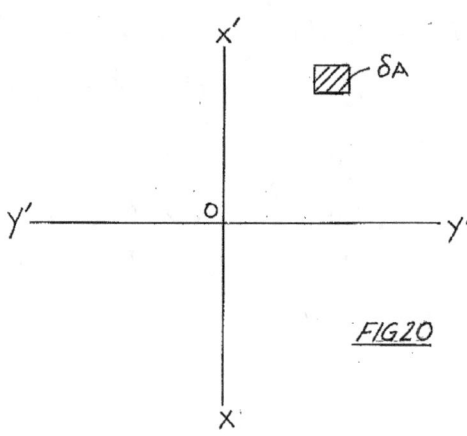

FIG 20

With rotation, $x = -y$ and $y = x$

$$\therefore \qquad P_{\text{new axes}} = -\int xy\, dA$$

If we had instead chosen an anti-clockwise rotation the result would have been the same. Because the sign of the product moment of area changes from positive to negative with rotation of the axes, it follows that for some value of angle within the angular range of rotation, $P_{XY} = 0$. This is so because you cannot move from a positive value to a negative one or vice versa without passing through zero! The perpendicular axes at the specific angles of rotation at which the value of P_{XY} vanishes are called the principal axes.

In Fig. 21 angle 'θ' is the rotation and $X''OX'''$, $Y''OY'''$ are the principal axes.

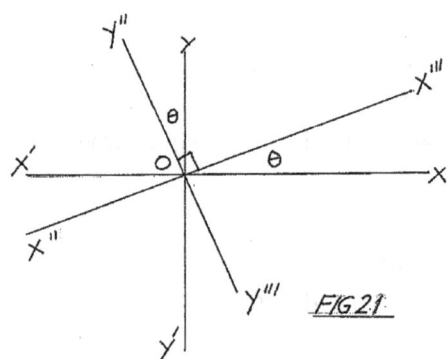

FIG 21

PARALLEL AXES THEOREM FOR PRODUCT MOMENT OF AREA

If $P_{X_c Y_c}$ is known for perpendicular axes say YOY_c and X_cOX' passing through the centroid (x_c, y_c) of a plane area, then the value of the product moment of area P_{XY} about parallel axes, say YOY' and XOX', distant 'a' and 'b' from the YOY_c and X_cOX' axes, are respectively given by :

~ 326 ~

$$P_{XY} = P_c Y_c + Aab$$

Where A = area of the figure.

This may be demonstrated by considering the rectangular area shown in Fig. 22.

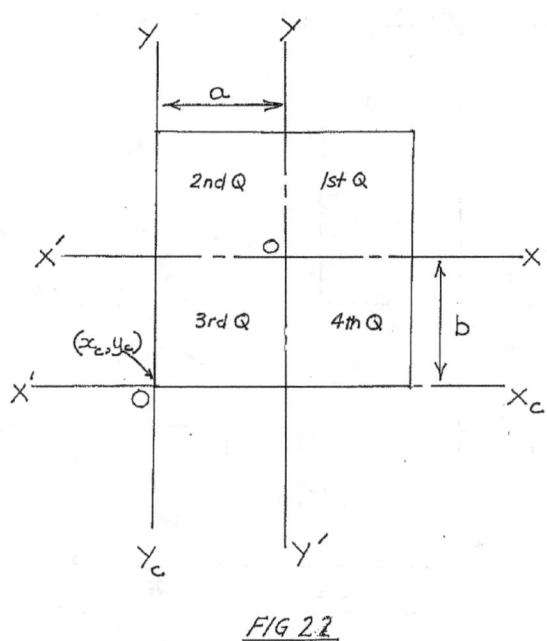

FIG 22

$$P_{XY} = \int x'y'dA$$

But
$$x' = x_c + a$$

and
$$y' = y_c + b$$

$\therefore$
$$\int xy\, dA = \int (x_c + a)(y_c + b)dA$$
$$= \int x_c y_c dA + \int ab\, dA + \int ay_c dA + \int bc_c dA$$

Observing that 'y_c' is positive in the 1st quadrant and negative in the 3rd quadrant $\int ay_c dA$ vanishes; likewise $\int bx\, dA$ x_c being positive in the fourth quadrant and negative in the second quadrant.

$\therefore$
$$P_{xy} = P_{x_x y_c} + Aab \qquad \dots \dots \dots (\alpha)$$

In words, if the product moment of area is known for axes passing through the centroid of the area, then the product moment of area for any other set of parallel axes say, YOY' and XOX' may be obtained from equation... (α), which may be regarded as the Parallel Axis Theorem for product of inertia. I have found the result to be quite useful in fact in the determination of product moment of area of a number of plane areas.

Illustrative Problem 10

Let us determine the product moment of area about perpendicular axes X_cO_cX, Y_cO_cY passing through the centroid for the quadrant of the circle shown in Fig. 23.

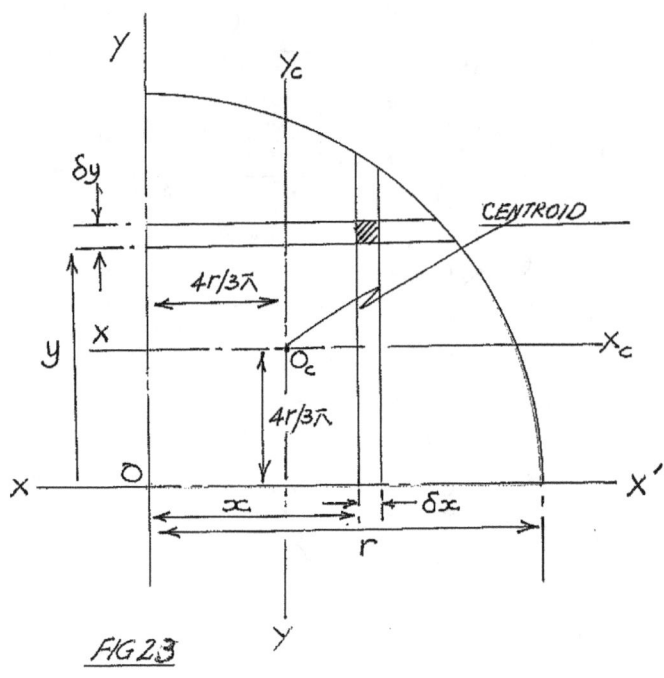

FIG 23

Quadrant of Circle of radius 'r'

$$P_{XY} = \iint xy\, dy\, dx, \text{ where } P_{XY} = \text{Product moment of area about } YOY',\ XOX'$$

$$= \int_0^r \int_0^{y=\sqrt{r^2-x^2}} [y\, dy]dx$$

$$= \int_0^r \left[\frac{y^2}{2}\right]_0^{\sqrt{(r^2-x^2)}} x\, dx$$

$$= \frac{1}{2}\int_0^r (r^2 x - x^3)dx$$

i.e. $$P_{XY} = \frac{r^4}{8}$$

By the Parallel Axis Theorem for Product Moment of area

~ 328 ~

$$\frac{r^4}{8} = P_{x_c y_c} + \frac{\pi r^2}{4}\left(\frac{4r}{3\pi}\right)\left(\frac{4r}{3\pi}\right)$$

$$= P_{x_c y_c} + \frac{4r^4}{9\pi}$$

$$\therefore \quad P_{x_c y_c} = \left(\frac{r^4}{8} - \frac{4r^4}{9\pi}\right)$$

$$\text{i.e.} \quad P_{x_c y_c} = \frac{r^4(9\pi - 32)}{72\pi}$$

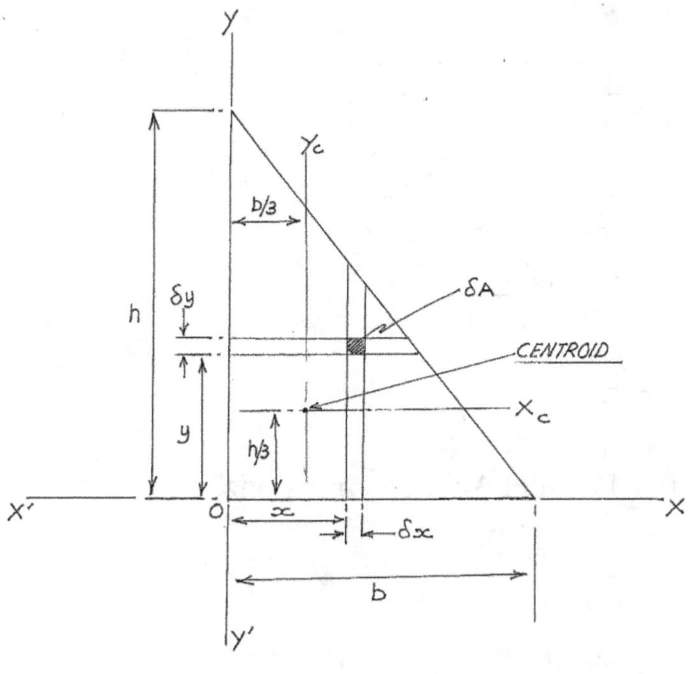

FIG 24.

Illustrative Problem 11

The problem of finding the product of inertia of a right-angled triangle such as that shown in Fig. 24, with reference to the triangle's centroidal axes may be solved by first determining the product of inertia about the axes OY and OX and then applying the Parallel Axis Theorem. Let us proceed to do that.

$$P_{xy} = \int xy\, dA$$

$$= \iint xy\, dy\, dx$$

The limits of integration are: For y : $y = 0$ to $y = h - \frac{hx}{b}$, and for $x : x = 0$ to $x = b$

$$\therefore \quad P_{XY} = \int_0^b \int_0^{(h - hx/b)} [y\, dy]x\, dx$$

$$= \int_0^b \left[\frac{y^2}{2} \right]_0^{(h-hx/b)} x \; dx$$

$$= \frac{1}{2} \int_0^b \left\{ h\left(\frac{1-x}{b}\right) \right\}^2 x \; dx$$

$$= \frac{h^2}{2} \int_0^b \left(1 - \frac{2x}{b} + \frac{x^2}{b^2} \right) x \; dx$$

$$= \frac{h^2}{2} \int_0^b \left(x - \frac{2x^2}{b} + \frac{x^3}{b^2} \right) dx$$

$$= \frac{h^2}{2} \left[\frac{x^2}{2} - \frac{2x^3}{3b} + \frac{x^4}{4b^2} \right]_0^b$$

$$= \frac{h^2}{2} \left(\frac{b^2}{2} - \frac{2b^2}{3} + \frac{b^2}{4} \right)$$

$$= \frac{b^2 h^2}{2} \left(\frac{1}{2} - \frac{2}{3} + \frac{1}{4} \right)$$

i.e. $\qquad P_{XY} = \dfrac{b^2 h^2}{24}$

Now we can apply the Parallel Axis theorem, viz:

$$I_{xy} = I_c + A \, \overline{xy}$$

i.e. $\qquad \dfrac{b^2 h^2}{24} = I_c + \dfrac{bh}{2}\left(\dfrac{b}{3} \cdot \dfrac{h}{3}\right) = I_c + \dfrac{b^2 h^2}{18}$

from which $\qquad I_c = \dfrac{b^2 h^2}{24} - \dfrac{b^2 h^2}{18}$,

$\qquad I_c = -\dfrac{b^2 h^2}{72}$, check.

The reader should note that if the gradient of the side of the triangle BC were

positive, in which case the equation of BC would be $y = \dfrac{h}{b}x + h$ then $P_{xy} = +\dfrac{b^2 h^2}{72}$

SECOND MOMENTS OF AREA WITH CHANGE OF DIRECTION OF AXES AND DETERMINATION OF THE PRINCIPAL AXES

As stated earlier, the values of the second moment of area of sections about a set of axes which are inclined to the conventional X- and Y-axes but which are themselves perpendicular to one another, are required in problems of asymmetrical bending, just to mention one area of application. In Fig. 25 it is required to find moments of inertia about the inclined orthogonal axes $Y''OY''$ and $X''OX''$. In order to avoid confusion in

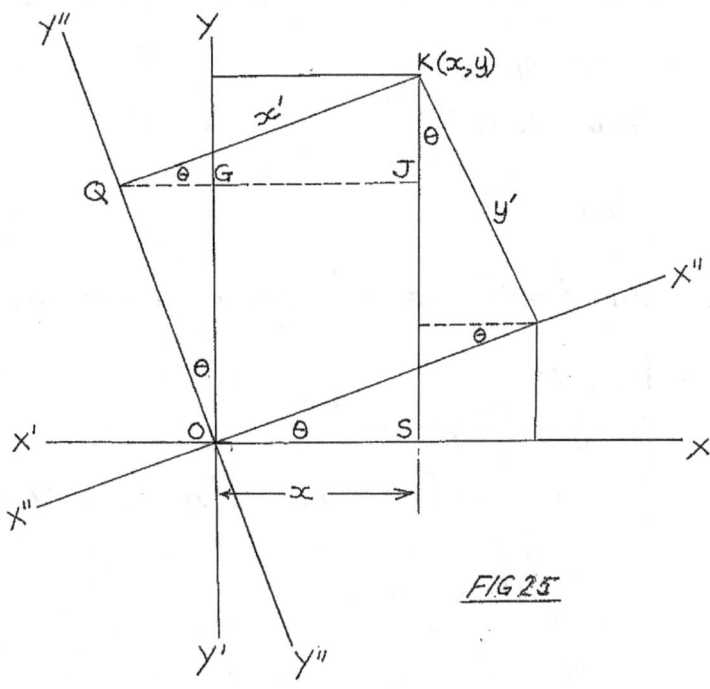

FIG 25

designating axes and the coordinates in reference to them, I shall write coordinates in the $YOY' - XOX'$ system as plain 'x' and 'y' and those in the $Y''OY'' - X''OX''$ system respectively, by x' and y'. With reference to Fig. 25

$$x = QJ - QG$$

But, $\dfrac{QJ}{x'} = Cos\theta$; $\dfrac{QG}{y'} = Sin\theta$

∴ $x = x'Cos\theta - y'Sin\theta$ . . . . (i)

Also $y = JK + JS$; $JS = OG$; $OQ = y'$

and, $\dfrac{JK}{x'} = Sin\theta$; $\dfrac{OG}{OQ} = \dfrac{OG}{y'} = Cos\theta$

∴ $y = x'Sin\theta + y'Cos\theta$ . . . . (ii)

Multiplying (i) by $Cos\theta$ and (ii) by $Sin\theta$ and adding the results we obtain

$$xCos\theta = x'Cos^2\theta - y'Sin\theta\,Cos\theta$$

and $$ySin\theta = x'Sin^2\theta + y'Sin\theta\,Cos\theta$$

from which

$$x' = ySin\theta + xCos\theta$$

Multiplying (i) by $Sin\theta$ and (ii) by $Cos\theta$ and subtracting (iii) from (iv):

$$xSin\theta = x'Cos\theta\, Sin\theta - y'Sin^2\theta \qquad \dots\dots\dots \text{(iii)}$$

and $\quad yCos\theta = x'Cos\theta\, Sin\theta + y'Cos^2\theta \qquad \dots\dots\dots \text{(iv)}$

$\therefore \qquad\quad y' = yCos\theta - xSin\theta$

By definition, the moment of inertia of an elemental area δA about the $X''OX''$ is

$$I_{X''OX''} = \int (y')^2 dA$$
$$= \int (yCos\theta - xSin\theta)^2 dA$$
$$\int y^2 Cos^2\theta\, dA + \int x^2 Sin^2\theta\, dA - 2\int xy\, Sin\theta\, Cos\theta\, dA$$

But since,

$$\int y^2 dA = I_X \; ; \; \int x^2 dA = I_Y \; ; \text{ and,}$$
$$\int xy\, dA = P_{XY}$$

we have

$$I_{X''OX''} = I_X Cos^2\theta + I_Y Sin^2\theta - P_{XY} Sin2\theta \qquad \dots\dots\dots \text{(v)}$$

Similarly, from

$$I_{YOY''} = \int (x')^2 dA$$

we obtain

$$I_{Y''OY''} = I_x Sin^2\theta + I_Y Cos^2\theta + P_{XY} Sin^2\theta \qquad \dots\dots\dots \text{(vi)}$$

Continuing, the product moment of area about the rotated orthogonal axes $Y''OY''$ and $X''OX''$ and $X''OX''$ may be determined from:

$$P_{x'y''} = \int x'y'dA$$
$$= \int (ySin\theta + xCos\theta)(yCos\theta - xSin\theta)dA$$
$$= \int y^2 Sin\theta\, Cos\theta\, dA + \int xyCos^2\theta\, dA$$
$$- \int xySin^2\theta dA - \int x^2 Sin\theta\, Cos\theta\, dA$$
$$= Sin\theta\, Cos\theta\int y^2 dA - Sin\theta\, Cos\theta\int x^2 dA$$
$$+ Cos^2\theta\int xy\, dA - Sin^2\theta\int xy\, dA$$
$$= I_X Sin\theta\, Cos\theta - I_Y Sin\theta\, Cos\theta + P_{xy} Cos^2\theta$$
$$- P_{XY} Sin^2\theta$$

$$= \frac{1}{2}(I_x - I_Y)Sin2\theta + P_{xy})(Cos^2\theta - Sin^2\theta)$$

i.e.
$$P_{X''Y''} = \frac{1}{2}(I_x - I_y)Sin2\theta + P_{XY}Cos2\theta \qquad \dots\dots\dots\dots \text{(vii)}$$

Recalling that $Cos^2\theta = (1 + Cos2\theta)/2$ and
$$Sin^2\theta = (1 - Cos2\theta)/2$$

we may rewrite the expression for $I_{X''}$ and $I_{Y''}$ in the form

$$I_{X''} = I_X\left(\frac{1 + Cos2\theta}{2}\right) + I_Y\left(\frac{1 - Cos2\theta}{2}\right) - P_{XY}Sin2\theta$$

i.e.
$$I_{X''} = \left(\frac{I_X + I_Y}{2}\right) + \frac{1}{2}((I_X - I_Y)Cos2\theta - P_{XY}Sin2\theta \qquad \dots\dots\dots\dots \text{(viii)}$$

and similarly,

$$I_{Y''} = \left(\frac{I_X + I_Y}{2}\right) + \frac{1}{2}((I_X - I_Y)Cos2\theta - P_{XY}Sin2\theta \qquad \dots\dots\dots\dots \text{(ix)}$$

It is important to note that upon adding the equations (v) and (vi) the following are obtained:

$$I_{X''} + I_Y = I_X(Cos^2\theta + Sin^2\theta) + I_Y(Sin^2\theta + Cos^2\theta)$$
$$= I_X + I_Y$$

or $\qquad I_{X''} + I_{Y''} = J,$ the polar moment of inertia for a plane area.

As you would recall, Equation (vi) may be re-arranged as follows:
$$\left\{I_{X''} - \left(\frac{I_X + I_Y}{2}\right)\right\} = \frac{1}{2}(I_X - I_Y)Cos2\theta - P_{xy}Sin2\theta$$

If now we square both sides of the latter expression, then,

$$\left\{I_{X''} - \left(\frac{I_X + I_Y}{2}\right)\right\}^2 = \left\{\frac{1}{2}(I_X - I_Y)Cos2\theta - P_{XY}Sin2\theta\right\}^2 \qquad \dots\dots\dots\dots \text{(x)}$$

Similarly squaring both sides of equation (v) yields
$$\{P_{x'y''}\}^2 = \left\{\frac{1}{2}(I_X - I_Y)Sin2\theta + P_{XY}Cos2\theta\right\}^2 \qquad \dots\dots\dots\dots \text{(xi)}$$

Adding (viii) and (ix) results in the following:

$$\left\{I_{X''}-\left(\frac{I_X+I_Y}{2}\right)\right\}^2+P^2_{x''y''}=\left\{\frac{1}{2}(I_X-I_Y)\right\}^2Cos2\theta+\left\{\frac{1}{2}(I_X-I_Y)\right\}^2Sin^22\theta$$

$$-2\left\{P_{xy}Sin2\theta\,Cos2\theta\left\{\frac{1}{2}(I_x-I_y)\right.\right.$$

$$+2\left\{P_{xy}Sin2\theta\,Cos2\theta\left\{\frac{1}{2}(I_x-I_y)\right.\right.$$

$$+P^2_{xy}(Sin^22\theta+Cos^22\theta)$$

$$=\left\{\frac{1}{2}(I_X-I_Y)\right\}^2+P^2_{XY}$$

i.e. $$\left\{I_{X''}-\left(\frac{I_X+I_Y}{2}\right)\right\}^2=\frac{1}{4}(I_X-I_Y)^2+P^2_{XY}\qquad\ldots\ldots\ldots\ldots\text{(xii)}$$

Recall your earlier encounter with co-ordinate geometry, in particular the equation of the circle, expressed in the form:

$$(x-h)^2+(y-k)^2=r^2,$$

its centre at $x=h,\ y=k$; and radius $'r''$

Upon examination of equation (xii) it will be deduced that it represents the equation of a circle in the $X''-Y''-$plane with centre at co-ordinates $x'=(I_X-I_Y)/2,\ y'=0$; and having a radius

$$'r'=\pm\sqrt{\frac{1}{4}\left\{(I_X+I_Y)^2+4P^2_{XY}\right\}}=\pm\frac{1}{2}\sqrt{(I_x+I_y)^2+4P^2_{XY}}$$

Thus, the equation (ix):

$$\left\{I_{X''}-\left(\frac{I_X+I_Y}{2}\right)\right\}^2=\frac{1}{4}(I_X-I_Y)^2+P^2_{XY}$$

may be represented graphically by a circle having product moments of area as ordinates and second moments of area as abscissae.   Before showing this representation, you may be interested to know that a similar construction is used in problems of stress analysis.   You will meet it later on in this book. Both representations are due to Professor Otto Mohr, a brilliant German engineer whose lifespan straddled the nineteenth and twentieth centuries.

To continue with the analysis, if we wish to determine the minimum (or maximum) value of $I_{X''}$ with respect to 'θ' then equation (vi), viz:

$$I_{X''} = \left(\frac{I_X + I_Y}{2}\right) + \frac{1}{2}(I_X - I_Y)\ Cos2\theta - P_{XY}\ Sin2\theta$$

must be differentiated with respect to 'θ' and the result equated to zero. That is to say,

$$\frac{dI_{X''}}{d\theta} = -2 \cdot \frac{1}{2}(I_X - I_Y)\ Sin2\theta - 2P_{XY}\ Cos2\theta = 0$$

from which: $2P_{XY}\ Cos2\theta = -(I_X - I_Y)\ Sin\theta$

i.e. $\qquad Tan2\theta = -\dfrac{2P_{XY}}{(I_X - I_Y)}$

or $\qquad Tan2\theta = \dfrac{2P_{XY}}{(I_Y - I_X)}$

Had we employed,
$$I_{Y''} = I_X Sin^2\theta + I_Y Cos^2\theta + P_{XY} Sin2\theta$$
and rewritten it for ease of computation in the form

$$I_{Y''} = I_X\left(\frac{1 - Cos2\theta}{2}\right) + I_Y\left(\frac{1 + Cos2\theta}{2}\right) + P_{XY} Sin2\theta$$

i.e. $\qquad I_{Y''} = I_X\left(\dfrac{I_X + I_Y}{2}\right) + \left(\dfrac{I_Y - I_X}{2}\right)Cos2\theta + P_{XY} Sin2\theta$

and then differentiated $I_{Y''}$ with respect to θ we would have obtained the same result as before, viz:

$$2P_{XY}\ Cos2\theta = (I_Y - I_X)Sin2\theta$$

i.e. $\qquad Tan2\theta \quad = 2P_{XY}(I_Y - I_X)$

The result for $Tan2\theta$ is reflected in Fig. 26

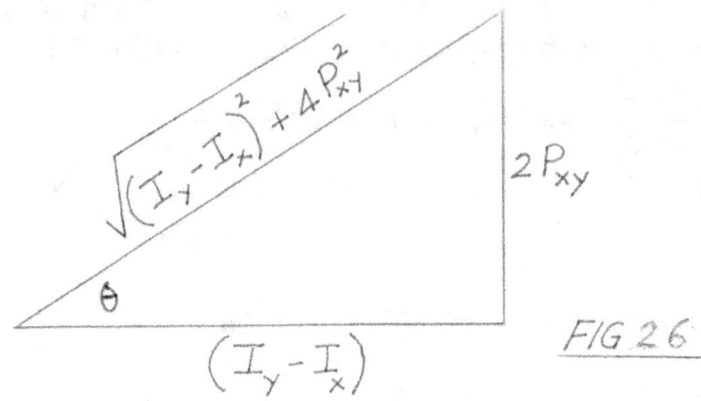

$$\sqrt{(I_Y - I_X)^2 + 4P_{XY}^2}$$

$2P_{xy}$

$(I_Y - I_X)$

θ

FIG 26

When the results, $Sin\theta = 2P_{XY}/\sqrt{(I_Y - I_x)^2 + 4P_{xy}^2}$ and $Cos2\theta =$ $(I_Y - I_X)/\sqrt{(I_X - I_X)^2 + 4P_{xy}^2}$ are substituted in equation (v) which is:

$P_{X'y'} = \dfrac{1}{2}(I_X - I_Y)Sin2\theta + P_{XY}Cos2\theta$, it is found that,

$$P_{x''y''} = \frac{1}{2}\frac{(I_X - I_Y)\cdot 2P_{XY}}{\sqrt{(I_Y - I_X)^2 + 4P_{XY}^2}} + \frac{P_{XY}(I_Y - I_X)}{\sqrt{(I_Y - I_X)^2 + 4P_{XY}^2}}$$

$$= \frac{P_{XY}(I_X - I_Y)}{\sqrt{(I_Y - I_X)^2 + 4P_{XY}^2}} - \frac{P_{XY}(I_Y - I_X)}{\sqrt{(I_Y - I_X)^2 + 4P_{XY}^2}}$$

$$= 0$$

Therefore, it is concluded that the product moment of area vanishes when the second moments of area are at their minimum and maximum values, i.e. when they are the principal second moments of area. Frame this in capital letters over your bed head: *"Remember always that the product moment of area of a plane section about its principal axes is zero."* Note also that the radii of gyration of an area with reference to the principal axes are called the principal radii of gyration.

The maximum and minimum values of the principal second moments of area may now be determined by substituting the values of $Sin2\theta$ and $Cos2\theta$ in $I_{x'}$ and $I_{y'}$. First, $I_{x'}$

$$I_{X''} = \frac{(I_x + I_y)}{2} + \frac{1}{2}\frac{(I_X - I_Y)(I_Y - I_X)}{\sqrt{(I_Y - I_X)^2 + 4P_{XY}^2}} - \frac{P_{XY}\cdot 2P_{XY}}{\sqrt{(I_Y - I_X)^2 + 4P_{XY}^2}}$$

$$= \frac{(I_X + I_Y)}{2} + \frac{(I_X - I_Y)(I_Y - I_X)}{2\sqrt{(I_Y - I_X)^2 + 4P_{XY}^2}} - \frac{2P_{XY}^2}{\sqrt{(I_Y - I_X)^2 + 4P_{XY}^2}}$$

$$= \frac{(I_X + I_Y)}{2} - \left[\frac{\left\{(I_Y - I_X)^2\right\}}{2\sqrt{(I_Y - I_X)^2 + 4P_{XY}^2}} + \frac{2P_{XY}^2}{\sqrt{\left(\frac{I_Y - I_X}{2}\right)^2 + 4P_{XY}^2}} \right]$$

$$= \left(\frac{I_X + I_Y}{2}\right) - \frac{\left[(I_Y - I_X)^2 + 4P_{XY}^2\right]}{2\left[(I_Y - I_X)^2 + 4P_{XY}^2\right]^{\frac{1}{2}}}$$

i.e. $$I_{min} = \left(\frac{I_X + I_Y}{2}\right) - \frac{1}{2}\sqrt{(I_Y - I_X)^2 + P_{xy}^2}$$

$$= \left(\frac{I_X + I_Y}{2}\right) - \frac{1}{2}\sqrt{4\left[\left\{\frac{I_Y - I_X}{2}\right\}^2 + P_{XY}^2\right]}$$

or $$I_{min} = \left(\frac{I_X + I_Y}{2}\right) - \sqrt{\left(\frac{I_Y - I_X}{2}\right)^2 + P_{xy}^2}$$

Substituting the same results for $Sin2\theta$ and $Cos2\theta$ in the equation for $I_{Y'}$, we obtain

i.e. $$I_{max} = \left(\frac{I_X + I_Y}{2}\right) + \sqrt{\left(\frac{I_Y - I_X}{2}\right)^2 + P_{xy}^2}$$

so that we may write

$$I_{max/min} = \left(\frac{I_X + I_Y}{2}\right) \pm \sqrt{\left(\frac{I_Y - I_X}{2}\right)^2 + P_{xy}^2}$$

Illustrative Example 12

Determine the principal moments of inertia with reference to the X and Y axes of the rectangle shown in Fig. 27

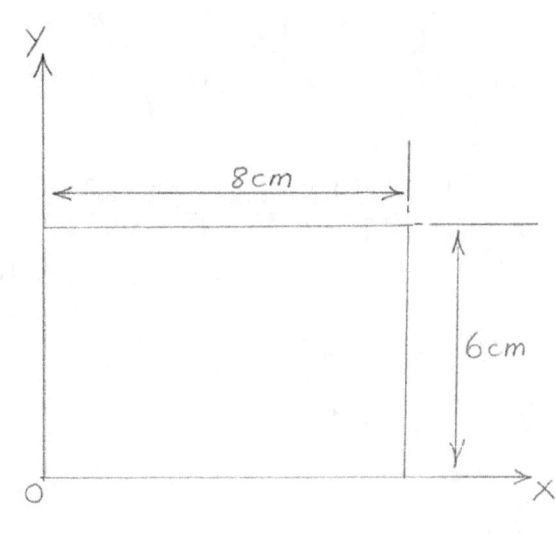

FIG.27

Commence by determining P_{XY}

By definition

$$P_{XY} = \int xy \ dA$$

$$= \iint xy \ \delta y \ dx$$

$$= \int_{0}^{8} \int_{0}^{6} y \ dy \ x dx$$

$$= \left(\frac{36}{2}\right)\left(\frac{64}{2}\right)$$

i.e. $P_{XY} = 576(cm)^4$

Next, we find I_X and I_Y

Accordingly $I_X = \dfrac{8(6)^3}{3} = 576(cm)^4$

and $I_Y = \dfrac{6(8)^3}{3} = 1024(cm)^4$

Assuming 'θ' to be angle between the X-Y axis and the principal axes OX' and OY' as shown in Fig. 28, then

$$\text{Tan } 2\theta = \frac{2P_{XY}}{I_Y - I_X}$$

$$= \frac{2(576)}{1024 - 576} = \frac{1152}{448} = 2.5714$$

~ 338 ~

or $\qquad 2\theta = Tan^{-1} 2.5714$

i.e. $\qquad 2\theta \approx 68.6^{o}$

$\therefore \qquad \theta = 34.3^{o}$

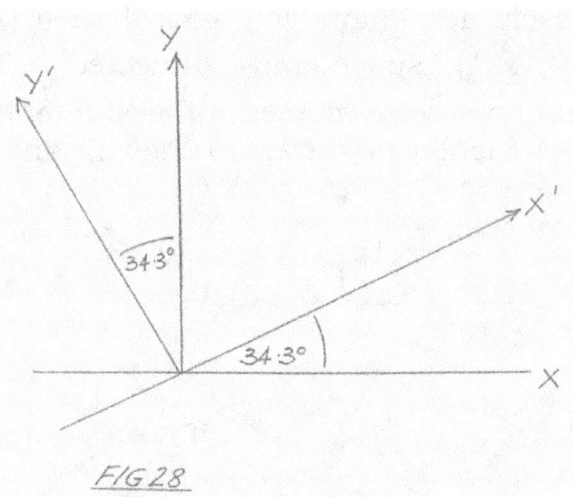

$FIG\ 28$

Now calculate I_{max} and I_{min}

$\therefore \qquad I_{max} = \dfrac{576 + 1024}{2} \pm \sqrt{(224)^2 + (576)^2}$

$$= (800 \pm 618)cm^4$$

so that $I_{max} \, 1418(cm)^4 ; \qquad I_{min} = 182(cm)^4$

MOHR'S GRAPHICAL SOLUTION FOR DETERMINATION OF PRINCIPAL AXES, I~MAX~ AND I~MIN~

Let us begin this article by restating some of the results derived in the preceding one:

$$I_{X'} = \frac{I_x + I_Y}{2} + \left(\frac{I_X - I_Y}{2}\right)Cos2\theta - P_{XY}Sin2\theta$$

$$I_{Y'} = \frac{I_X + I_Y}{2} - \left(\frac{I_X - I_Y}{2}\right)Cos2\theta + P_{XY}Sin2\theta$$

And,

$$Tan2\theta = +\frac{2P_{XY}}{(I_Y - I_X)} \qquad or \qquad Tan2\theta = -\frac{2P_{ay}}{I_x - I_Y}$$

The starting point of the graphical solution is choice of an appropriate scale for the moments and product of inertia so as to produce a reasonably-sized diagram. Next comes location of the origin of the axes of inertia and their erection: values of product of inertia as ordinates and those of second moments of area as abscissae. Points with coordinates $(I_X + P_{XY})$ and $(I_Y, -P_{XY})$ are located in (I_X, I_Y) vs P_{XY} space and connected. The point where the connecting line crosses the second moment of area line, is the centre of Mohr's Circle. The whole construction is shown in Fig. 29 and demonstrated in the following illustrative example

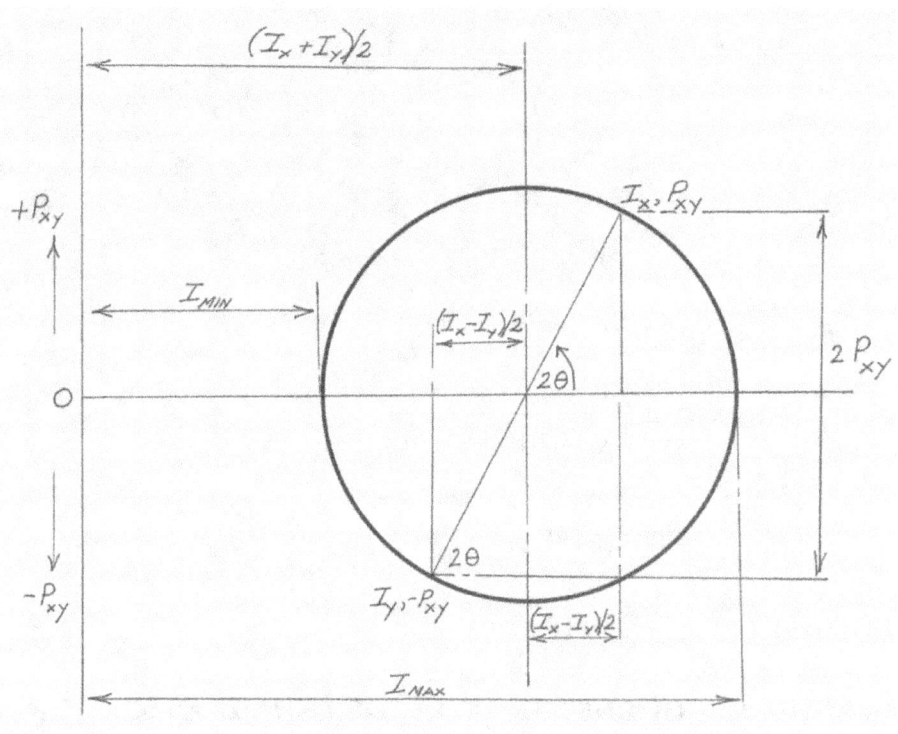

FIG 29: GRAPHICAL CONSTRUCTION OF MOHR'S CIRCLE FOR PRINCIPAL SECOND MOMENTS OF AREA AND THEIR ORIENTATION

Illustrative Example 13

Compute the principal moments of inertia with reference to the centroidal axes and the orientation of the principal axes of the plane composite area shown in Fig. 30. Also, construct the associated Mohr's Circle.

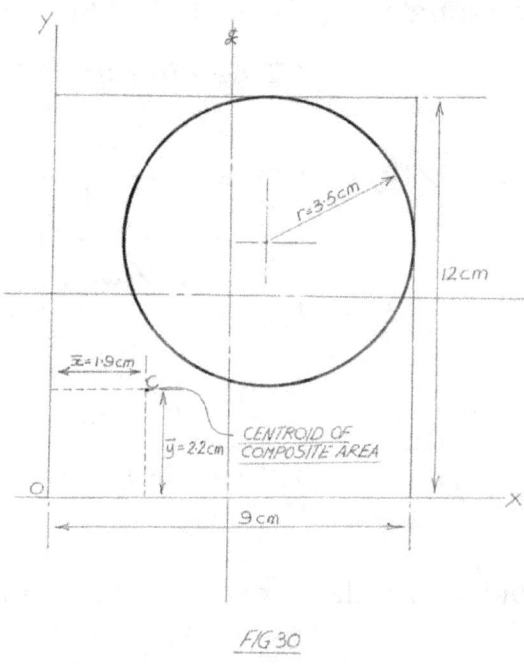

FIG 30

First, the centroid of the composite area is determined. Using the south-western most part of the area as the origin of the axes for finding $\bar{x}$ and $\bar{y}$, we have

$$\bar{x} = \frac{108(4.5) - \pi(3.5)^2(5.5)}{108 + \pi(3.5)^2}$$

$$= 1.9cm$$

$$\bar{y} = \frac{108(6) - \pi(3.5)^2(8.5)}{108 + \pi(3.5)^2}$$

$$= 2.2cm$$

Consider next the computation of product moment of area

For the rectangle = $P_{xy} = 0 + 12(9)(2.6)(3.8)$
$$= 1067(cm)^4$$

For the voided circular area of radius *3.5cm*:
$$P_{XY} = 0 + \pi (3.5)^2(3.6)(6.3)$$
$$= 873(cm)^4$$

~ 341 ~

$\therefore$ Nett P_{xy} = $(1067 - 873)(cm)^4$
  = $194(cm)^4$

and following this comes the computation of the second moments of area of the composite with reference to the centroidal axes:

$$I_X \text{ for the rectangle } = \frac{12(9)^3}{12} + 12(9)(2.6)^2$$

$$= (729 + 730)cm^4 = 1459\ cm^4$$

$$I_X \text{ for the voided circular area : } = \frac{\pi(3.5)^4}{4} + (3.5)^2(3.6)^2$$

$$= (118 + 499)(cm)^4$$
$$= 617(cm)^4$$

$\therefore$ Nett I_X = $(1459 - 617)(cm)^4 = 842(cm)^4$

Continuing

$$I_Y \text{ for the rectangle } = \frac{9(12)^3}{12} + 9(12)(3.8)$$

$$= (1296 + 1559)(cm)^4$$
$$= 2855(cm)^4$$

$$I_Y \text{ for the voided circular area } = \pi(3.5)^4 + \pi(3.5)^2(6.3)^2$$

$$= (118 + 1528)(cm)^4$$
$$= 1646(cm)^4$$

$\therefore$ Nett I_Y = $(2855 - 1646)(cm)^4$
  = $1209(cm)^4$

The principal moments of inertia are computed from the following:

$$I_{max/min} = \left(\frac{I_x + I_Y}{2}\right) + \sqrt{\left(\frac{I_X - I_X}{2}\right)^2 + P_{xy}^2}$$

$\therefore$
$$I_{max/min} = \left(\frac{842 + 1209}{2}\right) \pm \sqrt{\left(\frac{1209 - 842}{2}\right)^2 + (194)^2}$$

$$= 1025 \pm \sqrt{33672 + 37636}$$
$$= 1025 \pm 287$$

$\therefore$
$$I_{max} = 1312 cm^4$$
$$I_{min} = 738 cm^4$$

Now for the orientation of the axes. It was shown that:

$$Tan2\theta = \frac{2P_{xy}}{I_Y - I_X}$$

Hence
$$Tan2\theta = \frac{2(194)}{1209-842} = \frac{388}{367}$$

$$= 1.0572$$

i.e. $2\theta = Tan^{-1}1.0572$

$\therefore \qquad 2\theta = 46.6^o$

or $\qquad \theta = 23.3^o$

Refer to Fig. 31 which shows the orientation of the principal axes.

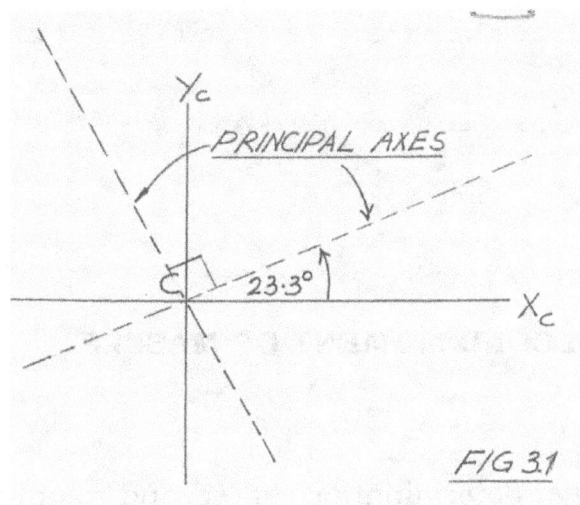

To construct Mohr's circle, we first set out the product of inertia and moment of inertia axes: P_{XY} as ordinates and I_X, I_Y as abscissae. Next we locate the centre of the circle at a distance equal *(842 + 1209)/2 = 1025cm⁴* from the origin and on the moment of inertia axis. The radius of Mohr's circle is computed from $\frac{1}{2}\sqrt{\left(\frac{I_X+I_Y}{2}\right)^2 + 4P_{XY}^2}$

i.e. $\frac{1}{2}\sqrt{\left(\frac{842+1209}{2}\right)^2 + 4(194)^2} = 548cm^4$

Points $(I_X,P_{XY}) = (842,194)$ and $(I_Y,-P_{XY}) = 1209-194)$ are plotted and joined and the angle of orientation of the principal axes measured. Here $2\theta = 46.6^o$ i.e. $\theta = 23^o$. Refer to Fig. 32.

Assignment: Construct Mohr's Circle for the immediately preceding problem.

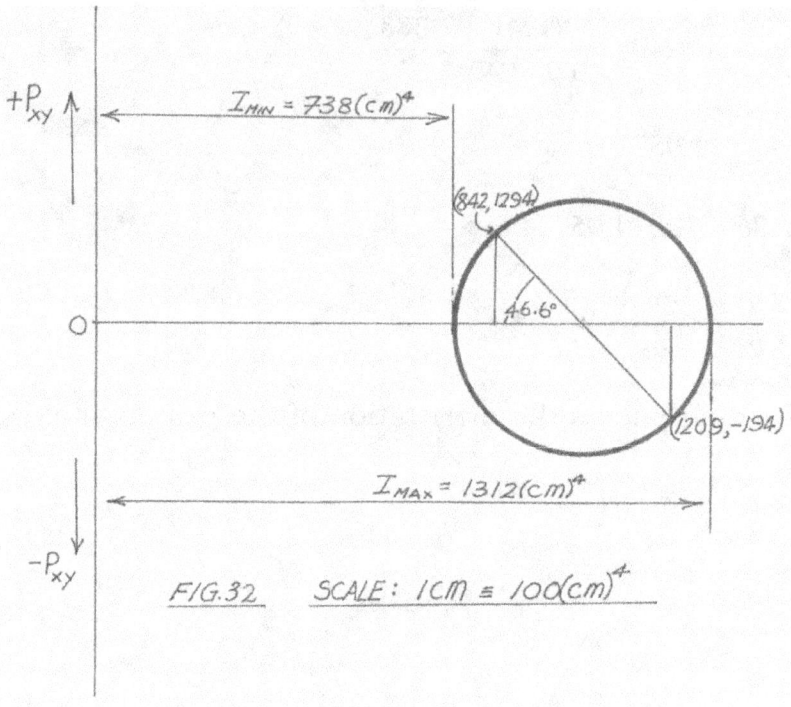

FIG.32   SCALE: $1CM \equiv 100(cm)^4$

INTRODUCTION TO SECOND MOMENT OF MASSES

Preliminaries

Before delving into the determination of second moment of masses it is considered useful to deal with a few preliminaries which will assist our future analyses. Accordingly, we shall derive general expressions for (i) a lamina of arbitrary shape, (ii) a rectangular lamina, and (iii) a circular lamina. A lamina is a thin plate; thin in the sense that its thickness is very small in comparison with its other dimensions.

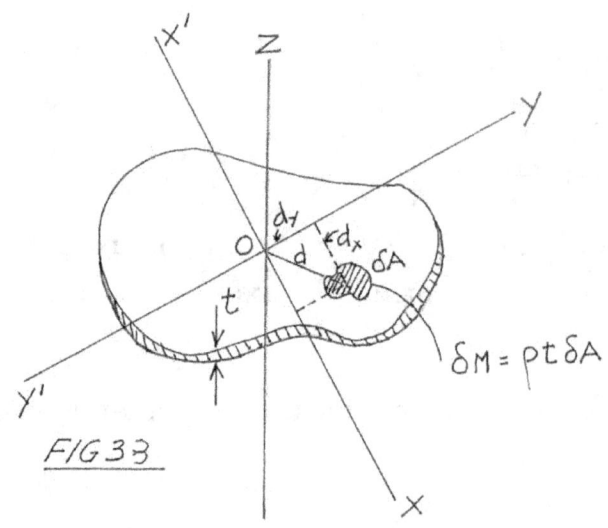

FIG 33

~ 344 ~

(i) Lamina of arbitrary shape

The lamina shown in Fig. 33 is in the X-Y plane. Axis Z' is perpendicular to this plane. The infinitesimal area of 'δA' has a mass '$\delta M' = \rho \delta A t$, where '$\rho$' is the density of the lamina, assumed constant throughout, and thickness 't' is likewise assumed constant. Accordingly, the second moment of mass of the entire lamina of mass M about, say, the Y-axis, is from our earlier definition given by:

$$I_Y = \int dM \; d_X^2$$
$$= \rho \, t \int dA \; d_X^2$$

Similarly,

$$I_X = \rho \, t \int dA \; d_Y^2$$

and $\quad I_Z = \rho \, t \int dA \; d^2 = \rho \, t \left[\int dA \left(d_X^2 + d_Y^2 \right) \right]$

Let us now turn attention to a rectangular shape such as that shown in Fig. 34

(ii) Laminar of rectangular shape

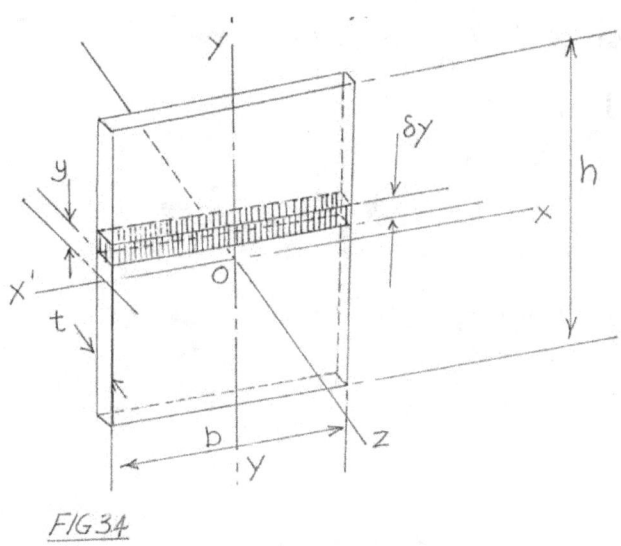

FIG 34

The origin of the axes is at the centroid of the plate of dimensions: height 'h' and width 'b'. Let t be its thickness and 'ρ' its density assumed constant throughout.

Considering the shaded strip of depth 'δy', distant 'y from X-axis. Its volume is $b \delta y \, t$ and its mass $\delta M = \rho$ bt δy.

By definition,

$$I_X = \int d_y^2 dM$$
$$= \int d_y^2 \rho\, bt_{dy}$$

The limits of integration are: $y = -h/2$ to $+h/2$ $\quad\therefore\quad$ $I_X = \int_{-h/2}^{h/2} y^2 \rho\, bt\, dy$

$$= \rho\, bt \frac{h^3}{12}$$

But the mass of the plate = $\rho\, bht$

$$\therefore \quad I_x = \frac{Mh^2}{12}$$

If we considered a strip parallel to the side with dimension 'h', then

$$\delta M = \rho ht\, \delta x$$

and

$$I_Y = \int d_x^2 dM$$
$$I_Y = \int d_x^2 \rho\, ht\, dx$$

This time the limits are: $x = -b/2$ to $+b/2$

i.e. $\qquad I_Y = \rho\, ht \int_{-b/2}^{b/2} x^2 dx$

from which $I_Y = \frac{Mb^2}{12}$

Since $\qquad J = I_X + I_Y$

J for the rectangular laminar = $\dfrac{M}{12}(h^2 + b^2)$

(iii) Circular Lamina

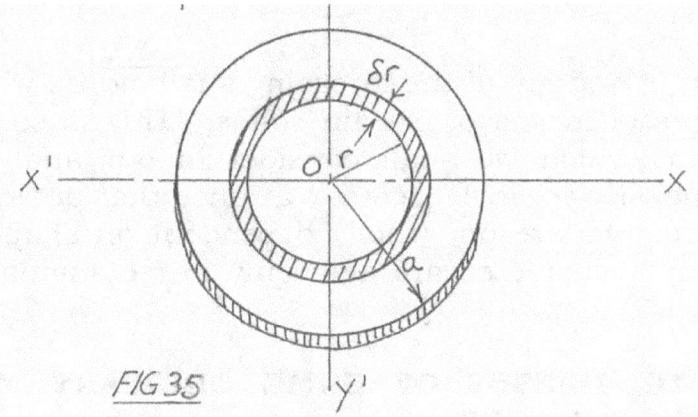

FIG 35

Proceeding in a similar fashion followed in the case of the rectangular lamina, we may straightway write down the relationship between mass of the elemental shaded portion shown in Fig. 35. It is

$$\delta M = 2\pi r \delta r \cdot t \cdot \rho$$

t and ρ being, respectively, the thickness and density of the lamina both assumed constant.

We shall adopt a short-cut in order to obtain the moments of inertia of the lamina about the X- and Y-axes. That is to say we shall first of all obtain I about an axis perpendicular to the plane of the lamina, i.e. J, and work backwards.

Now
$$J = \int_0^a r^2 dM$$

$$= \int_0^a 2\pi \, r^3 \rho t \; dr$$

$$= 2\pi\rho t \int_0^a r^3 dr = \frac{\pi\rho t \; a^4}{2}$$

and since the mass of the lamina = $\pi\rho t \; a^2$

$$J = \frac{Ma^2}{2}$$

Now
$$J = I_X + I_Y$$

But because of symmetry $\quad I_X = I_Y$

$$\therefore \quad I_X = \frac{Ma^2}{4} = I_y$$

The results obtained at (ii) and (iii) of the foregoing shall aid us when we come to consider the moments of inertia of certain solids. This is so because the basic infinitesimal strips which we shall consider in our analyses may be likened to these laminae. However, let me say at once that determinations of moments of inertia of masses are not wholly dependent on shapes for which the elemental strips to be considered are akin only to rectangular or circular laminae.

SECOND MOMENTS OF MASSES OF SOME ORDINARY MASSES OR MOMENTS OF INERTIA OF MASSES

In this section we determine second moments of mass of bodies or masses having geometric shapes for which the results previously obtained for the circular lamina and its rectangular counterpart, will be found to be useful points of departure. The familiar form

$$I = \int r^2 dM$$

may be rewritten

$$I = \int r^2 \rho \, dV$$

as was stated in our introduction to this chapter.

In these equations 'ρ' is density which may be constant or variable, and 'dV'' is the limit of the Infinitesimal volume 'δV'. Thus, when we write,

$$I = \rho \int r^2 \, dV$$

it is implied that the moment of inertia is a function only of geometric shape. If 'ρ' is not constant but varies throughout a region in X-, Y-, Z- space then,

$$I = \int r^2 \rho(x, y, z) dV$$

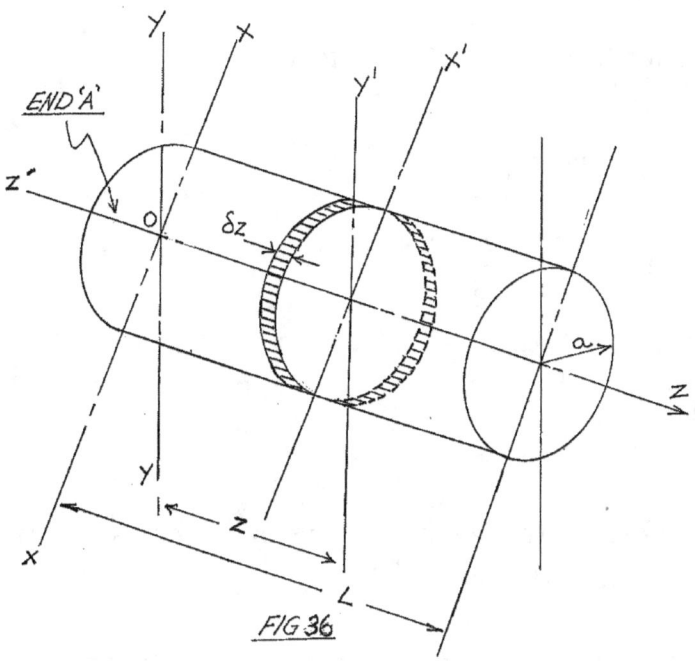

FIG 36

Consider now the solid right circular cylinder of mass M shown in Fig. 36. Its second moment of mass is required with reference to the end diameter at A.

Take the elemental disc of thickness 'δz' shown in the diagram. Let us designate its elemental moment of inertia about a diametral axis X' as $\delta I_{X'}$. Then $\delta I_{X'}$ =Mass of element multiplied by its radius and divided by 4 (Remember the moment of inertia of the circular lamina of radius 'a' about any diameter = $Ma^2/4$? Of course you do! We are here merely using a result derived in our preliminary analysis).

Therefore we may write

$$\delta I_{X_1} = \rho \pi a^2 \delta z) \left(\frac{a^2}{4} \right)$$

in which $\rho \pi a^2 \delta_z$ is the mass 'M' of the element.

Now, if we wish to transfer this moment of inertia to the end diameter with reference to the *X-axis*, through A which is distant 'z' away from axis X', then by the Parallel Axis Theorem:

$$\delta I_X = \delta I_{X'} + \delta M(z^2)$$

$$= \rho \pi a^2 dz \left(\frac{a^2}{4} \right) + \rho \pi a^2 dz(z^2)$$

i.e. $\quad \delta I_X = \rho \pi a^2 \left(\frac{a^2}{4} + z^2 \right)$

For the entire cylinder of length L, therefore,

$$\int_0^L dI_X = I_X = \rho\pi\, a^2 \int_0^L \left(\frac{a^2}{4} + z^2\right) dz$$

i.e. $\quad I_X = \rho\pi\, a^2\left(\frac{a^2 L}{4} + \frac{L^3}{3}\right)$

But mass of cylinder, $M = \rho\pi\, a^2 L$

$$\therefore \quad I_X = M\left(\frac{a^2}{4} + \frac{L^2}{3}\right)$$

Note that this I_X refers to the X-axis at the cylinder's end, not the one through its centroid which we here designate I_{CG}.

If we wish also to transfer this moment of inertia to an axis parallel to the X-axis but passing through the middle of the cylinder, i.e. through the cylinder's centroid, then again by the Parallel Axis Theorem,

$$I_x = I_{CG} + M\left(\frac{L}{2}\right)^2$$

centroid

$$\therefore \quad I_{CG} = M\left(\frac{a^2}{4} + \frac{L^2}{3}\right) - \frac{ML^2}{4}$$

from which $\quad I_{CG} = M\left(\frac{a^2}{4} + \frac{L^2}{12}\right)$

Observe that if we merely wanted the Moment of inertia of the solid cylinder about the axis of Z then we would have written

$$\delta I_z = (\rho\pi\, a^2 dz)\left(\frac{a^2}{2}\right)$$

Why $\left(\frac{a^2}{2}\right)$? Well that is because it was shown earlier in our consideration of laminae that the second moment of area of a thin disc about its polar axis is $\frac{Ma^2}{2}$ where M = mass of the disc, which in this case is $\rho\pi\, a^2 dz$. Hence the expression for δz. Therefore for the entire cylinder of length L'

$$I_Z = \rho\pi\, \frac{a^4}{2}\int_0^L dz$$

$$= \frac{\rho a^4 \pi L}{2}$$

and because the mass of the cylinder is $\rho a^2 \pi$,

$$I_z = \frac{M a^2}{2}$$

Evidently, for a hollow cylinder of external radius 'a' and internal radius, say, r

$$I_z = \frac{M}{2}(a^2 - r^2)$$

We could just as easily extend the application of the transfer equation given for an axis parallel to the X-axis, to one parallel to the Y-axis; Hence,

$$\delta I_X = \delta I_{x'} + \delta M(z^2)$$
$$\delta I_y = \delta I_{yX} + \delta M(z)$$

in which,

$\delta I_{x'}, \delta I_{y'}$ = moment of inertia of element about its own X-axis or
     Y-axis.
δM = mass of element; and
z = distance from the element to the axis with reference to
     which the moment of inertia is required.

Let us deal with some standard cases as illustrative examples.

Illustrative Example 14

To determine the moment of inertia of a solid right circular cone of base radius 'R', and height 'h', and constant density 'ρ' about the axis OX through its base, perpendicular to its own axis. Refer to Fig. 37

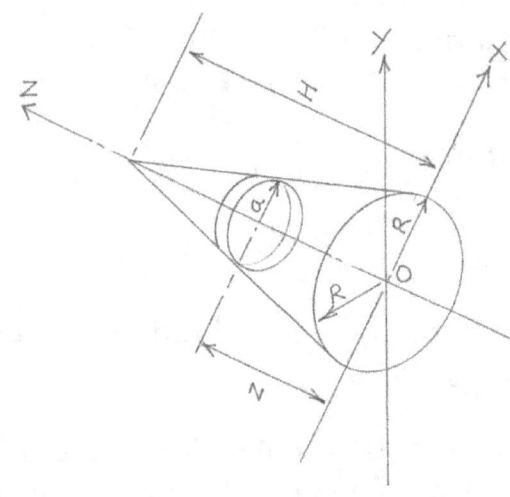

FIG 37

Employing,

$$\delta I_X = \delta I_{X'} + \delta M(z^2)$$

$\delta I_{X'}$ = moment of inertia of elemental disc about axis OX'

$$= \delta M\left(\frac{a^2}{4}\right)$$

$$= \text{density (area)(thickness)}\left(\frac{a^2}{4}\right)$$

$$= \rho\pi\, a^2 \cdot dz \cdot \left(\frac{a^2}{4}\right)$$

Now, $\delta M(z^2)$ = density(area)(thickness)(z^2)

$$= \rho\pi\, a^2 dz(z^2)$$

$\therefore$
$$\delta I_X = \rho\pi\, a^2 \delta z\left(\frac{a^2}{4}\right) + \rho\pi\, a^2 \delta z(z^2)$$

Referring again to Fig. 37

$$\frac{a}{R} = \frac{(H-z)}{H}$$

i.e.
$$a = \frac{R(H-z)}{H}$$

$\therefore$
$$\delta I_X = \rho\pi\left\{\frac{R(H-z)}{H}\right\}^2 dz \cdot \frac{1}{4}\left\{\frac{R(H-z)}{H}\right\}^2 + \rho\pi\left\{\frac{R(H-Z)}{H}\right\}^2 \delta z(z^2)$$

~ 352 ~

$$= \frac{\rho \pi R^4}{4H^4}(H-z)^4 dz + \frac{\rho \pi R^2}{H^2}(H-z)^2 z^2 dz$$

Let us substitute $u = (H-z)$, in which case $dz = du$ and the limits of integration will be as follows. When $z = H$; $u = 0$ and when $z = 0$; $u = H$.

Accordingly,

$$I_X = \frac{\rho \pi R^4}{4H^4}\int\limits_H^0 u^4(-du) + \frac{\rho \pi R^2}{H^2}\int\limits_H^0 u^2(H-u)^2(-du)$$

$$= \frac{\rho \pi R^4}{4H^4}\cdot\left(\frac{H^5}{5}\right) + \frac{\rho \pi R^2}{H^2}\int\limits_H^0 (u^2 H^2 - 2u^3 H + u^4)(-du)$$

$$= \frac{\rho \pi R^4 H}{20} + \frac{\rho \pi R^2}{H^2}\left(\frac{H^5}{30}\right)$$

or $\quad I_X = \frac{\rho \pi R^4 H}{20} + \frac{\rho \pi R^2 H^3}{30}$

Knowing that the mass 'M' of the cone is $\frac{1}{3}\rho \pi R^2 H$ we may write,

$$I_X = \frac{3MR^2}{20} + \frac{MH^2}{30} = \frac{M}{60}(9R^2 + 2H^2)$$

Illustrative Example 15

To determine the moment of inertia of the solid rectangular block shown in Fig. 38, with reference to the X-axis at its base. Assume density 'ρ' to be constant throughout.

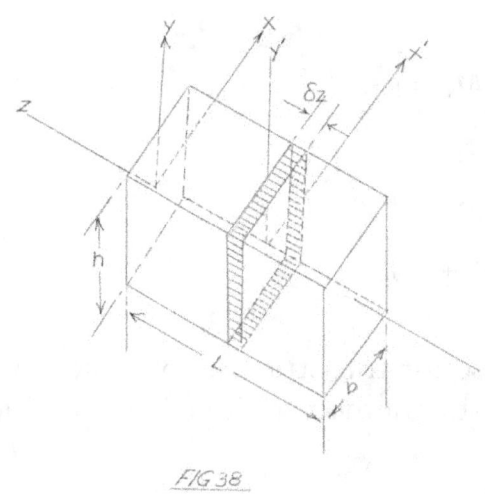

FIG 38

The basic relationship I am going to employ is:

$$\delta I_X = \delta I_{X'} + \delta M(z^2)$$

Noting that $\quad \delta I_{X'} = \delta M \cdot \dfrac{h^2}{12}$

where $\qquad \delta M = \rho b h \cdot \delta z \cdot \dfrac{h^2}{12}$

$$\delta I_X = \rho b h \cdot dz \cdot \frac{h^2}{12} + \rho b h \cdot \delta z(z^2)$$

$$= \rho b h \left(\frac{h^2}{12} \delta z + z^2 \delta z \right)$$

and $\qquad I_X = \rho b h \displaystyle\int_0^L \left(\dfrac{h^2}{12} dz + z^2 dz \right)$

$$= \rho b h \left[\frac{h^2 L}{12} + \frac{L^3}{3} \right]$$

$$= \rho b h L \left[\frac{h^2}{12} + \frac{L^2}{3} \right]$$

i.e. $\quad I_X = M \left[\dfrac{h^2}{12} + \dfrac{L^2}{3} \right]$

or $\quad I_X = \dfrac{M}{12}(h^2 + 4L^2)$

This is the moment of inertia with reference to the X-axis passing through 'A', that is an end diameter on the X-axis. To obtain the moment of inertia with respect to the X-axis through the centroid, we use the Parallel Axis Theorem.

Accordingly

$$I_X = I_{CG} + M \left(\frac{L}{2} \right)^2$$

$\therefore \qquad I_{CG} = \left(\dfrac{Mh^2}{12} + \dfrac{ML^2}{3} \right) - \dfrac{ML^2}{4}$

i.e. $\quad I_{CG} = \dfrac{M}{12}(h^2 + L^2)$

I leave it as an exercise for the student to show that with respect to the Y-axis through the centroid, the moment of inertia of the block is

$\frac{M}{12}(b^2 + L^2)$. [Hint: Consider an elemental strip running the whole length of the block and with thickness 'δy'..]

Illustrative Problem 16

To determine the moment of inertia of a uniform solid sphere of radius 'a' and density 'ρ' about a diameter. See Fig. 39

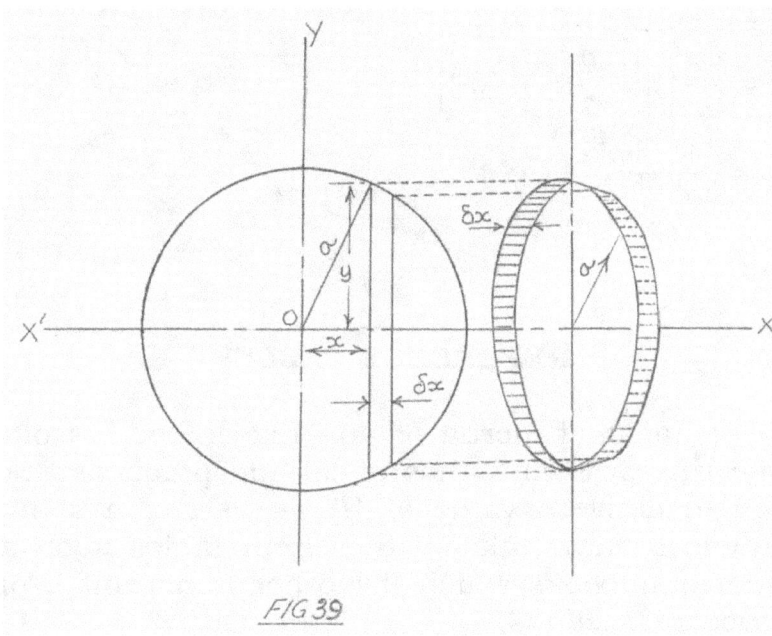

FIG 39

Visualise a solid disc of radius 'y' and thickness 'dx', at a distance x from 'O'; you are inside the sphere now. The volume of this disc is $\pi y^2 dx$. Referring to Fig. 39 it is deduced that

$$a^2 = x^2 + y^2$$
$$\therefore \quad y^2 = a^2 - x^2$$
$$\therefore \quad \text{Volume of elemental disc} = \pi(a^2 - x^2)\delta x$$

Assuming that 'ρ' is the density of the material of the sphere, the mass of this elemental disc, $\delta M = \rho\pi(a^2 - x^2)\delta x$

If M is the total mass of the entire sphere then $M = \rho \cdot \frac{4}{3}\pi\, a^3$

Therefore, in terms of 'M' the mass of our elemental disc 'δM' is

$$\delta M = \frac{\pi\rho(a^2 - x^2)\delta x}{\frac{4}{3}\rho\pi\, a^3}(M) \quad \text{or} \quad \frac{3M(a^2 - x^2)dx}{4a^3}$$

~ 355 ~

Reminding ourselves that the *X*-axis is perpendicular to the plane of the elemental disc, and recalling that the moment of inertia of a disc of mass '*M*' about an axis perpendicular to its plane is given by $\dfrac{M}{2}(radius)^2$, we may write down at once that the moment of inertia of our elemental disc of mass $3M\dfrac{(a^2-x^2)\delta x}{4a^3}$ is $3M\dfrac{(a^2-x^2)\delta x}{4a^3}\left(\dfrac{y^2}{2}\right)=\dfrac{3M(a^2-x^2)\delta x(a^2-x^2)}{8a^3}$

$$\therefore \quad I_X = \frac{3M}{8a^3}\int_{-a}^{+a}(a^2-x^2)^2\,dx$$

$$= \frac{3M}{4a^3}\left(a^5 - \frac{2}{3}a^5 + \frac{1}{5}a^5\right)$$

$$= \frac{3M}{4a^3}\cdot\frac{8a}{15}a^5$$

$$I_X = \frac{2Ma^2}{5}$$

REFERENCE TABLES

The value of the moment of inertia of an area or body is often required for solution of engineering problems, theoretical and practical; and unless one is specially required to derive a value of '*I*' as part of a solution, it is quite permissible to quote a value taken either from tables such as I have given hereunder or from handbooks which list other important properties as well, especially for structural sections.

Table 1 : I for some common Geometrical Sections and Areas

Name of Geometrical Section or Area	Figure	I_x, I_z	I_Y
Square		$I_X \dfrac{b^4}{12}$; $I_z = \dfrac{b^4}{6}$	$\dfrac{b^4}{12}$
		$I_X = \dfrac{b^4}{12}$; $I_z = \dfrac{b^4}{6}$	$\dfrac{b^4}{12}$
Rectangle		$I_X = \dfrac{bd^3}{12}$ $I_z = \dfrac{1}{12}\left(bd^3 + db^3\right)$	$\dfrac{db^3}{12}$
Triangle		$I_x = \dfrac{bh^3}{36}$ $I_z = \dfrac{1}{36}\left(bh^3 + hb^3\right)$	$\dfrac{hb^3}{36}$
Circle		$I_x = \dfrac{\pi r^4}{4}$ $I_z = J = \dfrac{\pi r^4}{2}$ $= \dfrac{\pi r^4}{2}$	$\dfrac{\pi r^4}{4}$
Quadrant of a Circle of radius 'r'		$\dfrac{\pi r^4}{16}$	$\dfrac{\pi r^4}{16}$

Ellipse of major axis 2a, minor axis 2b		$\dfrac{\pi \, ab^3}{4}$	$\dfrac{\pi \, a^3 b}{4}$
Semi-Circle		$\dfrac{\pi r^4}{8}$	$\dfrac{\pi r^4}{8}$
Tee		$I_x = \dfrac{(BD^2 - bd)^2}{12(BD - bd)} -$ $\dfrac{4BDbd(D-d)^2}{12(BD-bd)}$	$\dfrac{(D-b)B^3 + d(B-b)^3}{12}$
I(I-Section)		$I_x = \dfrac{BD^3 - bd^3}{12}$	$\dfrac{DB^3 - db^3}{12}$

In Table II the moments of inertia of some familiar masses or solid bodies are provided for ease of reference.

Table II : 'I' for some bodies of uniform density 'ρ' or homogeneous mass 'M' about axes of revolution X, Y, Z.

Body	Figure	I_x, I_z	I_Y
Thin Cylindrical Rod of length 'L'		$\dfrac{ML^2}{12}$	$\dfrac{ML^2}{12}$
Right Circular Cylinder of Radius 'a', length L		$I_X = M\left(\dfrac{a^2}{4}+\dfrac{L^2}{12}\right);$ $I_z = \dfrac{Ma^2}{2}$	$M\left(\dfrac{a^2}{4}+\dfrac{L^2}{12}\right);$
Rectangular Prism of width 'b', thickness 'h' and length 'L'		$\dfrac{M}{12}(h^2+L^2)$ $I_z = \dfrac{M}{12}(b^2+h^2)$	$\dfrac{M}{12}(b^2+L^2)$
Circular Disc of Radius 'a'		$I_x = \dfrac{Ma^2}{4}$ $I_z = \dfrac{Ma^2}{2}$	$M\dfrac{a^2}{4}$
Solid sphere		$2MA^2/5$ or $\dfrac{8}{15}\rho\pi r^2$	$2mA^2/5$ or $\dfrac{8}{10}\rho\pi r^3$

Right Circular Cone of base radius 'a', height 'h;'		$I_X = \dfrac{3}{20} M(a^2 + 4h^2)$ $I_Z = \dfrac{3Ma^2}{10}$	$\left(\dfrac{\rho \pi h^3 R^2}{30} + \dfrac{\rho \pi h^3 R^4}{20} \right)$ or $\left(\dfrac{m\ell}{10} + \dfrac{3MR^2}{20} \right)$
Hollow Right Circular Cylinder : Outside radius R; Inside radius r		$I_Z = \dfrac{M}{2}(R^2 - r^2)$	
Thin Rectangular Lamina		$I_x = \dfrac{Mb^2}{12}$ Polar M.I.(J) = $\dfrac{M}{12}(b^2 + a^2)$	$\dfrac{Ma^2}{12}$

Dimensional drawings of some popular structural sections and their properties are given in Appendix I. Density of material is assumed constant. Included in the data provided are: cross-sectional area, position of centroid (centre of mass), second moments of area with reference to conventional X-axis and Y-axis i.e. I_{xx}, I_{yy} and to principal axes U - U and V – V, i.e. I_{uu}, I_{vv}.

All axes pass through the centroid of the section; and, also the corresponding radii of gyration i.e. k_{xx}, k_{yy}, k_{uu} and k_{vv} are with reference to these axes.

SOLVED AND OTHER PROBLEMS

Q1. Using double integrals determine the second moment of area with reference to the *X-* and *Y-*axes of the region in the first quadrant defined by the parabola $y^2 = 9x$ and the lines $y = 0$ and $x = 3$. Check your results using single integrals or by any other method including the use of double integrals.

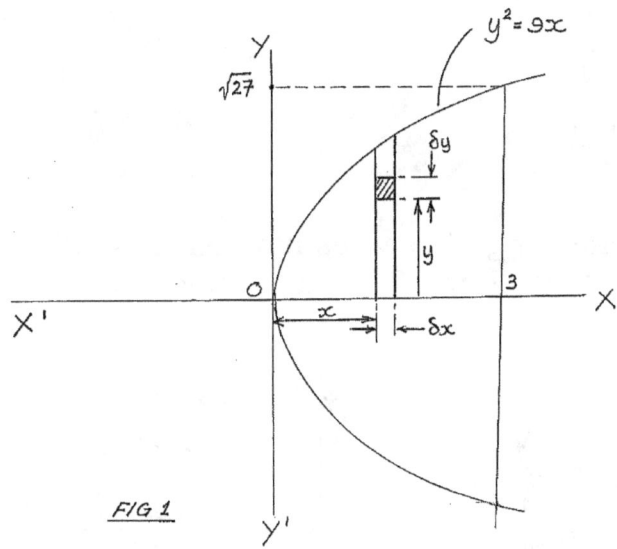

FIG 1

Refer to Fig. 1. Using double integrals, we have:

$$\delta I_{YY} = \delta y \, \delta x (x^2)$$
$$\delta I_{XX} = \delta y \, \delta x (y^2)$$

Evaluating I_{XX} first, and noting that the limits of integration are: in the case of *y:* from $y = 0$ to $y = 3\sqrt{x}$; and for *x*: from $x = 0$ to $x = 3$,

$$I_{XX} = \int_0^3 \int_0^y y^2 dy \; dx$$

i.e. $$I_{XX} = \int_0^3 \frac{y^3}{3} \; dx$$

$$= \frac{1}{3} \int_0^3 (3\sqrt{x})^3 \, dx$$

$$= \frac{9 \times 2 \times 9\sqrt{3}}{5}$$

or $$I_{XX} = \frac{162\sqrt{3}}{5}$$

Continuing,

$$I_{YY} = \int_0^3 \int_0^y x^2 \, dy \, dx$$

$$= \int_0^3 x^2 y \, dx$$

$$= \int_0^3 x^2 \, 3x^{1/2} \, dx$$

$$= 3 \int_0^3 x^{5/2} \, dx$$

$$= \frac{3.2}{7} (\sqrt{3})^7$$

$$I_{YY} = \frac{162\sqrt{3}}{7}$$

For the purpose of checking these results using single integrals two separate diagrams, viz. Figs. 2 and 3 are drawn. (q.v.).

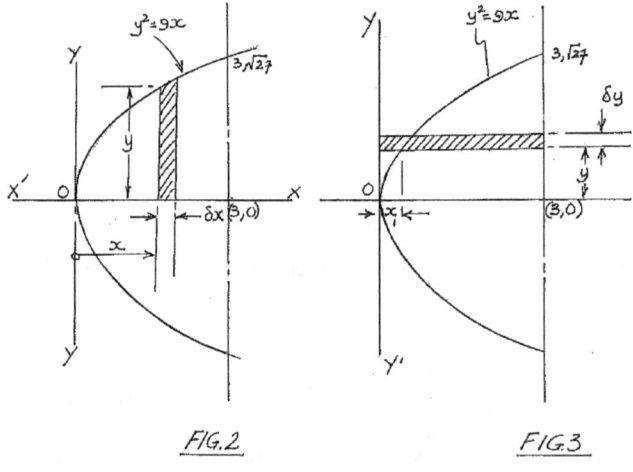

FIG.2   FIG.3

Referring to Fig. 2

$$\delta A = y \delta x$$

∴ $$i_{yy} = \int_0^3 x^2 \, y \, dx = \int_0^3 x^2 \cdot 3x^{1/2} \, dx$$

i.e. $$I_{YY} = 3 \left[\frac{2}{7}(\sqrt{3})^7 \right] = \frac{3.2}{7} \cdot 27\sqrt{3}$$

or $$I_{yy} = \frac{162\sqrt{3}}{7}, \text{ check.}$$

For I_{XX}, $\delta A = (3-x)\delta y$, and $I_{XX} = \int_0^{\sqrt{27}} dA \cdot y^2$

But

$$x_1 = \frac{y^2}{9}$$

$$\therefore \qquad \delta A = \left(3 - \frac{y^2}{9}\right)\delta y$$

$$\therefore \qquad I_{XX} = \int_0^{\sqrt{27}} \left(3 - \frac{y^2}{9}\right) y^2 \, dy$$

$$= \int_0^{\sqrt{27}} \left(3y^2 - \frac{y^4}{9}\right) dy$$

$$= \left[y^3 - \frac{y^5}{45}\right]_0^{\sqrt{27}}$$

$$= 81\sqrt{3}\left(1 - \frac{3}{5}\right)$$

i.e. $\qquad I_{XX} = \dfrac{162\sqrt{3}}{5}$, Check

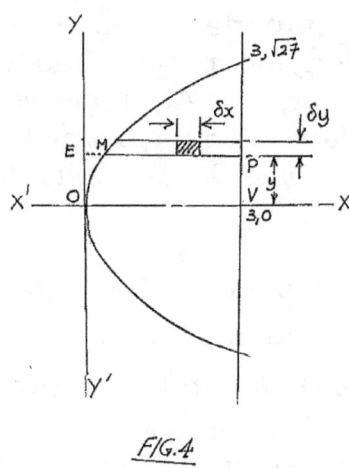

FIG.4

I could just have easily solved the problems using double integrals by writing

$$I_{YY} = \int_0^{\sqrt{27}} \int_{\frac{y^2}{9}}^3 y^2 \, dy \, dx$$

the limits of integration for 'x' being from $x = EM$ to $x = EP$ i.e. from $x = \dfrac{y^2}{9}$ to $x = 3$ and for y: from $y = 0$ to $y = VQ = \sqrt{27}$

$$\therefore \qquad I_{YY} = \frac{1}{3}\int_0^{\sqrt{27}} \left(9 - \frac{y^6}{2187}\right) dy$$

which works out to be

$$I_{YY} = 27\sqrt{3}\left(1 - \frac{1}{7}\right)$$

or $\quad I_{YY} = \dfrac{162\sqrt{3}}{7}, \quad$ as before

Similarly

$$I_{XX} = \int_0^{\sqrt{27}} \int_{y^{2/9}}^{3}$$

$$= \int_0^{\sqrt{27}} [x]_{y^{2/9}}^{3} = \int_0^{\sqrt{27}} \left(3y^2 - \frac{y^4}{9}\right) dy$$

$$= 81\sqrt{3} - \frac{243\sqrt{3}}{5} = 81\sqrt{3}\left(1 - \frac{3}{5}\right)$$

i.e. $\quad I_{XX} = \dfrac{162\sqrt{3}}{5} \quad$ as before

Q2. 'Vesica Piscis' is the name of the space shown hatched in Fig. 1. It is formed by the intersection of two circles of equal radii, the distance between centres being a common radius. This space was "... the basis of some of the most important geometric secrets of ancient operative masons. The underlying proportions of all the wonderful cathedral-architecture of the Middle Ages derived principally from the 'Vesica Piscis' and the equilateral triangle which is formed by use of this space". From *"Freemasonry: Some Deeper Considerations"* by Jack Bright.

(Questions):
(i)   Show that: the area of the 'Vesica Piscis' in Fig. 1 is given by $r^2(4\pi - 3\sqrt{3})/6$; and, also that,

(ii) $I_{XX} = r^4(8\pi - 3\sqrt{3})/16$; and $I_{YY} = (8\pi + 3\sqrt{3})/48$

Compare flow patterns across models having the shape of 'Vesica Piscis' with those of other models of different shapes in wind-and water-tunnels. Are there aerodynamic and naval architecture possibilities for air-craft and ship-design?

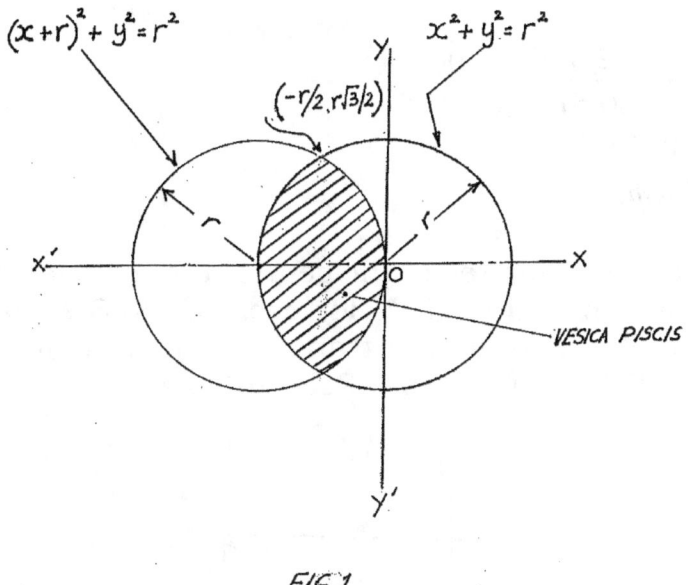

$(x+r)^2 + y^2 = r^2$ $x^2 + y^2 = r^2$

$(-r/2, r\sqrt{3}/2)$

VESICA PISCIS

FIG.1

Q3.   The cross-section of a steel channel beam is shown in Fig. 8a. Determine the moment of inertia of the cross-sectional area about (i) an axis through its centroid 'C' parallel to axis OX; and (ii) about the Y-axis.

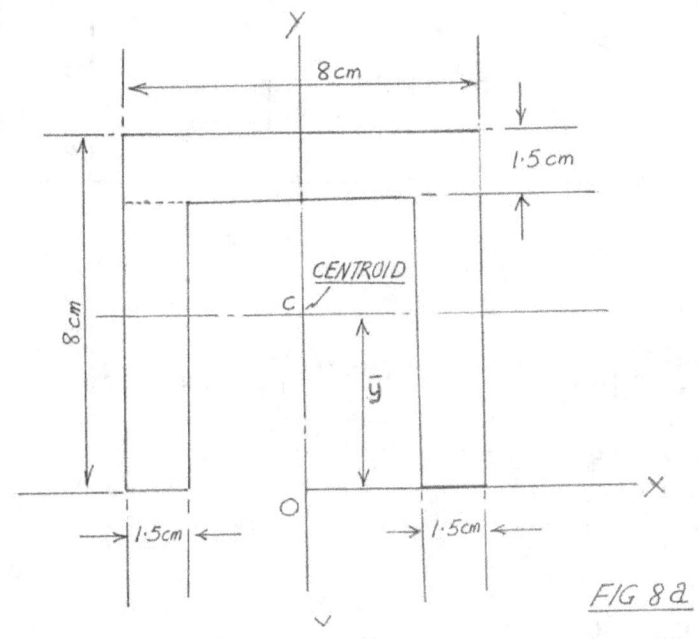

FIG 8a

The first thing to do is to determine the location of the centroid 'C', assumed to be distant $\bar{y}$ from the axis OX. Because the section is symmetrical about the axis OY, it follows at once that $\bar{x} = 0$

~ 365 ~

Referring to Fig. 1, dividing it into 3 separate rectangular areas and taking moments about OX,

$$\bar{y} = \frac{8(1.5)(7.25) + 2\{6.25)(1.5)(3.35)\}}{8(1.5) + 2(6.5 \times 1.5)}$$

$$= 150.375/31.5$$

$$\bar{y} = 4.77 cm$$

Accordingly the coordinates of the centroid of the channel are: *(0, 4.8)*. Next, we obtain the value of 'I' for each of the 3 separate areas chosen about each area's centroid and transfer each value to the centroidal axis using the parallel axis theorem. Refer to 8b.

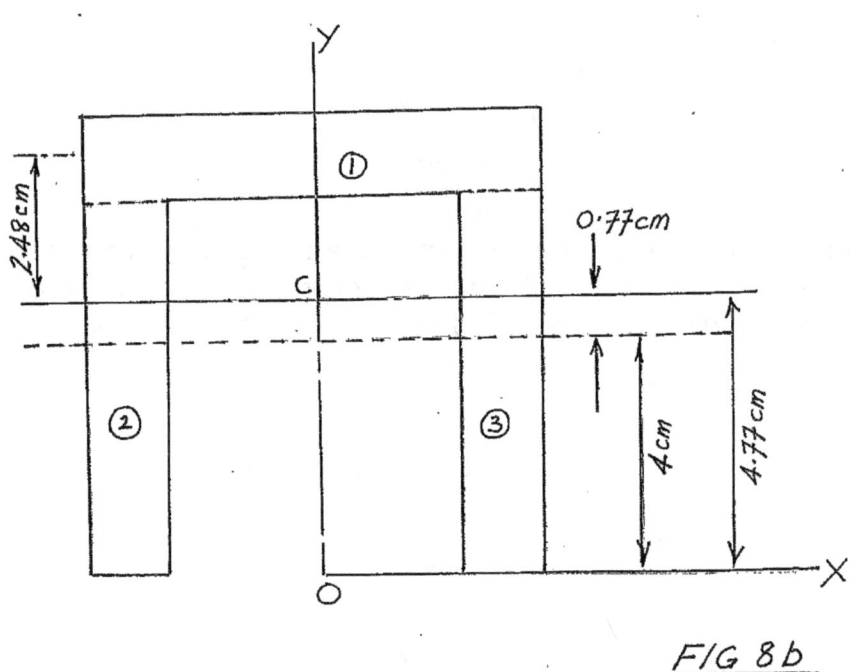

FIG 8b

$$I_C = \frac{8(1.5)^3}{12} + 8(1.5)(2.48)^2 + 2\left\{\frac{1.5(6.5)^3}{12} + 1.5(6.5)(0.77)^2\right\}$$

$$= 2.25 + 73.8 + 2(34.3 + 5.8)$$

$$= 75.05 + 80.2$$

$$\therefore I_C = 155.25(cm)^4$$

and

$$I_{OY} = \frac{1.5(8)^3}{12} + 2\left\{\frac{6.5}{12}(1.5)^3 + 6.5(1.5)(3.25)^2\right\}$$

$$= 64 + 2(1.8 + 103)$$

$$I_{OY} = 273.6(cm)^4$$

Q4. Determine the moment of inertia of the cross-sectional area of the I-beam shown in Fig. 1 about (i) *X-axis;* and (ii) *Y-axis,* these axes having their origin at the centroid of the area

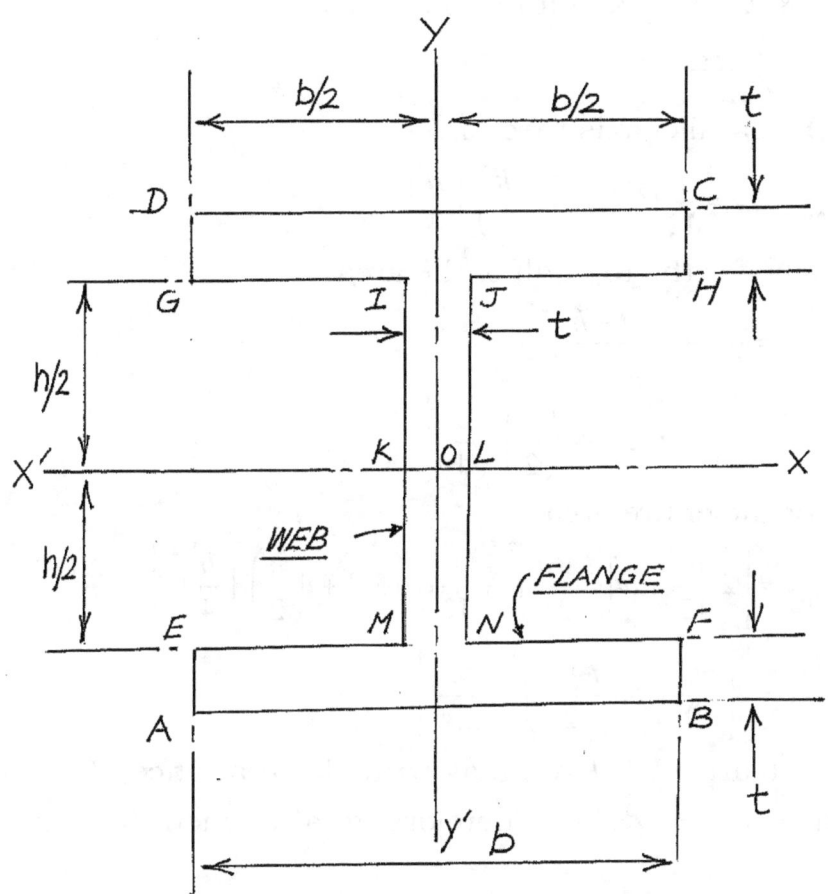

The area is symmetrical about both axes. Therefore its centroid is $h/2+t$ from the top *CD* of the upper flange and the same distance from the edge *AB* of the lower flange; hence the *XOX* -axis. Similarly, the centroid is $\left(\dfrac{b}{2}\right)$ from the imaginary lines *AEGD* and *BFHC*; hence the *YOY* -axis.

Again, because of symmetry I shall consider the top half of the area and multiply the result to obtain the whole.

For area *CDGH*, its I' about its own centroid is

$$I' = \frac{bt^3}{12}$$

Distance between its centroid and axis $\quad XOX = \left(\dfrac{t+h}{2}\right).$

Therefore by the parallel axis theorem

$$I'_{CDGH/XOX'} = \frac{bt^3}{12} + bt\left(\frac{t+h}{2}\right)^2$$

For area *IKLJ*, its I' about its centroid is

$$I'' = \frac{t(h/2)^2}{12}$$

And by the parallel axis theorem

$$I''_{IKLJ/XOX'} = \frac{t}{12}(h/2)^3 + t\left(\frac{h}{2}\right)\left(\frac{h}{4}\right)^2$$

Accordingly, for the top half of the area

$$I_{XOX'} = \frac{bt^3}{12} + bt\left(\frac{t+h}{2}\right)^2 +$$

$$\frac{t}{12}(h/2)^3 + t\left(\frac{h}{2}\right)\left(\frac{h}{4}\right)^2$$

So that for the entire area

$$I_{XOX'} = \left[2\frac{bt^3}{12} + bt\left(\frac{t+h}{2}\right)^2 + \frac{t}{12}(h/2)^3 + t\left(\frac{h}{2}\right)\left(\frac{h}{4}\right)^2\right]$$

$$= 2\left[\frac{bt^3}{12} + bt\left(\frac{t+h}{2}\right)^2 + \frac{th^3}{24}\right]$$

For the web area *IMNJ*, we may write down at once, $I_{YOY'} = ht^3/12$ and for each flange $I_{YOY'} = tb^3/12$. Therefore, total moment of inertia about the *Y-axis* is

$$I_{YOY'} = \frac{ht^3}{12} + 2\left(\frac{tb^3}{12}\right) = \frac{ht^3 + 2tb^3}{12}.$$

No one in his or her right mind would attempt to learn by rote the results just derived. My advice is always to work from first principles. As a consequence of frequent working you would easily recall the expressions for 'I' for basic figures such as rectangles and triangles about rectangular co-ordinate axes through their centroid; and, by applying the parallel axis theorem; readily work out the value of 'I' with respect to any other parallel axes.

Assignment:

Criticise the statement: "The term moment of inertia erroneously implies a quality of inertia."

Q5. Determine the polar moment of inertia, J_z and radius of gyration k_z of a right-angled triangle of height h and base b' shown in Fig. 19 with reference to the *Z-axis* passing through the centroid '0' at the intersection of the *X-axis* and *Y-axis* in the *X-Y plane* and perpendicular to it.

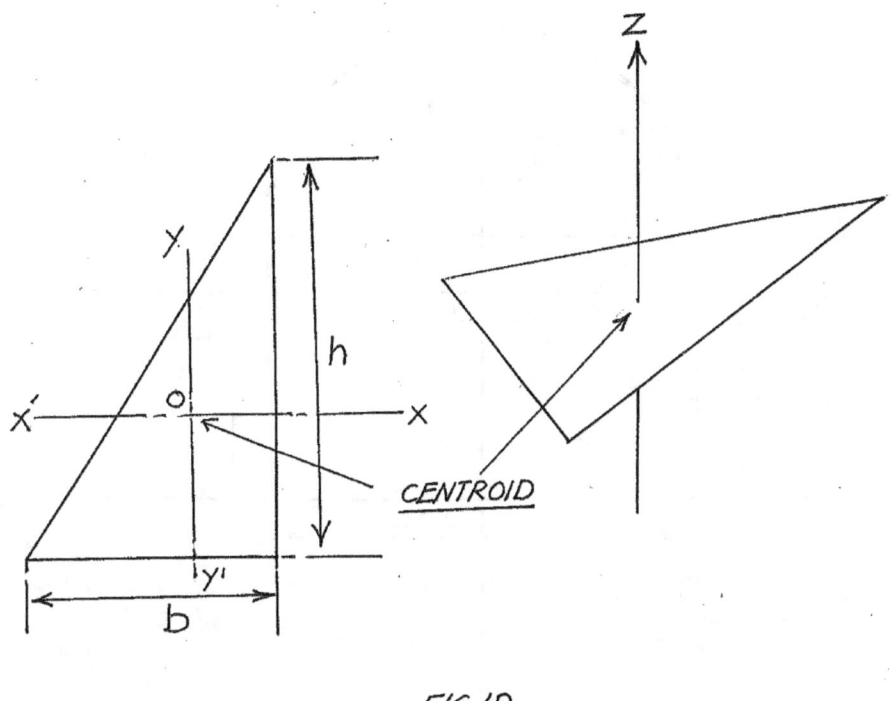

$$FIG\ 19$$

From previous work it is known that $I_{XX'}$ and $I_{YY'}$ with reference to orthogonal axes $X'OX$ and YOY' passing through the centroid '0' of the triangle are given by:

$$I_{XX'} = \frac{bh^3}{36} \quad ; \quad I_{YY'} = \frac{hb^3}{36}$$

Therefore, the polar moment of inertia, J_z, of the triangle with reference to the *Z-axis* is given by

$$J_z = \frac{bh^3}{36} + \frac{hb^3}{36} = \frac{bh}{36}\left(h^2 + b^2\right)$$

If k_z = radius of gyration, then

$$Ak_z^2 = \frac{bh}{36}\left(h^2 + b^2\right), \text{ where } A = \text{area of triangle}$$

i.e. $\quad \dfrac{1}{2}bh\,k_z^2 = \dfrac{bh}{36}(h^2 + b^2)$

from which

$$k_z = \frac{1}{3\sqrt{2}}\sqrt{h^2 + b^2}$$

Q6.   What is the polar moment of inertia of the area shown in Fig. 13 about its centroidal axis?

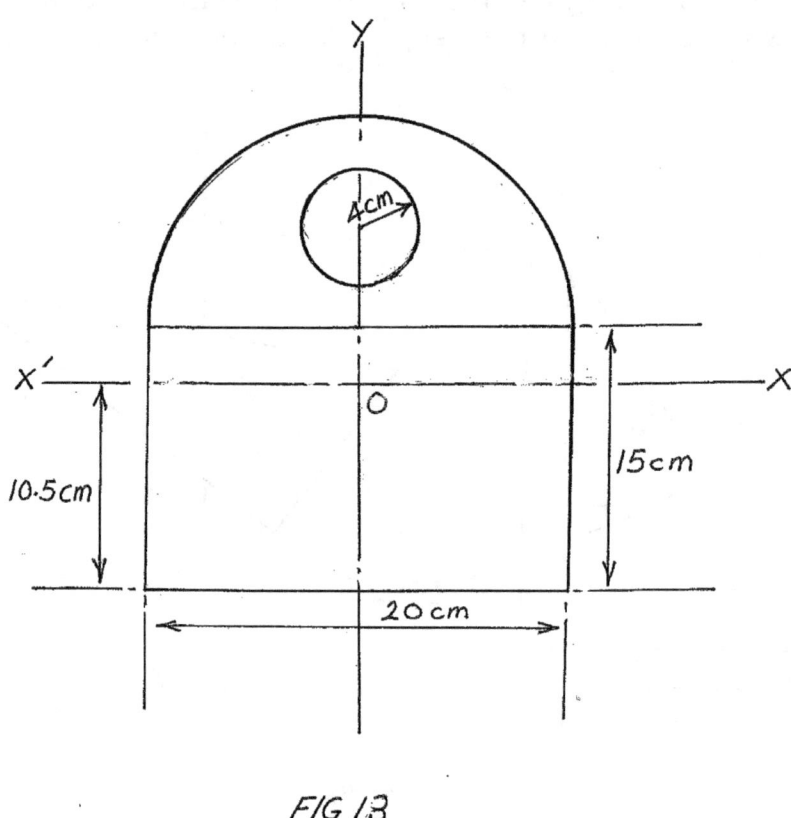

FIG 13

We already determined from an Illustrated Example that the value of the second moment of the entire area about the centroidal axis which is distant *10.5cm* from the base is *16,576.5cm⁴*. Because of symmetry the centroid of the composite area lies along the *YY'* - axis.

Here $\bar{x} = 0$.

Hence $I_{YY'}$ for semi-circular area $= \dfrac{\pi r^4}{8}$

$$= \dfrac{\pi (10)^4}{8}$$

$$= 3928 cm^4$$

$I_{YY'}$ for rectangular area $= \dfrac{15(20)^3}{12} cm^4$

$$= 10000 cm^4$$

$I_{YY'}$ for voided circular area $= \dfrac{\pi r^4}{4}$

$$= \frac{\pi(4)^4}{4}$$

$$= 219 cm^4$$

∴   Total $I_{YY'} = (3928 + 10000 - 219) cm^4$

$$= 13709 cm^4$$

∴ J_z for composite area about is centroid

$$= I_{XX'} + I_{YY'}$$

$$= 16576.5 cm^4 + 13709 cm^4$$

i.e. J_z = $30,285.5 cm^4$, say, $30,286 cm^4$

Q7.   Calculate the distance 'd' between the rectangular areas shown in Fig. 1 for the condition $I_{XX'} = I_{YY'}$

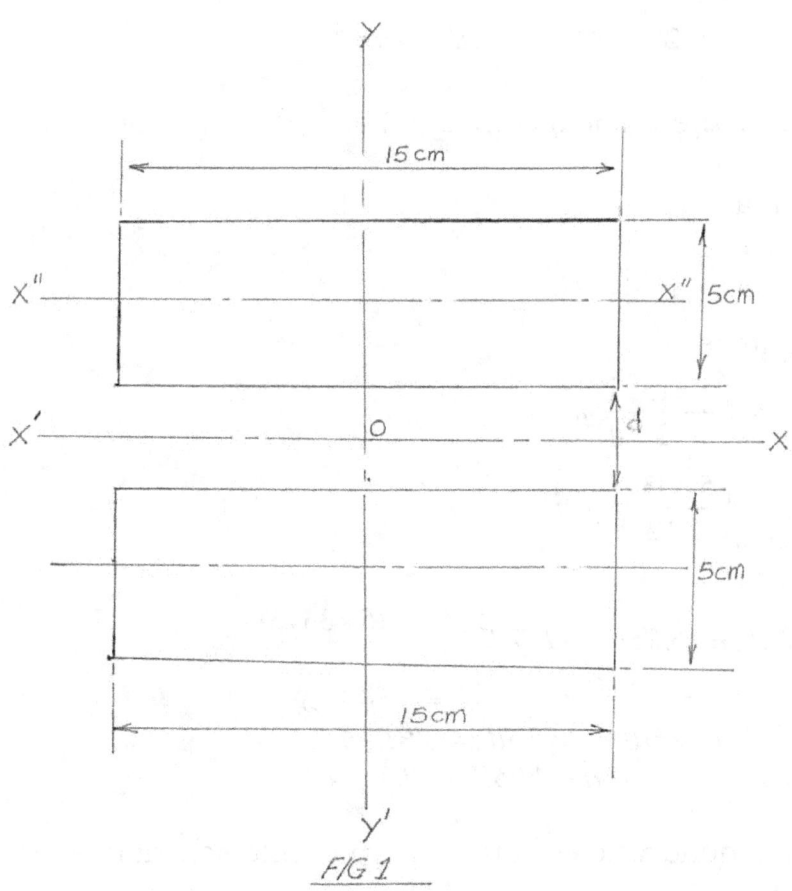

FIG 1

For $I_{X''X''}$; the horizontal centroidal axis through top area, we may straightway write

$$I_{X''X''} = \frac{15 \times (5)^3}{12} cm^4$$

Employing the Parallel-Axis theorem to transfer the axis to XOX'

$$I_{XX'} = I_{X'X'} + 75\left(\frac{d}{2} + 2.5\right)^2$$

$$= \frac{15(125)}{12} + 75\left(\frac{d^2}{4} + 2.5d + 6.25\right)$$

$$= \frac{625}{4} + 18.75d^2 + 187.5d + 468.75$$

i.e. $I_{XX'} = (625 + 18.75d^2 + 187.5d)cm^4$

For both areas (top and bottom):

$$I_{XX} = 2(625 + 18.75d^2 + 187.5d)cm^4$$

Now for $I_{YY'}$; the vertical centroidal axis through top area.

For top area
$$I_{YY'} = 5 \times \left(\frac{15^3}{12}\right)$$

For both areas
$$I_{YY'} = 2\left(\frac{5 \times 15^3}{12}\right)cm^4$$

$$= 2\left(\frac{5 \times 3375}{12}\right)cm^4$$

For $I_{XX'} = I_{YY'}$

$$2(625 + 18.75d^2 + 187.5d) = 2\frac{(5 \times 3375)}{12}$$

$$= 1406.25$$

∴ $18.75d^2 + 187.5d = 781.25$

or $d^2 + 10d - 41.67 = 0$

from which quadratic equation the possible solutions are:

$$d_{1,2} = \frac{-10 \pm \sqrt{100 + 166.7}}{2}$$

i.e. $d_{1,2} = \frac{-10 \pm 16.3}{2}$

Evidently the only possible solution is

$$d = \frac{6.3}{2}, \quad \text{say}$$

$$d = 3.2cm.$$

Q8. The enclosed S-shaped area in Fig. 1 may be assumed to be made up of semi-circular profiles. Given the dimensions shown find (i) the moment of inertia of the area about the horizontal axis through the centroid of the area; and (ii) the radius of gyration about the same axis

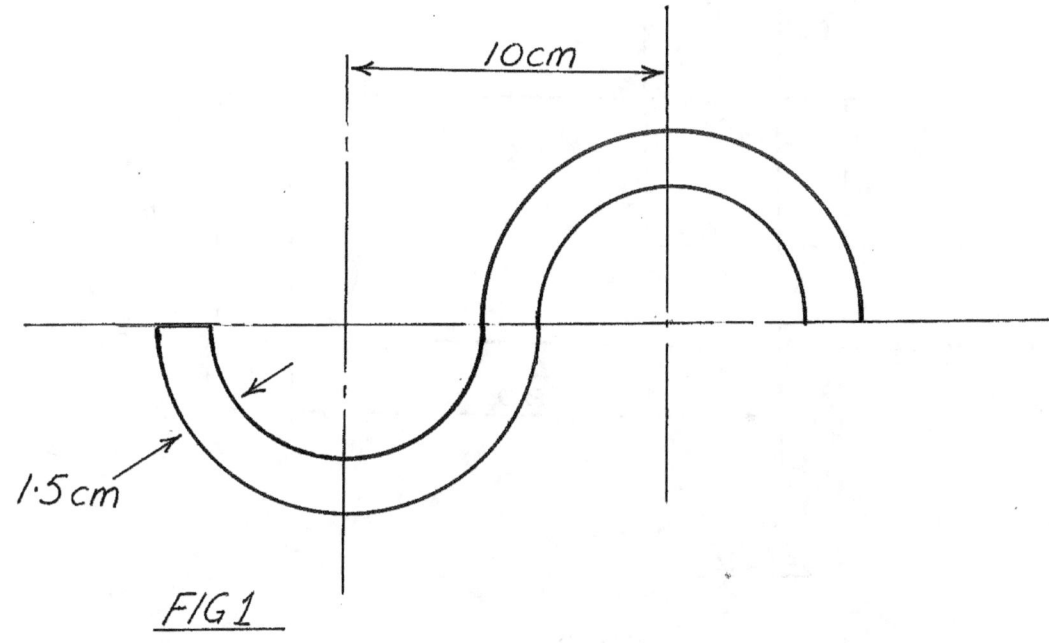

FIG 1

Ans = $602(cm)^4$; 7.2cm

Q9. State in your own words aided by appropriate sketch or sketches, the Parallel-Axis Theorem. The plane area having the dimensions shown in Fig 1 has the rectangular section *hcm wide* and *10cm deep* cut out of it as indicated. If the moment of inertia of the remaining section about the *XX'*-axis is 25320cm⁴ what is the value of *'h'*.

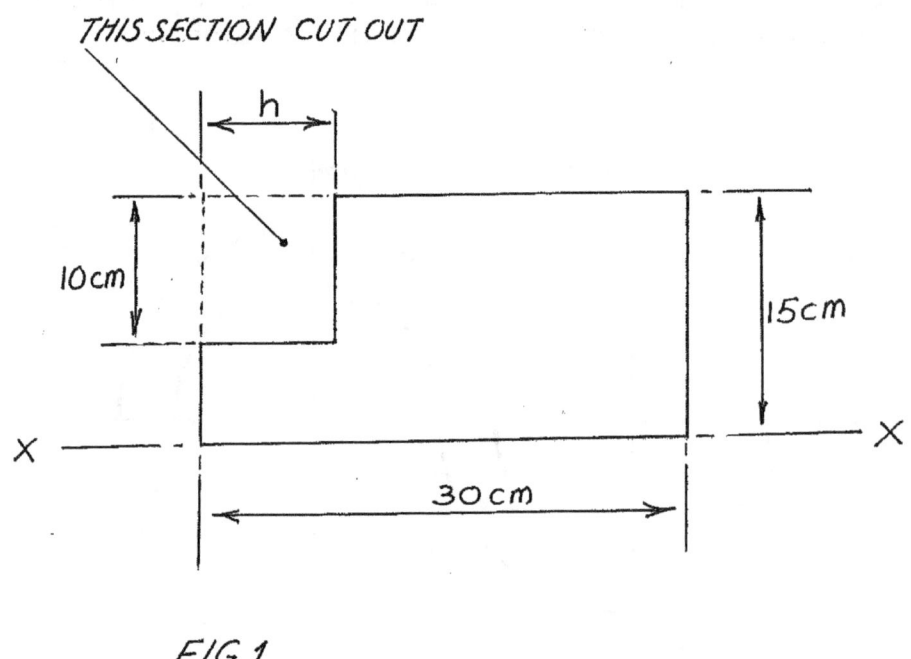

THIS SECTION CUT OUT

$FIG 1$

$I_{XX'}$ for the whole area $= \dfrac{30(15)^3}{3}$

$\qquad\qquad\qquad\qquad = 33750\ cm^4$

I_{CG} of area $h \times 10$ about its own centroid

$$= \frac{h(10)^3}{12}$$

To transfer this second moment of area use is made of the Parallel-Axis Theorem thus

$I_{XX'}$ for cutout $h \times 10$

$$= \frac{I}{CG} + \text{Area (Distance of centroid to } XX' - axis)^2$$

$$= \frac{h(10)^3}{12} + 10h(10)^2$$

$$= \frac{1000h}{12} + 1000h$$

~ 374 ~

Therefore

$I_{XX'}$ for whole area = $I_{XX'}$ for remaining part + $I_{XX'}$ for cut out

i.e.
$$33750 = 25320 + \frac{1000h}{12} + 1000h$$

$$8430 = \frac{500h}{6} + 1000h$$

from which $6500h = 50580$

i.e. $h = 7.8cm$

Q10. A steel stanchion has the cross-section shown in Fig. 1. Determine $I_{XX'}$ and $I_{YY'}$ in $(cm)^4$. What is product second moment of area of the section? Give the reasons for your answer, and state whether the values you obtained are the maximum and minimum product second moment of area. [See the compound struts on the building of an insurance company on upper St. Vincent Street, Port of Spain].

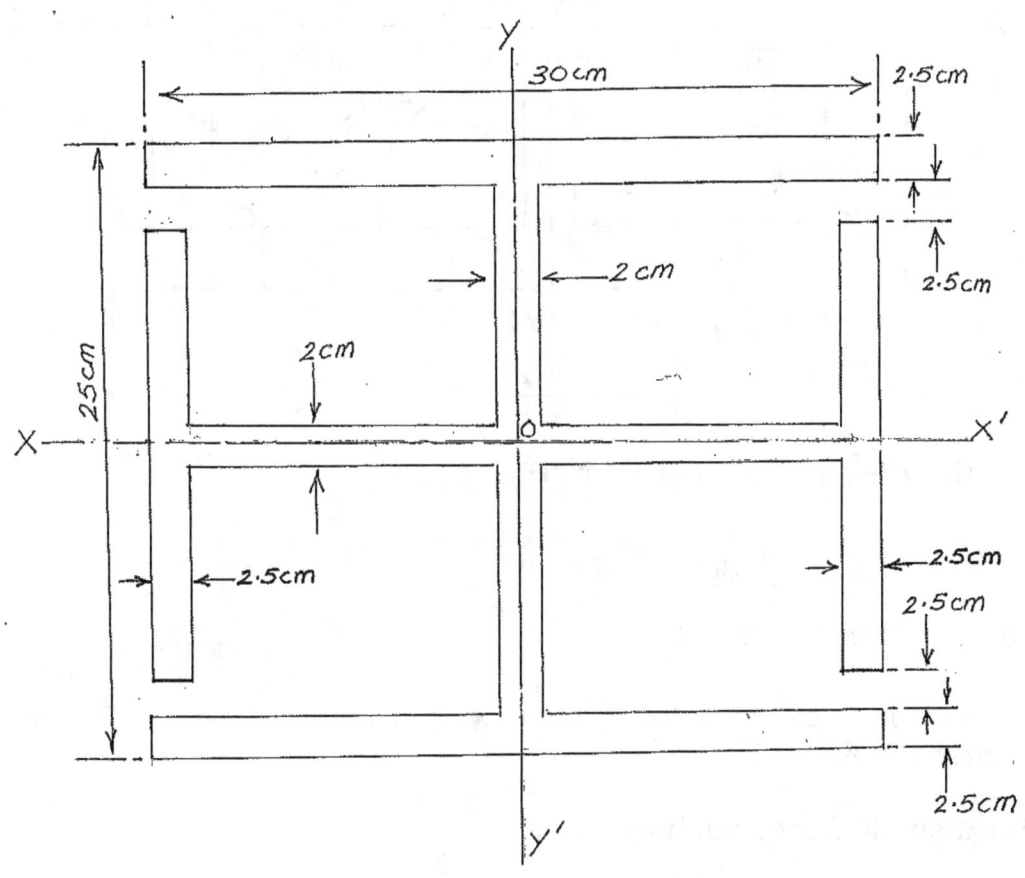

FIG. 1

In Fig. 2, the cross-sectional area is divided into 3 parts comprising: (i) the vertical I-beam comprising 2 flanges 30cm x 2.5cm and a web 2cm x 20cm; and (ii) 2 Tees, each comprising a flange : 15cm x 2.5cm and a web 11.5cm x 2 cm.

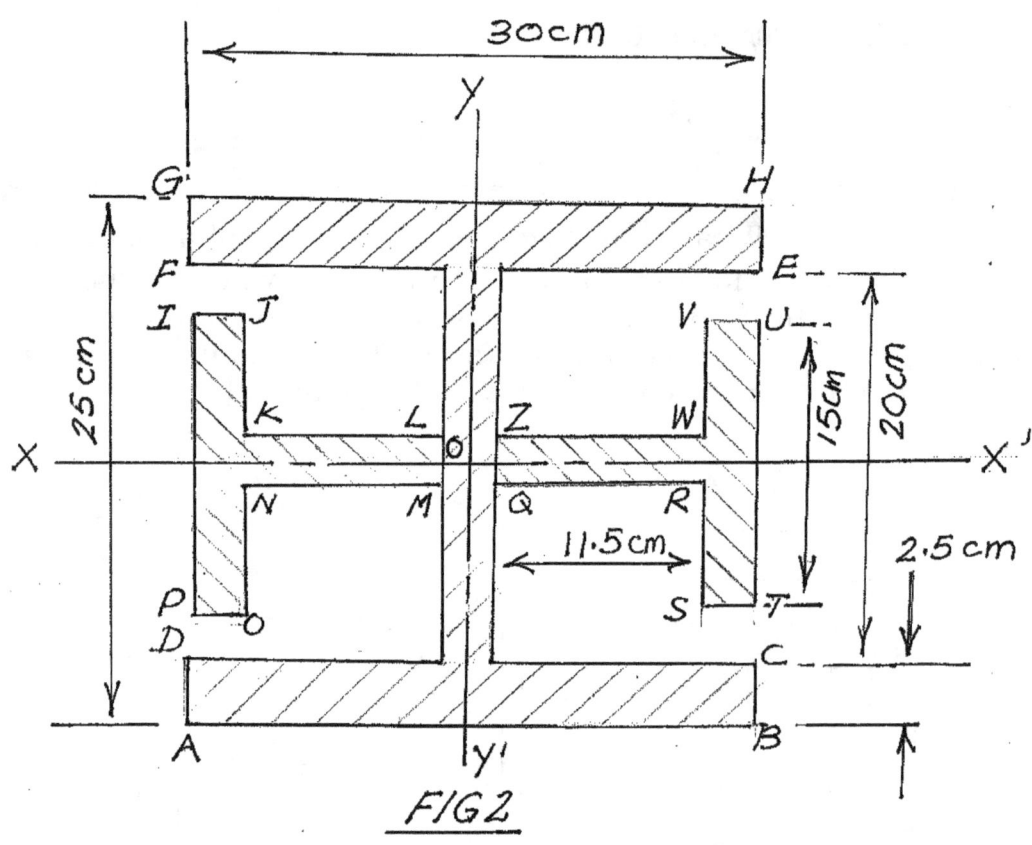

FIG2

For the *I-beam*, '*ABCDFEGH*' in Fig. 2,

$$I_{XX}(cm)^4 = \frac{1}{12}\left\{30(25)^3 - 28(20)^3\right\}$$

i.e. $I_{XX}(cm)^4 = 20395.8$

Turning now to the 2 Tees, each with a flange *15cm x 2.5cm* and a web *11.5cm x 2cm:*

For a single flange such as *STUV*

$$I_{XX}(cm)^4 = \frac{2.5}{12}\times(15)^3$$

and for a single web such as *QRWZ*

$$I_{XX'}(cm)^4 = \frac{11.5}{12} \times (2)^3$$

$\therefore$ For the latter flanges and web, the total I_{XX} is given by

$$I_{XX'} = (cm)^4 = 2\left\{\frac{2.5(15)^3}{12} + \frac{11.5(2)^3}{12}\right\}$$

$$= 2(703.125 + 7.66)$$

$$I_{XX}(cm)^4 \approx 1422$$

$\therefore$ *Total $I_{XX}(cm)^4$ for the I-section*

$$= 20395.8 + 1422$$

say, *21818*

Consider next Fig. 3

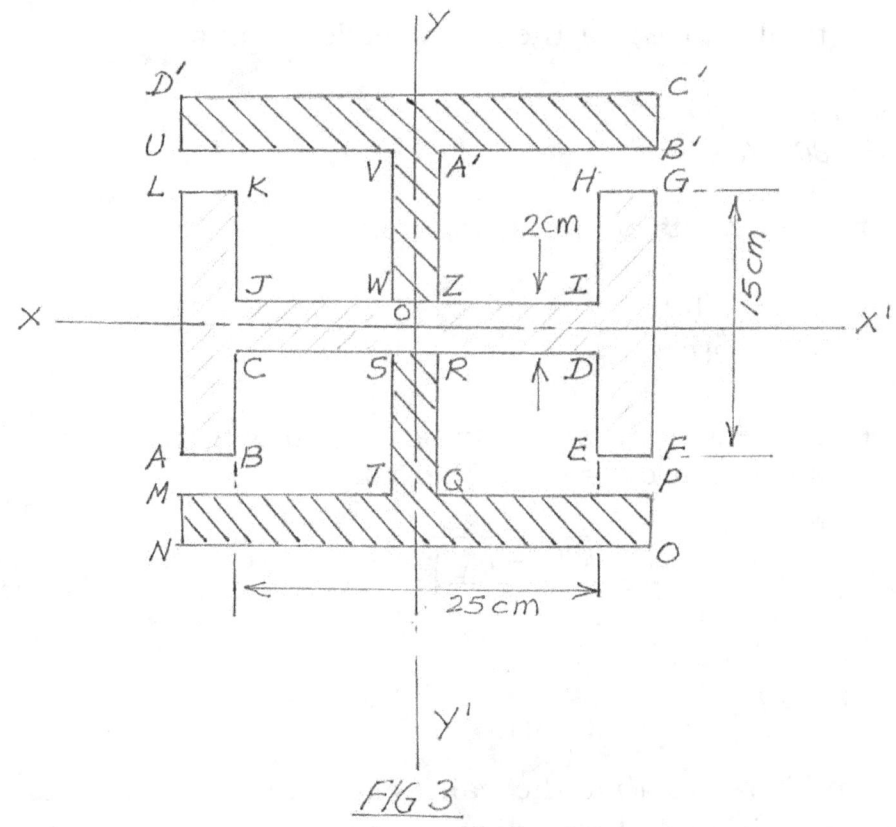

FIG 3

The horizontal I-beam 'LABCJWZIHGFED' consisting of 2 flanges 'LABK' and 'HEFG' and the web 'CSRDIZWJ' we have

$$\therefore \quad I_{YY'}(cm)^4 = \frac{1}{12}\left\{ 15(30)^3 - 13(25)^3 \right\}$$
$$= 16823$$

For the vertical Tees, we have for each of the flanges such as 'UB'C'D'

$$I_{YY'}(cm)^4 = \frac{1}{12}\left\{ 2.5(30)^3 \right\}$$

and for each web such as $WZA'V$

$$I_{YY'}(cm)^4 = \frac{9(2)^3}{12}$$

$\therefore$ $I_{YY'}(cm)^4$ for the flanges and webs such as 'U'B'CD' and 'WZA'V'

$$I_{YY'}(cm)^4 = 2\left[\frac{2.5(30)^3}{12} + \frac{9(2)^3}{12} \right]$$
$$= 11262$$

Therefore total $I_{XX'}(cm)$ for the cross-sectional area

$$= 16823 + 11262$$
$$= 28025 (cm)^4$$

Collecting the results of the calculation:

$$I_{XX'}(cm^4) = 21818$$
$$I_{YY'}(cm)^4 = 28025$$

Because the section is symmetrical about both axes, P_{xy} the product second moment of area is zero. Therefore

$$I_{max,min} = \frac{I_{YY} + I_{XX}}{2} \pm \sqrt{\left(\frac{I_{YY} - I_{XX}}{2} \right)}$$

i.e. $\quad I_{max} = I_{YY'}$ and

$$I_{min} = I_{XX'}$$

It is seen therefore that the values obtained for $I_{YY'}$ and $I_{XX'}$ in the foregoing are indeed the maximum and minimum second moments of area respectively

Q11. A slender rod of mass *5kg* and length *1m* has a metallic sphere of mass *15kg* screwed in at one end, the other end resting on a horizontal knife edge. The sphere which is solid has a diameter of *15cm*. Determine the moment of inertia of the rod-sphere combination about the horizontal knife-edge. See Fig. 1.

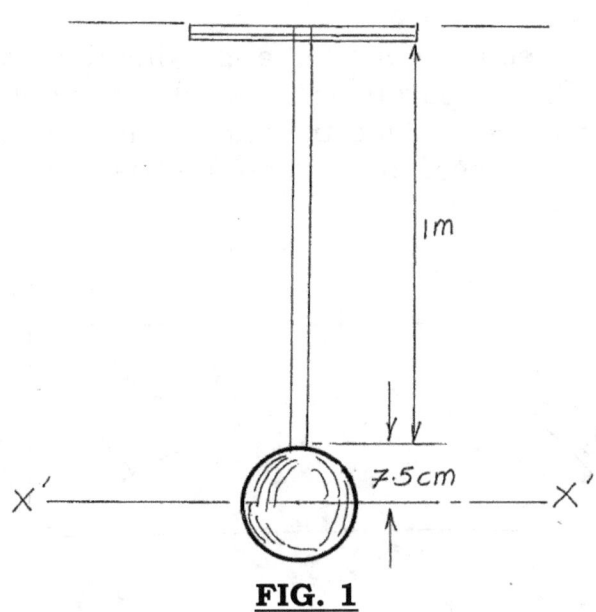

FIG. 1

For rod about knife-edge support $X''X$,

$$I_{X''X''} = \frac{m\ell^2}{12} + m\left(\frac{\ell}{2}\right)^2 = \frac{m\ell^2}{3}, \text{ where } M = \text{mass of rod assumed to be slender;}$$

and 'ℓ' its length.

Noting that $I_{XX'}$ for the sphere = $\frac{2}{5}Mr^2$, where M = mass of sphere; and r its radius and noting that this moment of inertia of the sphere has to be transferred to the knife edge '$X''X$', we have for the sphere.

$$I_{X'X} = \frac{2}{5}Mr^2 + M(107.5)^2 \text{ [Note that the distance from } XX' \text{ to } X''X \text{ is } (100 + 7.5)cm.]$$

Total $I_{X''X} = I_{X''X}$ for rod + $I_{X''X}$ for sphere

$$\therefore \quad I_{X''X} = \frac{m\ell^3}{3} + \frac{2}{5}Mr^2 + M(107.5)^2$$

Substituting the data provided

$$I_{X''X} = \frac{5(100)^2}{3} + \frac{2}{5}15(7.5)^2 + 15(107.5)^2$$

$$= 16666.7 + 337.5 + 173344$$
$$= 1903482 kg\ cm^2$$

and since

$$1m = 100cm$$
$$I_{x'x} = 190.3\ kg\ m^2$$

Q12. Figs. 1 and 2 represent an aerofoil approximation by aid of elliptical and parabolic arcs. If the parabolic arcs have their vertices on the *Y-axis,* determine the moment of inertia and polar moment of inertia of the enclosed area with respect to the *X-axis* and *Y-axis*

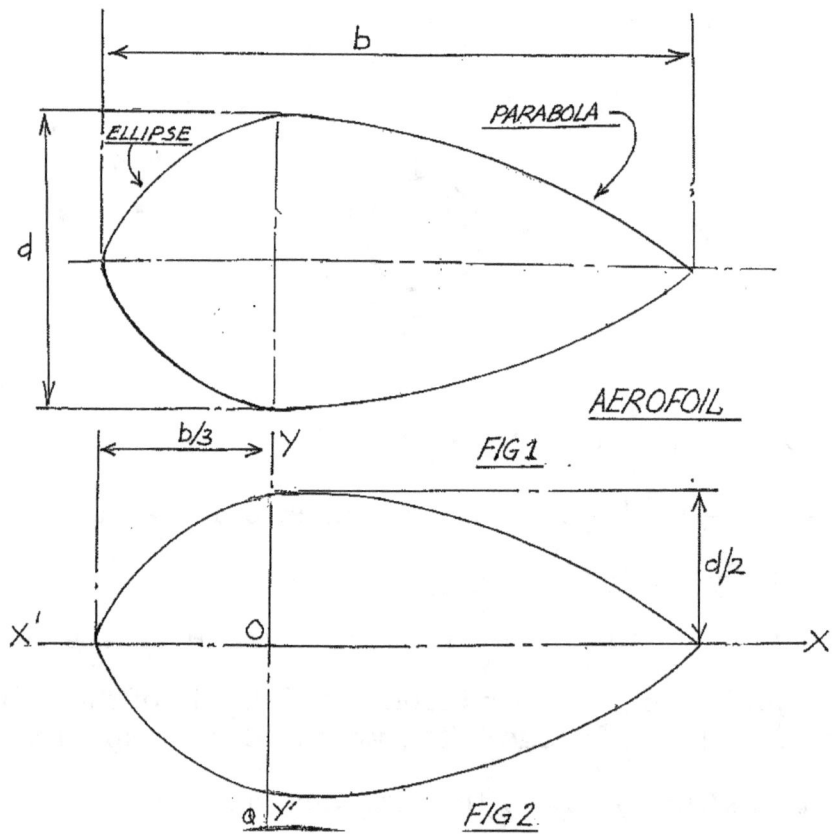

Observe that the parabolas were truncated. Therefore their vertices (their origins) were not shown. But we are told that they are on the *Y-axis*, at the origin.

Let equation of parabola be the form $y = ax^2 + cx + e$

when $x = 0$; $y = d/2$; and when $x = +\dfrac{2b}{3}, y = 0$

$$\therefore \quad \frac{d}{2} = o + o + e \qquad\qquad \therefore \quad e = \frac{d}{2}$$

Also
$$o = a\left(\frac{+2b}{3}\right)^2 + c\left(\frac{+2b}{3}\right) + \frac{d}{2};$$

i.e.
$$o = a\left(\frac{4b^2}{9}\right) + \frac{2bc}{3} + \frac{d}{2}$$

and
$$o = a\left(\frac{-2b}{3}\right)^2 + c\left(\frac{-2b}{3}\right) + \frac{d}{2};$$

i.e.
$$o = a\left(\frac{4b^2}{9}\right) - \frac{2b}{3}c + \frac{d}{2}$$

Accordingly, $\quad c = o,\ $ and
$$\frac{4b^2}{9}a = -\frac{d}{2}$$

or
$$a = -\frac{9d}{8b^2}$$

so that
$$y = \frac{d}{2} - \frac{9dx^2}{8b^2} \qquad\qquad \dots\dots\dots\ \text{(i)}$$

is the equation of the parabola

For the ellipse
$$\frac{x^2}{\left(\dfrac{b}{3}\right)^2} + \frac{y^2}{\left(\dfrac{d}{2}\right)^2} = 1$$

i.e.
$$\frac{9x^2}{b^2} + \frac{4y^2}{d^2} = 1$$

from which $\quad y^2 = \dfrac{d^2}{4} - 9\dfrac{d^2 x^2}{4b^2}$

or
$$y^2 = \frac{d^2}{4}\left(1 - \frac{9x^2}{b^2}\right)$$

or
$$y = \frac{d}{2}\left(1 - \frac{9x^2}{b^2}\right)^{\frac{1}{2}} \qquad\qquad \dots\dots\dots\ \text{(ii)}$$

is the equation of the ellipse

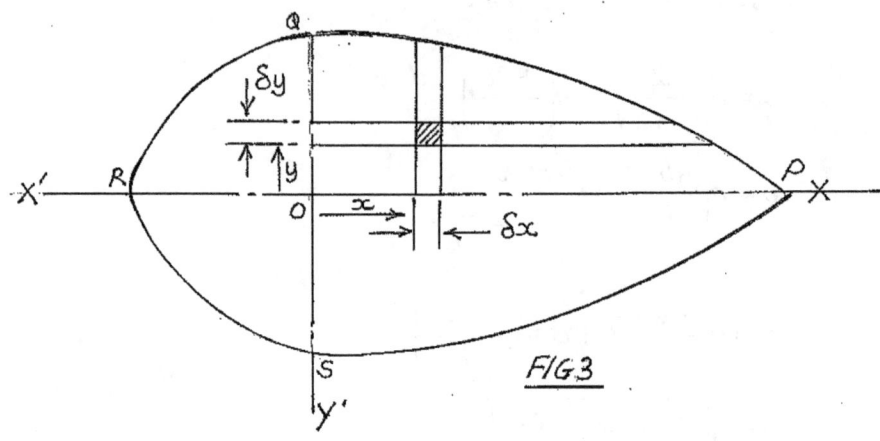

FIG.3

MI of area 'opq
w.r.t. X-axis

$$= \int_{3}^{\frac{2b}{3}} \int_{y_1}^{y_2} y^2 dy\ dx$$

$$= \int_{0}^{\frac{2b}{3}} \left[\begin{array}{c} y = \dfrac{d}{2} - \dfrac{9dx^2}{8b^2} \\[3mm] \\ y^2 dy \end{array} \right] dx$$

$$= \int_{0}^{\frac{2b}{3}} \left[\dfrac{y^3}{3} \right]_{0}^{y=\frac{d}{2}-\frac{9dx^2}{8b^2}} dx$$

$$= \int_{0}^{\frac{2b}{3}} \left[\dfrac{y^3}{3} \right]_{0}^{y=\frac{d}{2}-\frac{9dx^2}{8b^2}} dx$$

$$= \int_{0}^{\frac{2b}{3}} \dfrac{1}{3} \left(\dfrac{d}{2} - \dfrac{9dx^2}{8b^2} \right) dx$$

$$= \dfrac{d^3}{3} \int_{0}^{\frac{2b}{3}} \left(\dfrac{1}{2} - \dfrac{9dx^2}{8b^2} \right)^3 dx$$

$$= \dfrac{d^3}{3.2^3} \int_{0}^{\frac{2b}{3}} \left(1 - \dfrac{9x^2}{4b^2} \right) dx$$

$$= \dfrac{d^3}{24} \int_{0}^{\frac{2b}{3}} \left\{ 1 - \left(\dfrac{9x^2}{4b^2} \right)^3 - 3(1)\dfrac{9x^2}{4b^2} + 3(1)\left(\dfrac{81x^4}{16b^4} \right) \right\} dx$$

~ 382 ~

$$= \frac{d^3}{24} \int \left[1 - \frac{729x^6}{7(64)b^6} - \frac{27x^2}{4b^2} + \frac{243x^4}{16b^4} \right] dx$$

$$= \frac{d^3}{24} \left[x - \frac{729x^7}{7(64)b^6} - \frac{9x^3}{4b^2} + \frac{243x^5}{5(16)b^4} \right]_0^{\frac{2b}{3}}$$

$$= \frac{d^3}{24} \left[\frac{2b}{3} - \frac{729(128)b^7}{7(64)b^2(27)27(3)} - \frac{9(8b^3)}{27.4b^2} + \frac{243(32)b^5}{5(16)b^4 \, 243} \right)$$

$$= \frac{d^3}{24} \left[\frac{2b}{3} - \frac{128b}{7(64)(3)} - \frac{8b^3}{12b^2} + \frac{2b^5}{5b^4} \right]$$

$$= \frac{d^3}{24} \left[\frac{2b}{3} - \frac{128b}{21(64)} - \frac{2b}{3} + \frac{2b}{5} \right]$$

$$= \frac{d^3}{24} \left(\frac{2b}{5} - \frac{2b}{21} \right) = \frac{4bd^3}{315}$$

$\therefore$ MI of area 'opq
w.r.t. X-axis $= \dfrac{4bd^3}{315}$

$\therefore$ MI of area 'opqsp'
w.r.t. X-axis $= 2\left(\dfrac{4bd^3}{315} \right)$

$$= \frac{8bd^3}{315}$$

Considering the elliptical part of the area, we have

MI of area 'opr'
w.r.t. X-axis (See
Fig. 2)
$$= \int_0^{b/3} \int_{y=0}^{y=\frac{d}{2}\left(1-\frac{9x^2}{b^2}\right)^{1/2}} y^2 \, dy \, dx$$

$$= \int_0^{b/3} \left[\frac{y^3}{3} \right]_0^{y=\frac{d}{2}\left(1-\frac{9x^2}{b^2}\right)^{1/2}}$$

$$= \frac{1}{3} \int_0^{b/3} \frac{d^3}{8} \left(1 - \frac{9x^2}{b_2} \right)^{3/2} dx$$

Put $\dfrac{3x}{b} = Cos \, u$

$\therefore$ $MI = \dfrac{d^3}{24} \int_{24}^{b/3} (1 - Cos^2)^{3/2} \left(-\dfrac{b}{3} \right) Sinu \cdot du$

$$= \frac{d^3}{24} \int_0^{b/3} (Sin^2u)^{3/2} \left(-\frac{b}{3}\right) Sin u \ du$$

$$= \frac{bd^3}{72} \int_0^{b/3} Sin^4u \ du$$

Writing $I_n = \int Sin^4u \ du$, then

$$I_4 = -\frac{1}{4} Sin^3u Cos \ u + \frac{3}{4} \int \frac{1}{2}(1 - Cos2u)du$$

$$= -\frac{1}{4} Sin^3u \ Cos \ u + \frac{3}{8} \int \frac{1}{2}\left[u - \frac{1}{2}Sin2u\right]$$

$$= -\frac{1}{4} Sin^3u \ Cos \ u + \frac{3}{8}u - \frac{3}{16}Sin2u$$

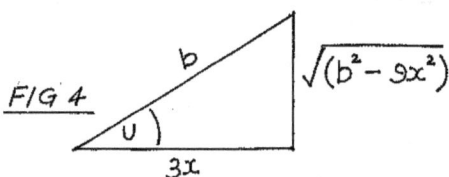

FIG 4

Referring to Fig. 4

$$Sin \ u = \frac{\sqrt{b^2 - 9x^2}}{b} = \sqrt{1 - \frac{9x^2}{b^2}}$$

$$\therefore \ I_4 = \frac{1}{4}\left\{\left(1 - \frac{9x^2}{b^2}\right)^{1/2}\right\}^3 \cdot \frac{3x}{b} + \frac{3}{8}Cos^{-1}\left(\frac{3x}{b}\right) - \frac{3}{16}(2)\left(1 - \frac{9x^2}{b^2}\right)\cdot\frac{3x}{b}$$

$\therefore$ MI of 'oqr' w.r.t.
X-axis

$$= -\frac{bd^3}{72}\left[-\frac{1}{4}\left(1 - \frac{9x^2}{b^2}\right)^{3/2}\cdot\frac{3x}{b} + \frac{3}{8}Cos^{-1}\left(\frac{3x}{b}\right) - \frac{3}{16}\cdot2\left(1 - \frac{9x}{b_2}\right)^{1/2}\cdot\frac{3x}{b}\right]_0^{b/3}$$

$$= -\frac{bd^3}{72}\left[-\frac{1}{4}\left(1 - \frac{9}{b^2}\cdot\frac{b^2}{9}\right)\cdot\frac{3}{b}\cdot\frac{b}{3} + \frac{3}{8}Cos^{-1}\left(\frac{3}{b}\cdot\frac{b}{3}\right)\right.$$

$$\left. -\frac{3}{16}\cdot2\left(1 - \frac{9}{b^2}\cdot\frac{b^2}{9}\right)\cdot\frac{3}{b}\cdot\frac{b}{3}\right]$$

$$-\left[-\frac{1}{4}(1-0)0 + \frac{3}{8}Cos^{-1}(0) - \frac{3}{16}\cdot2(1-0)(0)\right]$$

~ 384 ~

$$= -\frac{bd^3}{72}\left\{\frac{3}{8}Cos^{-1}(1)\right\} - \left\{\frac{3}{8}Cos^{-1}(0)\right\}$$

$$= -\frac{bd^3}{72}\left(\frac{3}{8}\cdot 0 - \frac{3}{8}\cdot\frac{\pi}{2}\right)$$

$$= \frac{bd^3}{72\cdot}\cdot\frac{3}{8}\cdot\frac{\pi}{2}$$

$$= \frac{\pi bd^3}{24(8)(2)}$$

MI of area 'oqr'
w.r.t. X-axis

$$\therefore \qquad = \frac{\pi bd^3}{384}$$

so that for total area 'oqrs' = $2\left(\dfrac{\pi bd^3}{384}\right)$

$$= \frac{\pi bd^3}{193}$$

$\therefore$ Total *MI* of areas enclosed by the parabolas and ellipses with respect to x-axis

$$= \frac{8bd^3}{315} + \frac{\pi bd^3}{192}$$

$$= \frac{2525.73}{315(192)}bd^3$$

$$\approx \frac{bd^3}{23.9}, \quad\text{say,}\quad \frac{bd^3}{24}$$

To determine *MI* of enclosed area of aerofoil about *Y-axis.*

Re-express equation of parabola in the form

$$x = \frac{2b}{3}\sqrt{(1-2y/d)}$$

Now for ½ parabolic area

MI of area 'opq' $\qquad = \displaystyle\int_0^{d/2}\int_0^{x=\frac{2b}{3}\left(1-\frac{2y}{d}\right)^{1/2}} x\,dx\,dy$

w.r.t. *Y-axis*

$$= \int_0^{d/2} \left[\frac{x^3}{3} \right]_0^{x=\frac{2b}{3}\left(1-\frac{2y}{d}\right)^{1/2}} dy$$

$$= \frac{1}{3} \int_0^{d/2} \left(\frac{2b}{3} \right)^3 \left(1 - \frac{2y}{d} \right)^{3/2} dy$$

$$= \frac{8b^3}{81} \int_0^{d/2} \left(1 - \frac{2y}{d} \right)^{3/2} dy \qquad \dots\dots\dots\dots \text{(i)}$$

Putting $Sin^2 u = \dfrac{2y}{d}$, expression (i) becomes

$$= \frac{8b^3}{81} \int_0^{d/2} \left(1 - Sin^2 u \right)^{3/2} \cdot \frac{d}{2} Sin 2u \; du$$

$$= \frac{8b^3 d}{81} \int_0^{d/2} \cdot Cos^3 u \cdot \frac{2}{2} \cdot Sin\, u \; Cos\, u \; du$$

$$= \frac{8b^3 d}{81} \int_0^{d/2} Cos^4 u \; Sin u \; du \qquad \dots\dots\dots\dots \text{(ii)}$$

Writing $\quad k = Cos u$
$$dk = -Sin u \cdot du$$

Therefore (ii) becomes

$$= \frac{8bd^3}{81} \int_0^{d/2} -k^4 dk$$

$$= \frac{8db^3}{81} \left[-\frac{k^5}{5} \right]_0^{d/2}$$

$$= -\frac{8db^3}{405} \left[Cos^5 u \right]_0^{d/2}$$

But $\quad Cos\, u = \sqrt{1 - Sin^2 u}$

$$= \sqrt{1 - \left(\frac{2y}{d} \right)^2} = \sqrt{1 - \frac{4y^2}{d^2}}$$

$\therefore \qquad Cos^5 u = \left(1 - \dfrac{4y^2}{d^2} \right)^{5/2}$

$\therefore \quad MI \text{ w.r.t } Y\text{-axis} = -\dfrac{8db^3}{405} \left[\left\{ 1 - \dfrac{4y^2}{d^2} \right\}^{5/2} \right]_0^{d/2}$

$$= -\frac{8db^3}{405} \left[1 - \frac{4}{d^2} \cdot \frac{d^2}{4} - 1 - 0 \right]$$

$$= -\frac{8db^3}{405}(1-1-1-0)$$

$$= +\frac{8db^3}{405}$$

so that *MI* about *Y-axis* for the total area enclosed by the parabolic arcs

$$2\left(\frac{8db^3}{405}\right) = \frac{16db^3}{405}$$

For determination of the *MI* of the elliptical area about *Y-axis* we write the equation of the ellipse in the form

$$x = \sqrt{\frac{b^2}{9} - \frac{4y^2 b^2}{9d^2}}$$

$$= \frac{b^2}{9}\left\{\left(1 - \frac{4y^2}{d^2}\right)\right\}^{1/2}$$

$$= \frac{b}{3}\left(1 - \frac{4y^2}{d^2}\right)^{1/2}$$

$\therefore$ MI of area 'oqr'
w.r.t. Y-axis

$$= \int\limits_{0} \int\limits_{3\left(\ \ d^2\ \right)} x^2 dx\ dy$$

$$= \int_{0}^{d/2} \left[\frac{x^3}{3}\right]_{0}^{b/3\left(1-\frac{4y^2}{d^2}\right)^{1/2}} dy$$

$$= \frac{1}{3}\int_{0}^{d/2} \left(\frac{b}{3}\right)^3 \left(1 - \frac{4y^2}{d^2}\right)^{3/2} dy$$

$$= \frac{b^3}{81}\int_{0}^{d/2} \left(1 - \frac{4y^2}{d^2}\right)^{3/2} dy$$

Let $\dfrac{2y}{d} = Cos\ u$ so that $\dfrac{2}{d}dy = -Sin\ u\ du$.

$\therefore$ Integral $= \dfrac{b^3}{81}\int^{d/2} \left(1 - Cos^2 u\right)^{3/2}\dfrac{d}{u}\cdot(-Sin\ u\ du)$

$$= \frac{b^3}{81}\ \frac{d}{2}\int_{0}^{d/2} -Sin^3 u\ Sinu\ du$$

$$= -\frac{db^3}{162} \int_0^{d/2} Sin^4u \; du$$

$$= -\frac{db^3}{162} \left[-\frac{1}{4} Sin^3u \; Cos \; u \right]_0^{d/2} + \frac{3}{4} \int_0^{d/2} Sin^2u \; du$$

But $Sin^2u = \dfrac{1 - Cos2u}{2}$

$$\therefore \text{Integral} = -\frac{db^3}{162} \left[-\frac{1}{4} Sin^3u \; Cos \; u + \frac{3}{4} \int_0^{d/2} \left(\frac{1 - Cos2u}{2} \right) du \right.$$

$$= -\frac{db^3}{162} \left[-\frac{1}{4} Sin^3u \; Cos \; u + \frac{3}{8}u - \frac{3}{8} \int_0^{du} Cos2u \; du \right.$$

$$= -\frac{db^3}{162} \left[-\frac{1}{4} Sin^3u \; Cos \; u \; \frac{3}{8}u - \frac{3}{16} Sin2u \right]_0^{d/2}$$

Substituting $Cosu = 2y/d$ which means that

$$Sinu \quad = \frac{\sqrt{d^2 - 4y^2}}{d}$$

$$= \sqrt{1 - 4y^2 / d^2}$$

$$\therefore \quad \text{Integral} = -\frac{db^3}{162} \left\{ -\frac{1}{4} \left(\sqrt{1 - \frac{4y^2}{d^2}} \right)^3 \cdot \frac{2y}{d} + \frac{3}{8} Cos^{-1} \frac{2y}{d} \right.$$

$$-\frac{3}{16} \cdot 2 \left(\sqrt{1 - \frac{4y^2}{d^2}} \right) \frac{2y}{d} - \left\{ -\frac{1}{4} \left(\sqrt{1 - \frac{4y^2}{d^2}} \right)^3 \frac{2y}{d} \right.$$

$$\left. +\frac{3}{8} Cos^{-1} \frac{2y}{d} - \frac{3}{16} \cdot 2 \left(\sqrt{1 - \frac{4y^2}{d^2}} \right) \cdot \frac{2y}{d} \right]_0^{d/2}$$

$$= -\frac{db^3}{162} \left[\left\{ 0 + \frac{3}{8} Cos^{-1}(1) - 0 - \right\} - \left\{ -0 + \frac{3}{8} Cos^{-1}(0) - 0 \right\} \right]$$

$$= -\frac{db^3}{162} \left[0 - \frac{3}{8} \cdot \frac{\pi}{2} \right]$$

MI of area
'oqr' w.r.t.

Y-axis

$$= +\frac{db^3}{162} \cdot \frac{3\pi}{16}$$

$$= \frac{\pi \, db^3}{864}$$

∴   MI for whole
area 'oqrs'
w.r.t. Y-axis

$$= 2\left\{\frac{\pi \, db^3}{864}\right\}$$

∴ Total MI w.r.t. Y-axis of

area enclosed by aerofoil

$$= \frac{\pi db^3}{432}$$

$$= \frac{16db^3}{405} + \frac{\pi \, db^3}{432} = 0.0395 + 0.0073$$

$$= 0.0468$$

$$\approx \frac{db^3}{21}$$

Accordingly, polar moment of inertia of aerofoil (about perpendicular axis through intersection of *X- and Y-axis*
say

$$I_Z = I_{XX} + I_{YY}$$

$$= \frac{bd^3}{24} + \frac{db^3}{21} = \left(\frac{7bd^3 + 8db^3}{168}\right)$$

$$= bd(7d^2 + 8b^2)/168$$

Polar moment of inertia $\approx \dfrac{bd(7d^2 + 8b^2)}{168}$

In Table 1, I have tabulated the values of Second Moments of Area of some common geometrical shapes. This is mostly for reference purposes.

Q13. Determine the product moment of area about its centroidal axis of the composite plane area shown in Fig. 1.

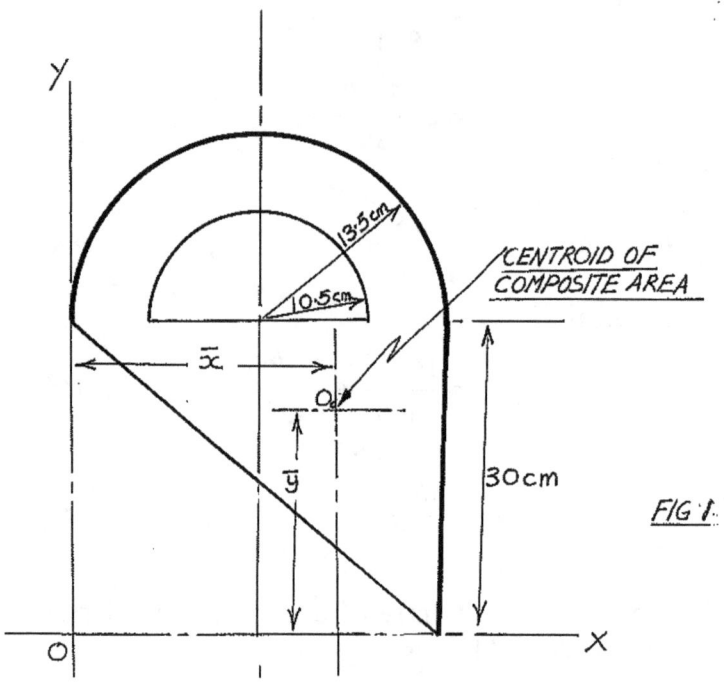

FIG 1.

Let $\bar{x}$ and $\bar{y}$ be the coordinates of the centroid of the composite area:

$$\therefore \qquad \bar{x} = \frac{\dfrac{27(30)}{2}(18) + \dfrac{\pi}{2}(13.5)^2(13.5) - \dfrac{\pi}{2}(10.5)^2 10.5}{27(15) + \dfrac{\pi}{2}(13.5)^2 - \dfrac{\pi}{2}(10.5)}$$

$$= \frac{7290 + 3865 - 1819}{528}$$

$$= 20.96cm, \text{ say, } 21cm$$

$$\text{and} \qquad \bar{y} = 27\frac{(30)20}{2} + \frac{\pi}{2}(13.5)^2\left\{\frac{4}{3}(13.5) + 30\right\}$$

$$\frac{-\dfrac{\pi}{2}(10.5)\left\{\dfrac{4}{3\pi}(10.5) + 30\right\}}{528}$$

$$= 23.3cm$$

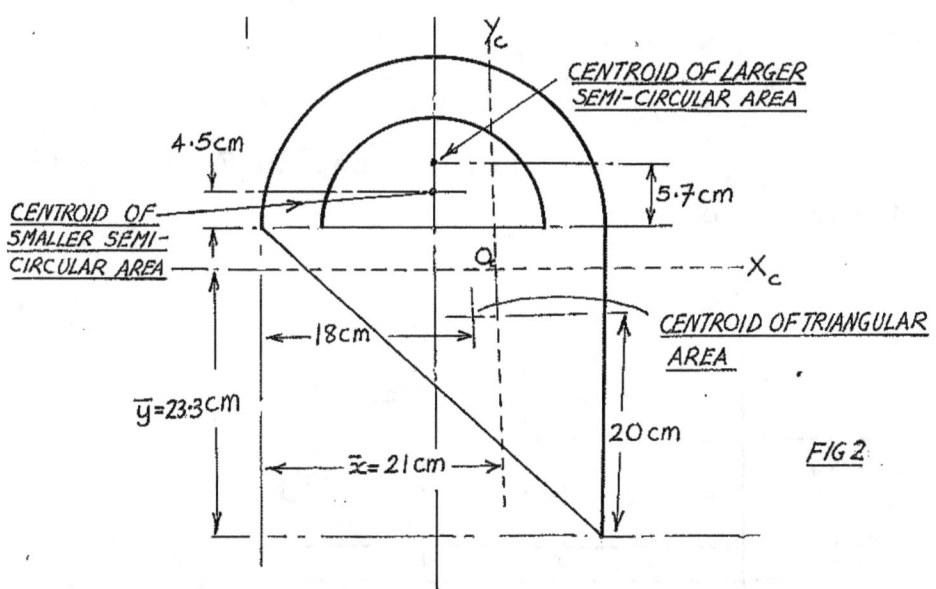

FIG 2

'x','y' distances of centroid of component areas from $\bar{x}$ and $\bar{y}$ of the composite area, shown again as Fig. 2 :

For (i):   x = 18 - 21 = -3cm   ; y = 20 - 23.5 = -3.3cm
"   (ii): x = 13.5 - 21 = -7.5cm ; y = 35.7 - 23.3 = 12.4cm
"   (iii) x = 13.5 - 21 = -7.5cm ; y = 34.5 - 23.3 = +11.2cm

These data and all other relevant information required to complete the calculation are shown in the following tabulation:

Component Area	Area Sq.cm	x	y	Axy	P_{xy}
(1)	405	-3	-3.3	+4009.5	$\dfrac{-(30)^2(27)^2}{72} = -9112.5$
(2)	286	-7.5	12.4	-26598	0
(3)	173	-7.5	11.2	-14532	0

For all areas: $P = \sum \left(P_{\overline{xy}} + Axy \right)$

Let me write this out in full to avoid any confusion

$P = \sum \left(P_{\overline{xy}} + Axy \right)$ = $(P_{xy(\text{its own centroid})} + A_{(1)}\, x'y')$ for the triangle +

$(P_{xy(\text{its own centroid})} + A_{(2)}\, x''y'')$ for the semi-circle) -

$(P_{xy(\text{its own centroid})} + A_{(3)}\, x'''y''')$ for the voided area

= -9112.5 + 4009.5 + 0 - 26598)

~ 391 ~

$-\{0+(-14532)\}$

i.e. $P_{xy} = -17,169cm^4$

Q14. (1) Show that the $P_{x'y'}$ for the rectangle shown in Fig.1 is $b^2h^2/4$, with reference to the *XY-axis*.

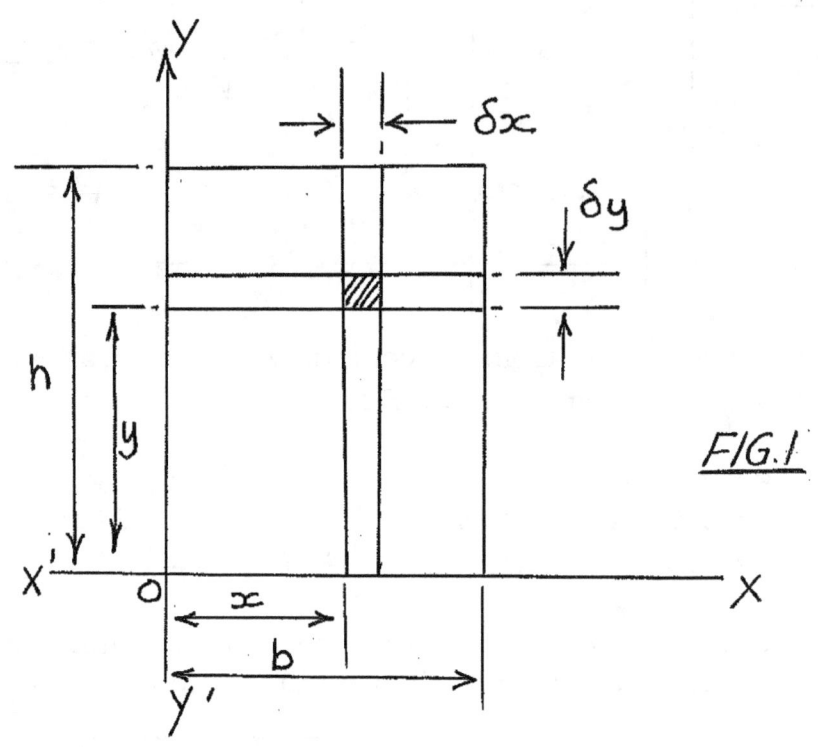

FIG.1

$$P_{XY} = \int xy \; dA$$

Putting $\quad dA = \delta y \; \delta x,$

$$P_{XY} = \iint xy \; dy \; dx$$

$$= \int_0^b \int_0^h \left[y \; dy\right] x \; dx$$

$$= \int_0^b \left[\frac{y^2}{2}\right]_0^h x \; dx$$

$$= \frac{h^2}{2} \int_0^b x \; dx$$

i.e. $\quad P_{XY} = \dfrac{b^2 h^2}{4}$

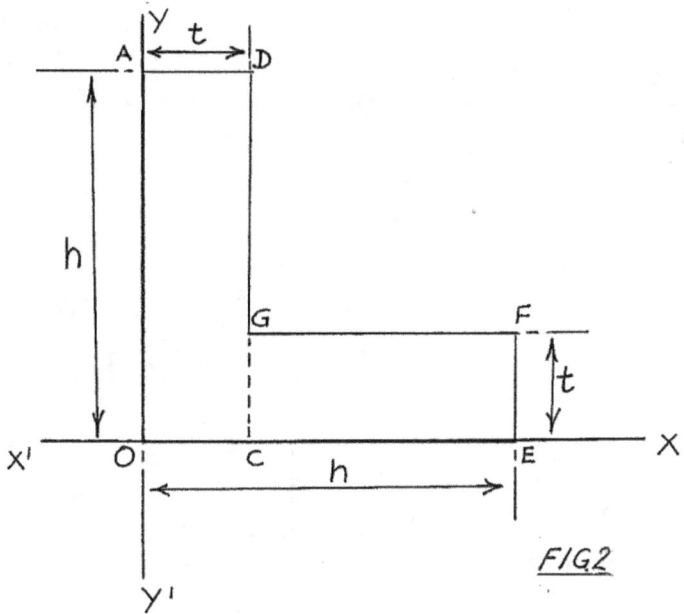

FIG2

Q14. (2) What is the product moment of area of the angle section shown in Fig. 28 with reference to the *X-Y axis?*

The section is divided into the two rectangles *AOCD* and *CEFG* as indicated.

Using the result obtained in Q14 we may write

$$P_{XY} = \frac{t^2 h^2}{4}$$

By inspection we deduce on the grounds of symmetry that the product moment of area of rectangle *CEFG* about its own centroid is zero. Applying with reference to the X-Y axis the parallel axis theorem for product of moments of area:

$$P_{XY} = 0 + (h-t)t\left\{t + \left(\frac{h-t}{2}\right)\right\} \cdot \frac{t}{2}$$

$$\therefore \qquad P_{x'y'_{CFNG}} = (h-t)t\left(\frac{h+t}{2}\right) = \frac{t^2}{4}(h^2 - t^2)$$

so that for both rectangles

$$P_{x'y'} = \frac{t^2 h^2}{4} + \frac{t^2}{4}(h^2 - t^2)$$

~ 393 ~

Q15. Determine the principal moments of inertia with reference to the centroidal axes of the plane section shown in Fig. 1. Also determine the orientation of the principal axes and draw Mohr's Circle using your derived results.

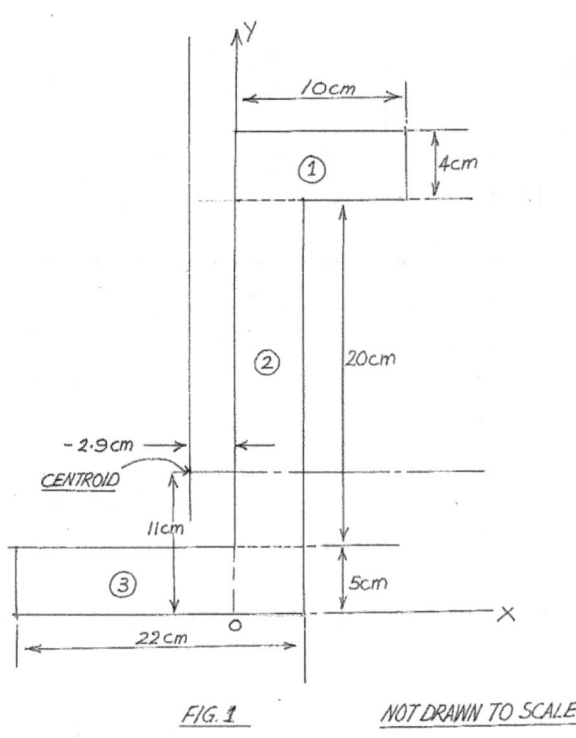

FIG. 1   NOT DRAWN TO SCALE

Area (1): $10 \times 4 = 40(cm)^2$
"   (2): $20 \times 4 = 80(cm)^2$
"   (3): $22 \times 5 = 110(cm)^2$

Note the origin of the X-Y axes for the computation of the coordinates $(\bar{x}, \bar{y})$ of the centroid of the composite area, divided into three elements: (1), (2) and (3).

The data necessary for the calculation of $\bar{x}, \bar{y}$ are entered in the table following:

Element	Area (A) (cm)²	x cm	Ax	y cm	Ay
(1)	40	5	200	27	1080
(2)	80	2	160	15	1200
(3)	110	-9	-990	2.5	275
Σ	230		-630		2555

~ 394 ~

Based on these data

$$\bar{x} = -\frac{630}{230}$$
$$\approx -2.9cm$$

$$\bar{y} = \frac{2555}{230}$$
$$\approx 11cm$$

Next, another tabulation is drawn up, this time showing: the distances of the centroid of each of the elemental areas from the axes marked Y' and X' passing through the centroid of the composite area, these distances being designated y and x respectively. Also tabulated are the second moments of area for each elemental area about its *X- and Y-axis* transferred to the parallel axes through the centroid of the composite area; and, lastly the product moment of area for each of the elemental areas with reference to the centroidal axes for the composite area.

Element	Distance of Element's Centroid from Centroid of Composite Area		$I_Y + Ax^2$ (cm)⁴	$I_X + Ay^2$ (cm)⁴	$P_{xy} = I_{cc} + Axy$ (cm)⁴
(1)	+7.9	+16	$\dfrac{4(10)^3}{12} + 40(7.9)^2$ = 2830	$\dfrac{10(4)^3}{12} + 40(16)^2$ = 10293	0+40(7.9)(16) = 5056
(2)	+4.9	+4	$\dfrac{20(4)^3}{12} + 80(4.9)^2$ = 2027.5	$\dfrac{4(20)^3}{12} + 80(4)^2$ = 3947	0+80(4.9)(4) = 1568
(3)	-6.1	-8.5	$\dfrac{5(22)^3}{12} + 230(-6.1)^2$ = 12995	$\dfrac{22(5)^3}{12} + 230(-8.5)^2$ = 16846	0+110(-6.1)(-8.5) =5703
			$\sum 17852$	$\sum 31086$	$\sum 12327$

Using the tabulated results and noting that $P_{XY} = +12327$

$$Tan2\theta = -\frac{P_{xy}}{(I_X - I_Y)/2}$$

$$2\theta = Tan^{-1} - \frac{12327}{(31086 - 17852)2}$$

$$= Tan^{-1}(-12327/6617)$$

i.e. $\qquad 2\theta = Tan^{-1} - 1.8689$

so that remembering your previous encounter with trigonometrical equations and recalling

$$2\theta = -61.8° + n360$$

and that the tangent is negative in the second and fourth quadrant.

$$2\theta = 118.2° + n \cdot 360$$

$$\theta = 58.1° + n \cdot 180$$

where 'n' is any positive or negative integer including zero. Proceed now to compute $I_{max,min}$.

$$I_{max} = \frac{1}{2}(I_X + I_Y) + \sqrt{\left(\frac{I_X - I_Y}{2}\right)^2 + P_{xy}^2}$$

$$= \frac{1}{2}(31086 + 17852) + \sqrt{(6617)^2 + (12327)^2}$$

$$= 24469 + 13991$$

$$I_{max} = 38460 cm^4$$

and

$$I_{min} = 24469 - 13991$$

$$= 10478 \ cm^4$$

Let us check these calculations by constructing Mohr's Circle. Using $I_X = 31086(cm)^4$; $I_Y = 17852(cm)^4$; $P_{XY} = 12327$, Mohr's Diagram is drawn as in Fig. 2.

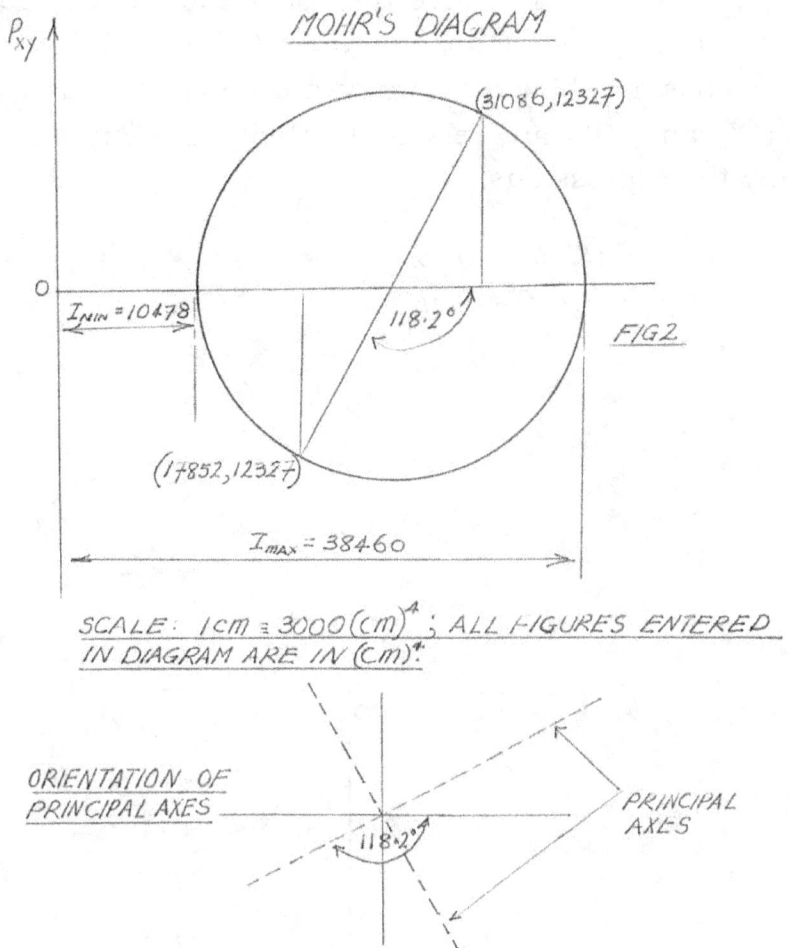

MOHR'S DIAGRAM

P_{xy}

(31086, 12327)

$I_{MIN} = 10478$

118.2°

FIG 2

(17852, 12327)

$I_{MAX} = 38460$

SCALE: $1cm = 3000 (cm)^4$; ALL FIGURES ENTERED IN DIAGRAM ARE IN $(cm)^4$.

ORIENTATION OF
PRINCIPAL AXES

118.2°

PRINCIPAL
AXES

Q16. In Fig.1, *XOX'* and *YOY'* are the conventional orthogonal axes for a two dimensional region. Orthogonal axes *OX$_1$* and *OY$_1$* are at θ° to *XOX'* and *YOY'* as shown. Show that the co-ordinates x_1 and y_1 with respect to the axes *OX$_1$* and *OY$_1$* are related to their counterparts on the *XOX'* and *YOY'* by the expressions:

$$x_1 = ySin\,\theta + xCos\,\theta$$
$$y_1 = yCos\,\theta - xSin\,\theta$$

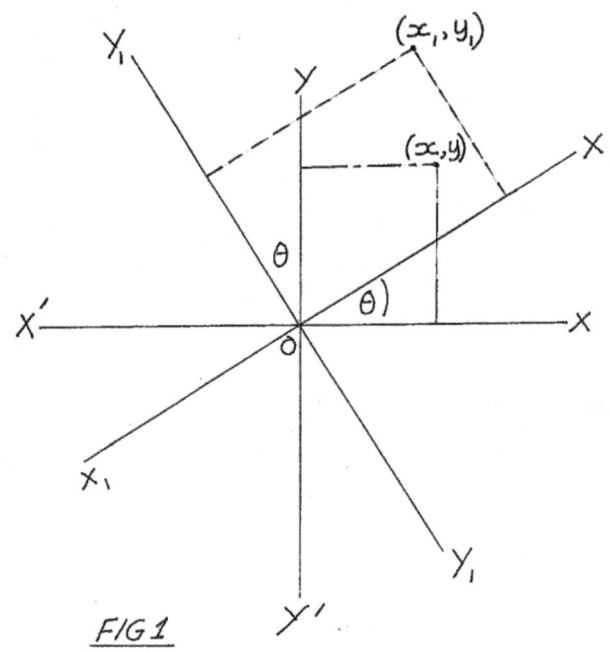

FIG 1

Q17. Referring to the diagram given in Q. 15, show that the radii of gyration on the X$_1$- and Y-axes are given by:

$$k_{Y_1}^2 = k_X^2 Sin^2\theta + k_Y^2 Cos^2\theta$$
$$k_{Y_1}^2 = k_X^2 Cos^2\theta + k_Y^2 Sin^2\theta$$

Q18. The thin rectangular plate *ABCD* of mass '*M*'kg shown in Fig. 1 is 8cm long and 6cm wide. Determine the moment of inertia of the plate about an axis passing through the diagonal *BD*.

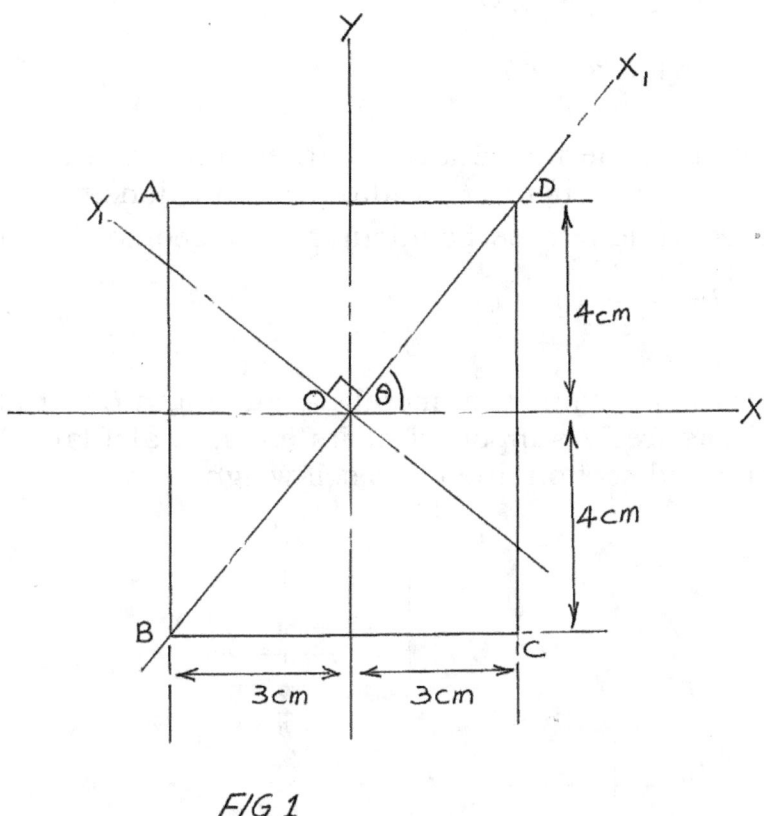

FIG 1

Let angle between axis *OX* and diagonal *BD*, the axis OX_1 be 'θ'. Straightway we may write

$$I_{OX} = M\frac{(8)^2}{12} = \frac{16}{3}M \; kg \; cm^2$$

and

$$I_{OY} = M\frac{(6)^2}{12} = 3M \; kg \; cm^2$$

Now,

$$I_{OX_1} = I_{OX}Cos^2\theta + I_{OY}Sin^2\theta$$

Here, $Cos\theta = \dfrac{3}{5}; \quad Sin\theta = \dfrac{4}{5}$

$$\therefore \qquad I_{OX_1} = M\left\{\frac{16}{3}\left(\frac{3}{5}\right)^2 + 3\left(\frac{4}{5}\right)^2\right\} kg\ cm^2$$

$$= M\left(\frac{48}{25} + \frac{48}{25}\right) kg\ cm^2$$

i.e. $\qquad I_{OX_1} = 3.84 M (kg\ cm^2)$

Q19. Define moment of inertia. Show that the moment of inertia of a uniform circular cylinder of length 'ℓ', radius 'r' and density 'ρ' about an axis through its centroid and perpendicular to its geometrical axis is given by

$$I = \frac{\pi \rho r^2 \ell}{12}(3r^2 + \ell^2)$$

A uniform circular cylindrical metal bar of length 6m, radius 30mm and weight 250N is freely supported at its ends. Calculate the deflection of the bar at its mid section due to its self weight.

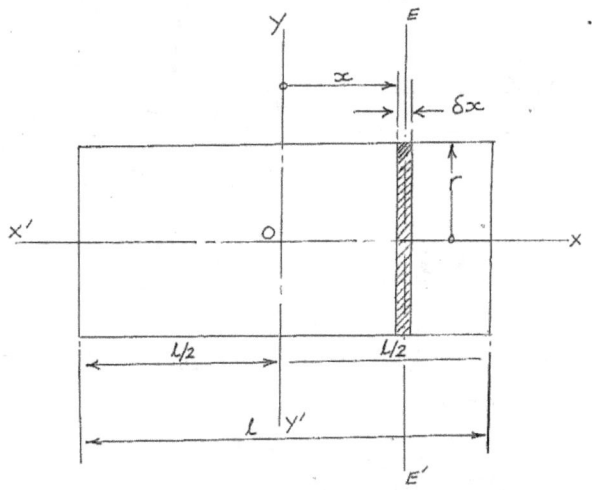

FIG.1   UNIFORM CIRCULAR CYLINDER OF
LENGTH 'ℓ' AND RADIUS 'r'

Moment of Inertia is always stated with reference to a particular axis and the concept is best explained by considering a system of particles of masses $m_1, m_2, m_3, ... m_n$, with perpendicular distances of these masses from the particular axis of $r_1, r_2, r_3 .... r_n$ respectively. By definition, Moment of Inertia, I is given by:

$$I \equiv m_1 r_1^2 + m_2 r_2^2 + m_3 r_3^2 + ... m_n r_n^2$$

~ 400 ~

Now if one were to combine all the masses into a single mass (say) 'M' and if M were placed at a perpendicular distance 'k' from the same particular axis, then

$$I = Mk^2$$

where 'k' is the radius of gyration.

In Fig. 1 axis YY' is the axis passing through the centroid about which the moment of inertia of the cylinder is to be determined.

Now the moment of inertia of the elemental disc of thickness 'dx' about axis EE' passing through its own centroid and parallel to the YY'-axis is

$$I_{EE'} = \frac{1}{4} Mr^2$$

where mass of elemental disc $\delta M = \rho \, \pi r^2 \delta x$

$$\therefore \quad I_{EE'} = \frac{1}{4} \rho \, \pi r^4 \delta x$$

By the parallel-axis theorem, δI_{YY} of the elemental disc is

$$\delta I_{YY'} = I_{EE'} + \delta M$$

$$\therefore \quad \delta I_{YY'} = \frac{1}{4} \rho \, \pi r^4 \delta x + \rho \, \pi r^2 \delta x \cdot x^2$$

$$= \frac{1}{4} \rho \pi r^4 \delta x + \rho \, \pi r^2 x^2 \delta x$$

$$= \pi \rho r^2 \left(\frac{r^2}{4} + x^2 \right) \delta x$$

$$I_{YY'} = \pi \rho r^2 \int_{-l}^{l/2} \left(\frac{r^2}{4} + x^2 \right) dx$$

the limits of integration being from $x = -\ell/2$ to $x = +\ell/2$

Integrating

$$I_{YY'} = \pi \rho r^2 \left[\frac{r^2 x}{4} + \frac{x^3}{3} \right]_{-\ell/2}^{+\ell/2}$$

$$= \pi \rho r^2 \left[\frac{1}{8} r^2 \ell + \frac{\ell^3}{24} + \frac{1}{8} r^2 \ell + \frac{\ell^3}{24} \right]$$

$$= \pi \rho \, r^2 \left(\frac{1}{4} r^2 \ell + \frac{\ell^3}{12} \right)$$

$$= \pi \rho \, r^2 \left(\frac{3 r^2 \ell + \ell^3}{12} \right)$$

$$I_{YY'} = \frac{\pi \rho \, r^2 \ell}{12}(3r^2 + \ell^2)$$

You must be careful to note that in the numerical part of the question you have to recall the well-known formula for central deflection of a freely supported beam carrying a uniformly distributed load viz:

$$\delta = \frac{5WL^3}{384EI},$$

That the 'I' in this formula is the moment of inertia of a circular area about its diameter, i.e. $I = \dfrac{\pi r^4}{4}$. At this juncture you probably do not know about this relationship for deflection as yet but you can take it on trust. Accordingly,

$$\delta = \frac{5WL}{384EI} = \frac{5WL}{384E\dfrac{\pi r^4}{4}}$$

$$\delta = \frac{4 \times 5WL^3}{384 \, \pi \, Er^4}$$

According to the data

$$
\begin{aligned}
W &= 250N \\
L &= 6m \\
E &= 208 \times 10^9 N/m^2 \\
r &= 30mm = \left(\frac{30}{1000}\right)m
\end{aligned}
$$

$$\therefore \quad \delta = \frac{4 \times 5 \times 250 \times (6)^3 \times 10^{12}}{384 \times \pi \times 208 \times 10^9 \times 81 \times 10^4}$$

$$\delta = 0.0053m, \quad \text{say}$$

$$\delta = 5.3mm$$

Q20. Determine the moment of inertia of the steel prism shown in Fig. 1 about the *X-and Y-axes* passing through the centre of the base as indicated. Take density of steel = *7850kg/m³ = 0.007850kg/(cm)³*

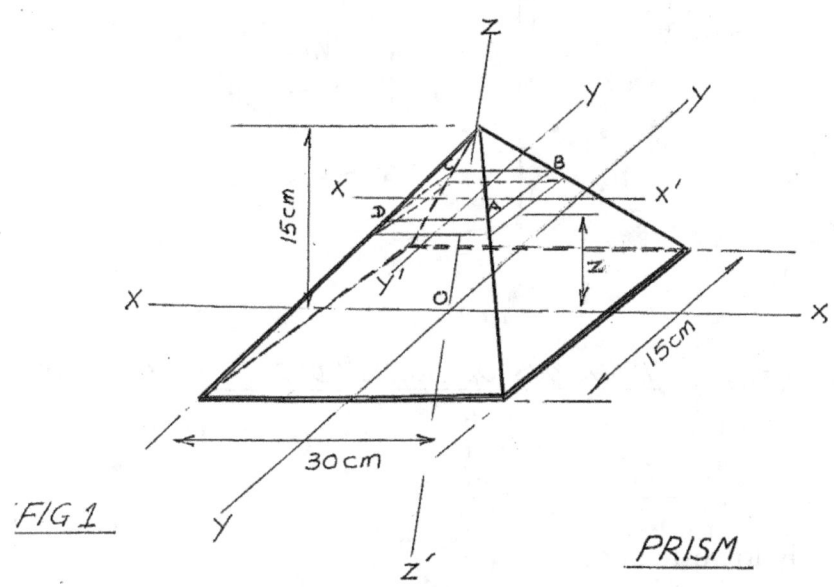

FIG 1

PRISM

Consider a section 'z' *cm* from the base of the prism.

Therefore the cross-sectional area of this section is: *(30-2z)(15-z)*. The thickness of the infinitesimal volume is 'δz', so that the volume 'δV' of this infinitesimal slice or element of the prism is given by

$$\delta V = 2(15-z)(15-z)\delta z \quad cm^3$$

and its mass

$$\delta M = 2\rho(15-z)^2 \delta z \quad kg$$

where ρ = density, assumed constant and in units of *kg/cm³*

Now, for a thin rectangular plate as in Fig. 2

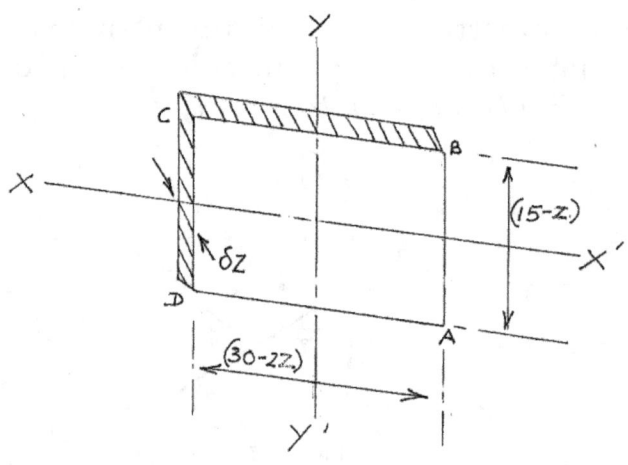

$$I_{XX} = \frac{Mh^2}{12}; \quad I_{YY} = \frac{M\ell^2}{12}$$

Accordingly for our slice

$$\delta I'_{XX'} = \delta M \cdot \frac{(15-z)^2}{12}$$

and

$$\delta I'_{YY'} = \delta M \left\{ \frac{2(15-z)}{12} \right\}^2$$

Remember these axes XX' and YY' are those at the slice. In order to obtain, say, I_{XX} and I_{YY} at the base we have to employ the Parallel-Axis Theorem. The axes with respect to which the second moments are required, being distance 'x' away:

$$\delta I_{XX} = \left\{ \frac{(15-z)^2}{12} + z^2 \right\} \delta M$$

$$= \left\{ \frac{(15-z)^2}{12} + z^2 \right\} 2\rho(15-z)^2 \, \delta z$$

$$I_{XX} = \frac{2\rho}{10^6} \int \left\{ \frac{(15-z)^2}{12} + z^2 \right\} (15-z)^2 \, dz$$

The limits of integration are $z = 0$ to $z = 15$

$$I_{XX} = 2\rho \int_0^{15} \left\{ \frac{(15-z)^2}{12} + z^2 \right\} (15-z)^2 \, dz$$

By a similar token

$$\delta I'_{YY} = \left[\frac{\{2(15-z)\}^2}{12} + z^2\right] 2\rho(15-z)^2 \, dz$$

and

$$I_{YY} = 2\rho \int_0^{15} \left\{4\frac{(15-z)^2}{12} + z^2\right\} (15-z)^2 \, dz$$

$$= 2\rho \int_0^{15} \left\{\frac{(15-z)^2}{3} + z^2\right\} (15-z)^2 \, dz$$

$$I_{YY} = 2\rho \int_0^{15} \left\{4\frac{(15-z)^2}{12} + z^2\right\} (15-z)^2 \, dz$$

Let us evaluate these integrals

$$I_{XX} = \frac{\rho}{6} \int_0^{15} \left\{ (15z)^2 + 12z^2 \right\}(15-z)^2 \, dz$$

$$= \frac{\rho}{6}\left[\int_0^{15} (15-z)^4 \, dz + 12\int z^2(15-z)^2 \, dz \right]$$

Now $\int_0^{15} (15-z)^4 \, dz$ may be transformed to

$\int_0^{15} u(-du)$ when we substitute *15-z* for u because with $u = 15 - x$; $dz = -du$

Integrating

$$\int_0^{15} u^4(-du) = \left[-\frac{u^5}{5}\right]_0^{15}$$

To change the limits of integration:

When $z = 0$; $u = 15$ and when $z = 15$, $u = 0$

$$\therefore \quad \int_0^{15} u^4(-du) = \left[-\frac{u_5}{5}\right]_{15}^{0}$$

$$= \left[\frac{u^5}{5}\right]_0^{15} = \frac{(15)^5}{5} = 151875$$

Continuing $\int_0^{15} z^2(15-z)^2 \, dz = \int_0^{15} 225z^2 - 30z^3 + z)^4 \, dz$

$$= \left[75z^3 - 7.5z^4 + \frac{z^5}{5}\right]_0^{15}$$

$$= 75(3375) - 7.5(50625) + 15\,1875$$

~ 405 ~

$$= 25312.5$$

$$I_{XX} = \frac{\rho}{6}\left[151875 + 12(25312.5)\right] = \frac{\rho}{6}(455625)$$

$$= \frac{0.007850 \times 455625}{6} \, kg \; cm^2 = 596 kg.cm^2$$

Therefore the value of the moment of inertia of the prism about the *XX-axis* at its base is *596 kg.cm²*.

I leave to the reader evaluation of the integral to give the value of the moment of inertia of the prism about the *YY-axis* at its base.

Q21. The *45 cm* square steel plate *2 cm* thick and made from material having a uniform density of *7850 kg/m³* is to be connected to a shaft. The plate has a hole of diameter *10 cm* at its centre and four equi-spaced holes each *5 cm* in diameter drilled through the plate on a pitch circle of *30 cm* diameter. Find the mass moment of inertia of the plate with respect to a perpendicular axis passing through its centre. See Fig. 1

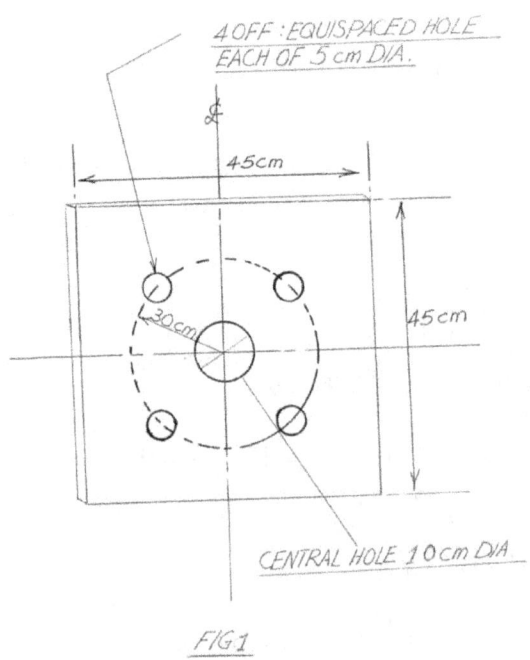

FIG 1

Let M = mass of whole plate i.e. without any holes

      m_1 = mass of material drilled out of the hole at the centre of the plate

      m_2 = mass of material from each equi-spaced hole

Considering the plate as a thin rectangular laminar, its thickness being negligible in comparison with its other dimensions *(2 cm <<< 45 cm)* we recall that its moment of inertia about an axis passing through its centroid and perpendicular to its plane is, working in *kg* and *metres*

$$\frac{M}{12}\left\{\left(\frac{45}{100}\right)^2 + \left(\frac{45}{100}\right)^2\right\} = \frac{0.4050}{12}M = 0.03375M$$

For the central circular hole, its moment of inertia about an axis passing through its centre and perpendicular to the plate:

$$\frac{m_1}{2}\left(\frac{5}{100}\right)^2 = 0.00125m_1$$

Dealing now with one of the equi-spaced holes, its moment of inertia about an axis passing through its centre and perpendicular to the plate:

$$\frac{m_2}{2}\left(\frac{2.5}{100}\right)^2 = 0.0003125m_2$$

At this stage let us evaluate M, m_1 and m_2.

$$M_1 = \left(\frac{45}{100}\right)^2 \cdot \frac{2}{100}(7850)kg$$

$$= 0.2025 \times \frac{2}{100} \times 7850$$

$$= 31.8kg$$

$$m_1 = \frac{\pi}{4}\left(\frac{10}{100}\right)^2 \times \frac{2}{100} \times 7850$$

$$= \frac{\pi}{4} \cdot \frac{100}{1000} \cdot \frac{2}{100} \times 7850$$

$$= 12.33kg$$

Similarly

$$m_2 = \frac{\pi}{4}\left(\frac{2.5}{100}\right)^2 \cdot \frac{2}{100} \times 7850$$

$$\approx 0.8kg$$

∴ Moment of Inertia of plate = *31.8(0.03375) kg m²*
                            = *1.07 kg m²*

Moment of Inertia of central circular hole of diameter 10 cm,
                            = *0.0125m₁*
                            = *0.00125(12.3) kg m²*
                            = *0.015 kg m²*

and for one of the four equi-spaced circular holes, the corresponding moment of inertia is *0.0003125(0.8) = 0.00025 kg m²*. But we have to transfer this moment of inertia to the plate's centre. Therefore the Parallel Axis Theorem is invoked. That is to say I_{centre} = *0.00025 + 0.8*

$\left(\dfrac{15}{100}\right)^2$ = *0.00025 + 0.018 = 0.01825 kg m²*. For the *4* holes I_{centre} = *0.073 kg m²*.

Therefore the total *M.I.* about an axis through the centre of the plate and perpendicular to its plane is:

(1.075 – 0.015 – 0.073) kg m² = 0.987 kg m².

Q22. The hollow steel block shown in Fig. 1 has a density of *7850 kg* per cubic metre. The concentric hole of diameter extends throughout the length of the block. Calculate the polar moment of inertia of the mass of the block about its longitudinal centroidal axis.

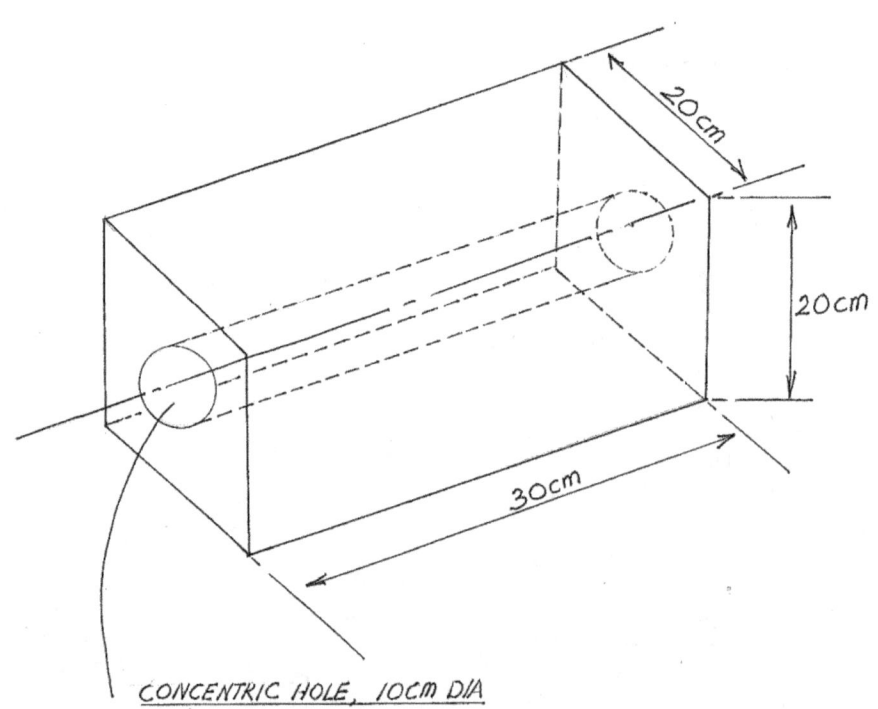

CONCENTRIC HOLE, 10cm DIA

FIG 1

Volume of solid block, i.e. without hole

$\qquad V = 20 \times 20 \times 30$ *cubic centimetres (cc)*
$\qquad\qquad = 12000\ cc$
Now $\qquad 1m = 100\ cm$

or $\qquad 1m^3 = 10^6\ cc$

$\therefore \qquad Mass\ of\ block \quad = \dfrac{12000}{10^6}(7850)kg$

$\qquad\qquad\qquad = 94.2\ kg$

We must now determine the mass of steel which was removed to produce the concentric hole of *10cm* diameter. Accordingly:

$$Vol.\ of\ material\ removed = \dfrac{\pi d^2}{4}(30)cm^3$$

$$= \dfrac{\pi(100)(30)}{4}$$

$$= 2355.9\ cm^3,\ \ say$$
$$2356\ cm^3$$

$\therefore \qquad$ Mass of this material

$$= \dfrac{2356(7850)}{10^6}kg$$

$$= 18.5\ kg$$

Let I_z be the polar axis. Therefore for the solid block (and working in *kg* and metres):

$$I_Z = M\dfrac{\left[\left(\dfrac{20}{100}\right)^2 + \left(\dfrac{20}{100}\right)^2\right]}{12}$$

$$= \dfrac{94.2(0.04+0.04)}{12}$$

$$= 0.628\ kg\ m^2$$

For the cylindrical material extracted

$$I_z = \dfrac{m}{2}\left(\dfrac{10}{100}\right)^2$$

$$= \dfrac{18.5}{2}(0.01)$$

i.e. I_z for the cylindrical material extracted

$$= 0.0925\ kg\ m^2$$

so that moment of inertia of the hollow block about its horizontal centroidal axis is given by :

$$I_{z(hollow\ block)} = 0.628\ kg\ m^2 - 0.0925\ kg\ m^2$$
$$= 0.5355\ kg\ m^2,\ \ say = .0.54\ kg\ m^3$$

Q23. A flywheel made of carbon steel having a density of *7850 kg/m³* has the dimensions shown in Fig. 1. Assuming the rim and hub to be hollow cylinders and the web to be a thin disc, calculate the moment of inertia of the flywheel about its polar axis in *kg(mm)²*.

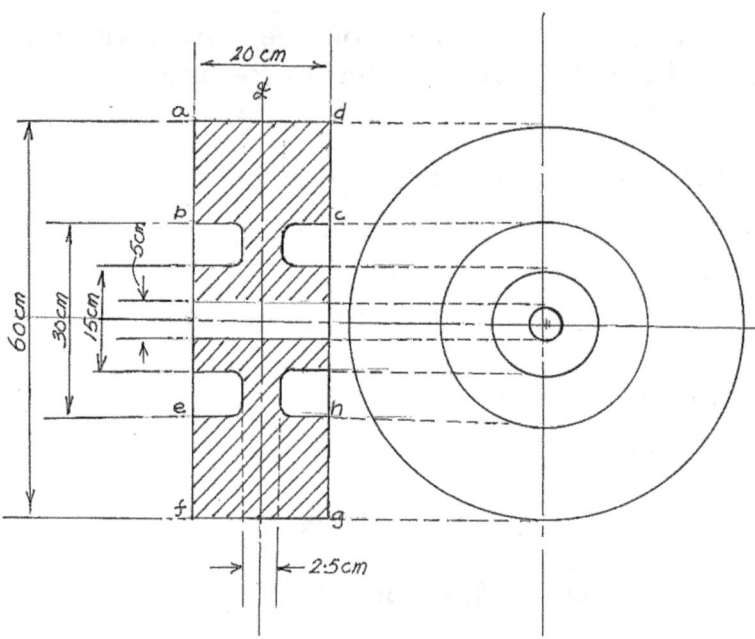

FIG. 1

Considering the rim, with cross-section *abcd/efgh* as shown in the diagram as a hollow cylinder, then its outer diameter is *60 cm* and its inner diameter *30 cm*. Therefore its polar moment of inertia is:

$$I_z = \frac{1}{2} m(R^2 - r^2)$$

But $m = \frac{\rho \pi}{2}(R^2 - r^2)L$, in which ρ = density of material assumed uniform throughout; R,r, the external and internal radii, respectively and L,, the length of the cylinder.

Working in millimeters and noting that the density of the material which is given as *7850 kg/m³*, is equivalent to *7.85 x 10⁻⁶kg/(mm)³*, we have

$$m = 7.85 \times 10^{-6} \times \frac{\pi}{2} \left\{ (600)^2 - (300)^2 \right\} 200 kg$$

$$= 7.85 \times 10^{-6} \times \pi \times 900 \times 300 \times 200$$
$$= 7.85 \times \pi \times 54 kg$$
$$= 1332\ kg$$

$$\therefore \qquad I_z = \frac{1}{2}(1332)\left\{ (600)^2 - (300)^2 \right\} kg(mm)^2$$

$$= 666(900)(300) kg(mm)^2$$
$$= 17982 \times 10^4 \ kg(mm)^2$$

For the hub which must also be treated as a hollow cylinder, its mass, say, m_1 is given by

$$m_1 = \frac{7.85}{2} \times 10^{-6} \times \pi \left\{ (150)^2 (50)^2 \right\} 200$$

$$= \frac{7.85}{2} \times 10^{-6} \times \pi \times 200 \times 100 \times 200$$

or $\qquad m_1 = 53.7 kg$

from which

$$I'_z = \frac{53.7}{2} \left\{ (150)^2 - (50)^2 \right\}$$

$$= \frac{53.7}{2} (200)(100)$$

$$= 53.7 \times 10^4 \ kg(mm)^2.$$

Now to the web. We are to treat it as a thin hollow disc It was shown earlier that the polar moment of inertia of a thin disc of radius 'a'$= \frac{ma^2}{2}$. Evidently for a thin hollow disc with internal radius, say, b,

$$I_z = \frac{1}{2} m(a^2 - b^2)$$

Here, mass of web, $m_2 \quad = \quad \rho \frac{\pi}{2} \left\{ (300)^2 - (150)^2 \right\} 25 kg$

$$= 7.85 \times \frac{10^{-6}}{2} \times 450 \times 150 \times 25 kg$$

$$= 132469 \times 10^{-4} kg$$

$$m_2 = 13.2 kg$$

$\therefore \qquad I''_2 = \frac{1}{2} 13.2 \left\{ (300)^2 - (150)^2 \right\} kg(mm)^2$

$$= \frac{1}{2} 13.2 \times 450 \times 150 kg(mm^{2)}$$

$$= 445500 \ kg(mm)^2$$

$$I''_2 = 44.5 \times 10^4 kg(mm)^2$$

Therefore for the complete flywheel, the polar moment of inertia = $(17983 + 53.7 + 55.5)10^4 \ kg(mm)^2 = 18080 \times 10^4 \ kg(mm)^2 = 180.8 \ kg(m^2)$.

Q24. Using cylindrical co-ordinates show that the moment of inertia with reference to the centroidal axis passing through its vertex and centre of its base, of a right circular cone of height 'h' and base radius 'a' is given by the expression $3Ma^2/10$ in which M = mass of cone. Assume density 'ρ' to be constant throughout.

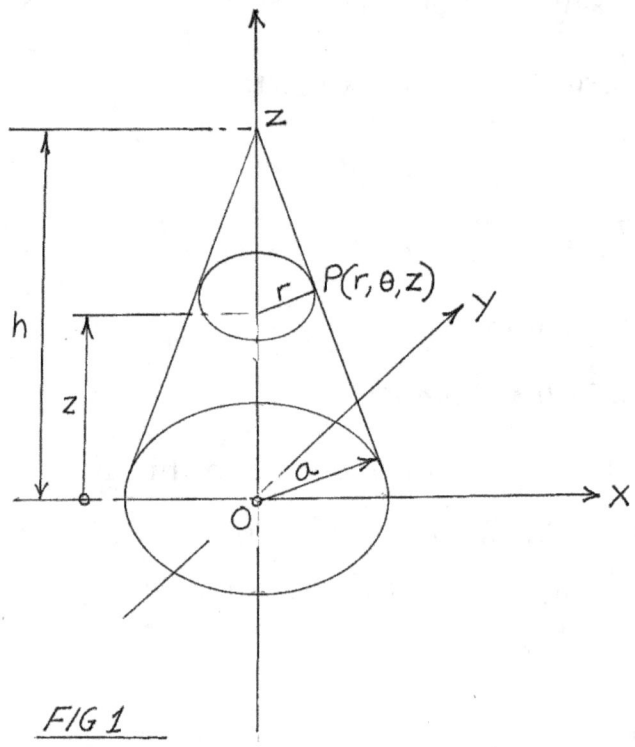

<u>FIG 1</u>

Mass of cone = $\dfrac{\rho \pi a^2 h}{3}$

With reference to Fig. 1, let 'P', be the point on the surface of the cone; its cylindrical coordinates are r, θ, z. Let I_z be the required moment of inertia with reference to the *Z-axis*.

Accordingly

$$I_Z = \int_0^h \int_0^{2\pi} \int_0^r (\rho dV)r^2$$

$$= \rho \int_0^h \int_0^{2\pi} \int_0^r r^2 r\, dr\, d\theta\, dz$$

By geometrical consideration

$$\frac{r}{a} = \frac{h-z}{h}$$

i.e. $r = \dfrac{a(h-z)}{h}$

$$\therefore \qquad I_z = \rho \int_0^h \int_0^{2\pi} \int_0^{a(h-z)/h} r^3\, dr\, d\theta\, dz$$

$$= \rho \int_0^h \int_0^{2\pi} \frac{r^4}{4}\, d\theta\, dz$$

$$= \rho \frac{a^4}{4h^4} \int_0^h \int_0^{2\pi} (h-z)^4\, d\theta\, dz$$

$$= \rho \frac{2\pi\, a^4}{4h^4} \int_0^h (h-z)^4\, dz$$

Substituting $u = h - z$

$$-du = dz$$

$$\therefore \qquad I_z = \frac{\rho\pi\, a^4}{2h^4} \int_h^0 u^4 (-du)$$

$$= \frac{\rho\pi\, a^4}{2h^4} \int_0^h u^4\, du$$

$$= \frac{\rho\pi\, a^4}{2h^4} \cdot \frac{u^5}{5} = \frac{\pi a^4 h^5}{10h^4}$$

$$= \frac{\rho\pi\, a^2 h}{10}$$

But $\qquad M = \dfrac{\rho\pi\, a^2 h}{3}$

$$\therefore \qquad I_z = \frac{3Ma^2}{10} \qquad\qquad\qquad \text{Q.E.D}$$

Q25. If the density 'ρ' at any point '$P'(r,\theta,z)$ in the cone considered in Question 24 is directly proportional to its distance 'r' squared from the axis of the cone, then determine (i) the mass of the cone, and (ii) its centroidal axis through the vertex.

Using cylindrical co-ordinates, the mass 'M' of the cone is, referring to Fig. 1 of Question 24:

$$M = \int_0^h \int_0^{2\pi} \int_0^r \rho\, dv$$

and because we are told : $\rho = Kr^2$

$$M = \int_0^h \int_0^{2\pi} \int_0^{\frac{a(h-z)}{h}} Kr^2 r \, dr \, d\theta \, dz$$

$$= K\int_0^h \int_0^{2\pi} \int_0^{\frac{a(h-z)}{h}} r^3 \, dr \, d\theta \, dz$$

Following the steps taken in the solution of Question 23, we find that:

$$M = \frac{\pi K \, ha^a}{10}$$

Now for I_Z.

In cylindrical coordinates this is expressed by:

$$I_Z = \int_0^h \int_0^{2\pi} \int_0^{\frac{a(h-z)}{h}} \rho \, r^2 r \, dr \, d\theta \, dz$$

Putting $\rho = Kr^2$

$$I_Z = \int_0^h \int_0^{2\pi} \int_0^{\frac{a(h-z)}{h}} Kr^2 r^2 r \, dr \, d\theta \, dz$$

$$= \int_0^h \int_0^{2\pi} \int_0^{\frac{a(h-z)}{h}} Kr^5 \, dr \, d\theta \, dz$$

$$= K\int_0^h \int_0^{2\pi} \left[\frac{r^6}{6} \right]_0^{\frac{a(h-z)}{h}} d\theta \, d\theta$$

$$= \frac{Ka^6}{h^6} \int_0^h \int_0^{2\pi} (h-z)^6 \, dz$$

i.e.
$$I_z = \frac{2\pi Ka^6}{h^6} \cdot \int_0^h (h-z)^6 \, dz$$

Putting $(h-z)u$

$$I_Z = \frac{2\pi \, Ka^6}{h^6} \int_h^0 u^6(-du) = \frac{2\pi \, Ka^6}{h^6} \int_0^h u^6 \, du$$

$$= \frac{2\pi \, Ka}{h^6} \frac{h^7}{7}$$

$$I_Z = \frac{2\pi \, Ka^6 h}{7}$$

CHAPTER 5

ELASTIC TENSILE, COMPRESSIVE, BEARING AND SHEAR STRESSES AND THERMAL STRESSES; ELASTIC STRAIN; ALLOWABLE STRESS, POISSON'S RATIO; AND, PLANE STRESS AND PLANE STRAIN

The stresses and strains we are about to define and describe are said to be elastic in the sense that once the agents causing them are removed the bodies return to their unstressed and unstrained state in every particular.

The concept of stress in engineering science is defined as force per unit area. The symbol commonly used to designate it is 'σ'. To illustrate, consider a prismatic[3] bar of steel affixed vertically to a rigid support and having an external force 'F' applied vertically downwards at its free end as shown in Fig. 1.a.

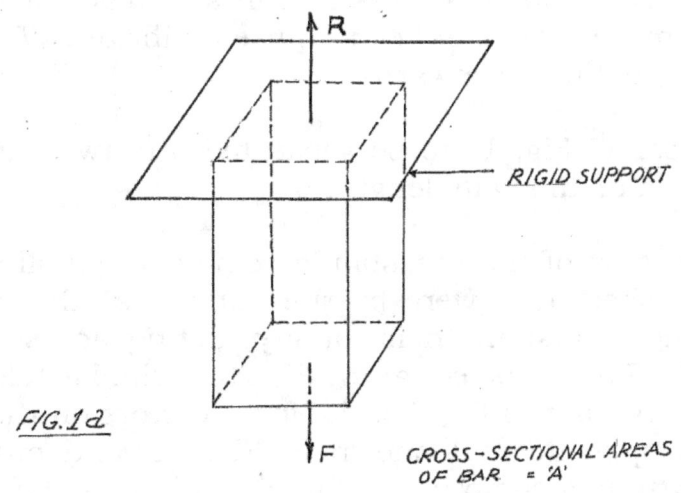

FIG. 1a

RIGID SUPPORT

CROSS-SECTIONAL AREAS
OF BAR = 'A'

It is assumed force 'F' act through the point of intersection of the diagonals of all cross-sections of the bar. The bar is being stretched or elongated by 'F' and is said to be in a state of tension. The bar is also in a state of equilibrium because force 'F' which is acting downwards is balanced by force 'R' acting upwards at the support, i.e. $R = F$

---

[3] "prismatic" refers to "a solid whose ends are equal and parallel plane figures and whose lateral faces are parallelograms"

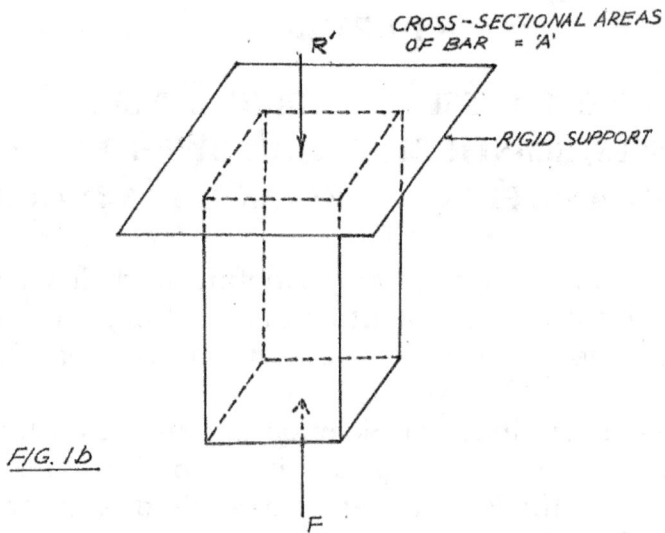

CROSS-SECTIONAL AREAS
OF BAR = 'A'

R'

RIGID SUPPORT

F

FIG. 1b

If as shown in Fig. 1b the force 'F' acts upwards instead, then the bar is said to be in a state of compression. As before, for equilibrium, 'F' acting upwards is balanced by 'R' = F' acting downwards.

Now imagine the bar in Fig. 1a to be separated into two parts by a horizontal cut at any cross-section along its length.

In order that each part of the bar should remain in equilibrium, the internal force vectors unleashed as it were by the cut across the two cross-sectional areas at any section must in their totality, acting across surfaces such as 'abcd' and 'efgh' in Fig. 1c, be equal to 'F'. By a similar token when a similar cut is made across the bar in Fig. 1b, the force vectors on the exposed surfaces act in a direction opposite that shown in Fig. 1c, and must be equal to 'F' which in such a case is a compressive force. All in compliance with one of the axioms of statical equilibrium, in this case: $\sum Force_Y = 0$.

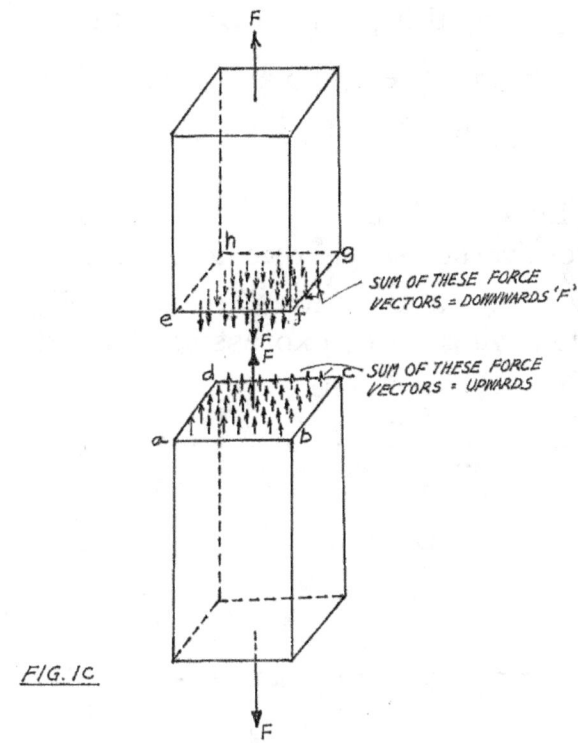

FIG. IC

STRESS

Elastic Tensile and Compressive Stresses

By definition just stated, stress is force per unit area. Thus with reference to Fig. 1a.

$$\text{Tensile stress, } \sigma_T = \frac{F}{A}$$

'A' being the cross-sectional area of the bar perpendicular to the axis of the bar. Likewise, with reference to Fig. 1b,

$$\text{Compressive stress, } \sigma_c = -\frac{F}{A}$$

In this book, tensile stresses are treated as positive and compressive stresses as negative, i.e.

$$\sigma_T : +ve \quad ; \quad \sigma_c : -ve$$

Now, while on the basis of statical considerations it is evident that in order to satisfy vertical equilibrium at every horizontal cross-section of the bars chosen as examples, an upward or downward force F, must be balanced by respectively a downward or upward F, how in fact is stress distributed across different cross-sections of the bar? The principle named after the famous mathematician and stress analyst Adhémar Barrè de Saint-Venant, insofar as it relates here to tension and compression, provides the answers. In short, at a distance not less than one bar-width, or in the case of a circular cylindrical

~ 417 ~

bar, one bar-diameter, from the point of application 'F', localized effects are attenuated, resulting in the average values given by $\sigma_T = \dfrac{F}{A}$ and $\sigma_c = -\dfrac{F}{A}$. Should we take this on trust?

In the "Theory of Elasticity" by Timoshenko and Goodier (1951), the distribution of compressive stress σ_y due to a compressive force 'P' across a prismatic bar of length '2c' and width '2ℓ' as in Fig. 1d where $c >>$, ℓ is approximated by the following series expression derived on the basis of Fourier analysis :

$$\sigma_y = -\frac{P}{2\ell} - \frac{P}{\ell} \sum_{m-1}^{\infty} \left[\frac{m\pi}{\ell}(c-y)+1 \right] e^{-\frac{m\pi}{\ell}(c-y)} \, Cos \, \frac{m\pi x}{\ell}$$

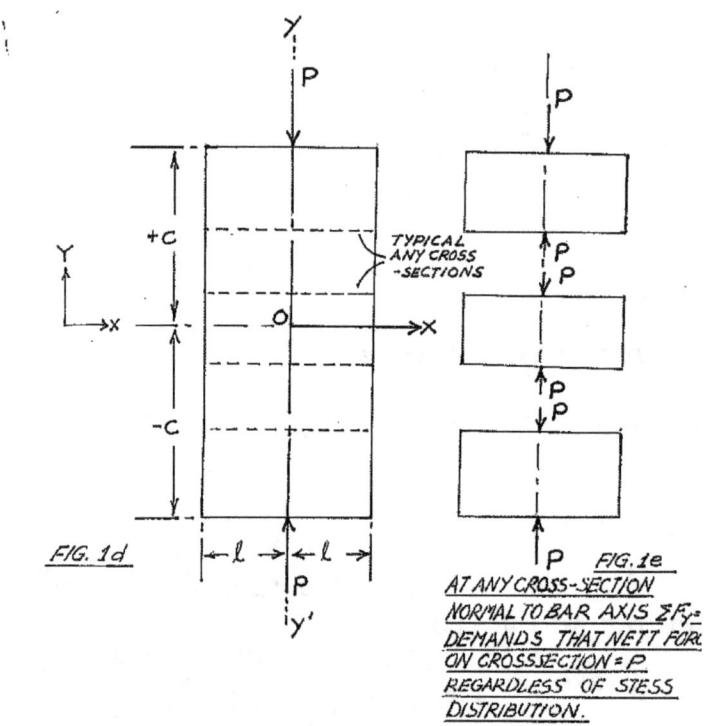

FIG. 1d

TYPICAL ANY CROSS-SECTIONS

FIG. 1e

AT ANY CROSS-SECTION NORMAL TO BAR AXIS ΣF_y DEMANDS THAT NETT FORC ON CROSSSECTION = P REGARDLESS OF STESS DISTRIBUTION.

σ_y, being a compressive stress is negative. When $y = c - 2\ell$,

$$\sigma_y = -\frac{P}{2\ell} - \frac{P}{\ell} \left\{ \left(\frac{2\pi+1}{e^{2\pi}} \right) Cos \frac{\pi x}{\ell} + \left(\frac{4\pi+1}{e^{4\pi}} \right) Cos \frac{2\pi x}{\ell} + \right.$$

$$\left(\frac{6\pi+1}{e^{6\pi}} \right) Cos \frac{3\pi x}{\ell}$$

I have shown stress distributions in Fig. 1f(a), 1f(b) for $y = c - \ell/2$ and $y = c - \ell$ respectively. Fig. 1f(c) is a diagram showing the distribution of compressive stress across a section where $y = c - 2\ell$ i.e. $y = c -$ width of bar: a section at a width-of-the-bar's length from the point of application of 'P'. The maximum value of σ_y occurs when $Cos\, m\pi x / \ell = 1$, i.e. when $x = 0$ (the centre-line of the bar).

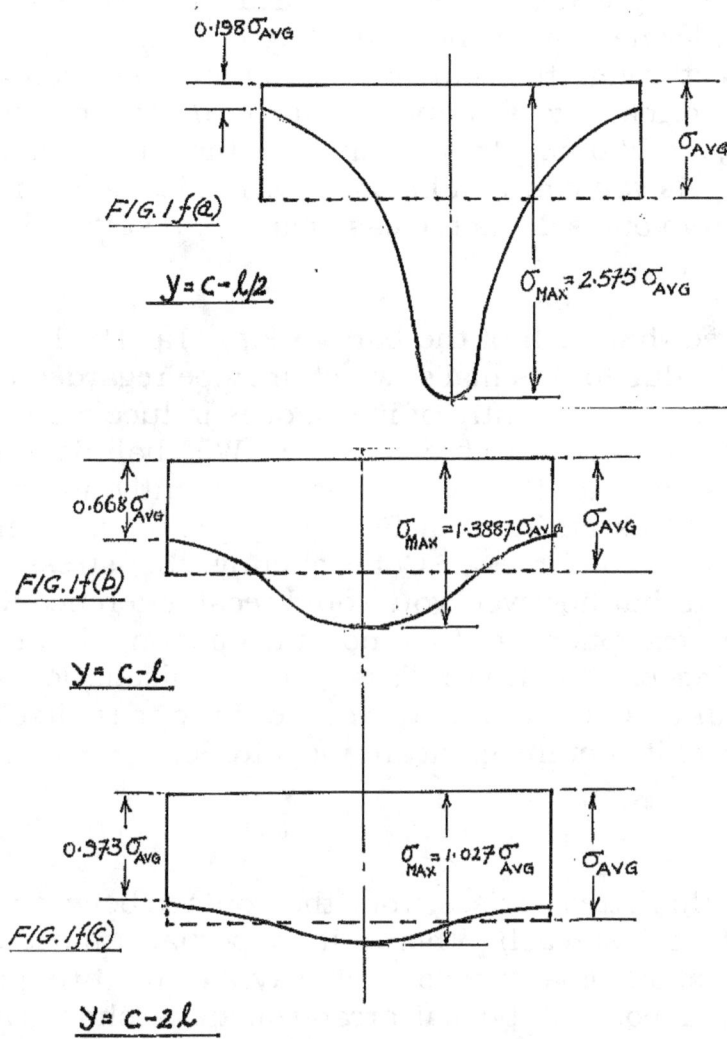

FIG.1f(a)

$y = c - \ell/2$

$0.198\,\sigma_{AVG}$

σ_{AVG}

$\sigma_{MAX} = 2.575\,\sigma_{AVG}$

FIG.1f(b)

$y = c - \ell$

$0.668\,\sigma_{AVG}$

$\sigma_{MAX} = 1.3887\,\sigma_{AVG}$

σ_{AVG}

FIG.1f(c)

$y = c - 2\ell$

$0.973\,\sigma_{AVG}$

$\sigma_{MAX} = 1.027\,\sigma_{AVG}$

σ_{AVG}

Thus, it may be deduced that at a distance equal to the width of the bar from the point of application of P, the stress distribution is virtually uniform right across the section.

Let me emphasize again that in accordance with the laws of statics, irrespective of the actual stress distribution across any cross-section such as '$abcd$'/'$efgh$' in Fig. 1c, the integration of that distribution over the entire cross-sectional area i.e. stress times area must equal the applied load, viz 'F' in Fig. 1c or 'P' in Fig. 1d, so that the equilibrium condition $\sum F_y = 0$ is satisfied at every cross-section. See Fig. 1e.

The student's attention is also drawn to the emphasis placed on having the line of action of 'forces 'F' in Fig. 1a, F' in Fig. 1b 'F″' in Fig. 1c and 'P' in Fig. 1d pass through the points of intersection of the diagonals of the cross-sectional areas. These points of intersection where the whole area of the cross-section may be imagined as being concentrated are called the centroids of the cross-sections. We treated formally with them in Chapter 3. If the lines of action of 'F' and 'P' do not pass through the centroids of all the cross-sections of the bar, then apart from the normal stresses due to tension (F) and compression (P) as in Figs. 1c and 1e, additional stresses due to the displacement from the centroid would arise. As you can well imagine such misalignment would cause a greater pull or push to one side as the case may be, resulting in a bending of the bar.

It should also be noted that each of the bars in Figs. 1a, 1b, 1c, 1d has mass. Each bar's own weight due to this mass, which may be regarded as an internal force acting through each bar's centre of mass, does induce a stress. However, this effect was ignored in this introductory note. We shall deal with it in due course. Let me just say at this stage that frequently in many practical applications, self weight as an internal force is negligible in comparison with external and other applied forces. Consequently the stress due to it is neglected. Having said that however, you would recall reference was made to a good example of an exception to this not uncommon situation when we considered the loading on the Howe Truss in our discussion of Free-Body Diagrams earlier on in this text, the weight of the Howe truss itself being taken into account because of its not insignificant magnitude.

Bearing Stress

As its name implies this stress arises from the contact between two surfaces pressing or bearing against each other. It is perhaps most appropriately exemplified by the action of a tensile pull, say, 'F' on two plates of steel fastened together by a bolt. A typical arrangement is shown in Fig. 2, the thickness of each plate being $t/2$ and the bolt which fits snugly in the hole is of diameter 'd'.

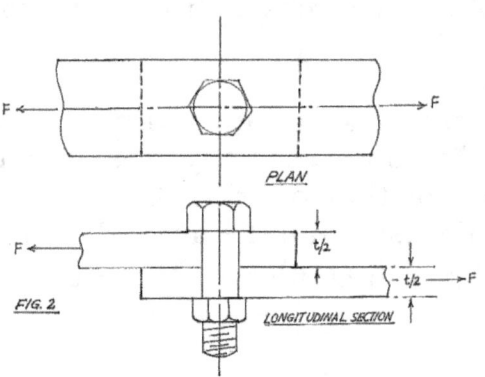

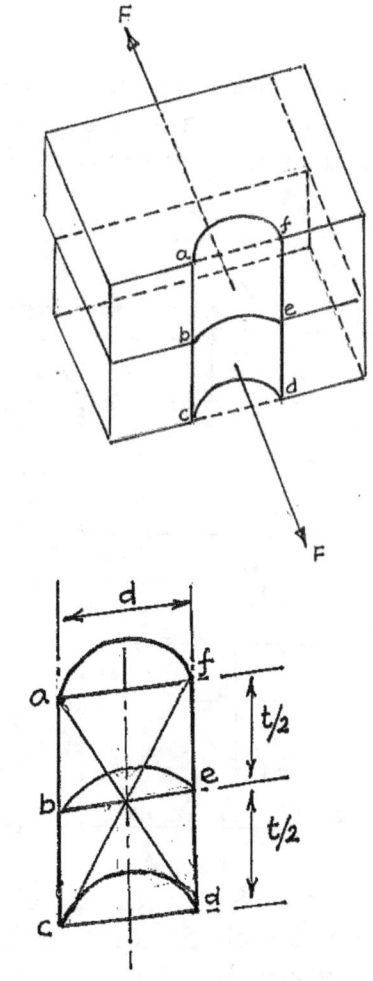

<u>FIG. 2a</u>

By the pulling-apart action of 'F', the surfaces of the plate and shank of the bolt bear against one another. The projected contact area between bolt and plate perpendicular to F is $2(t/2)d = td$, i.e. total thickness of connected plates times the diameter of the bolt. This projected area is represented by the plane 'abcdef' in Fig. 2a. The bearing stress, say, σ_B is given by $\sigma_B = F/td$. In the foregoing analysis it is assumed that the bearing force 'F' passes through the centre of the projected area. If instead of only one bolt there are 'n' bolts of the same diameter at the section, then the bearing stress would be $\dfrac{F}{ntd}$.

~ 421 ~

SHEAR STRESS

Consider another kind of stress. To illustrate it, take a rectangular block of stiff, solid rubber of cross-sectional area 'A' and stick it firmly to a rigid horizontal surface. Such a block is shown in Fig. 3.

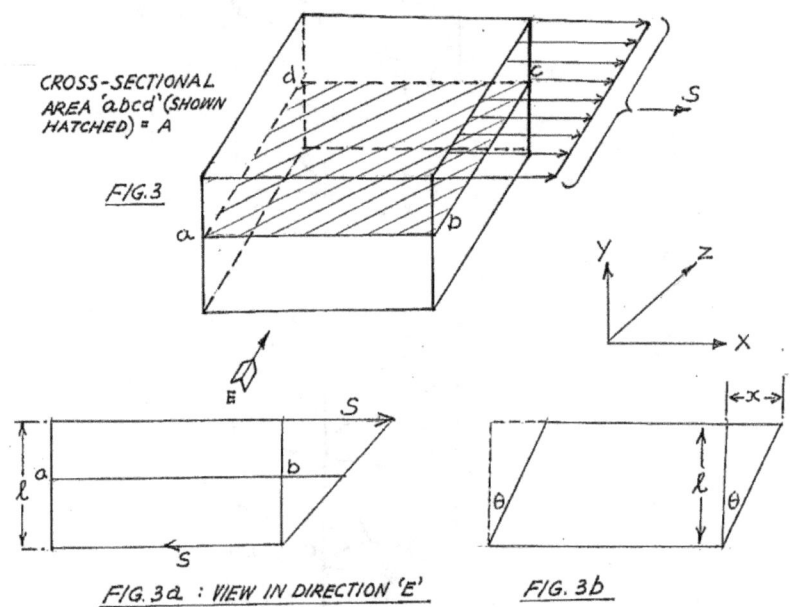

CROSS-SECTIONAL AREA 'abcd' (SHOWN HATCHED) = A

FIG.3

FIG. 3a : VIEW IN DIRECTION 'E'

FIG. 3b

Then apply a horizontal force 'S' uniformly along the top edge of the block as shown in Figs 3a, 3b. The block distorts. The deformed state of the block is shown in Fig. 3b. In this state the top edge of the block extends by a distance, say, x. Were we to take a horizontal section through the block, a force 'S' on the lower side of the upper portion would have to act in a direction opposite to that of 'P' as in Fig. 3a and 3b in order to maintain equilibrium. Likewise 'S' on the upper surface of the lower section would have to be in the same direction as 'S' in Figs. 3a and 3b; and a force 'S' would similarly be required at the junction between the bottom surface of the lower section and the surface to which the block is fixed. Each infinitesimal layer of the block slides over the next from top to bottom, the view of the block in the direction of arrow 'E' being a shown in Fig. 3b. It is the sliding or shearing action that gives rise to shear stress. Shear stress is generally denoted by either 'q' or 'τ' and is defined thus:

$$q \ (\ or \ \tau) = \frac{S}{A}$$

Before dealing with an important property of shear stress it is well to note that stress is a vector quantity. That is to say, stress is a quantity with magnitude and direction. But it is a vector of a different kind. How so? Because stress by definition is always associated with a given area. Whereas force vectors

~ 422 ~

obey the mathematical laws of vector analysis, stress vectors do not.   To illustrate, refer to Fig. 4 which shows an object

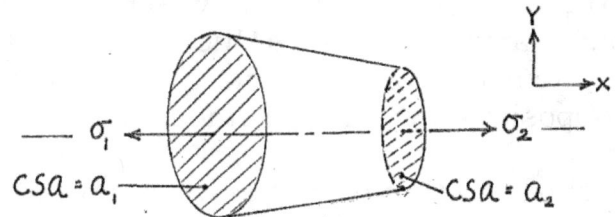

IS THE CONDITION $\Sigma F_x = 0$, SATISFIED BY $\sigma_1 = \sigma_2$?
THE ANSWER IS: NO! BUT, $\sigma_1 a_1 = \sigma_2 a_2$?: YES!

FIG. 4

having stresses σ_1 and σ_2 acting in opposite directions in the same line of action on surfaces of different cross-sectional area, each surface perpendicular to the line of action σ_1 and σ_2.   With reference to Fig. 4, it is incorrect to write $\sigma_1 = \sigma_2$.   The stresses must first of all be multiplied by the cross-sectional areas over which they act in order to convert them to forces.   Consequently with reference to Fig. 4, we may write correctly, $\sigma_1 a_1 = \sigma_2 a_2$.   You will come across these stress vectors a great deal more if you stay with Mechanics of Solids throughout your academic studies in engineering.   In the mathematical theory of elasticity you will meet these special vectors there.   They are called tensors.

Now to the important property of shearing stresses.

The loading of the block in Fig. 3 causes a shear stress, say, 'q' to act on its top surface and a stress 'q' in the opposite direction on its bottom surface.   While these forces satisfy $\sum F_X = 0$, they do not satisfy $\sum M_{ZZ} = 0$.

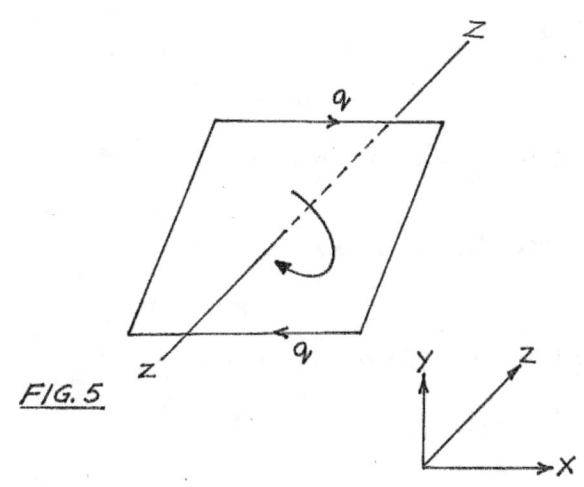

FIG. 5

Referring to Fig. 5, it should be clear that the shear forces due to 'q' acting together would cause the block to rotate clockwise about the Z-axis as shown in the diagram . Note that the plane on which the shear stresses act is perpendicular to the Z-Z axis. The forces due to 'q' constitute a couple about the Z-Z axis. The system can only be brought into equilibrium by another couple acting in the opposite sense in the same plane.

In Fig. 6 the block is shown with shear stresses q_1 on top and bottom surfaces and q_2 on the sides of the deformed block. Assume unit thickness of block and $AB=CD=a$, and $BC=AD=b$. Also, let the perpendicular distances from side DC to B and from AD (extended to E) to point C 'h' and 'k' respectively.

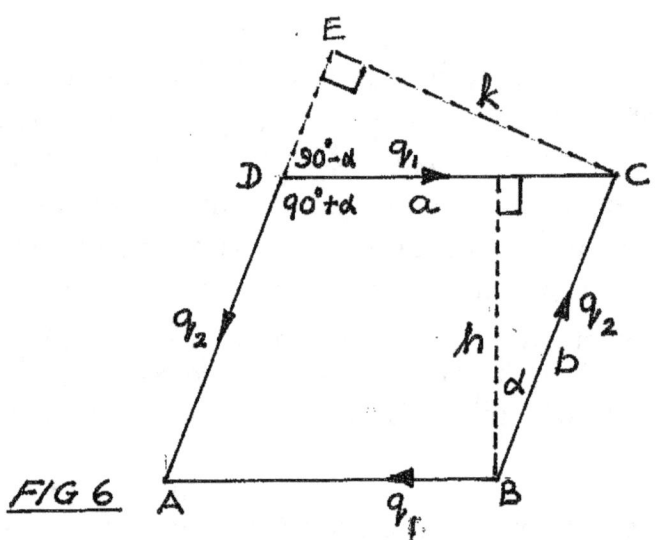

FIG 6

For $\quad \sum M_{ZZ} = 0$, couple due to 'q_1' must equal couple due to 'q_2'.

Couple due to q_1 = Force due to $q_1 \times h$

$\qquad = q_1 \times$ *Area on which* q_1 *acts* $\times h$

$\qquad = q_1 (1 \times a) \times h$

But, $\qquad Cos\alpha = \dfrac{h}{b}$

$\therefore \qquad$ Couple due to $q_1 = q\ ab\ Cos\alpha$

Similarly,

Couple due to $q_2 \quad =$ Force due to $q_2 \times k$

$\qquad = q_2 (1 \times b) \times k$

But $\qquad \dfrac{k}{a} = Sin\ (90-\alpha) = Cos\alpha$

$\therefore \qquad q_2 = q_2\ ab\ Cos\alpha$

For $\sum M_{zz} = 0$

Couple due to   'q_1' = Couple due to 'q_2'

$\therefore$ $q_1 \, ab \, Cos\alpha = q_2 \, ab \, Cos\alpha$

i.e. $q_1 = q_2$

The stresses q_2 acting on the other pair of opposite sides of the block are q_1's complementary shearing stresses.

In general, shear on one surface is always associated with shear of equal intensity acting in a direction perpendicular to that of the original shear and giving rise to a turning moment of equal magnitude but of opposite sense. Remember it well!

Thermal Stress

Temperature stresses are induced in practically all materials due directly to changes in the temperature and constraining conditions. Consider an unconstrained mild steel rod of original length 'l_o' at, say, temperature 't_o'. If the rod's temperature is increased by, say, 'Δt' throughout its length, then the expanded length of the rod would be $l_o + \alpha l_o \Delta t$, '$\alpha$' being the coefficient of linear expansion of mild steel, assumed constant throughout range of 'Δt.

Fig. 6a refers.

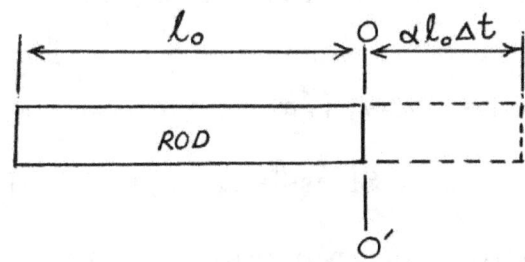

FIG. 6a: ROD EXPANDING FREELY FOR TEMPERATURE INCREASE 'Δt'

Evidently, if the rod at its original length 'l_o' at temperature t_o is now fitted between two rigid supports and its temperature increased by Δt to $t_o + \Delta t$, then the expansion $\alpha l_o \Delta t$ which would have occurred, were the rod free to expand, would be prevented. Such a situation would be analogous to one in which a force, say 'F', acts to compress the rod back to position 0-0' as shown in Fig. 6b. Thus, a compressive strain is induced in the rod. If the associated stress is, say, 'σ' then 'σ' = F/A, in which A is the cross-sectional area of the rod, assumed uniform throughout.

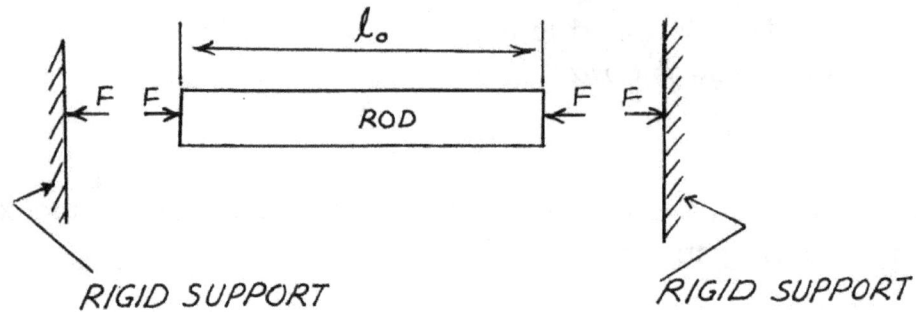

FIG. 6b: FREE-BODY DIAGRAM OF ROD AT TEMPERATURE $t_o + \Delta t$ AND CONSTRAINED BY RIGID SUPPORTS AT ITS ENDS. FORCE 'F' ACTING ON THE ROD HAS IN EFFECT COMPRESSED IT BY AN AMOUNT $\alpha l_o \Delta t$ BACK TO ITS ORIGINAL LENGTH 'L$_o$'. EQUAL AND OPPOSITE FORCES 'F' ACT ON SUPPORTS AS SHOWN. THE SYSTEM IS IN EQUILIBRIUM.

If the rod were free to expand, its original length at temperature $t_o + \Delta t$ would have been $l_o + \alpha l_o \Delta t = l_o (1 + \alpha \Delta t)$. but as depicted in Fig. 6b the force 'F' has compressed it from length, $l_o (1 + \alpha \Delta t)$, by an amount equal to $\alpha l_o \Delta t$, back to l_o. now, compressive strain = Amount of compression/original length from which compressed, i.e."

$$\text{Compressive Strain} = \alpha l_o \Delta t / (l_o + \alpha l_o \Delta t) = \alpha \Delta t / (1 + \alpha \Delta t)$$

$$\therefore \text{Thermal Compressive Stress} = E \alpha \Delta t / (1 + \alpha \Delta t)$$

and Compressive Force 'F' on rod and supports $= AE \alpha \Delta t / (1 + \alpha \Delta t)$. 'E' is Young's modulus of elasticity for the mild steel rod.

Consider next the case where the temperature of the same rod is decreased from 't_o' to $t_o - \Delta t$.

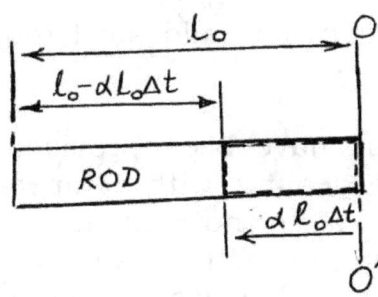

FIG.6c: ROD CONTRACTING FREELY FOR TEMPERATURE DECREASE 'Δt'

The rod is free to contract. 'α' is the coefficient of linear thermal contraction. The rod contracts to a length $= l_0 - \alpha l_0 \Delta t$.

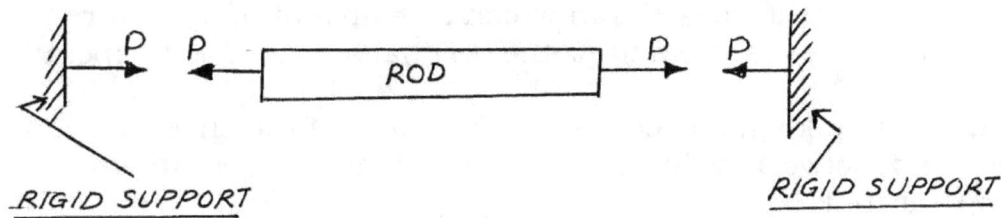

FIG. 6d: FREE-BODY DIAGRAM OF ROD AT TEMPERATURE $t_0 - \Delta t$ AND CONSTRAINED BY RIGID SUPPORTS AT ITS ENDS. TENSILE FORCES 'P' ARE NECESSARY TO PULL THE ROD OF LENGTH $l_0 - \alpha l_0 \Delta t$, BACK A DISTANCE $\alpha l_0 \Delta t$ to l_0. EQUAL AND OPPOSITE FORCES 'P' ACT ON THE SUPPORTS AS SHOWN. THE SYSTEM IS IN EQUILIBRIUM

When the rod is placed between rigid supports, it follows that in order to retain the length 'l_0' after its contraction a force say, P would be required to pull the rod back from its original contracted length $l_0 - \alpha l_0 \Delta t$, back to '$l_0$'. Thus the rod in effect is extended a distance $\alpha l_0 \Delta t$ back to l_0.

$$\therefore \text{Tensile Strain} = \alpha l_0 \Delta t / (l_0 - \alpha l_0 \Delta t) = \alpha \Delta t / (1 - \alpha t)$$

~ 427 ~

Accordingly, thermal tensile stress, say, $\sigma_T = \frac{E\sigma\Delta t}{(1-\alpha t)}$ and $P = \frac{AE\alpha\Delta t}{(1-\alpha\Delta t)}$ in which, as before E = Young's modulus of elasticity for mild steel rod, and A = cross-sectional area of rod.

Worked examples and other problems have been provided later on in this Chapter to illustrate the method of dealing with temperature stresses in composite bars. The important point to note at this stage is that by constraining thermal expansion and contraction, stresses are induced in materials. Unions, bends and expansion bellows are often used to mitigate these in steam pipes.

Here is an easy problem to refresh your memory of your early introduction to Physics.

Show, using an appropriate diagram, that an analogous expression for determination of the area of a thin rectangular sheet of material with a coefficient of superficial expansion 'α', when its temperature is increased throughout by Δt, is given by the approximate expression: $A_t = A_o (1 + 2\,\alpha\Delta t)$ in which A_t = area at temperature $t_o + \Delta t$ in which A_t = area at temperature $t_o + \Delta t$, 't_o' being the original temperature and A_o the original area at this temperature.

Assuming, you had to take into account the thickness of the sheet, what would be the relationship between volume V_t (at temperature $t_o + \Delta t$) and V_o the volume at temperature t_o?

UNITS OF STRESS

In the System International (SI), the fundamental unit for stress is Newton per square metre : symbolically N/m^2. It is what is called a derived unit because it is formed from three of the seven basic units of this system, viz. mass (kg), length(m) and time(s). N/m^2 has the units $kg\ m^{-1}\ s^{-2}$. Because it is so small, being roughly equivalent to 0.000145 lbf/sq.in, it is necessary to use multiples and sub-multiples of it. Such are: GN/m^2, MN/m^2, kN/m^2, $N/(mm)^2$ and $kN(mm)^2$. Pressure also has the same units as stress but again because of the small size of the unit N/m^2 which incidentally is called a Pascal in France, a permitted unit, the bar, which is equal to $10^5 N/m^2$ or $10^5 Pa$ is preferred. 1 bar is approximately 1 atmosphere $\approx$ 14.7 lb/sq.in.

Illustrative Example 1

Each of the two mild steel plates in Fig. 7, is 20mm thick and 15cm wide. They are fastened together by three 24mm diameter bolts. The bolts are a close fit in the holes drilled in the plates. The combination is subjected to a tensile force of 100kN. Calculate the following : (i) the average normal stress in the plates at a section well away from the points of application at the ends where the tensile forces act (ii) the stress across the section AA', neglecting the effects of stress concentration, (iii) the average stress in the bolts, and (iv) the bearing stress between the plate and the bolts.

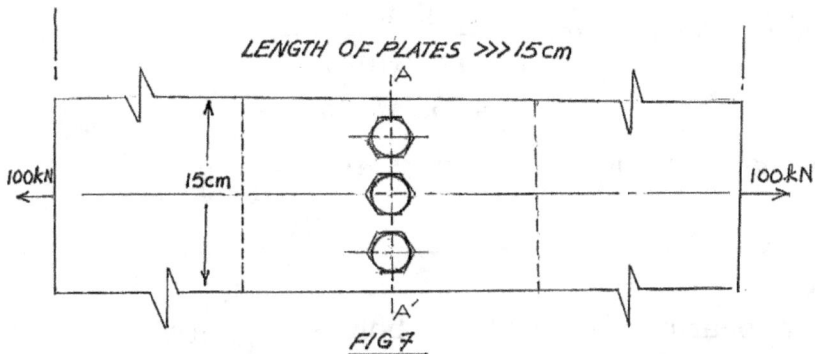

FIG 7

It is assumed that at a cross section of the fastened plates distant not less than say, one and one-half times the width of a plate from the ends where each tensile force of 100kN is applied, the average tensile stress 'σ' is given by $\sigma =$ Force/Area.

Working in newtons and metres:

$$\sigma = \frac{100 \times 10^3}{\left(\frac{15}{100} \times \frac{20}{1000} \right)} N/m^2$$

$$= \frac{10^8}{3} N/m^2, \quad \text{say}$$

$$\sigma = 33.3 MN/m^2$$

In order to determine the average tensile stress, σ' across the section $A-A'$, the projected area of the three bolts, i.e. $3(24)(20)^2 (mm^2)$ must be subtracted from the entire cross-sectional area of a plate, viz $(20 \times 150)(mm)^2$. Therefore the effective cross-sectional area = 1560(mm)².

Accordingly

$$\sigma' = \frac{100 \times 10^3}{1560/10^6} = 64 \times 10^6 N/m^2$$

i.e. $\sigma' = 64 MN/m^2$

Note that the effects of stress concentration due to the bolt holes have been neglected here. Bear in mind that such effects generally account for a higher value than that obtained for σ'.

(iii)   Shear area of a single bolt $= \dfrac{\pi}{4} \times \left(\dfrac{24}{1000}\right)^2 (m)^2 = 452.3 \times 10^{-6} (m)^2$;

$1357 \times 10^{-6} (m)^2$ for 3 bolts.

$\therefore$ Average shear stress $= \dfrac{100 \times 10^3\, N}{1357 \times 10^{-6}\, m^2}$

$= 73.7\, MN/m^2$

(iv)   Each bolt has a (projected) bearing area $= \dfrac{24}{1000} \times \dfrac{20}{1000} (m)^2$

$= \dfrac{480}{10^6}\, m^2.$

$\therefore$ Total bearing area for three bolts $= \dfrac{1440}{10^6} (m)^2$

$\therefore$ Average bearing stress $= \dfrac{100 \times 10^3}{1440 \times 10^{-6}}$

$= 69.4\, MN/m^2.$

STRAIN

Tensile and Compressive Strain

If the initial length, that is the unloaded length of the bar in Fig. 1a is 'ℓ'_0 and that as a consequence of the application of force 'F' the bar assumes a new length, say, ℓ, then the direct engineering tensile strain 'e' is given by the ratio:

$$e = \dfrac{increase\ in\ length}{initial\ length}$$

Evidently, $\ell > \ell_0$ and 'e' is positive for tensile 'F'. Similarly, if 'F' is a compressive force $\ell < \ell_o$ and 'e' is negative. Summarizing,

Tensile Strain : $\ell > \ell_0$:

$$e_T = \dfrac{\ell - \ell_0}{\ell_0} = \left(\dfrac{\ell}{\ell_0} - 1\right),\ \text{positive}$$

Compression Strain : $\ell < \ell_0$

$$e_c = \dfrac{\ell - \ell_0}{\ell_0} = \left(\dfrac{\ell}{\ell_0} - 1\right),\ \text{negative}$$

By its very definition, strain is dimensionless. You may however come across such references to it expressed as say millimetres per millimetre. Such however do not alter its dimensionlessness.   Sometimes strain may be expressed as a percent, e.g.

$$e\% = \frac{100(extension \ or \ contraction)}{original \ \dim ension}$$

The definition of direct engineering strain implicitly assumes that throughout the length of the material under either tension or compression, tensile strain or compressive strain as the case may be is the same for each part of the material, in our example in bar in Fig. 1a.   That is to say strain is uniform throughout.   Thus for one-quarter of the bar, the strain is one-quarter of the strain for the whole bar; for one-half length of the bar, the strain is one-half the strain of the whole bar, and so on.

You may however also come across in the literature what is described in some texts as "true engineering strain." This strain designated $'e_T'$ is derived in the following manner, it being assumed that strain, far from being uniform varies throughout the bar, logarithmically as we shall show presently in the following:

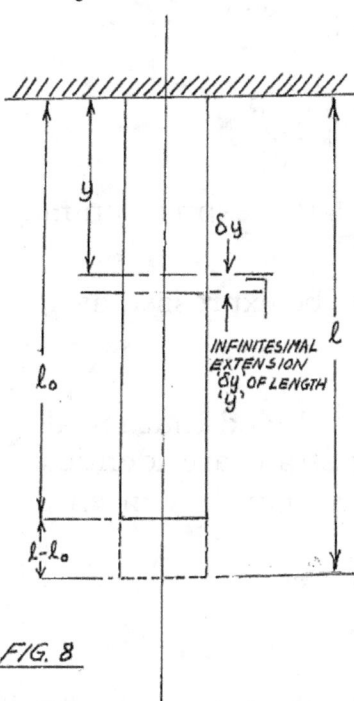

FIG. 8

In Fig. 8, the bar of original length $'\ell_0'$ is stretched to length $'\ell'$. The strain of the elemental length of bar 'y' is δe_T. Accordingly

$$\delta e_T = \frac{\delta y}{y}$$

so that

$$\int de_T = \int_{\ell_0}^{\ell} \frac{dy}{y}$$

from which $\quad e_T = Log_e \dfrac{\ell}{\ell_0}$ $\qquad \dotsb$ (i)

It was shown before that :

$$e = \frac{\ell}{\ell_0} - 1 \quad \text{or} \qquad \frac{\ell}{\ell_0} = e + 1$$

$$\therefore \qquad Log\left(\frac{\ell}{\ell_0}\right) = Log(e+1)$$

But $\quad Log\left(\dfrac{\ell}{\ell_0}\right) = e_T$

$$\therefore \qquad e_T = Log(e+1) \quad \text{or} \quad e^{e_T} = e + 1 \qquad \dotsb (ii)$$

Remembering the series expansion for e^x, viz

$$e^x = 1 + x + \frac{x^2}{2!} + \frac{x^3}{3!} + \dotso$$

we may write

$$e^{e_T} = 1 + e_T + \frac{(e_T)^2}{2!} + \frac{(e_T)^3}{3!} \dotso$$

and because e_T is small the latter expression may be simplified thus :

$$e^{e_T} = 1 + e_T$$

which on the basis of (ii) may be expressed as

$$e + 1 = 1 + e_T$$

It may therefore be safely concluded that for all practical purposes engineering strain and true engineering strain are identical. Note that the final length of the elongated bar in tension may be written as $\ell_0(1+e)$; and in compression $\ell_0(1-e)$.

Shearing Strain

In Fig. 9 the same block of rubber used in the illustration for shear stress is acted upon by two pairs of shearing stress on four of its sides. The block is thereby distorted but it is in equilibrium because the sum of the moments about the Z-axis of the forces due to the shearing stress is zero.

In Fig. 9a, a section through the distorted tube parallel to the X-Y plane is shown and in Fig. 9b, the section is re-oriented to show the total distortion as one angle to one side. As pointed out before this angle is always small

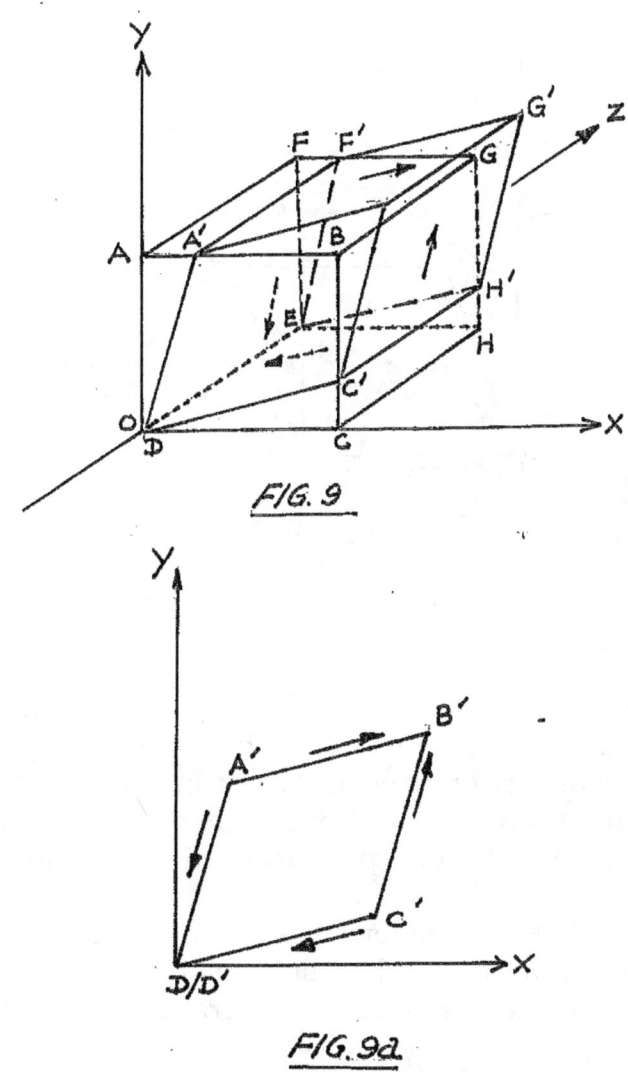

FIG. 9

FIG. 9a

so small, in fact that $\tan \gamma$ may be taken as equal to $\dfrac{u}{y}$ as in Fig. 9b. As every schoolboy knows when an angle is very small as is the case with 'γ', $\underset{\gamma \to 0}{Lt} \dfrac{\tan \gamma}{\gamma} = 1$, that is to say the tangent of 'γ' may be replaced by the angle γ itself, in radians. Thus shear strain is the angle of distortion 'γ', given by u/y with reference to Fig. 9b.

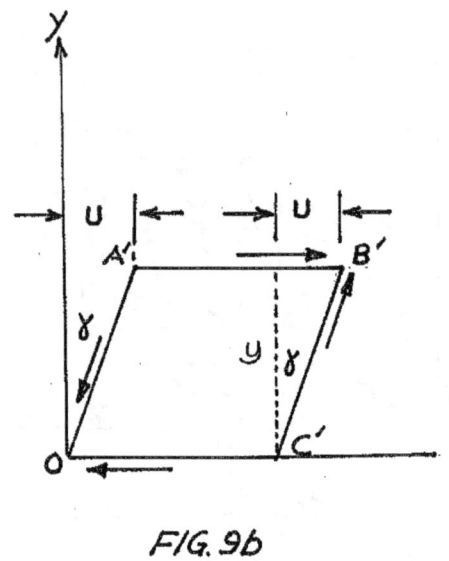

<u>FIG. 9b</u>

Thermal Strain

When thermal stress was considered, the free expansion and free contraction of the metal rod was $\alpha l_o \Delta t$: $+\alpha l_o \Delta t$ in the case of expansion and $-\alpha l_o \Delta t / l_o$ in the case of contraction. Clearly such 'free' changes in the length of the rod had no stresses associated with them. By definition strain in these circumstances is numerically $\alpha \Delta t$, i.e. strain is directly proportional to temperature change.

If the same rod we have been considering is fixed to a rigid support at one end and free to expand at the other, and is subjected to an elastic strain 'e_x' and also a thermal strain, say, $\alpha \Delta t$, then the total strain = $e_x + \alpha \Delta t$.

Before dealing with the illustrative examples on thermal stresses and strains, I shall treat with Hooke's law for uniaxial stress, a tool required from our engineering kit to aid the demonstration.

Volumetric Strain

The volumetric strain which engineers are interested is that due to the action of compressive forces acting on a volume as for example in the case a cylinder subjected to internal pressure, hydrostatic or otherwise or simply that caused by hydrostatic pressure acting on a submerged body such as a submarine or an undersea oilwell well-head.

If 'δV' is the change in the original volume V_o' then volumetric strain e_v is given by:

$$e_v = Lt \frac{\delta V}{V_o} = \frac{dV}{V_o}$$

~ 434 ~

To illustrate, take the case of the cuboid. A cuboid is merely a particular case of a right prism, all its faces being rectangular. One such cuboid is shown in Fig. 10.

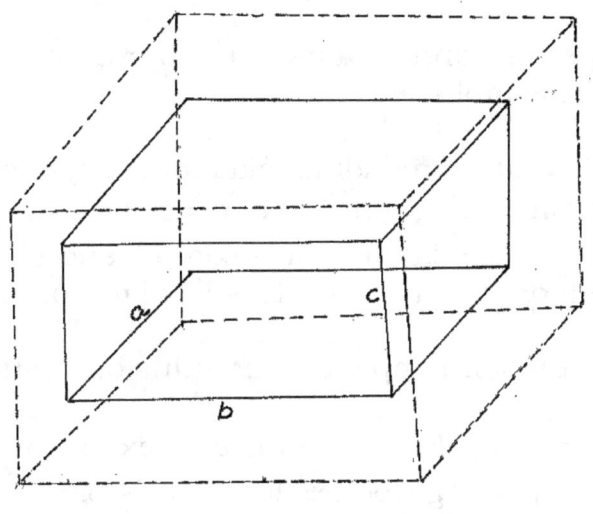

FIG. 10: ILLUSTRATING VOLUMETRIC STRAIN

---- ORIGINAL OUTLINE OF CUBOID
——— FINAL   "   "   "

Assume the cube to be submerged and that as a result of hydrostatic pressure acting on its faces, the new lengths of its sides are: $(a-\delta a)$, $(b-\delta b)$ and $(c-\delta c)$. Consequently the change in volume is:

$$abc-(a-\delta a)\ (b-\delta b)\ (c-\delta c)$$
$$= abc-\{abc-ac(db)-bc(da)+c(da)(db)-ab(dc)$$
$$\quad + a(dc(db)+b(da)(dc)-(dc)da)(db)\}$$

Neglecting second order quantities since the contractions are small, we discard:

$$c(da)(db), a(dc)(db), b(da)(dc) \text{ and } (dc)(da)(db)$$
$$\therefore \quad \text{Change in volume} \ = \ abc-abc+ac(b)+bc(da)+ab(dc)$$
$$= ac(db)+bc(da)+ab(dc)$$

so that

$$\text{Volumetric strain} \ = \ \frac{ac(db)+bc(da)+ab(dc)}{abc}$$

say

$$e_v = \frac{dV}{V_o} = \frac{db}{b}+\frac{da}{a}+\frac{dc}{c}$$

i.e. volumetric strain of the cuboid in Fig. 20 is given by the sum of the individual linear strains of its edges.

If the cuboid were a cube then, e_v = three times the linear strain of the edge.

THE MODULLI OF ELASTICITY

Having defined elastic stress and strains it is appropriate that the connection between them should be established.

The connection between uni-directional elastic tension and compression as described in the foregoing is given by : $E = \sigma_T / e_T$; $E = \sigma_c / e_c$. These two expressions symbolize Hooke's law for uni-axial elastic stress and strain. We shall have a great deal more to say about this law later on in this text.

E is called Young's modulus or simply the modulus of elasticity.

For the bar in Fig. 1a, $\sigma_T = F/A$ and strain, e_T = extension/original length, i.e. $(\ell - \ell_o)\ell_o$. Therefore, by applying Hooke's law to this bar

$$E = \frac{F}{A}\frac{1}{e_T} \quad or \quad \frac{F}{A(\ell - \ell_0)}$$

where ℓ = final length, and ℓ_o = original length.

For pure elastic shear; i.e. an elastic state of stress with only shear stress acting as was represented in the foregoing article on this type of stress, the connection is:

$$C = \frac{q}{\gamma}$$

Some authors use the symbol 'G' instead of 'C' but in either case it is referred to as the shearing modulus or modulus of rigidity.

Finally, the Volume or Bulk modulus which is generally denoted by the symbol 'K' is defined thus:

$$K = \frac{P}{Volumeteric\ Strain}$$

'P' being the hydrostatic pressure which causes stresses in all directions on the body. Therefore we may write:

$$K = \frac{P}{dV/V_o}$$

You may also come across in the technical literature 'volumetric strain' which is really a hydroscopic strain caused by shrinkage due to moisture loss. Two materials in which this occurs are wood and concrete although conceptually they are the same. There is no connection between strain due to loss of moisture and engineering volumetric strain due to stress.

STIFFNESS OF A BAR

Stiffness is defined as load per unit deflection or force (say, F) per unit deflection. The deflection may be an extension or compression. Young's modulus may therefore be expressed in terms of the stiffness of, say, a rod or a bar. Stiffness is generally denoted by 'k' or 'S'. By Hooke's law:

$$E = \frac{Load\ (or\ Force)}{Cross-sectional\ area\ (A)} \times \frac{length\ (\ell)}{(extension\ or\ compression)}$$

and with a little rearrangement, this relationship becomes

$$E = \frac{Load\ (or\ Force)}{(Extension\ or\ deflection)} \times \frac{length}{cross-sectional\ area}$$

$$E = \frac{S\ell}{A}$$

from which $\quad S = \dfrac{AE}{\ell}$

Consider the component bar in Fig. 11. It consists of 'n' parts in series, each part having a different length and different cross-sectional area from the others. The compound bar could be considered as comprising a set of individual bars connected in series. The material composition of the bar may also vary from section to section. The bar carries a load 'W' at its free end.

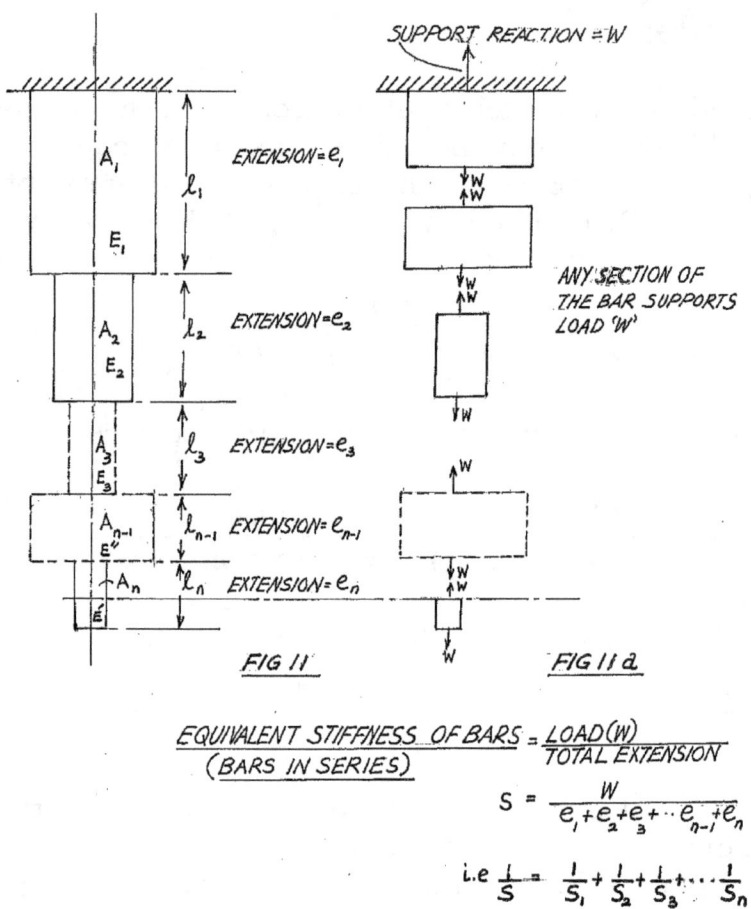

SUPPORT REACTION = W

EXTENSION = e_1

EXTENSION = e_2

ANY SECTION OF THE BAR SUPPORTS LOAD 'W'

EXTENSION = e_3

EXTENSION = e_{n-1}

EXTENSION = e_n

FIG 11

FIG 11 a

$$\frac{\text{EQUIVALENT STIFFNESS OF BARS}}{\text{(BARS IN SERIES)}} = \frac{\text{LOAD (W)}}{\text{TOTAL EXTENSION}}$$

$$S = \frac{W}{e_1 + e_2 + e_3 + \cdots e_{n-1} + e_n}$$

$$\text{i.e } \frac{1}{S} = \frac{1}{S_1} + \frac{1}{S_2} + \frac{1}{S_3} + \cdots \frac{1}{S_n}$$

The FBDs for each part of the bar are shown in Fig. 11a. Evidently the load, in this case effecting tensile stress, in each part is 'W'. But it should also be clear that the extension or deflection of each part would be unique to it. Let the extension of the top portion = e_1; and that of the lowest, e_n.

Designate 'S' as the equivalent stiffness of the complete system and the stiffness of the upper and lower portions of the bar by S_1 and S_n respectively. Remember 'W' is the load on each component part of the bar. Accordingly, by definition

$$S = \frac{W}{\sum e}$$

$$S = \frac{W}{e + e_2 + \ldots e_n}$$

i.e. $$S = \frac{1}{\dfrac{e_1}{W} + \dfrac{e_2}{W} + \dfrac{e_n}{W}} = \frac{1}{1/S_1 + 1/S_s + \ldots 1/S_n}$$

which may be rewritten as

$$\frac{1}{S} = \frac{1}{S_1} + \frac{1}{S_2} + \frac{1}{S_3} + \ldots \frac{1}{S_{n-1}} + \frac{1}{S_n}$$

~ 438 ~

Thus, we may generalize and say that the reciprocal of the equivalent stiffness of a system comprising any number of elastic bars in series, is equal to the sum of the reciprocal of the stiffness of each component bar.

Consider next a number of parallel bars of the same length each made of different materials as before and all having different cross-sectional areas. See Fig. 11b. The bars are attached to a rigid rod supporting a load 'W' at its mid-point.

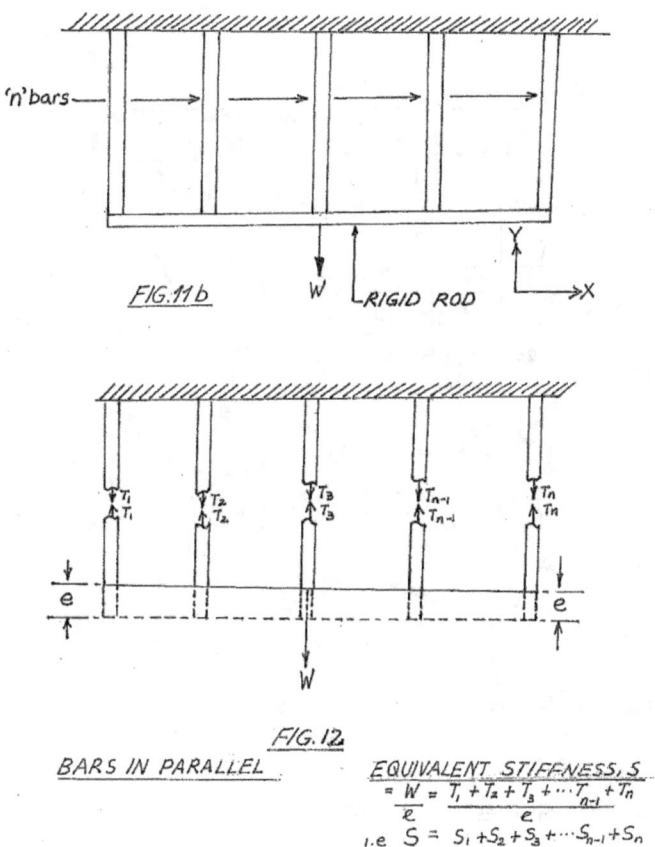

FIG.11b

RIGID ROD

FIG. 12

BARS IN PARALLEL

EQUIVALENT STIFFNESS, S

$$= \frac{W}{e} = \frac{T_1 + T_2 + T_3 + \cdots T_{n-1} + T_n}{e}$$

i.e $S = S_1 + S_2 + S_3 + \cdots S_{n-1} + S_n$

For vertical equilibrium $\sum F_Y = 0$

i.e. $T_1 + T_2 + T_3 + \ldots T_{n-1} + T_n = W.$

If each bar is stretched the same amount 'e' by 'W' then dividing the latter equation throughout by 'e' we obtain:

$$\frac{T_1}{e} + \frac{T_2}{e} + \frac{T_3}{e} + \ldots \frac{T_{n-1}}{e} + \frac{T_n}{e} = \frac{W}{e}$$

The equivalent stiffness of the system is $\frac{W}{e}$, denoted by 'S'.

$$\therefore \quad S = S_1 + S_2 + S_3 + \ldots S_{n-1} + S_n$$

Thus, the equivalent stiffness of a system comprising any number of elastic bars in parallel is equal to the sum of the individual stiffnesses.

Illustrative Example 2

A bar is made up of the three sections shown in Fig. 12. Draw the Free-Body diagram for each section and hence show the state of stress there. Then compute the extension or compression of each part. Neglect stress concentration at the junctions and take $E = 210GN/m^2$.

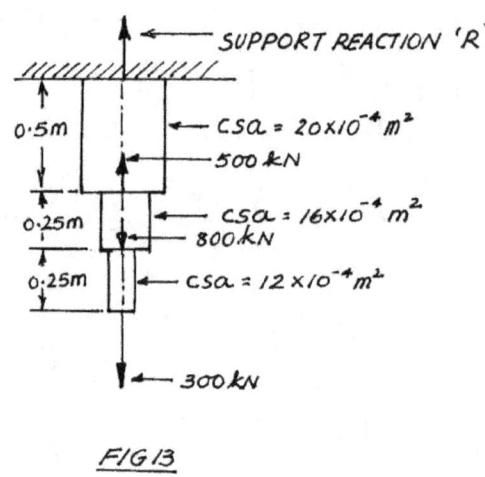

FIG 13

Let the support reaction be 'R'.

For vertical equilibrium:
$$R + 500 - 800 - 300 = 0$$
$$\therefore \quad R = 600kN$$

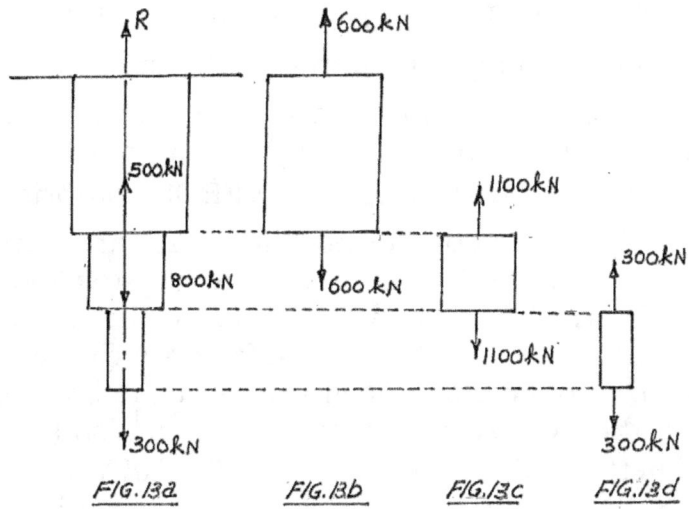

FIG.13a   FIG.13b   FIG.13c   FIG.13d

FBDS for the 3 parts are shown in Figs 13b, 13c and 13d. As can be seen all are in tension. By Hooke's law we may write:

$$extension = \frac{Force}{Area} \cdot \frac{length}{E}$$

The condition represented by Fig. 13b

$$extension = \frac{600 \times 10^3 \times 0.5}{20 \times 10^{-4} \times 210 \times 10^9}$$

$$= \frac{6 \times 500}{20 \times 210}$$

$$0.71mm$$

Similarly for Fig. 13c

$$extension = \frac{1100 \times 10^3 \times 0.25}{16 \times 10^{-4} \times 210 \times 10^9}$$

$$= \frac{11 \times 250}{16 \times 210} mm = 0.82mm$$

and for Fig. 12d

$$extension = \frac{300 \times 10^3 \times 0.250}{120 \times 10^{-4} \times 210 \times 10^9}$$

$$= \frac{3 \times 250}{12 \times 210} = 0.3mm$$

$$\therefore \quad \text{Nett deflection} = (0.71 + 0.82 + 0.30)mm$$

$$= 1.83mm.$$

In the following illustrative example the bar in Fig. 14 is restrained between two fixed rigid supports. In the preceding example it was restrained at one end only.

The first thing to do is to write down the equation of equilibrium for $\sum F_Y = 0$ or $\sum F_X = 0$, for the system of loading depending on the orientation of the bar. Reactions are designated, their directions having been assumed arbitrarily. When reactions R_1 and R_2 are substituted for the rigid supports as in Fig. 14a, $\sum F_Y = 0$, translates to $R_1 + R_2 = 90k$. Now refer to the FBDS comprising Fig. 14b. Because each uniform section of the compound bar must carry either a purely tensile or compressive force, it should be evident that the top-most section is under tension, because R_2 upwards at its top must be balanced by R_2 downwards at its bottom. At the top of the next section there has to be a force R_2 acting upwards at its top to balance R_2 downwards at the bottom of the first section. And at the top of the next section down, there is also a force of 20kN acting downwards. The nett result there : a force $R_2 - 20kN$ acting upwards. This is balanced by $R_2 - 20kN$, downwards at the bottom of this section. The free-body diagrams and the additional explanations provided in the solution should assist. Observe that when you add forces at the same cross-sections across diagrams, the original loading configuration is obtained. Examine the drawings carefully. The detailed solution of Illustrative Example 3 now follows:

Illustrative Example 3

The prismatic bar of varying cross-sectional area shown in Fig. 1 is loaded as indicated in the diagram. Draw appropriate free-body diagrams for each section of the bar indicating the forces acting and compute the corresponding strains and elongations or compressions. Assume Hooke's law for stress and strain. Take E = 210 x 10⁶N/m².

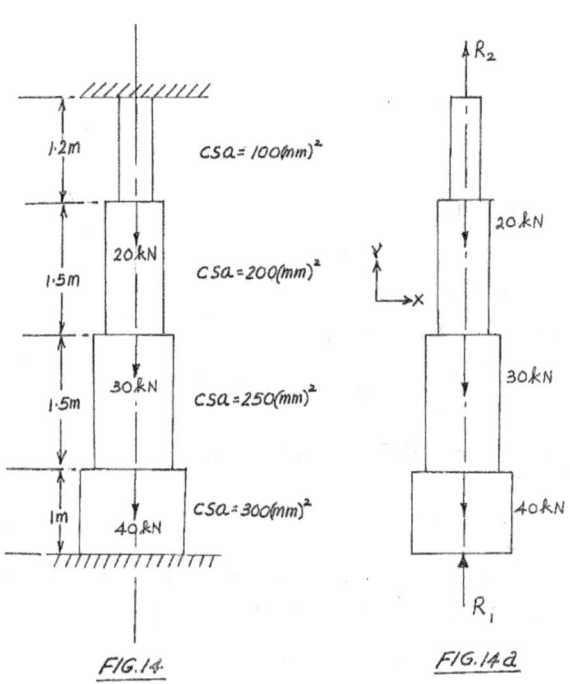

FIG.14.   FIG.14a.

It is assumed that the supports are rigid. Let R_1 = reaction at end E and R_2 = reaction at A. These reactions are entered in Fig. 14a.

Applying $\sum P_y = 0$, we have :

$$R_1 + R_2 = (20 + 30 + 40)kN$$

i.e. $\quad R_1 + R_2 = 90kN \qquad \qquad \cdots \cdots \cdots \cdots$ (i)

The free-body diagrams are prepared starting at the top.

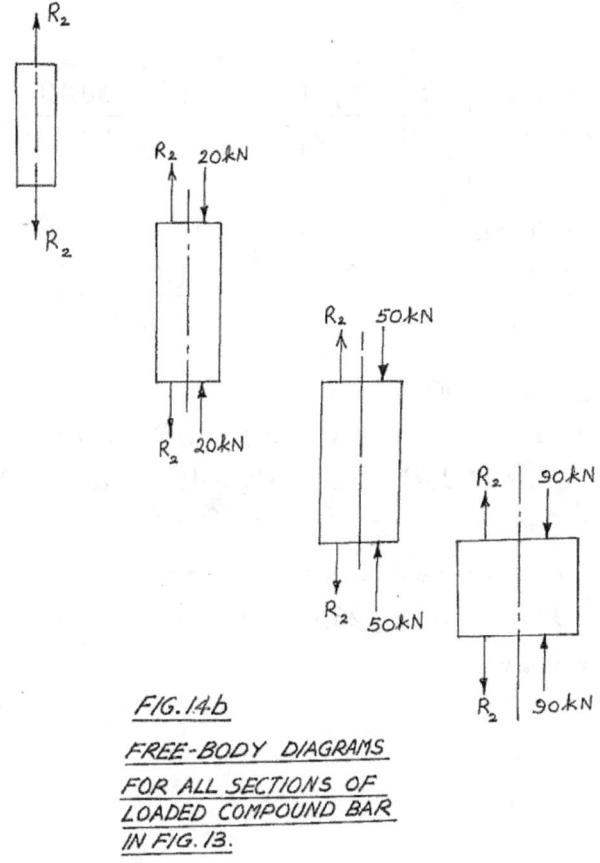

FIG. 14b

FREE-BODY DIAGRAMS
FOR ALL SECTIONS OF
LOADED COMPOUND BAR
IN FIG. 13.

The first section is under a tensile force R_2. When the top face of section 2 is merged with the bottom face of section 1, the R_2s cancel, leaving 20kN acting downwards at the junction of the two sections. Similarly, when the top face of section 2 is merged with the bottom face of section 2, the R_2s cancel, leaving a nett force of 30kN acting downwards ; and so on for the other section. Note that the bottom face of section 4 has force R_2 acting downwards, i.e. $-R_2$ and 90kN acting upwards; the result being $(90-R_2)kN$, which by equation (i) is R_1.

Now by Hooke's law

$$E = \frac{Stress}{Strain}$$

~ 443 ~

$$= \frac{P}{csa} \cdot \frac{length}{ext}$$

or, in short $\quad ext = \frac{P}{E} \cdot \frac{length}{csa}$ $\qquad \cdots \cdots \cdots \cdots \cdots$ (ii)

Because of the rigidity of the supports the sum of all extensions and compressions must be zero, i.e. $\sum (extensions + compressions) = 0$. Accordingly, if we apply (ii) to the whole bar taking each section at a time, then assuming the reaction to be in kN

$$\therefore \quad \sum ext = 0 = \frac{R_2}{E \cdot} \frac{10^3 \times 1.2}{100 \times 10^6} + \frac{(R_2 - 20)10^3 \times 1.5}{E \cdot 200 \times 10^{-6}} + \frac{(R_2 - 50)10^3 \times 1.5}{E \cdot 250 \times 10^{-6}}$$

$$+ \frac{R_2 - 90)10^3}{E.300 \times 10^{-6}}$$

$$\therefore \quad 0 = \frac{1.2R_2}{100} + \frac{1.5(R_2 - 20)}{200} + \frac{1.5(R_2 - 50)}{250} + \frac{(R_2 - 90)1}{300}$$

Multiplying throughout by 1500 we get
$$0 = 15(1.2R_2) + 11.25(R_2 - 20) + 9(R_2 - 50) + 5(R_2 - 90)$$

i.e. $\qquad 0 = 18R_2 + 11.25R_2 - 225 + 9R_2 - 450 + 5R_2 - 450$

from which $\quad 43.25R_2 = 1125$

$\qquad$ or $\qquad R_2 = 26.01 kN$, say, $26 kN$

Therefore by (i) $\qquad R_1 = 90 - R_2 = (90 - 26)kN$

$\qquad \therefore \qquad R_1 = 64 kN$

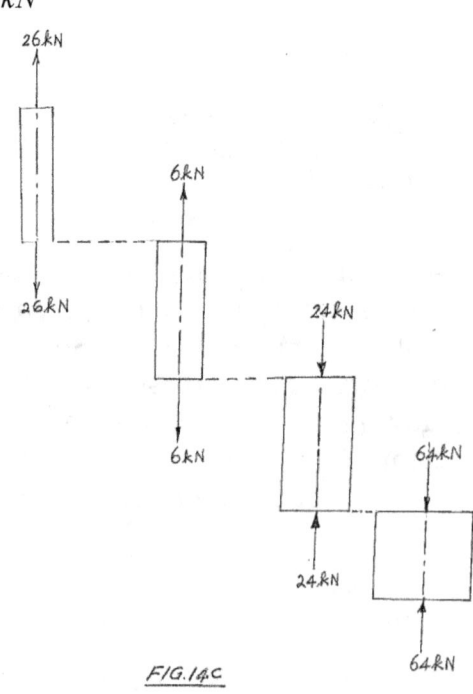

FIG.14C

Combine these diagrams to arrive at the original load configuration. Under that loading it is now know that:

(i)   Section 1 is under a Tensile load of 26kN;
(ii)   Section 2 is under a tensile load of 6kN;
(iii) Section 3 is under a Compressive load of 24kN; and
(iv)   Section 4 is under a Compressive load of 64kN. Refer to Fig. 14c

Tensile strain in section 1, $e_1 = \dfrac{26 \times 10^3}{100 \times 10^{-6}} \times \dfrac{1}{210 \times 10^9}$

$$e_q = 0.001238$$

so that actual extension in this section, say, $y_1 = 1.2(0.001238)$

$$y_1 = 0.0014857m$$

say $y_1 = 0.001486m = 1.486 \times 10^{-3} m$ or $1.486mm$

Similarly $e_2 = \dfrac{6 \times 10^3}{200 \times 10^{-6}} \cdot \dfrac{1}{210 \times 10^9}$

$$e_2 = 0.000142857$$

$\therefore$ $y_2 = 1.5(0.000142857)m$

i.e. $y_2 = 0.000214285 = 2.143 \times 10^{-4} m$ or $0.2143mm$

Further $e_3 = \dfrac{24 \times 10^3}{250 \times 10^{-6}} \cdot \dfrac{1}{210 \times 10^9}$

$$= -0.0004571$$

and $y_3 = -1.5(0.0004571)m$

say, $y_3 = -0.000686m = 0.686mm$

Finally

$$e_4 = -\dfrac{64 \times 10^3}{300 \times 10^{-6}} \cdot \dfrac{1}{210 \times 10^9}$$

$$= -0.00101587$$

$\therefore$ $y_4 = -1(0.00101587)m$

$$= y_4 = -0.001016m \text{ or } -1.102mm$$

Total elongation $= y_1 + y_2$

$$= 0.001486 + 0.000214$$

$$= 0.0017m$$

Total compression $= -(y_3 + y_4)$

$$= -0.000686 - 0.001016$$

$$= -0.001702m$$

Close enough I would say! That is, elongation + compression = 0.

Notice that while the nett elongation and compression add up to zero, the same cannot be said of the strains, the nett tensile strain being 0.001381 and nett compressive strain equal – 0.001473.

Illustrative Example 4

A composite laminate is made of three different materials: 2 mm thick cloth; 6 mm thick mat; and 3 mm woven-roving. What is the maximum carrying load capacity in tension per metre width of this composite laminate?   You are provided with the following data.

	Cloth	**Mat**	**Woven Roving**
Tensile Strength (σ):	123.5MN/m²	46MN/m²	164MN/m²
Tensile Modulus (E)	9.7GN/m²	6.2GN/m²	9.3GN/m²

The strains in the layers must be the same so we may write

$$\frac{\sigma_c}{E_c} = \frac{\sigma_m}{E_m} = \frac{\sigma_{wr}}{E_{wr}} \qquad \dots\dots\dots\dots\dots \quad (i)$$

from which

$$\sigma_c = E_c \cdot \frac{\sigma_m}{E_m} = \frac{E_c \sigma_{wr}}{E_{wr}} \qquad \dots\dots\dots\dots \quad (ii)$$

Also, $$\sigma_m = E_m \cdot \frac{\sigma_c}{E_c} = \frac{E_m \cdot \sigma_{wr}}{E_{wr}} \qquad \dots\dots\dots\dots \quad (iii)$$

and $$\sigma_{wr} = \frac{E_{wr}}{E_c} \sigma_c = \frac{E_{wr} \sigma_m}{E_m} \qquad \dots\dots\dots\dots \quad (iv)$$

Considering the tensile strength and tensile modulus of cloth using equations (ii)

$$123.5 = \frac{9.7\sigma_m}{6.2}$$

$$\therefore \qquad \sigma_m = 78.9 MN/m^2$$

Also $$123.5 = \frac{9.7 \times \sigma_{wr}}{9.3}$$

$$\therefore \qquad \sigma_{wr} = 118.4 MN/m^2.$$

Secondly, considering the tensile strength and tensile modulus of mat and using equations (iii),

$$46 = \frac{6.2 \times \sigma_c}{9.7}$$

$$\therefore \qquad \sigma_c = 72 MN/m^2$$

Also $\qquad 46 = \dfrac{6.2 \times \sigma_{wr}}{9.3}$

from which $\quad \sigma_{wr} = 69 MN/m^2$

Thirdly, considering the tensile strength and tensile modulus of woven roving and using equations (iv)

$$164 = \dfrac{9.3 \times \sigma_c}{9.7}$$

$$\therefore \qquad = 171 MN/m^2$$

Finally $\qquad 164 = \dfrac{9.3 \sigma_m}{6.2}$

i.e. $\quad \sigma_m = 109 MN/m^2$

The results of the computation are tabulated as follows:

Max. Allowable Stress MN/m²	Computed Stresses MN/m²		
In Cloth = 123.5	Mat = 78.9		WR = 118.4
In Mat = 46	Cloth = 72		WR = 69
In WR = 64	Mat = 109		C = 171

Inspection of the table reveals that the best combination of working stresses are as follows :
Mat @ 46 MN/m²; Cloth @72MN/m²; and woven roving @ 69MN/m².

Therefore maximum loading capacity in tension <u>per metre width</u> of material for the given thickness is given by:

Max Tensile load per metre width
$= \sigma_m A_m + \sigma_c A_c + \sigma_{wr} A_{wr}$

$$= 46\left(\dfrac{6}{1000}\right) + 72\left(\dfrac{2}{1000}\right) + 69\left(\dfrac{3}{1000}\right)$$
$$= 0.627 MN$$
$$= 627 kN$$

Illustrative Example 5

Consider steel rod AB shown in Fig. 15. Its length is ℓ_o at temperature 'T_o' and it is heated so that its new temperature, assumed uniformly distributed throughout its length is 'T'. The rod is free to expand; it is not constrained. Explain what happens when the rod is (i) free to expand (ii) constrained. Evidently the rod would extend by an amount $\alpha(T - T_o)\ell$ when free to expand in which α = coefficient of linear thermal expansion. Refer to Fig. 15a. Based on our definition of strain, the rod's strain is $\alpha(T - T_o)\ell / \ell = \alpha(T - T_o)$. There is no stress here! Thus, strain is not necessarily associated with stress.

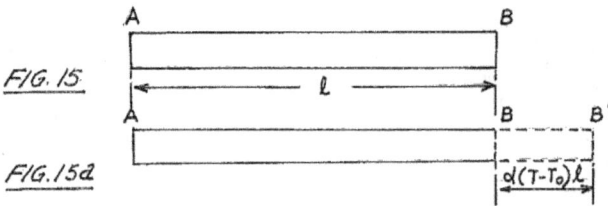

FIG. 15

FIG. 15a

If in its state of elevated temperature it is necessary to get end B' back to B, then a force, say, F shown in Fig. 15b would be required to compress the rod a distance $= \alpha(T - T_o)\ell$. See the Free Body diagram immediately below.

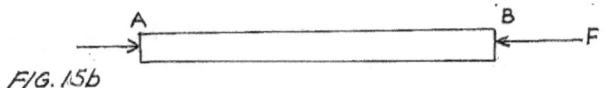

FIG. 15b

Putting

$$E = \frac{Stress}{Strain}$$

$$\sigma = stress$$

$$E = \frac{\sigma}{\alpha \dfrac{(T - T_o)\ell}{\ell}} = \frac{\sigma}{\alpha(T - T_o)} \quad \text{or} \quad \sigma = E\alpha(T - T_o)$$

or

$$E = \frac{F}{csa\{\alpha(T - T_o)}$$

where csa = cross-sectional area of rod, and E = Young modulus.

It follows therefore that if the rod were constrained by rigid supports at A and B at the original temperature T_o and then heated to temperature 'T', then a compressive stress $\sigma = E\alpha(T - T_o)$ would be induced in the rod, although the rod has not physically changed its length; a perfect example of stress (thermal stress) without strain.

The expansion of metals is a matter of some engineering importance. It has to be taken into account in the design and erection of steel trusses, as was pointed out in the previous chapter, and also steel bridges. Also advantage is taken of the significant differences in the magnitude of coefficients of linear expansion to produce bi-metallic strips (or compound strips) employing, say, brass with a coefficient of approximately $19 \times 10^{-6} / C$, and iron with a coefficient of roughly $12 \times 10^{-6} /^{\circ} C$ for use in thermostatic controls. Some of you would have encountered these applications in your earlier studies in Physics.

Illustrative Example 6

A manifold attached to a steam line in a refinery distillation plant is firmly fixed to a rigid support. A straight length of lagged steam line of 150mm and 120mm outside and inside diameter respectively connects this line to a valve fitted to a condenser. If there is no allowance for expansion of the length of lagged line what is the magnitude of the stress induced in the wall of this line for a temperature rise of 189°C? Calculate also the thrust on the valve. Take coefficient of linear thermal expansion for steel, $\alpha_s = 120 \times 10^{-6} /^{\circ} C$ and E_s = 210GN/m².

The strain in steam line due to temperature rise $= \alpha_s \Delta T$ in which $\Delta T =$ temperature rise.
Accordingly, strain $= 12 \times 10^{-6} \times 180 = 2160 \times 10^{-6}$.

Let $\sigma =$ stress induced in the steam line due to the temperature rise. By Hooke's law:

$$E = \frac{\sigma}{strain}$$

$\therefore \qquad 210 \times 10^9 = \frac{\sigma}{2160 \times 10^{-6}} = \frac{\sigma \times 10}{2160}$

i.e. $\qquad \sigma = 2160 \times 210 \times 10^3 = 216 \times 21 \times 10^5$

$\qquad\qquad = 4536 \times 10^5$

$\qquad \sigma = 453.6 MN / m^2$

The thrust on the valve is in fact the force in pipe-wall due to this stress.

Cross-sectional area of pipe (csa)$_p$ $= \pi \dfrac{\left\{ (150)^2 - (120)^2 \right\}}{4} (mm)^2$

$\qquad\qquad = \pi \dfrac{(270)(30)}{4} mm^2$

$(csa)_p = \dfrac{6361}{10^6} m^2$

$\therefore$ Thrust on condenser valve $\sigma(csa)_p = 453.6 \times 10^6 \times 6361 \times 10^{-6} N$

$\qquad\qquad = 2885 kN$

Illustrative Example 7

Two cylinders are connected by a straight length of steel pipe 5m long, at room temperature. Determine the stress induced in this pipe for a temperature rise of 180°C assuming that thermal expansion is limited to 0.5mm. Take $\alpha_s = 12 \times 10^{-6}/{}^{\circ}C$ and E = 206GN/m².

The fact that thermal expansion is limited to 0.5mm means that the line can expand only by an amount = 0.5mm without causing any thermal stress. But the full expansion of the pipe's length due to the temperature rise $= \alpha_s \Delta T \ell$. Therefore, employing the data provided:

$$\text{Full expansion} = 12 \times 10^{-6}(180)5m$$
$$= 12 \times 10^{-6} \times 180 \times 5000mm$$
$$= 10.8mm$$

Expansion allowance $= 0.5mm$

so that actual expansion $= 10.8 - 0.5 = 10.3mm$

Employing Hooke's law

$$E = \frac{\sigma}{strain} = \frac{\sigma}{\dfrac{10.3}{5000}}$$

where σ = stress in pipe.

$$\therefore \quad 206 \times 10^9 = \frac{\sigma}{10.3} \times 5000$$
$$\sigma = 424 \times 10^6 \, N/m^2$$

or $\quad \sigma = 424MN/m^2$

Illustrative Example 8

Consider the case of two different materials permanently bonded together to form a composite which is then subjected to an increase in temperature. How do we determine the resulting forces and stresses acting in each component part of the composite arising from the change in temperature?

Take the composite bar comprising a uniform brass strip of length 'ℓ' and cross-sectional area (a_b) bonded to a uniform steel strip also of length 'ℓ' but cross-sectional area 'a_{fe}'. See Fig. 16.

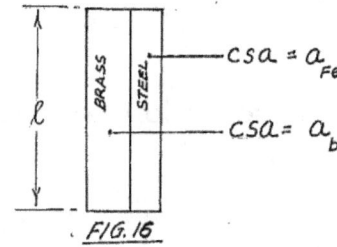

FIG. 16

The composite is initially at room temperature 'T_o'. Its temperature is then increased by an amount ΔT, to $T_o + \Delta T$ assumed uniformly distributed throughout. It is assumed that the heated composite does not bend or twist. The coefficient of expansion of brass $\alpha_b = 18.5 \times 10^{-6}/^\circ C$ and that of steel, $\alpha_{fe} = 12.5 \times 10^{-6}/^\circ C$.

In order to make the determination we shall consider first the effect of the temperature increase on the strips as separate elements and following that the outcome of their combination as a single composite.

It follows at once that since $\alpha_b > \alpha_{fe}$, when the strips are separate and therefore can each expand freely, the brass strip's thermal extension would be greater than the iron's, i.e. $\alpha_b \Delta T \ell > \alpha_{fe} \Delta T \ell$. This is depicted in Figs 16a and 16b.

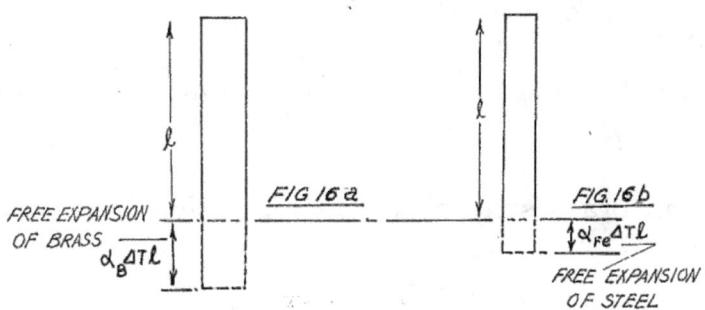

Consider next the condition when the strips are bonded to form the composite. As one would expect each strip cannot behave in quite the same manner as explained previously. The strips extend alright to be sure but the brass strip's full extension is hampered by the iron bar's lower quantum of expansion, in consequence of which the brass strip can only extend to level $E - E'$ shown in Fig. 16c. In this diagram although the strips are bonded together they are shown separately for the purpose of clarification. That is to say the effect of the iron strip is to produce a compression from level $P - P'$ to level $E - E'$ in the brass strip; in effect $P - P'$ is pushed back to $E - E'$. It should be clear that level $E - E'$ is the final equilibrium position of the heated compound composite strip. This being the case, it follows that the iron strip of which the free expansion is $\alpha_{fe} \Delta T \ell$, must experience a pull from level $Q - Q'$ to level $E - E'$.

That is to say the effect of brass strip is the creation of a tension in the steel strip.

The free-body diagrams for the condition are shown in Fig. 16c.

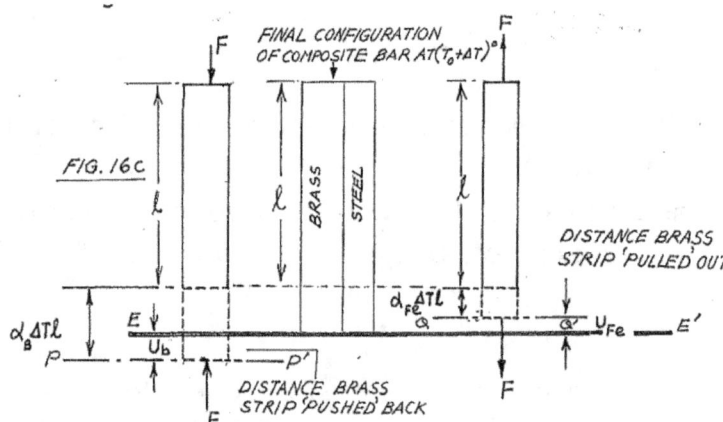

Evidently for equilibrium, the compressive force 'F' in the brass strip must equal the tensile force 'F' in the iron strip. Note that we are up to this point dealing with forces here not stresses. Because the cross-sectional areas of the strips are different, it follows that for equal forces 'F', the stress in the strips must be different. Let σ_b = stress in the brass strip and σ_{fe} = stress in iron strip.

$$\therefore \qquad \sigma_b = \frac{F}{(csa)_b} \quad ; \quad \sigma_{fe} = \frac{F}{(csa)_{fe}}$$

so that, $\qquad \sigma_b (csa)_b = \sigma_{fe} (csa)_{fe}$ .... (i)

If u_b = compression in brass strip and u_{fe} = extension in iron strip, both caused by 'F' then by Hooke's law

$$E_b = \frac{\sigma_b}{u_b} \cdot \ell \quad \text{and} \quad E_{fe} = \frac{u_{fe} \ell}{u_{fe}}$$

from which

$$u_b = \frac{\sigma_b \ell}{E_b} \quad ; \quad u_{fe} = \frac{\sigma_{fe} \ell}{E_{fe}}$$

Referring again to Fig. 16c it is seen that

$$\alpha_b \Delta T \ell - u_b = \alpha_{fe} \Delta T \ell + u_{fe}$$

i.e. $\qquad \alpha_b \Delta T \ell - \frac{\sigma_b \ell}{E_b} = \alpha_{fe} \Delta T \ell + \frac{\sigma_{fe} \ell}{E_{fe}}$

or $\qquad \alpha_b \Delta T \ell - \frac{\alpha_b}{E_b} = \alpha_{fe} \Delta T + \frac{\sigma_{fe}}{E_{fe}}$ .... (ii)

By equation (i) $\qquad \sigma_b = \sigma_{fe} \dfrac{(csa)_b}{(csa)_{fe}}$

$\qquad\qquad$ or $\qquad \sigma_{fe} = \sigma_b \dfrac{(csa)_{fe}}{(cas)_b}$

Depending on which stress is required, the appropriate substitution is made in (ii). For example if the stress in the brass strip is required, equation (ii) with the aid of equation (i) is manipulated to obtain

$$\frac{\sigma_b}{E_b} + \frac{\sigma_{fe}}{E_{fe}} = \alpha_b \Delta T - \alpha_{fe} \Delta T = \Delta T (\alpha_b - \alpha_{fe})$$

i.e. $\qquad \sigma_b \left\{ \dfrac{1}{E_b} + \dfrac{(csa)_b}{(csa)_{fe}} \cdot \dfrac{1}{E_{fe}} \right\} = \Delta T (\alpha_b - \alpha_{fe})$ $\qquad$ . . . . (iii)

Having determined the stress σ_b and σ_{fe}, the force 'F' may be readily obtained by multiplying the stress by the relevant cross-sectional areas.

As a general rule you might find it a useful step in problems involving thermal stresses and strains in composite bodies to consider first the effect of temperature rise on the individual components and after that the action that is necessary to obtain a final configuration as was done in the foregoing.

Illustrative Example 9

A vertical rectangular reinforced concrete column $500mm \times 200m$ and $2m$ long is subjected to a compressive force of 350kN. Neglecting the effects of buckling, the length of column being less than 15 times its diameter, and assuming maximum permissible working stresses in concrete and reinforcing bars (rebars) to be 5600kN/m² and 84000 kN/m² respectively, calculate the total cross-sectional area of steel required.

The rebars are bound together by transverse reinforcement to resist the tendency of the rods to bulge outwards and to prevent the concrete from splitting along planes of maximum shear stress (of which more later). Take E_c = 21GN/m² and E_s = 210GN/m².

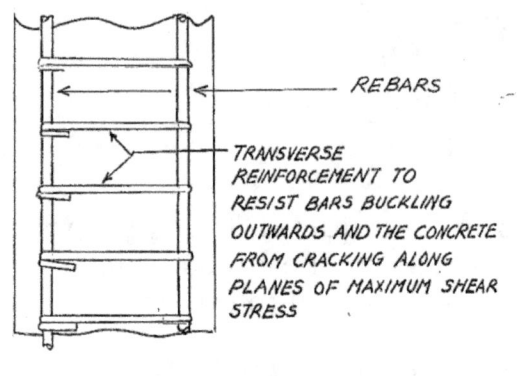

REBARS

TRANSVERSE REINFORCEMENT TO RESIST BARS BUCKLING OUTWARDS AND THE CONCRETE FROM CRACKING ALONG PLANES OF MAXIMUM SHEAR STRESS

FIG. 17

Compressive force shared by steel and concrete.

Let σ_c and σ_s be the actual stresses in concrete and steel respectively.

$$\therefore \qquad \sigma_C a_C + \sigma_s a_s = 350kN \qquad \dots\dots\dots\dots\dots\dots \text{(i)}$$

Where a_c and a_s are cross-sectional areas of concrete and steel.

Now, because the strains must be equal

$$\frac{\sigma_c}{E_c} = \frac{\sigma_s}{E_s} \qquad \dots\dots\dots\dots\dots\dots\dots \text{(ii)}$$

$$\text{or} \qquad \sigma_c = \frac{E_c}{E_s} \cdot \sigma_s \qquad \dots\dots\dots\dots\dots\dots \text{(iii)}$$

Similarly,

$$\sigma_s = \frac{E_s}{E_c} \cdot \sigma_c \qquad \dots\dots\dots\dots\dots\dots \text{(iv)}$$

Substituting permissible stress of 5600 kN/m² in equation (iii), we get

$$5600 = \frac{21}{210} \cdot \sigma_s$$

$$\therefore \qquad \sigma_s = 56000kN/m^2$$

Similarly for a permissible stress in the steel of 84000kN/m², we have from equation (iv)

$$84000 = \frac{210}{21} \cdot \sigma_c$$

$$\therefore \qquad \sigma_c = 8400kN/m^2$$

Clearly this value of σ_c will cause failure of the concrete, its allowable stress being only 5600kN/m². However with $\sigma_c = 5600kN/m^2$, the stress in the total steel reinforcement $\sigma_s = 56000kN/m^2$ and failure will not result, its allowable stress being $84000kN/m^2$.

Using equation (i)

~ 454 ~

$$350kN = \sigma_c \text{ (Total cross-sectional area of column minus}$$
$$\text{Area of steel) } + \sigma_s \text{ (area of steel)}$$

Let σ = total cross-section area of column

i.e. $\sigma_c(a - a_s) + \sigma_s a_s = 350kN$

But $a_c = (300 \times 200)(mm)^2$ or $50000/10^6 (m)^2$

$$\therefore \quad 56000 \times 10^3 \left\{ \frac{60000}{10^6} - a_s \right\} + 56000 \times 10^3 a_s = 350 \times 10^3 N \qquad \dots \dots \text{(v)}$$

where a_s = cross-sectional area of steel in sq. metres

From (v) $50400 a_s = 14(m)^2$

$\therefore$ $a_s = 0.000278(m)^2$

or $a_s = 278(mm)^2$.

Thus the total area of steel $= 278(mm)^2$. Reference to a Table of US Metric sizes, one (1) # 10 Metric size rebar of nominal diameter 9.525 mm has a nominal area of 71(mm)². Therefore four (4) such bars with total $a_s = 284(mm)^2$ should suffice. A longitudinal section of the column showing the bars and transverse reinforcement is shown in Fig. 17. A transverse section of the bar is given in Fig. 17a.

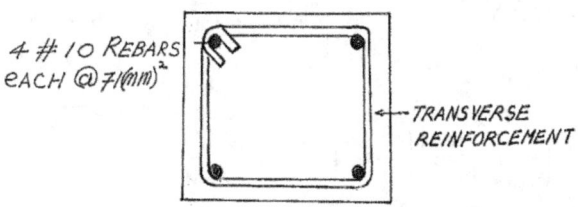

4 # 10 REBARS eACH @ 71(mm)²

TRANSVERSE REINFORCEMENT

FIG.17a:CROSS-SECTION OF COLUMN

To determine the deflection 'δ' in the steel we employ Hooke's law :

$$E_s = \frac{\sigma_s}{\delta} = 2000$$

$$\therefore \quad \delta = \frac{\sigma_s \times 2000}{E_s}$$

$$= \frac{56000 \times 2000}{210 \times 10^9}$$

$$= \frac{112}{210 \times 10^3}$$

$$= 0.0005 mm$$

For practical purposes, compression is zero.

Illustrative Example 10

Three identical steel cables each 1 metre apart support a rigid bar and a concentrated force of 15kN. The force is 250mm to the right of the middle wire as shown in Fig. 1. Neglecting the weight of the bar, calculate the force carried by each wire.

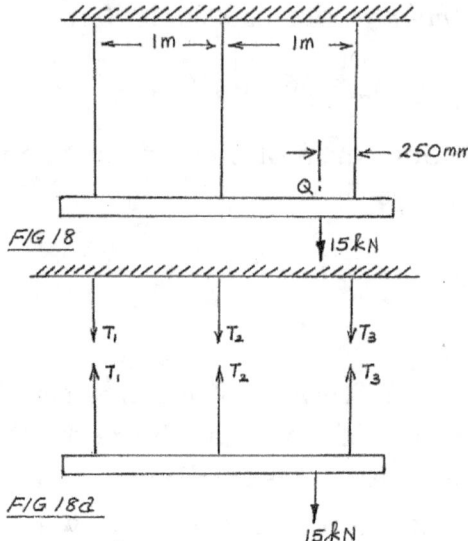

For vertical equilibrium:
$$T_1 + T_2 + T_3 = 15kN \qquad \dots\dots\dots\dots\dots\dots\dots\dots\dots \text{(i)}$$

Moments about 'Q':
$$T_1\left(1\frac{1}{4}m\right) + T_2\left(\frac{1}{4}m\right) = T_3\left(\frac{3}{4}m\right)$$

i.e.
$$5T_1 + T_2 = 3T_3 \qquad \dots\dots\dots\dots\dots\dots\dots\dots\dots \text{(ii)}$$

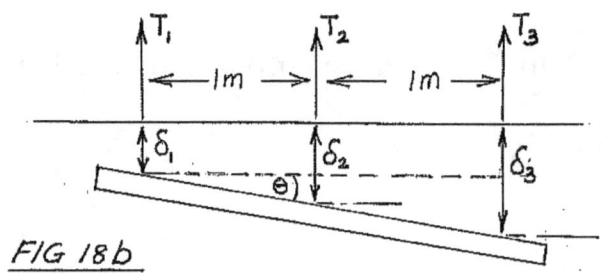

δ_s, δ_2, δ_3 are the vertical extensions of the rigid bar under the influence of the force of 15kN.

Employing the relationship, $E = \dfrac{T}{A} \cdot \dfrac{\ell}{\delta}$ which may be rewritten as $\delta = T\ell/AE$; and, noting that for each cable ℓ/AE is a constant, say, 'k', the cables being identical : $\delta_1 = kT_1$; $\delta_2 = kT_2$; $\delta_3 = kT_3$.

Since the rod is rigid, it is deduced from Fig. 18b that,

$$Tan\theta = \frac{\delta_2 - \delta_1}{1} = \frac{\delta_3 - \delta_2}{1}$$

or $\quad kT_2 - kT_1 = kT_3 - kT_2$

i.e. $\quad T_2 - T_1 = T_3 - T_2$

$\therefore \qquad T_3 = 2T_2 - T_1$ . . . . (iii)

Substituting this result in (ii) and (i)

$$5T_1 + T_2 = 6T_2 - 3T_1$$

from which

$$T_1 = \frac{5}{8} T_2$$ . . . . (iii)

and

$$3T_2 = 15kN$$

$\therefore \qquad T_2 = 5kN$

so that

$$T_1 = \frac{25}{8} kN$$

From (i) $\quad T_3 = \dfrac{55}{8} kN$

Ans : $\qquad T_1 = \dfrac{25}{8} kN$; $T_2 = 5kN$; $T_3 \dfrac{55}{8} kN$

The following Illustrative Example deals with volumetric strain which derives from hydroscopic action and not from the application of any mechanical stress. Later on when we come to consider the application of complex stress we shall consider volumetric strain as one of the effects of such application. In the example the strains are caused by shrinkage due to loss of moisture but the calculations are similar to those followed later on, except that whereas here the strains are obtained by direct tri-axial stress analysis the strains must first of all be computed. But we shall take all that in our stride.

Illustrative Example 11

A specimen of 'two-by-four green and dressed Laurier *(Nectandra spp.)* (a local timber used for furniture) was 5m long by 12cm wide by 6cm thick before drying in a solar kiln. At the end of the drying process, the length of the rod had shortened by 10mm; the width was reduced to 11.7cm and the thickness was measured to be 5.5cm. Compute the hygroscopic strain.

The 'green' specimen is shown in dotted outline in Fig. 19. Its original dimensions c, b, and a are respectively in the direction of the X-, Y- and Z-coordinate axes. The solid outline in Fig. 19 represents the specimen after kiln-drying.

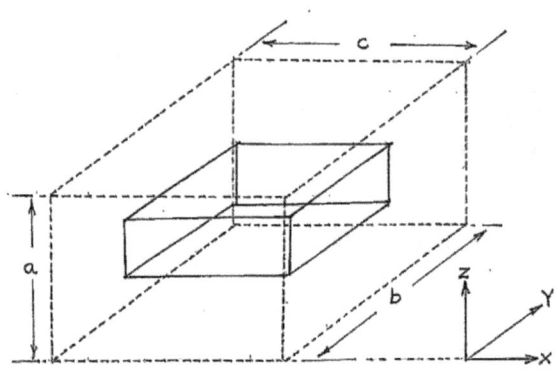

OUTLINE OF SOLID BEFORE SHRINKAGE IN
DOTTED LINES THROUGHOUT
SHRUNKEN OBJECT IN SOLID LINES THROUGHOUT

FIG.19

By definition : $Strain = \dfrac{Final\ dimension - Original\ dimension}{Original\ dimension}$

$\therefore$ $Final\ dimension = Original\ dimension \times Strain + Original\ dimension$

Final dimension = Original dimension × Strain + Original dimension
In this case, all strains are negative, Denoting

e_X = strain in direction of X-axis
e_Y = "   "   "   " Y-axis
e_Z = "   "   "   " Z-axis

Here $e_X = -\dfrac{10}{5000} = -0.002$

$e_Y = -\left(\dfrac{12-11.7}{12}\right) = -0.025$

$e_Z = -\left(\dfrac{6-5.5}{6}\right) = -0.0833$

The final volume, say V_F may be expressed as
$$V_F = c(1-e_x)b(1-e_y)a(1-e_z)$$ . . . . (I)

This is so because : Final dimension = original dimension – reduction in dimension. Thus in relation to the X-coordinate

Final dimension $= -ce_X + c = c(1 - e_X)$

Rewriting (i)

$$V_F = abc \ (1 - e_X) \ (1 - e_Y) \ (1 - e_Z)$$

or $\quad V_F = V(1 - e_X) \ (1 - e_Y)(1 - e_Z) \qquad \dots \dots \dots \dots \dots \dots$ (ii)

where $\quad V$ = original volume, which in this case $= 5 \times \left(\dfrac{12}{100}\right) \times \left(\dfrac{6}{100}\right) = 0.036 (m)^3$

Substitution of the relevant data in (ii) produces :

$$V_F = 0.036(1 - 0.002) \ (1 - 0.025) \ (1 - 0.0833)$$

$$= 0.036(0.892)$$

$$= 0.0321 (m)^3$$

$\therefore \quad$ Reduction in volume $\quad = 0.036 - 0.0321$

$$= 0.0039 m^3$$

from which

$$\text{Volumetric Strain} \quad = \dfrac{-0.0039}{0.036}$$

$$= -0.108$$

This may be expressed as a percent. Thus volumetric strain of specimen of Laurier $= -10.8\%$. A rather neat way of arriving at an excellent approximation of volumetric strain is to write the original volume 'V' in a format such as $V = xyz$ in which x, y, and z are respectively the dimensions in the X-, Y-, and Z-coordinate system. Then the logarithm of both sides of the expression is taken and the result differentiated, thus:

$$V = xyz$$

$$Log V = Log x + Log y + Log z$$

and

$$\frac{dV}{V} = \frac{dx}{x} + \frac{dy}{y} + \frac{dz}{z}$$

or $\quad \dfrac{dV}{V} = e_x + e_Y + e_Z$

Therefore,

Volumetric Strain = sum of strains in the direction of the 3 coordinate axes. Using the foregoing computed values of strains in Illustrative Example 11

$$\frac{dV}{V} = -0.002 - 0.025 - 0.0833$$

i.e. $\quad \dfrac{dV}{V} = -0.1103$

or $\quad \dfrac{dV}{V} \approx -11\%$

This method gives an excellent approximate result because in the computation
$$V + \delta V = (x + \delta x)\ (y + \delta y)\ (z + \delta z)$$
$$= xyz + yz\,\delta x + xz\,\delta y + z\,\delta x\,\delta y + xy\,\delta z + x\,\delta y\,\delta z + y\,\delta x\,\delta z + \delta x\,\delta y\,\delta z$$

We can justifiably ignore 2nd and 3rd order infinitesimal such as $z\,\delta x\,\delta y$, $x\,\delta y\,\delta z$, $y\,\delta x\,\delta z$, and $\delta x\,\delta y\,\delta z$ because strains are such small quantities anyway.

$\therefore \quad V + \delta V = xyz + yz\,\delta x + xz\,\delta y + xy\,\delta z$

or $\quad dV = yz\,dx + zd\,y + xy\,dz$

Dividing both sides by $V = xyz$ we get
$$\frac{dV}{V} = \frac{dx}{x} + \frac{dy}{y} + \frac{dx}{z}$$
which is the result we get by writing
$$V = xyz$$
and taking the logarithm of both sides and differentiating as was done in the foregoing.

The same result may be obtained in a slightly different manner. The derivation also yields an approximate result and we shall explain why.

When, $V = xyz$ we obtain by partial differentiation:

$$\frac{\partial V}{\partial x} = yz \qquad (y \text{ and } z \text{ are assumed constant})$$

$$\frac{\partial V}{\partial y} = xz \qquad (x \text{ and } z \text{ are assumed constant})$$

$$\frac{\partial V}{\partial Z} = xy \qquad (x \text{ and } y \text{ assumed constant})$$

Now $\quad dV = \dfrac{\partial V}{\partial x} \cdot dx + \dfrac{\partial V}{\partial y} \cdot dy + \dfrac{\partial V}{\partial z} \cdot dz$

i.e. $\quad dV = yz \cdot dx + xz \cdot dy + xy \cdot dz$

Dividing throughout by $V = xyz$
$$\frac{dV}{V} = \frac{yz}{xyz} \cdot dx + \frac{xzdy}{xyz} + \frac{xydz}{xyz}$$

$$\therefore \qquad \frac{dV}{V} = \frac{dx}{x} + \frac{dy}{y} + \frac{dz}{z}$$

The resulting volumetric strain is approximate because in performing the partial differentiations it was assumed that certain variables were constant. Although the assumptions were unquestionably valid because the strains occurring are so small as to allow us to treat two of the three independent variables as constant for each of the partial differentiations, the reality is in fact that strain along each of the three coordinate axes is occurring simultaneously. Strictly speaking nothing is constant!

Illustrative Example 12

A length of 500mm of sucker rods is suspended in the tubing of an oil well. Compute the elongation of the first 250 metres at the top assuming all the rods to constitute a single uniform rod of constant cross-sectional area and density. Take density of the sucker-rod grade of steel as 7800kg/m². Assume g = 10m/s².

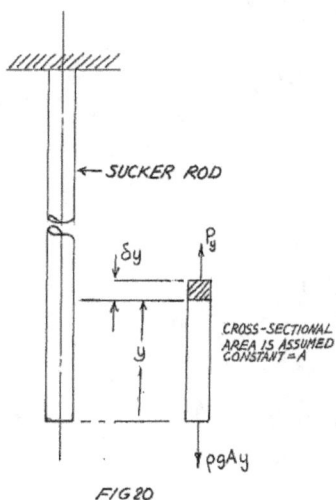

FIG 20

In Fig. 20 it is assumed the rods are attached to a rigid support at the well-head.

Let $\delta\Delta$ = infinitesimal deformation of an element of sucker rod of length 'δy'

then
$$\frac{\delta\Delta}{\delta y} = e$$

and because the stress on element $= \dfrac{P_y}{A_y}$

$$\frac{\delta\Delta}{\delta y} = \frac{P_y}{A_y} \cdot \frac{1}{E}$$

or $\qquad \delta\Delta = \dfrac{P_y \delta_y}{A_y E}$ .... (i)

$\therefore$ Total deformation of rod from $y=0$ to $y=250$ (i.e. origin of Y-axis : $y=0$ at depth of 500m.

$$\int_0^{250} d\Delta = \dfrac{1}{E} \int_0^{250} \dfrac{P_y}{A} dy \qquad \qquad .... \text{(iii)}$$

Remember that P_y is the force due to the weight of sucker rod of length 'y'.

i.e. $\qquad P_y = \rho\, g\, A\, y$

where $\qquad \rho$ = density of rod in kg/m^2

$\qquad g$ = acceleration due to gravity = 10m/s^2 approximately.

Equation (iii) may now be rewritten as

$$\Delta_{250} - \Delta_0 = \Delta = \dfrac{1}{E} \int_0^{250} \rho\, g\, \dfrac{A}{A} dy$$

$$= \dfrac{1}{E} \int_0^{250} \rho\, g\, y\, d\,y$$

i.e $\qquad \Delta_1 = \dfrac{\rho g}{2E} \left[y^2\right]_0^{250}$ .... (iv)

Substituting values in result (iv)

$$\Delta = \dfrac{7800 \cdot 10}{210 \times 10^9 \times 12} (250)^2$$

$$= \dfrac{7800}{210 \times 10^9 \times 2} \times 62500$$

$\therefore \qquad \Delta = 0.0116m$

Check units in equation (iv)

$\qquad LHS = m$

$\qquad RHS = \dfrac{kg}{m^3} \cdot \dfrac{m}{s^2} \cdot \dfrac{s^2 \cdot m}{kg\, m} \cdot m^2 = m \qquad \qquad$ OK

$\therefore$ Elongation of the first 250m of the rod from the top end is 0.0116m or 11.6 millimetres.

ALLOWABLE STRESS AND FACTOR OF SAFETY

Determination of the size of an engineering component which is to undergo stress of any kind, be it of a nature static, dynamic, cyclical, variable or of a combination involving any or all of these, is directly dependent on the probable value of the maximum stress and strains induced by the loads and displacements deriving therefrom.

Remember well that it is not only by the application of force that stresses and strains are caused; displacements also cause strains and stresses; in short, stress can cause strain and strain can cause stress. Based on determination of the maximum strains and stresses and a knowledge of the mechanical and other properties of the materials likely to be chosen for manufacture, at the same time contemplating all kinds of imponderable scenarios i.e. the worst that can happen, the designer estimates a component's size.

Evidently, if the estimated maximum stress exceeds the elastic limit of the material chosen, there will be some permanent deformation; the component may even fail outright. The designer therefore insures against the risk of such an outcome by sizing the component in such a way that an estimated in-service maximum stress is fixed at yield point. This in-service maximum stress is generally referred to as the allowable stress, σ_A i.e. the highest stress that can be safely tolerated by the component. This is engineering-design risk management.

The designer's conservation and caution is reflected in what is normally referred to as a factor of safety, which is as some authors have stated quite appropriately, in truth and fact, a factor of ignorance. Factor of safety (FOS) is related to allowable stress by the following expression:

$$\text{Factor of Safety} = \frac{\textit{Ultimate Strength of a material}}{\max \textit{imum allowable stress}}$$

The ultimate design load may be elastic limit or yield strength of the material or it may otherwise be specified. For example, if the ultimate strength of a certain wood is, say, $500kN/m^2$ and the factor of safety chosen is 5, then the maximum allowable working load is $100kN/m^2$. Take the case of the portion of the wooden frame shown immediately below.

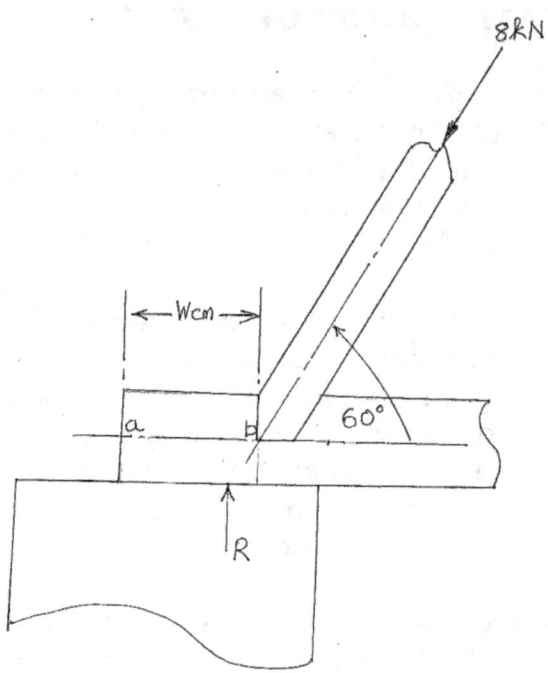

Suppose the ultimate strength of the wood in shear parallel to the grain is 400kN/m² and the designer chooses a factor of safety of, say, 8, then the maximum allowable shear stress parallel to the grain is 4000kN/m²/8=500kN/m².

If the wood is $10cm \times 10cm$, then for a horizontal shear force of $8Cos60° kN$:

$$500kN/m^2 \times \left(\frac{10cm \times Wcm}{10^4} \right) m^2 = 8Cos60° = 4kN$$

$$\therefore \quad 5W = 40$$
$$\therefore \quad W = 8cm$$

Illustrative Example 13

One end of the steel crankshaft of a lawnmower is 20mm diameter and carries a flywheel fastened to it by a square key 5mm by 5m made of an aluminium alloy. The key is also to protect the shaft when the blade jams suddenly, as for example, when in operational mode it strikes, say, a tree stump. The cost of replacing a key is a great deal less than that of changing a crankshaft. Shaft speed in the instant case is 3600 rpm and is provided by a 2625W engine. If the key is to fail by shear what is its length, assuming a maximum allowable working shear stress of 7MN/m²?

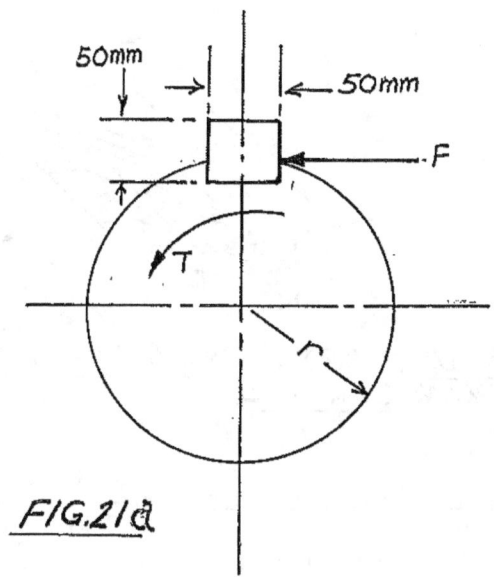

FIG.21a

In Fig. 21a a section through the shaft fitted with the square key is shown
Let $\qquad T$ = shaft torque in Nm.

$\qquad r$ = radius of shaft end, m.

Assuming that the entire torque transmitted is located at the surface of the shaft as shown, then
$$T = Fr \qquad \dots\dots\dots\dots\dots\dots\dots\dots\dots\dots\dots\dots\dots\dots \quad \text{(i)}$$
Now,
$\qquad$ Work done per revolution $= T \cdot \omega \qquad \dots\dots\dots\dots\dots\dots \quad \text{(ii)}$
where $\qquad \omega$ = angular velocity in radians/s
$$\therefore \qquad T \times \frac{2\pi(3600)}{60} = 2625$$
from which $\quad T = \dfrac{2625 \times 60}{2\pi \times 3600}$
$$= 6.96 Nm.$$
From (i) $\qquad F = \dfrac{T}{r}$
$$= \frac{6.96}{\left(\dfrac{10}{1000}\right)}$$
$$= 696 N$$

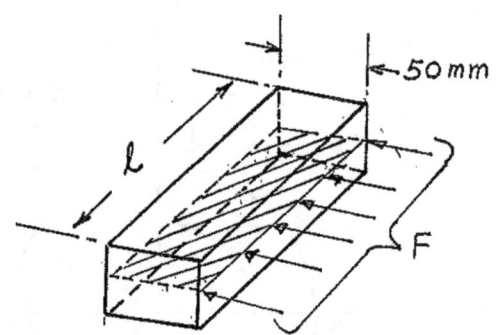

FIG.21b : SQUARE KEY SHOWING TYPICAL SHEARING PLANE (HATCHED)

If the key is to fail by shear, then shear area = $\dfrac{5}{1000} \times \dfrac{\ell}{1000}$

where ℓ = length of key in millimeters.

But maximum allowable working shear stress = $7 \times 10^6 \, N/m^2$

$$\therefore \quad 7 \times 10^6 = \dfrac{F}{\dfrac{5}{1000} \times \dfrac{\ell}{1000}}$$

or $\quad 7 \times 10^6 = \dfrac{696 \times 10^6}{5\ell}$

i.e. $\quad 35\ell = 69\ell$

$\qquad 45\ell = 696$

$\qquad \ell = 19.9mm, \quad$ say $\quad 20mm$

Illustrative Example 14

The bell-crank lever shown in Fig. 22 is pivoted to the support 'A' by means of a steel pin of diameter 'd' mm. If the allowable shear stress in the pin is limited to $0.11MN/m^2$ what is the minimum value of 'd'

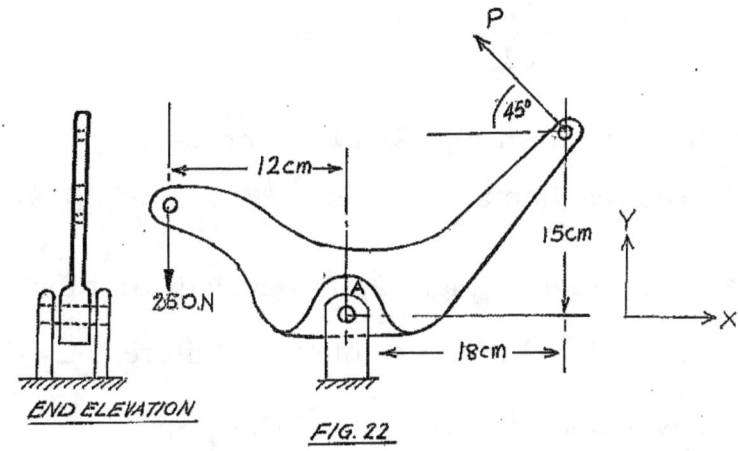

FIG. 22

END ELEVATION

The Free-body diagram is given in Fig. 22a in which R_H and R_V are respectively the horizontal and vertical components of the reaction at the pin at 'A'.

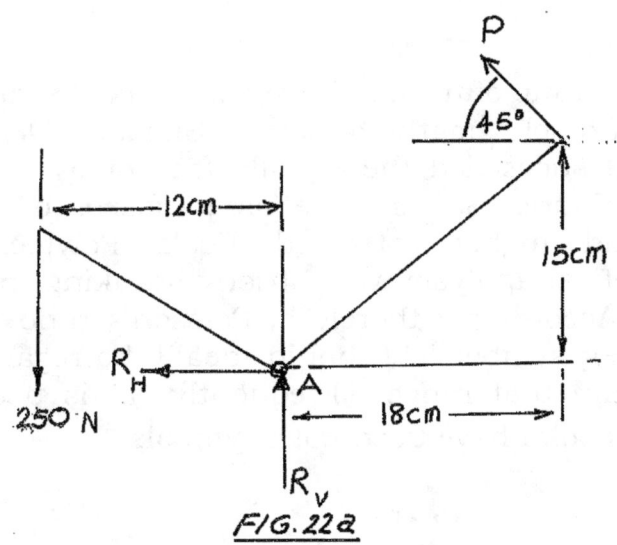

FIG. 22a

By moments at A (and working in N and cm)

$$250(12) + PSin45°(18) + PCos45°(15) = 0$$

i.e. $$3000 + \frac{P18}{\sqrt{2}} + \frac{P15}{\sqrt{2}} = 0$$

from which $$P = 128.5N$$

Also, for $\sum F_x = 0$

$$R_H - PCos45° = 0$$

i.e. $$R_H = \frac{128.5}{\sqrt{2}} = 91N$$

and for $\sum F_y = 0$

$$R_V - 250 + PCos45° = 0$$

i.e. $\quad R_V = 250 - \dfrac{128.5}{\sqrt{2}} = 91N = 159N$

The assumed direction of R_V in Fig. 22a is incorrect : R_V acts downwards; its magnitude = 5kN. The reaction at A, viz $R = \sqrt{R_v^2 + R_H^2} = \sqrt{(159)^2 + (91)^2}$ 183.2N.

Referring to the end elevation of Fig. 22, it is evident that for a pin diameter 'd' the total area to be considered in a case of shear failure is $2\left(\dfrac{\pi d^2}{4}\right)$ i.e. $\dfrac{\pi d^2}{2}$.

For a maximum allowable shear stress of 0.11MN/m²,

$$0.11 \times 10^6 \times \frac{\pi d^2}{2} = 183.2$$

$$\therefore \quad d^2 = 0.00106m^2 \quad d = 0.0326m \quad \text{or} \quad d = 32.6mm$$

POISSON'S RATIO

This is defined as the ratio of transverse strain to axial strain. It is named after the outstanding French mathematician Simeon Dennis Poisson who, incidentally in 1808 succeeded the equally fabulously-brilliant Jean Baptiste Joseph Fourier of Fourier-Series fame as Professor of Mathematics at the world-renowned and highly esteemed Ecole Polytechnique. Poisson's undergirding theoretical analysis was, strictly-speaking, restricted to strain in the elastic range. Accordingly therefore, Poisson's ratio which is universally denoted by the Greek symbol 'v', should ideally be regarded as being validly applicable only within that range. Frequently 'μ' is used as an alternative symbol. In this textbook I have used both symbols.

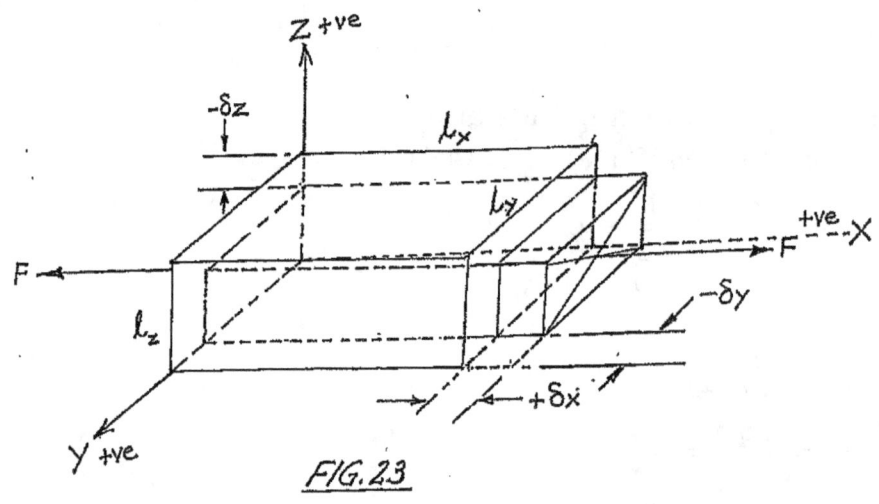

FIG. 23

In Fig. 23, I have shown a rectangular block of material, say, steel of dimensions ℓ_x, ℓ_y and ℓ_z. When a tensile force 'F' is applied at the mid-points of each y-z plane at the ends of the block, the deformations measured in the direction of the X-, Y- and Z-axes are respectively, δx, $-\delta y$ and $-\delta z$, the negative signs attaching to strains in the directions of the Y-, and Z-axes because the dimensions of the body in these directions are reduced as a result of the application of 'F' along the X-axis. Observe that the extension δx and contractions δy and δz are for the sake of illustration not distributed symmetrically. It is as though the Z-Y plane, the Z-X plane and the X-Y plane were all rigid surfaces and the extensions and contractions were measured in the direction of the +ve X-axis in the case of 'δx' and in the direction of –ve Y-axis and –ve Z-axis in the case of δy and δz respectively. Thus, we may write with reference to Fig. 22

$$e_x = +\frac{\delta_X}{\ell_X}$$

$$e_Y = -\frac{\delta_Y}{\ell} \quad \text{or} \quad e_Y = \frac{\delta Y}{\ell_Y}$$

$$e_z = -\frac{\delta Z}{\ell_z} \quad \text{or}$$

By definition, $v \text{ or } \mu = \dfrac{-e_Y}{e_Z}$; also $v \text{ or } \mu = -\dfrac{-e_Z}{e_X}$

Notice that the transverse strains are of sign opposite to the direct strain to which they are related. Therefore, if the direct strain is tensile, the transverse strains are compressive, i.e. of opposite sign, and vice-versa. When a value for Poisson's ratio is given one is concerned only with magnitude and not with sign. Thus from a strictly mathematical convention point of view, Poisson's ratio should be expressed thus:

$$v = \left| \frac{transverse \ strain}{direct \ axial \ strain} \right|, \quad \text{eg.} \quad v = \frac{-e_y}{e_x} = -\frac{e_x}{e_x}$$

Poisson's ratio is never given a negative value; it is the strain that is negative not Poisson's ratio. It is well to note that for so-called isotropic materials, that is to say, materials ideally having uniform properties in all directions, Poisson's theoretical analysis, based as it was on the assumption of elastic conditions of stress and strain throughout, yielded a value of 'v' or 'μ' $= \dfrac{1}{4}$. While this is a result which accords closely with experimental determinations using amorphous materials such as glass, it is also a matter of experimental record that for most engineering metals and alloys the value of 'v' or 'μ' is closer to 1/3 than it is to ¼. Do not confuse a ratio of strains with a ratio of deformations. The other crucial fact to recall always is that transverse or lateral strain is of opposite sign to the direct axial strain causing it. This would perhaps be more readily understood in the Chapter on Hooke's generalized law.

Illustrative Example 15

A cubical block of steel of side 10cm is subjected to a compressive force of 20kN whose line of action is along the X-axis passing through the centre of the sides *abcd* and *efgh* as shown in Fig. 24.

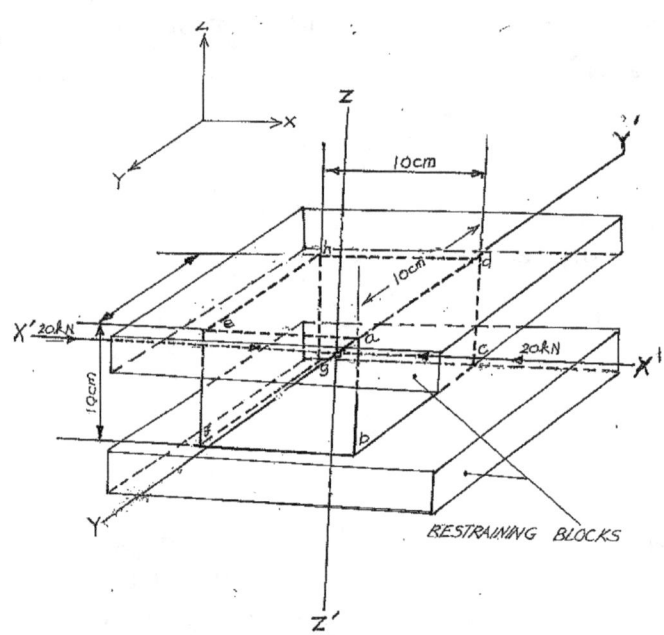

FIG. 24

In the Z-direction the block is restrained, but not so along the Y-axis. Determine total strain along the Y-axis and calculate the force acting on the restraining blocks. i.e. along the Z-axis. Assume isotropic material and take E = 200GN/m².

Because the material is assumed isotropic the value of 'μ' is the same throughout its matrix.

Solution

Area of face of cubical steel $= 10 \times 10 (cm)^2 = \dfrac{100}{10^4} m^2 = 10^{-2} m^2$

∴   Stress 'σ_x' in the direction of X-axis $= -\dfrac{200 \times 1000}{10^{-2}} N/m^2 = -2 \times 10^6 N/m^2$ or

$-2 MN/m^2$; the negative sign to indicate the stress is compressive.

Therefore strain $e_X = \dfrac{\sigma_X}{E} = -\dfrac{2 \times 10^6}{200 \times 10^9}$

$$= -1 \times 10^{-5}$$

Now, transverse strain e_Y due to strain e_X is $-0.3\ell_X$, from $e_Y = ve_X$. Similarly transverse strain e_Z in the direction of the Z-axis, associated with this same e_X, is given by $e_Z = -0.3e_X$, the material being isotropic. Remember that transverse or lateral strains are always of opposite sense to the direct axial strains.

Further, as in evident from the process of compression along the X-axis by the compressive force of 20kN, a stress σ_Z develops in the positive direction of the Z-axis and thereby causing a direct strain of $\dfrac{\sigma_Z}{E} = \sigma_Z / 200 \times 10^9$. The lateral strains associated with this direct strain are $e_Y = -0.3e_Z$ and $e_X = -0.3e_Z$.

i.e. $e_Y = -0.3\left\{\dfrac{\sigma_Z}{200 \times 10^9}\right\}$

and $e_Z = -0.3\left\{\dfrac{\sigma_Z}{200 \times 10^9}\right\}$

∴ Total strain in the Z-direction is therefore

$$= \dfrac{\sigma_Z}{200 \times 10^9} - 0.3e_X$$

$$= \dfrac{\sigma_Z}{200 \times 10^9} - 0.3(-1 \times 10^{-5}) = \dfrac{\sigma_Z}{200 \times 10^9} + 0.3 \times 10^{-5}$$

However, because the block is restrained in the direction of the Z-axis, this means that

$$\dfrac{\sigma_Z}{200 \times 100^9} + 0.3 \times 10^{-5} = 0$$

from which

$$\sigma_Z = -60 \times 10^4 \, N/m^2$$

or $\sigma_Z = -0.6 MN/m^2$

To find the force acting on the restraining blocks we multiply σ_Z by the surface area of a block. Accordingly,

Restraining Block Force $= -0.6 \times 10^6 \times \dfrac{100}{10^4} N = -6kN$

In order to determine the total strain along the Y-axis we have to consider the two components of strain viz. (i) the transverse strain e_Y already calculated; and (ii) the transverse component along the Y-axis due to σ_Z i.e. $-ve_Z$

(i) e_Y due to $e_X = -0.3e_x$

$$= +0.3 \times 10^{-5}$$

(ii) e_Y due to $e_Z = -0.3 \left\{ \dfrac{-60 \times 10^4}{200 \times 10^9} \right\}$

∴   Total $e_Y = +0.3 \times 10^{-5} + 0.3(3 \times 10^{-6})$

$$= 0.3 \times 10^{-5} + 0.09 \times 10^{-5}$$

$$e_Y = +0.39 \times 10^{-5}$$

Later on when we come to consider the generalized Hooke's law; we shall check these results.

NOTE ON PLANE STRESS AND PLANE STRAIN

Before wrapping up this Chapter it is perhaps opportune to consider the concepts of Plane Stress and Plane Strain. Many students ask me : what is meant by the terms plane stress and plane strain? I respond by saying that plane stress is a condition which, in its most generalized state, consists of two normal stresses in combination with one shearing stress; plane strain by two components of direct strain with a single shear strain.

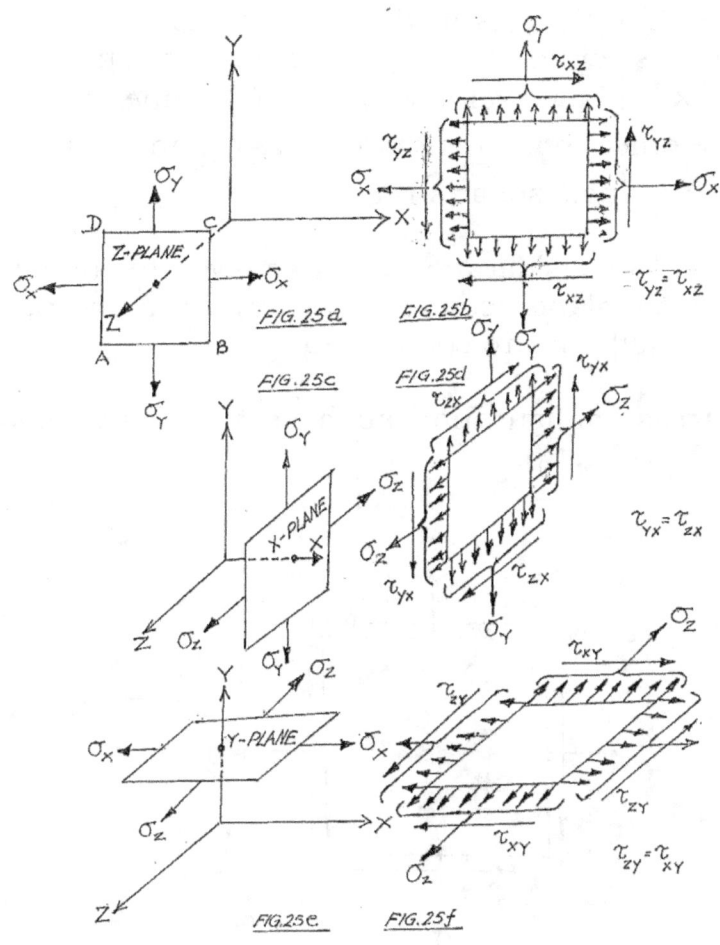

FIG. 25a   FIG.25b

FIG.25c   FIG.25d

FIG.25e   FIG.25f

This is first exemplified in Figs. 25a and 25b. The geometrical plane *ABCD* in Fig. 25a is a Z-plane i.e. all points on this plane have the same Z-coordinate. The subscripts for shear stress are labelled in accordance with the following convention : the first subscript refers to the axis in which direction it is pointing and the second to the plane on which it is acting. Thus τ_{XZ} refers to shearing stress acting in the X-direction on the Z-plane. The two normal stresses are σ_X and σ_Y and the complementary shear stress $\tau_{XZ}(=\tau_{YZ})$. Plane stress is also most appropriately referred to as two-dimensional stress. The stress depicted in Fig 25a is X-Y plane stress. The other possible configurations are X-Z and Y-Z.

The cases of plane stress for stresses in the X-Z or Y-plane) and Y-Z (or X-plane) are shown in Figs 25c, 25d and 25e, 25f respectively. If one chooses to work exclusively in the X-Y plane, then plane stress implies $\sigma_{ZZ} = 0$; and so too the shearing stresses τ_{XZ} and τ_{YZ} which is another way of saying that any component of stress perpendicular to the X-Y plane is zero.

In plane strain the condition, as in the case of plane stress, is defined by two components of direct strain and a single one of shear strain. And if we choose to work say in the X-Y plane exclusively then for plane strain in this plane, e_Z, γ_{XZ} and $\gamma_{YZ} = 0$; the other components in the X-Y plane being e_X, e_Y, $\gamma_{XY} (= \gamma_{XY})$: 2 direct strains and a single shear strain.

In Chapters 12 and 13 consideration shall be given to compound stresses and strains respectively in plane sections. In advanced mechanics of materials more complicated conditions are considered.

A typical plane stress configuration such as that which shall be considered later is represented by Fig. 26.

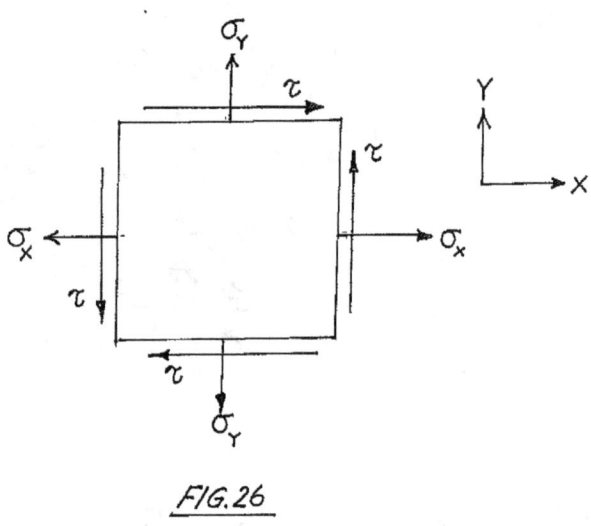

<u>FIG. 26</u>

SOLVED AND OTHER PROBLEMS

Q1. The wire *OA* in Fig. 1 is tied to the fixed point *'O'* and carries a mass of 150kg at its end. The length of the wire is 1.5m. The mass moves in a horizontal circle with constant angular velocity $\omega \, rad/s^2$ such that *OA* makes an angle of 45° with the vertical. If the allowable stress in the wire is 410 N/(mm)² what is the design-diameter of the wire? Take g = 9.81 m/s².

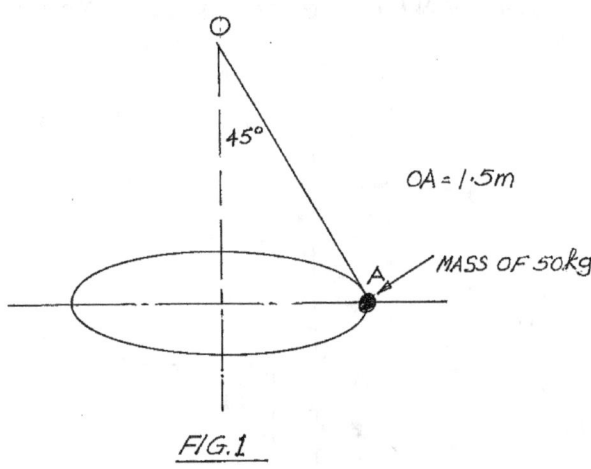

FIG.1

Ans ≈ approx 2.5mm

Q2. A vertical circular cylindrical composite bar of uniform cross-section has a diameter of 6cm and is one metre long. A tensile force of 150kN acts through the perpendicular axis of the bar. The two parts of the bar are each made of steel and aluminium of length $'y_s'$ metres and $'y_A'$ metres respectively. Assuming the steel and aluminium portions of the bar are firmly connected together and elongate by the same amount, determine (a) the value of $'y_s'$ and $'y_A'$; and, (b) the total elongation of the bar. Take E_s = 210 GN/m² and E_A = 70 GN/m².

Ans : $y_A = 25cm$; $y_s = 75cm$; Total Elongation = 0.5mm

Q3. Assuming an empirical relationship $\sigma^n = Ee$ instead of Hooke's law for the relationship between stress $'\sigma'$, strain $'e'$ and Young's modulus E where *'n'* is a dimensionless number depending on the property of the material, determine the total extension of the composite bar described in Question 2.

Ans : $\left(\dfrac{0.75E_A \sigma^{n_s} + 0.25E_s \sigma^{n_A}}{E_S E_A} \right) m$

Q4. The lifting device shown in Fig. 1 comprises a frame *ABC* of which the two steel bars *AC* and *BC*, both of uniform cross-sectional area throughout, are attached to rigid supports at *A* and *B*. At 'C' a two-sheave pulley is fixed to the frame. A load equivalent to 30kN is suspended from a hook attached to a single sheave pulley linked to the pulley at 'C'. Draw the free-body diagrams for the pulley at 'C' and for the frame; and determine the required cross-sectional areas of bars *AC* and *BC*, given that the allowable stresses in tension and compression are 140 MN/m² and 100 MN/m² respectively. Neglect the weight of the pulleys.

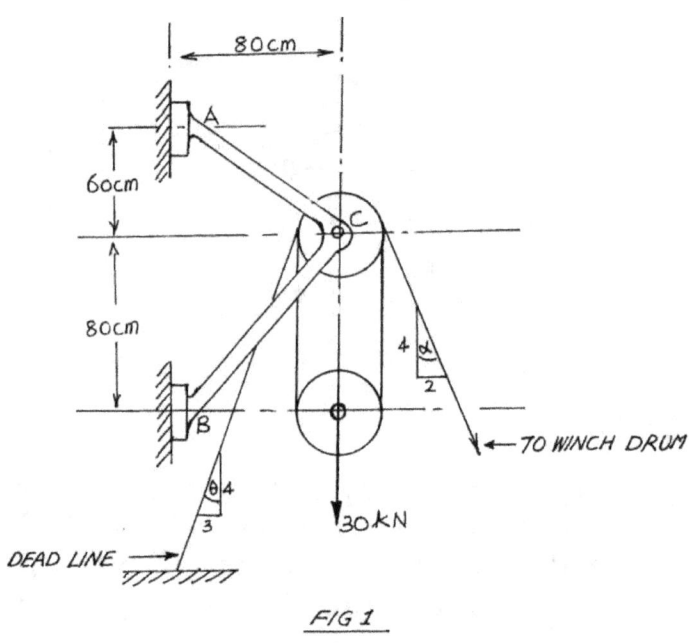

FIG 1

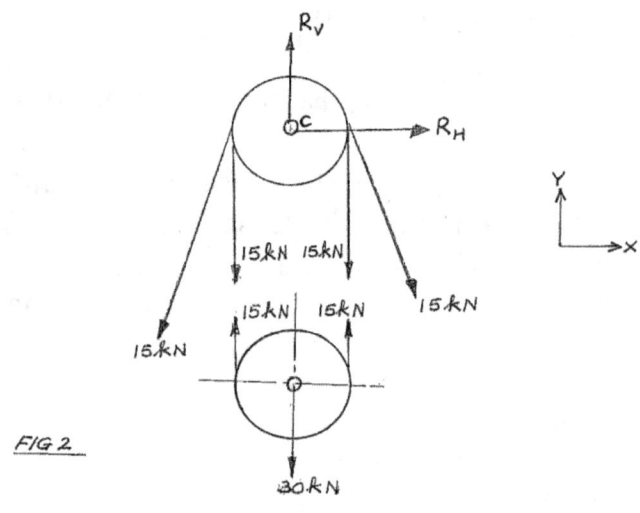

FIG 2

~ 476 ~

Fig. 2 is the *FBD* for the pulley at *C*. R_V and R_H are the vertical and horizontal reactions respectively at the pulley support.

Considering equilibrium at 'C'.

For $\quad \sum F_Y = 0 \ : \ R_V - 2(15) - 15 Cos\theta - 15 Cos\alpha = 0$

i.e. $\qquad R_V - 30 - 15\left(\dfrac{4}{5}\right) - 15\left(\dfrac{2}{\sqrt{5}}\right) = 0$

or $\qquad R_V - 42 - 13.4 \quad = 0$

$$\therefore \qquad R_V = 55.4 kN$$

Next, consider also at C, $\sum F_X = 0$

$$R_H - 15 Sin\theta + 15 Sin\alpha \qquad\qquad = 0$$

or $\qquad R_H - 15\left(\dfrac{3}{5}\right) + 15 \cdot \left(\dfrac{2}{2\sqrt{5}}\right) = 0$

$$R_H - 9 + 6.69 \qquad = 0$$

$$\therefore \qquad R_H \approx 2.3 kN$$

Fig. 3 is the *FBD* for the frame *ABC*

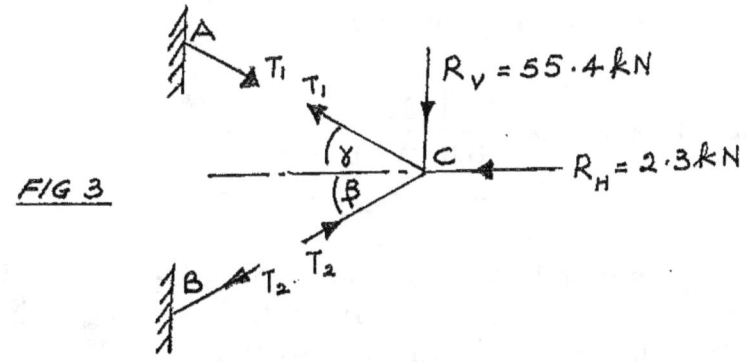

FIG 3

In Fig. 3, T_1 is assumed tensile ; T_2, compressive.
For at C, $\sum F_Y = 0$

$$T_1\ Sin\gamma + T_2\ Sin\beta - R_V = 0$$

i.e. $\quad T_1\left(\dfrac{60}{100}\right) + T_2\left(\dfrac{80}{80\sqrt{2}}\right) - 55.4 = 0$

$$0.6 T_1 + 0.71 T_2 = 55.4 \qquad\qquad \dots\dots\dots\dots\dots \ \text{(i)}$$

and also at C, for $\sum F_X = 0$

$$T_1\ Cos\gamma - T_2\ Cos\beta + 2.3 \quad = 0$$

i.e. $\quad T_1\left(\dfrac{80}{100}\right) - T_2\left(\dfrac{80}{80\sqrt{2}}\right) + 2.3 = 0$

or $\quad 0.8T_1 - 0.71T_2 = -2.3$ $\qquad \qquad \ldots \ldots \ldots \ldots \ldots$ (ii)

Adding (i) and (ii)

$$1.4T_1 = 53.1 kN$$

$\therefore \qquad T_1 = 37.9 kN$

so that by (ii) $\qquad 30.3 - 0.71T_2 = -2.3$

$$0.7T_2 = 32.5$$

$\therefore \qquad \qquad T_2 = 45.8 kN$

Our assumption as to the nature of the forces is proven correct. Member AC is in tension and BC in compression.

Because stress $\sigma =$ Force/area, we have in the case of AC, its cross-sectional area is given by:

$$(area)_{AC} = \frac{55.4 kN}{140 MN/m^2} = \frac{55.4 \times 10^3 \, m^2}{140 \times 10^6 10^3}$$

$$= \frac{55.4}{14 \times 10^4} m^2 = \frac{55.4}{14} cm^2$$

$$= 3.957 (cm)^2, \text{ say, } 4 (cm)^2$$

Similarly, for BC $(area)_{BC} = \dfrac{45.3 kN}{100 MN/m^2} = \dfrac{45.3 \times 10^3}{100 \times 10^6} m^2$

$$= 4.53 cm^2, \text{ say, } 4.6 cm^2$$

Q5. A symmetrically-shaped block of mass 25000kg is placed on top of three vertical cylinders placed in a straight line and equidistant from each other. Each cylinder is 30mm in diameter and 40mm in height. The block and the base upon which the cylinders rest may be assumed to be rigid, and the centre of gravity of the block is directly over the middle cylinder. Find the load supported by each cylinder when the middle cylinder is (i) : 0.030mm shorter, and (ii) 0.030mm longer than the others. Take E = 210GN/m² and g = 10m/sec².

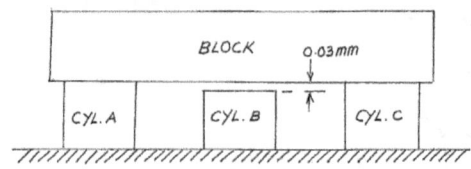

FIG. 1

~ 478 ~

Case (i) In order that the block may touch the top cylinder B, both cylinders A and C must be compressed 0.03mm.

From
$$E = \frac{W \cdot 40mm}{A \cdot 0.03}$$

where
E = Young's modulus, N/m^2
A = Cross-sectional area of cylinder in m^2.
W = load on each of cylinders A and C in N

$$W = 210 \times 10^9 \times \frac{\pi}{4} \left(\frac{30}{1000}\right)^2 \times \frac{0.03}{40}$$

$$\therefore \quad W = \frac{21000 \times \pi \times 27}{16}$$

$$= 111345N$$

Total mass of block = 25000kg.
$\therefore$ " weight " " = 250,000N

So, of the total weight of 250,000N, an amount equal to 2(111345)N, i.e. 222,690N, was necessary in order to compress cylinders A and C to the top surface of cylinder B.

$\therefore$ Balance of load equally distributed on A, B and C after the compression

$$= \frac{250,000 - 2(111345)}{3}$$

$$= 9103N.$$

Accordingly the load on each of the cylinders is given by:

Cyl. A : $111,345 + 9103$

$$= 120,448N$$

Cyl. B = $9,103N$, and

Cyl. C : $111,345 + 9103$

$$= 120,448N$$

Case (ii). Here, it is cylinder B that must be compressed 0.03mm before load sharing begins.

$\therefore$ Balance of load equally distributed on cylinders A, B and C after compression

$$= \frac{250,000 - 1(111345)}{3} = 46,218N$$

$\therefore$ Loads are : *Cyl. B* = $46,218N$

Cyl. B : $111,345 + 46,218$

$$= 157,563N$$

Cycl. C : $46,218N$

Q6. During the drilling of an oil-well the drilling string got stuck in the hole (as sometimes occurs). The hole was vertical. In order to estimate the depth at which the string was stuck, the drilling engineer marked off a length of drill-pipe, 25 cm on the part of the string above the derrick floor. Employing the draw-works, the string was slowly stretched a distance of 0.25 m. The marked off gauge length of 25 cm had increased to 25.0015 cm. When the pull on the string was released the marked off gauge length returned to its original length and the elastic 'pull-up' was reduced to zero. Estimate the depth at which the string was stuck.

Ans : 4167 m

Q7. The cross-sectional area of a vertically suspended steel bar 15 cm long subjected to an axial load of 100kN, varies along its length according to the relationship $\left(1+\dfrac{x^2}{225}\right) cm^2$, '$x$', being the distance in cm from the bar's rigid support. Determine the total extension of the bar. Neglect the self weight of the bar. Take E = 210 x 10⁵ N/(cm)².

The objective is to determine the total extension of the bar. If one merely substituted $x = 15 cm$ in the expression : $(1+x^2/225)$, and obtained thereby the cross-sectional area at the end of the bar, and then proceeded to use $E = \dfrac{Load}{Area} \times \dfrac{length}{extension}$, one would simply have found therefrom the "spot" extension at that cross-sectional area. What is wanted is the sum of all such extensions over the length of the bar. Evidently, to obtain such an integrated sum means a resort to the mathematical fiction of the infinitesimal.

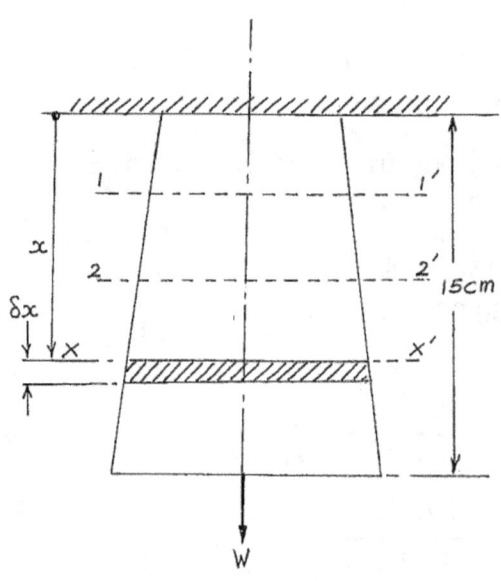

Suppose the bar has the configuration shown in the above diagram. It should be clear that at any section, e.g. $1-1'$ or $2-2'$ the load is W; in short $W_x = W = $ constant, throughout.

Considering a section at $X - X'$, 'x' from the support and of length δx, let its infinitesimal extension be $\delta\Delta$. Therefore

Tensile strain $\ 'e' = \dfrac{\delta\Delta}{\delta x}$

so that by Hooke's law

$$E = \frac{W}{\left(1 + \dfrac{x^2}{225}\right)} \cdot \frac{\delta x}{\delta\Delta}$$

In the limit $\ \delta\Delta = d\Delta$ and $\delta x = dx$
so that we may write

$$d\Delta = \frac{W}{E} \cdot \frac{dx}{(1 + x^2/225)}$$

The limits of integration are : $x = 0$ to $x = 15$. Accordingly

$$\int_0^{15} d\Delta = \Delta = \frac{225W}{E} \int_0^{15} \frac{dx}{\{(15)^2 + x^2\}}$$

$I = \int \dfrac{dx}{a^2 + x^2}$ is a standard form which when evaluated is $= \dfrac{1}{a}\tan^{-1}\dfrac{x}{a} +$ Constant

$$\therefore \quad \frac{225W}{E} \int_0^{15} \frac{dx}{\{(15^2 + x^2)\}} = \frac{225W}{E} \cdot \frac{1}{15}\tan^{-1}\frac{15}{15}$$

Substituting $\ W = 100kN$ and $E = 210 \times 10^5 \, Nm/(cm)^2$ [Note E is here expressed as N/cm²]

$$\therefore \quad \Delta = \frac{225 \times 100 \times 10^3 \times \pi}{210 \times 10^5 \times 15 \times 4} cm$$

$$= \frac{3\pi}{168} cm \equiv \frac{30\pi}{168} mm$$

i.e.   Total extension $= \Delta \approx 0.056 mm$.

Q8. A hexagon head bolt 20 mm nominal diameter, 130 mm long between shoulder of head and nut, and having a pitch of 2.5 mm is used to fasten together two steel plates each 5 cm square. Because the torque wrench is not considered a very reliable device for obtaining the force induced in the bolt, a procedure is followed whereby the nut is first tightened to ensure proper seating and then loosened. The nut is then again made snug against the plate and turned through an additional 15º. How such stress is caused in the unthreaded portion of the bolt and what is the pressure between the plates if they are assumed to be rigid? Take E = 210 x 109 N/m².

When the nut is turned through 15º, the bolt is stretched by an amount

$$= \left(\frac{15}{360}\right) \; times \; the \; pitch$$

$$= \frac{15}{360} \times 2.5mm$$

$$= 0.1042mm$$

∴   Strain in bolt $= \left(\dfrac{0.1042}{130}\right)$

By Hooke's law :

$$E = \frac{Stress}{Strain}$$

∴ $210 \times 10^9 = \dfrac{Stress}{\dfrac{0.1042}{130}}$

Stress in bolt $= \underline{168 \times 10^6 \, N/m^2}$

See Fig. 1

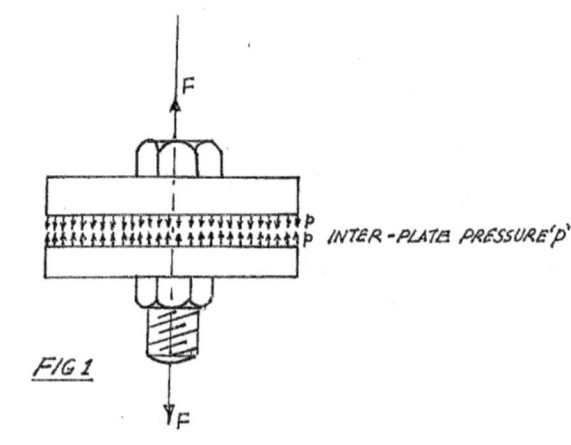

FIG 1

<u>BOLT UNDER TENSION ;</u>
<u>PLATE UNDER PRESSURE.</u>

Let F = force in bolt caused by tightening of nut.
Accordingly,

$$168 \times 10^6 = \frac{F}{\dfrac{\pi}{4}\left(\dfrac{20}{1000}\right)^2}$$

$$= \frac{F \times 4 \times 10^6}{\pi \times 400}$$

i.e. $\quad F = \dfrac{168 \times 10^6 \times \pi \times 400}{4 \times 10^6}$

$$= 100\pi \times 168$$

$$= 52802.4 \; Newtons, \text{ i.e. } 52802.4N$$

so that pressure 'p' between plates is given by

$$p = \frac{F}{A}$$

where $\qquad$ A $\;=\;$ area of plates

$$= \left(\frac{5}{100} \times \frac{5}{100}\right) m^2$$

$\therefore \qquad p = \dfrac{52802.4 \times 10^4}{25}$

or $\qquad p = 2.1 MN/m^2$

Q9. A composite laminate is made up of 3 different materials namely, mat, woven roving and cloth each having respectively a modulus of elasticity designated E_m, E_{wr} and E_C. Show that the composite modulus of elasticity (E_R) is given by:

$$E_R = \frac{E_m a_m + E_{wr} a_{wr} + E_c a_c}{a_m + a_{wr} + a_c}$$

where a_m, a_{wr} and a_c are respectively the cross sectional areas of the mat, woven roving and cloth. Calculate the modulus of elasticity of a composite laminate for which the following data apply:

	Modulus of Elasticity	Area
Mat Laminate	6.2G N/m²	$6 \times 10^{-3} m^2$ / metre width
Woven Roving Laminate	9.3G N/m²	$3 \times 10^{-3} m^2$ / metre width
Cloth Laminate	9.7G N/m²	$2 \times 10^{-3} m^2$ / metre width

From the general expressions

$$E = \frac{Stress}{Strain}$$

and

$$Stress = \frac{Load}{Area}$$

we may write $E_R = \frac{Total\ Load}{Total\ area} \cdot \frac{1}{Strain}$ . . . . (i)

i.e.

$$E_R = \frac{\sigma_m a_m + \sigma_{wr} a_{wr} + \sigma_c a_c}{(a_m + a_{wr} + a_c) \cdot (strain)}$$

$$= \frac{f_m a_m (strain) + E_{wr} a_{wr} (strain) + E_c a_c (strain)}{(a_m + a_{wr} + a_c) strain}$$

$$= \frac{(strain)\ (E_m a_m + E_{wr} a_{wr} + E_c a_c)}{(strain)\ (a_m + a_{wr} + a_c)}$$

$$\therefore \quad E_{composite} = \left(\frac{E_m a_m + E_{wr} a_{wr} + E_c a_c}{a_m + a_{wr} + a_c} \right) \qquad \text{Q.E.D}$$

This result may also be rewritten as

$$E_c \sum_{}^{i=m,wr,c} a_i = \sum_{}^{i=m,wr,c} E_i a_i$$

or

$$E_c = \frac{\sum_{}^{i=m,wr,c} E_i a_1}{\sum_{l=m,wr,c}^{} a_i}$$

Considering 1 metre width of the laminates

$$E_{comp} = \frac{(6.2 \times 10^9 \times 6 \times 10^{-3} + 9.3 \times 10^9 \times 3 \times 10^{-3} + 9.7 \times 10^9 \times 2 \times 10^{-3})}{6 \times 10^{-3} + 3 \times 10^{-3} + 2 \times 10^{-3}}$$

$$= \frac{(37.2 + 27.9 + 19.4)10^9 \cdot 10^{-3}}{11 \times 10^{-3}}$$

$$\therefore \qquad E_{comp} = 7.7 GN/m^2$$

Q10. A cube of aluminium of side 35mm is subjected to shearing stress such that the shear strain 'γ' is 0.0008. Refer to Fig. 1. Calculate the diagonal strain.

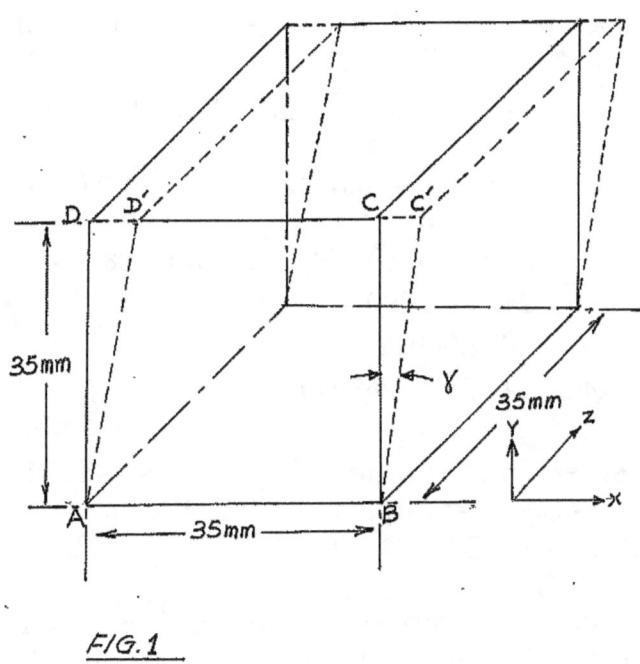

FIG.1

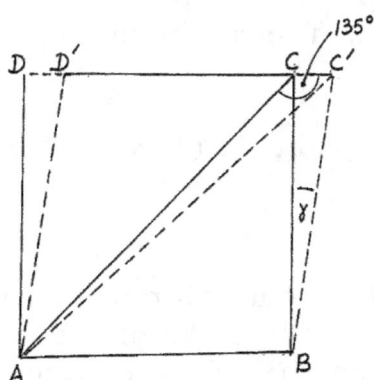

FIG. 2 : END VIEW OF CUBE
IN DIRECTION OF Z-AXIS

Because $\gamma = 0.0008$ radians, i.e. γ is very small, we may write $\gamma = \tan\gamma$. Accordingly with reference to Fig. 2

$$\frac{CC'}{35} = 0.0008$$

which makes $CC' = 0.028mm$

Also

$$AC^2 = (35)^2 + (35)^2$$
$$= 2(35)^2$$
$$\therefore \quad AC = 35\sqrt{2}$$

But extension of diagonal AC is $AC' - AC$. Employing the cosine law in triangle ACC',

$$(AC')^2 = (AC)^2 + (CC')^2 - 2(AC)\ (CC')\ Cos135°$$

$$= (35\sqrt{2})^2 + (0.028)^2 - 2(35\sqrt{2})(0.028)\left(\frac{-1}{\sqrt{2}}\right)$$

$$= (35)^2(2) + (0.028)^2 + 2(35)(0.028)$$
$$= 2450 + 0.000784 + 1.96$$
$$(AC')^2 = 2451.960784$$
$$\therefore \quad AC' = 49.51727763mm$$

so that the extension of the diagonal is 49.51727763 = $35\sqrt{2}$ = 49.51727763-49.434 = approx. 0.08327mm. Accordingly, diagonal strain = 0.08327/49.434 = 0.001684 = 16.8 x 10^{-4}.

Q11. In a glueability test the strength of the bond between the plain cylindrical surfaces of a PVC pipe of external diameter, say, 2.5 cm which is fitted into a PVC adapter plugged at its threaded end, is tested by applying water pressure to the glued combination. The length of the glued portion of the assembly is 4 cm. The combination failed at a water pressure of 3 bars.

What was the shear stress in the glued joint at the point of failure?

Ans: 48.7 kN/m².

Q12. The short circular cylindrical column shown in Fig. 1 is firmly fixed between rigid supports. It has two different but uniform cross-sections for each half of its length. At the junction of the two sections a horizontal platform is rigidly affixed to the column as shown. Equal loads of 150kN each act vertically downwards at each end of the platform and at a distance of 0.2m from the middle of the column. Determine the load on each portion of the column.

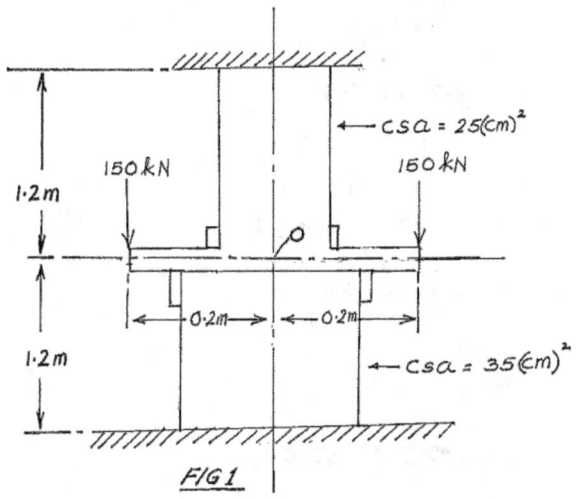

FIG 1

The 150kN loads produce opposing moments of magnitude 150kN (0.2m) ≡ 30kNm at the middle of the column. Result zero moment at centre-line. But the total vertical load at the middle column = 300kN. This is shown in Fig. 2

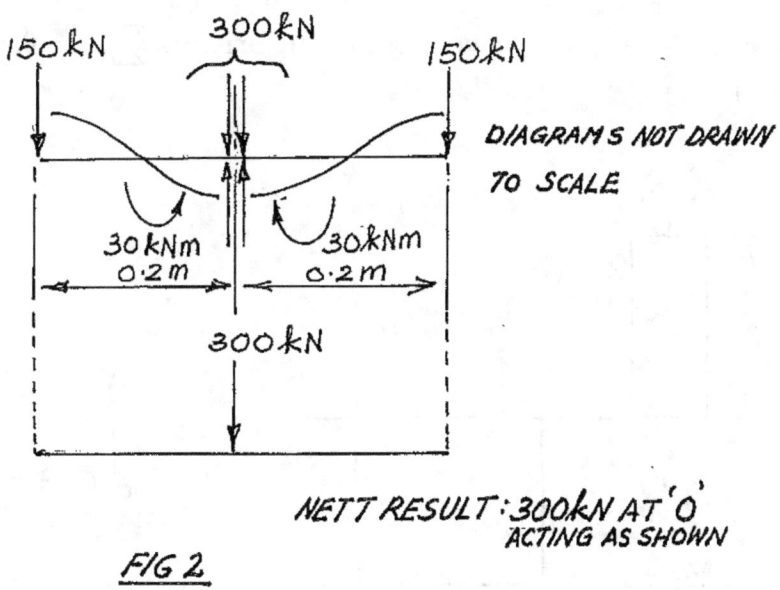

NETT RESULT: 300kN AT 'O' ACTING AS SHOWN

FIG 2

The loading on the column may now be represented by the diagram shown as Fig. 3.

~ 487 ~

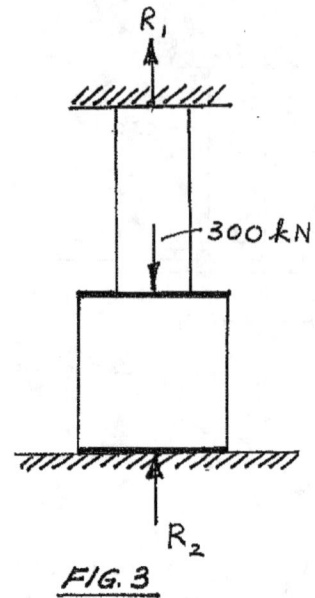

FIG. 3

The reactions at the supports R_1 and R_2 are shown in this figure.

With reference to Fig. 3, for vertical equilibrium, $\sum F_Y = 0,$ we have

$R_1 + R_2 = 300$

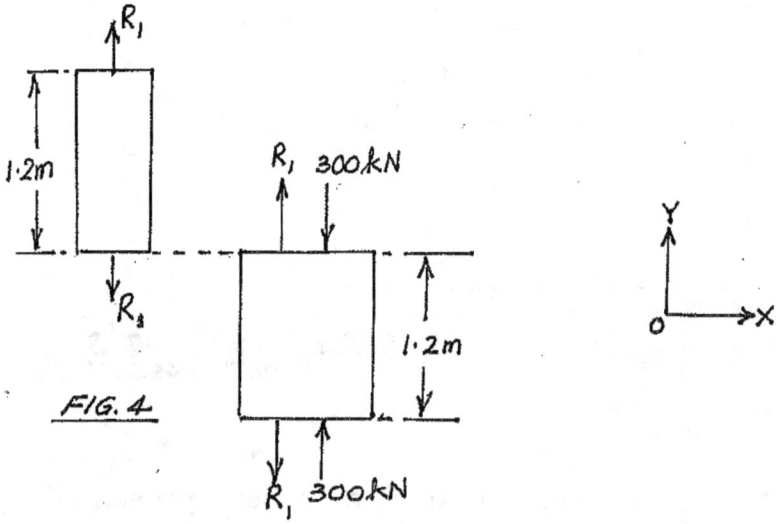

FIG. 4

Fig. 3 is now divided into 2 parts as shown in Fig. 4

Let u_1 = extension of upper section of bar

" u_2 =   "   " lower   "   "   "

Evidently, $u_1 + u_2 = 0$

i.e. $\left\{ u_1 = \dfrac{R_1(1.2)}{EA_1} \right\} + \left\{ u_2 = \dfrac{(R_1 - 300)}{EA_2}(1.2) \right\} = 0$

or $\dfrac{R_1}{A_1} + \dfrac{(R_1 - 300)}{A_2} = 0$

or $\dfrac{R_1}{25} + \dfrac{R_1 - 300}{35} = 0$

or $35R_1 + 25R_1 - 7500 = 0$

i.e. $60R_1 = 7500$

∴ $R_1 = 125 kN$

and $R_2 = 175 kN$

See Fig. 5

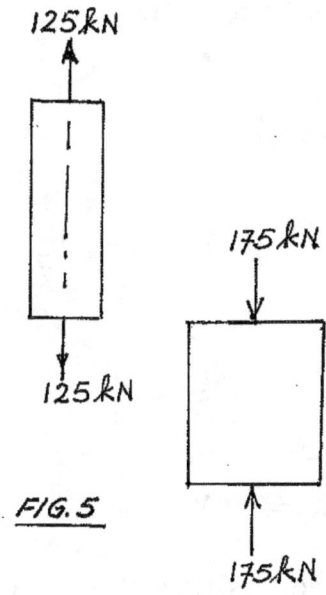

125 kN

125 kN

175 kN

175 kN

FIG. 5

Q13. Four stainless steel rods each of identical uniform cross-section throughout and jointly supporting a load of 75kN at their ends are connected to a rigid support as shown in Fig. 1. Determine the load borne by each rod.

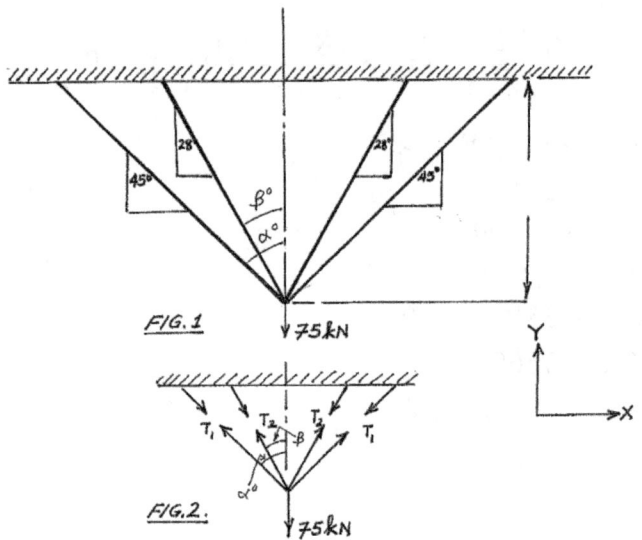

The FBD for the system is shown as Fig. 2. Because of the symmetrical nature of the alignment of the rods, the forces in the outermost two are designated T_1 ; and , the innermost two : T_2.

For $\quad \sum F_Y = 0$

$$2T_1\ Cos\alpha + 2T_2\ Cos\beta = 0 \qquad \dots\dots\dots\dots\dots \quad \text{(i)}$$

and because

$$Cos\alpha = \frac{4}{4\sqrt{2}} = \frac{1}{\sqrt{2}}; \text{ and, } Cos\beta = \frac{15}{17}$$

equation (i) becomes :

$$2T_1\left(\frac{1}{\sqrt{2}}\right) + 2T_2\left(\frac{15}{17}\right) = 75kN$$

or $\quad \sqrt{2}T_1 + 1.7647T_2 = 75kN$

or $\quad 1.414T_1 + 1.7647T_2 = 75kN \qquad \dots\dots\dots\dots\dots \quad \text{(ii)}$

By inspection of Fig. 2 we deduce that a consideration of horizontal equilibrium by application of $\sum F_X = 0$ will get us nowhere because

$$T_1 Sin\alpha + T_2 Sin\beta - T_2 Sin\beta - T_1 Sin\alpha = 0.$$

The problem is thus indeterminate : two unknowns T_1 and T_2 and only one equation, viz. (ii). Therefore we have to consider the geometry of deformation of the rods.

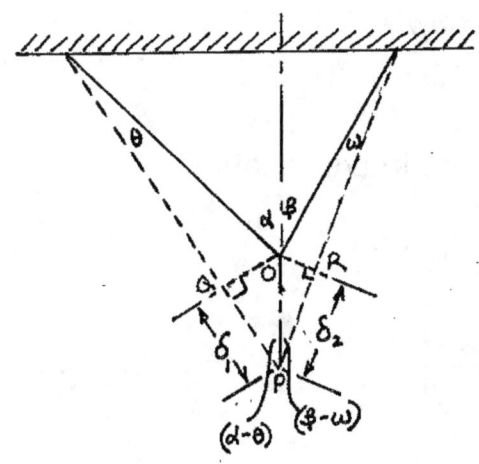

FIG.3

$$\alpha = Cos^{-1}\frac{1}{\sqrt{2}} \qquad \therefore \qquad \alpha = 45^{\circ}$$

$$\beta = Cos^{-1}\frac{15}{17} \qquad \therefore \qquad \beta \approx 62^{\circ}$$

Let A = cross-sectional area of each rod.

Referring to Fig. 3 we deduce that angle OPQ is $(\alpha - \theta)^{0}$ and angle $OPR = (\beta - \omega)^{0}$. Because angles θ and ω are small, we approximate these angles to α and β respectively. Hence,

$$\frac{\delta_1}{OP} = Cos\alpha \qquad \therefore \qquad OP = \frac{\delta_1}{Cos45^{\circ}}$$

Also

$$\frac{\delta_2}{OP} = Cos\beta \qquad \therefore \qquad OP = \frac{\delta_2}{Cos\beta}$$

$$\therefore \qquad \frac{\delta_1}{\delta_2} = \frac{Cos\alpha}{Cos\beta} \qquad \dots\dots\dots\dots\dots\dots\dots\dots\dots\dots\dots \text{(iii)}$$

$$E = \frac{T_1}{A} \cdot \frac{\ell/Cos\alpha}{\delta_1} \qquad \dots\dots\dots\dots\dots\dots\dots\dots \text{(iv)}$$

$$E = \frac{T_2}{A} \cdot \frac{\ell/Cos\beta}{\delta_2} \qquad \dots\dots\dots\dots\dots\dots\dots \text{(v)}$$

From (iv) and (v)

$$\delta_1 = \frac{T_1}{AE} \cdot \frac{\ell}{Cos\alpha}$$

$$\delta_2 = \frac{T_2}{AE} \cdot \frac{\ell}{Cos\beta}$$

from which,

$$\frac{\delta_1}{\delta_2} = \frac{T_1 \, Cos\beta}{T_2 \, Cos\alpha} \qquad \dotsb \qquad \text{(vi)}$$

Substituting result (iii) in (vi) we obtain

$$\frac{Cos\alpha}{Cos\beta} = \frac{T_1 \, Cos\beta}{T_2 Cos\alpha}$$

$$\therefore \qquad \frac{T_1}{T_2} = \frac{Cos^2\alpha}{Cos^2\beta} \qquad \dotsb \qquad \text{(vii)}$$

$$= \left(\frac{1}{\sqrt{2}}\right)^2 \cdot \frac{1}{(15/17)^2}$$

$$\frac{T_1}{T_2} = \frac{289}{250} = 0.64 \qquad \dotsb \qquad \text{(viii)}$$

Substituting the latter result in (i)

$$1.414(0.64T_2) + 1.76T_2 = 75kN$$
$$\text{i.e.} \qquad 2.674T_2 = 75kN$$
$$\therefore \qquad T_2 = 28.1kN$$
$$T_1 = 0.6T_2 \approx 18kN$$

Q14. A travelling block together with accessories including casing strings is lifted by eight wire ropes, each 60 metres long, diameter 40 mm and mass of 10kg per metre. With buoyancy effects taken into account, the total mass of the block, accessories and casing is 250,000kg when the string is stationary and "off bottom" in a shallow hole. See Fig. 1. Calculate the maximum stress and extension in each wire rope at this stage of the lift. Take E = 300GN/m² and assume $g \approx 10m/s^2$.

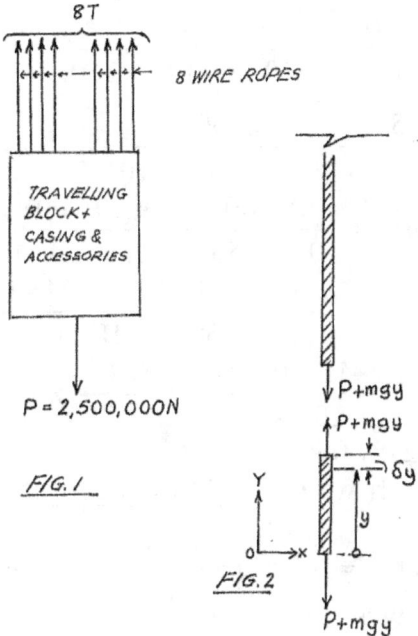

FIG. 1

FIG. 2

Mass of wire rope $= m = 10 kg/m$

$\therefore$ Force due to mass $'m' \equiv 100 N/m.$

Also, $P \equiv 2,500,000 N$

Cross-Sectional Area of 1 wire rope

$$= \frac{\pi d^2}{4}$$

$$= \frac{\pi \times 40 \times 40}{4} sq.\ mm$$

$$= \frac{\pi \times 40 \times 40}{4 \times 10^6} sq.\ metre$$

$$= 1.26 \times 10^{-3} m^2$$

Combining cross sectional areas of the 8 wire ropes, total c.s.a.

$$= 8 \times 1.26 \times 10^{-3} m^2$$

i.e. $csa = 10.1 \times 10^{-3} m^2$

Consider infinitesimal extension 'δe' of infinitesimal element of length 'δy'. See Fig. 2, and apply Hooke's law treating the eight ropes as a single unit.

$$\therefore \quad E = \frac{P + mgy}{csa} \cdot \frac{\delta y}{\delta e}$$

from which $\displaystyle\int de = \int_0^{60} \frac{P + mgy}{10.1 \times 10^{-3}} \cdot \frac{dy}{E}$ **[Note : Each rope = 60m long]**

i.e. $\displaystyle e = \int_0^{60} \frac{(2,500,000 + 48000y)dy}{10.1 \times 10^{-3} \times 350 \times 10^9}$

~ 493 ~

$$= \left[\frac{2.5 \times 10^6 \times 10^3 \times y}{10.1 \times 350 \times 10^9} + \frac{24000y^2}{10.1 \times 350 \times 10^9} \right]_0^{60}$$

$$\approx \frac{2.5y}{10.1 \times 350} + \frac{24000y^2}{10.1 \times 350 \times 10^9}$$

$$\therefore \quad \text{Extension} = \frac{2.5 \times 60}{10.1 \times 350} + \frac{24000 \times 60 \times 60}{10.1 \times 350 \times 10^9} \, m$$

$$= \frac{2.5 \times 60 \times 1000}{10.1 \times 350} + \frac{864 \times 1000}{10.1 \times 350 \times 10^4} \, mm$$

$$= \frac{15 \times 1000}{10.1 \times 35} + \frac{864}{10.1 \times 350} \, mm$$

$$= 42.4 + 0.24 \, mm$$

$\therefore \quad$ Extension of each wire $= \underline{42.6 \text{ millimetres}}$

Maximum stress at $x = 60$ $metres = \sigma_{60}$. Noting that each wire rope extends 42.6mm, we write

$$\sigma_{60} = \frac{E \cdot e}{\ell}$$

$$= \frac{350 \times 10^9}{60} \times \frac{42.6}{1000}$$

$$\therefore \quad \sigma_{60} = \underline{248.5 \times 10^6 \, N/m^2}$$

which is the stress in each wire.

Collecting results : $e = 42.6 mm$
$$\sigma_{60} = 248.5 MN/m^2$$

Q15. A slim cylindrical rod of constant cross-sectional area and length 'L' is rotated about a vertical axis through one of its ends with a constant angular velocity $\omega rad/s^2$. The density of the rod is 'ρ'. Assuming the rod to be elastic with a Young's modulus 'E' determine its total extension. Neglect any effects due to the boss and bearing at the end about which the rod rotates

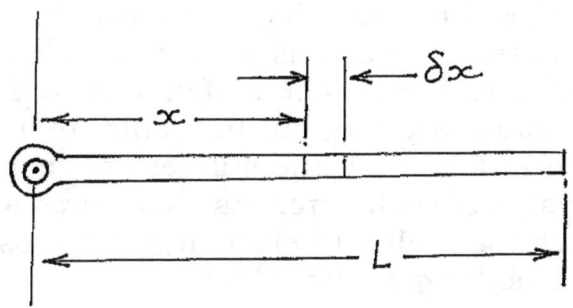

FIG.1

Consider the infinitesimal element of the rod in Fig. 1.

$$\text{Mass of element} = \text{Density} \times \text{Volume}$$
$$= \rho \cdot A \cdot \delta x$$

in which A = cross-sectional area.

Centrifugal force on element $= \rho \cdot A \cdot \delta x \omega^2 x$

Applying Hooke's law:

$$E = \frac{Force}{Area} \cdot \frac{Length}{Extension}$$

to the infinitesimal element

$$E = \frac{\rho A \cdot \delta x \cdot \omega^2 x}{A} \cdot \frac{x}{\delta e}$$

in which

δe = elastic extension of element over length 'x'.

$$\therefore \quad \delta e = \frac{\rho A \, x^2 \, \delta x \cdot \omega^2}{AE}$$

so that

$$\int_0^L de = \int_0^L \frac{\rho \omega^2 x^2 \, dx}{E}$$

i.e.

$$e = \frac{\rho \omega^2 L^3}{3E}$$

Q16. A rigid platform $2m \times 2m$ on which several heavy spotlights and accessories are mounted is to be suspended by four identical steel wires from rigid structural members in the roof of a church. The total mass of platform and accessories is 200kg but the centre of mass of the whole lot is located as shown in Fig. 1. If the stiffness of each wire is $1960N/mm$ then find the extension of each wire. Assume wires are stretched within the elastic limit. E = modulus of elasticity; a = cross-sectional area of each wire and L = length ; $g \cong 10m/s^2$.

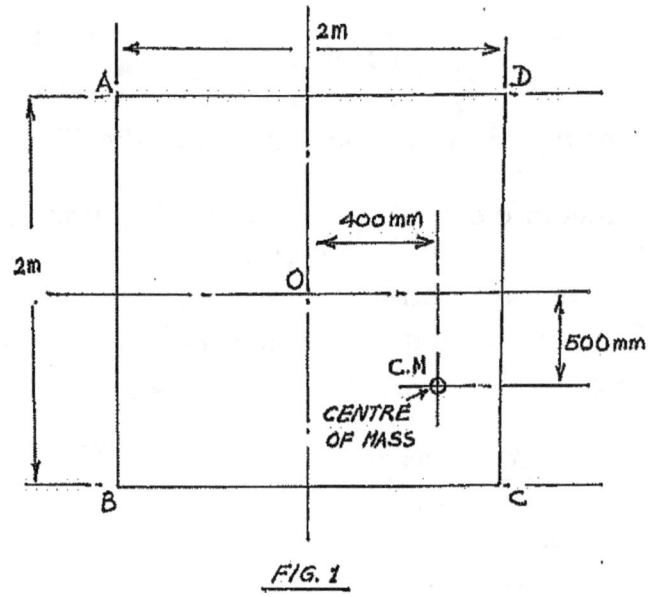

FIG. 1

Let tensions of wires at A, B, C, D be designated T_A, T_B, T_C and T_D respectively

$$\therefore \quad T_A + T_B + T_C + T_D = 2000N \quad \dots\dots\dots\dots\dots\dots \quad (i)$$

Considering moments about AB, we have

$$T_C(2) + T_D(2) = 2000(1.4)$$

$$\text{or} \quad T_C + T_D \quad = 1400 \quad \dots\dots\dots\dots\dots \quad (ii)$$

Likewise, moments about AD,

$$T_B(2)(+T_C(2) \quad = 2000(1.5)$$

$$\text{i.e.} \quad T_B + T_C \quad = 1500 \quad \dots\dots\dots\dots\dots\dots \quad (iii)$$

Also, let δ_A, δ_B, δ_C, δ_D be the extensions of the wires at A, B, C and D. [Remember the centre of mass of the loaded platform is not at the centre of the platform]. Now since the wires are identical in length ℓ, cross-sectional area and modulus of elasticity, the tension in each wire divided

~ 496 ~

by its deflection, i.e. each wire's stiffness 'S' is a constant, say, k. [Note

$: S = \dfrac{Force}{Extension} = EA / \ell$

Accordingly, $\delta_A = T_A / k$; $\delta_B = \dfrac{T_B}{k}$; $\delta_C = \dfrac{T_c}{k}$; $\delta_D = \dfrac{T_D}{k}$

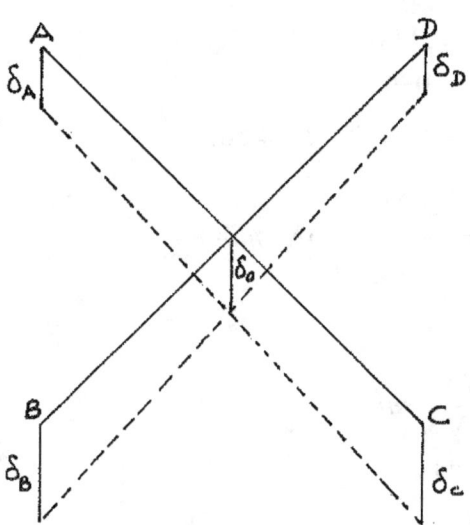

FIG.2: DISPOSITION OF DEFLECTIONS OF WIRES AT CORNERS AND AT CENTRE OF PLATFORM

If δ_0 is the deflection at the centre of the platform, then

$$\delta_0 = \dfrac{\delta_B + \delta_D}{2}$$

$$= \dfrac{\delta_A + \delta_C}{2}$$

so that

$$\delta_B + \delta_D = \delta_A + \delta_C$$

or $\quad \dfrac{T_B}{k} + \dfrac{T_D}{k} = \dfrac{T_A}{k} + \dfrac{T_c}{k}$

i.e. $\quad T_B + T_D = T_A + T_C \qquad \dots\dots\dots\dots\dots\dots\dots\dots$ (iv)

From (ii) and (iii)
$$T_D = 1400 - T_C$$
$$T_B = 1500 - T_C$$

Substituting in (iv)
$$1500 - T_C + 1400 - T_C = T_A + T_C$$

i.e. $2900 - 2T_C = T_A = +T_C$

or $T_A = 2900 - 3T_C$

Equation (i) now becomes
$$2900 - 3T_C + 1500 - T_C + T_C + 1400 - T_C = 2000N$$
$$5800 - 4T_C = 2000$$
$$4T_C = 3800$$
$$\therefore \quad T_C = 950N$$

so that $T_D = 450N; \quad T_B = 550N; \quad T_A = 50N$

Since stiffness, $S = \dfrac{T}{\delta} = \dfrac{Ea}{L} = 1960N/mm$

$$\delta_A = \frac{T_A}{1960}$$

$$= \frac{50N}{1960N} \cdot mm$$

i.e. $\underline{\delta_A = 0.025mm}$

Also, $\delta_B = \dfrac{550N}{1960N} \cdot mm$

$\underline{\delta_B = 0.275mm}$

$$\delta_C = \frac{950N}{1960N} \cdot mm$$

$\underline{\delta_C = 0.475mm}$

and, $\delta_D = \dfrac{450N}{1960N}, mm$

$\underline{\delta_D = 0.225mm}$

Q17. The frame shown in Fig. 1 is supported at *A* and *B* by two circular cylindrical steel bars of equal length and 6mm diameter and hinged at *C*. The load effect due to a block and tackle arrangement attached at '*D*' is a downward vertical force of 5kN. Determine the tension in each of the bars and hence the resultant reaction at the hinge. What would be the new tensions in the bars assuming a uniform temperature increase in them of 40°C? Use a coefficient of thermal expansion of $20 \times 10^{-6}/°C$ for the bars. At what temperature will the tension in bar '*B*' be reduced to two-fifths of its value before the temperature increase? Take E = 200GN/m².

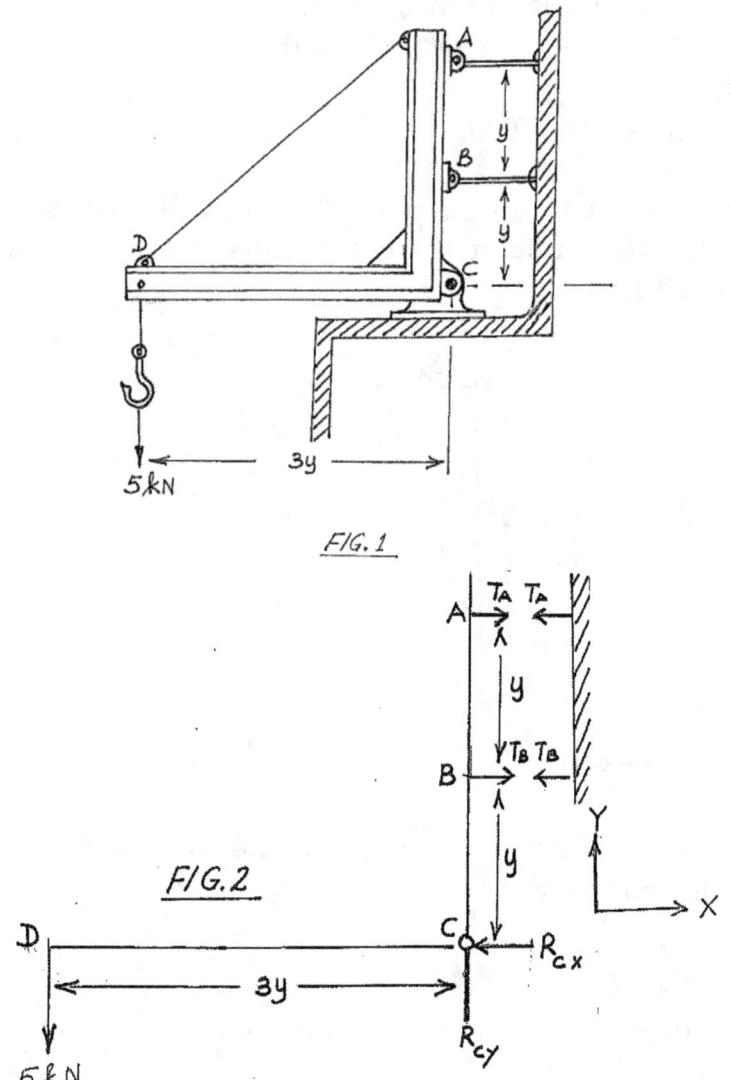

FIG. 1

FIG. 2

Fig. 2 is the free-body diagram.

Referring to this diagram and applying
$$\sum F_y = 0$$
we have,

$$R_{CY} - 5kN = 0$$
$$\therefore \quad R_{CY} = 5kN$$

Applying $\sum F_X = 0$

$$T_A + T_B - R_{CX} = 0$$

or
$$R_{CX} = T_A + T_B \qquad \dots\dots\dots\dots\dots\dots \text{(i)}$$

Taking moments about the hinge at C

$$T_A(2y) + T_B(y) - 50000(3y) = 0$$

from which

$$2T_A + T_B = 15000 \qquad \dots\dots\dots\dots\dots\dots\dots \text{(ii)}$$

This is as far as we can go with the laws of statics. We have to look elsewhere. Let us consider the extensions of the bars due to the tensions in them. See Fig. 2a.

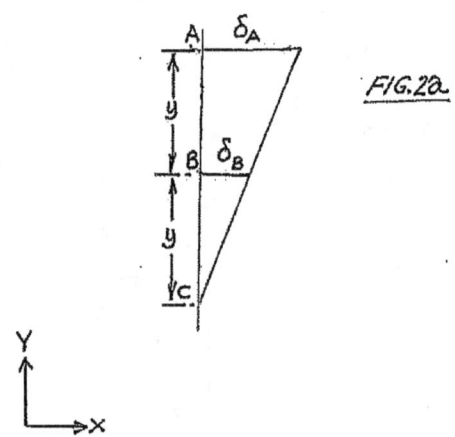

FIG.2a.

Let the extension of bar at $A = \delta_A$; and that at $B = \delta_B$
Evidently, by geometry

$$\frac{\delta_B}{\delta_A} = \frac{y}{2y}$$

i.e. $\qquad \delta_A = 2\delta_B \qquad \dots\dots\dots\dots\dots\dots\dots \text{(iii)}$

Hence, by Hooke's law

$$\delta_A = \frac{T_A}{A_a} \cdot \frac{\ell}{E}; \quad \delta_B = \frac{T_B}{A_b} \cdot \frac{\ell}{E}$$

in which A_a, A_b are respectively the cross-sectional areas of bars A and B; $\ell =$ length of bar A = length of bar B; and E = Young's modulus. Therefore

$$\frac{\delta_A}{\delta_B} = \frac{T_A}{T_B} \cdot \frac{A_b}{A_a} = 2$$

The wires have also the same cross-sectional areas, so that in effect

$$T_A = 2T_B$$

Substituting this result in (ii),

$$5T_B = 15000$$

or $\qquad T_B = 3000N = 3kN$

$$\therefore \quad T_A = 6000N = 6kN$$

By (i) $R_{CX} = 9kN$

Accordingly, the resultant reaction at C, say 'R' is obtained from:
$$R = \sqrt{(R_{CY}^2 + R_{CX}^2)}$$
$$= \sqrt{(5)^2 + (9)^2} = \sqrt{106}$$
$$R = 10.3kN$$

and its direction inclined at α^0 to the horizontal.
$$\alpha^0 = Tan^{-1} R_{CY} / R_{CX}$$
$$= Tan^{-1} 5/9$$
$$\therefore \quad \alpha \cong 29.1^o$$

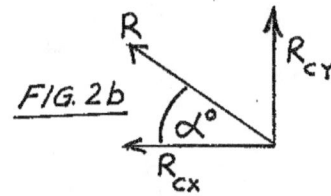

FIG. 2b

When the bars are subjected to the temperature increase let the tensions in them be T'_A and T'_B at A and B respectively.

In each bar the total strain is made up of two parts viz. a strain due to the stress in it and a thermal strain. This may be written thus:

Total strain = strain due to stress caused by the tension in the bar + pure thermal strain

If therefore the total strains in the bars at A and B are now designated δ'_A and δ'_B respectively, then
$$\frac{\delta'_A}{\ell} = \frac{T'_A}{A_a E} + \alpha \Delta t$$

where the new symbols α and Δt stand for coefficient of thermal expansion and increase in temperature, respectively.
Similarly,
$$\frac{\delta'_B}{\ell} = \frac{T'_B}{A_b E} + \alpha \Delta t$$

As before however
$$\delta'_A = 2\delta'_B$$
so that with 'ℓs' cancelling out

~ 501 ~

$$\frac{T'_A}{A_a E} + \alpha \Delta t = 2\left(\frac{T'_B}{A_B E} + \alpha \Delta t\right)$$

$$\frac{T'_A}{A_a E} = \frac{2T'_B}{A_B E} + \alpha \Delta t \qquad \ldots\ldots\ldots\ldots\ldots\ldots\ldots \text{(iv)}$$

Taking moments about C before
$$2T'_A + T'_B = 15000$$

i.e $\qquad T'_A = 7500 - \frac{T'_B}{2} \qquad \ldots\ldots\ldots\ldots\ldots\ldots\ldots \text{(v)}$

From (iv)

$$T'_A = 2T'_B \cdot \frac{A_a E}{A_b E} + A_a E \alpha \Delta t$$

Because $A_a E = A_B E$
$$T'_A = 2T'_B + A_a E \alpha \Delta t$$

or

$$7500 - \frac{T'_B}{2} = 2T'_B + A_a E \alpha \Delta t$$

i.e. $\qquad 2.5 T'_B = 7500 - A_a E \alpha \Delta t \qquad \ldots\ldots\ldots\ldots\ldots \text{(vi)}$

Given $A_a = \dfrac{\pi(6)^2}{4 \times 10^6} m^2 = 28.3 \times 10^{-6} m^2; \quad \Delta t = 40^\circ C$ and

$E = 200 \times 10^9 \, N/m^2$

$\therefore \qquad 2.5 T'_B = 7500 - 28.3 \times 10^{-6} \times 200 \times 10^9 \times 20 \times 10^{-6} \times 40$
$$= 7500 - 4528 = 2972$$

i.e. $\qquad T'_B = 1188.8 N, \quad \text{say} \quad 1189 N$

so that by (v)
$$T'_A = 7500 - 594.4, \quad \text{say} \quad 6906 N$$

If $(\Delta t)'$ be the temperature at which $T'_B = \dfrac{2}{5} T_B = \dfrac{2}{5}(3000)N$, viz, 1200N,

then by (vi)
$$2.5(1200) = 7500 - 28.3 \times 10^{-6} \times 200 \times 10^9 \times 20 \times 10^{-6} \times (\Delta t)'$$

i.e. $\qquad 3000 = 7500 - 113.3(\Delta t)$

from which $(\Delta t)' = 39.7^\circ C$.

Q18. The partly tapered bar shown in Fig. 1 is made up of three sections. It is affixed vertically to a rigid support at its top. The first section *'AB'* varies uniformly from a diameter of 100mm at the top to a diameter of 60mm at a section at *B*. Sections *BC* and *CD* are uniform circular cylindrical sections. *BC* is 60mm diameter and *BC* 50mm diameter. The bar carries the three loads shown. Determine the nett extension of the free end. The bar is of uniform density throughout, but you may neglect its weight. Take E = 210GN/m².

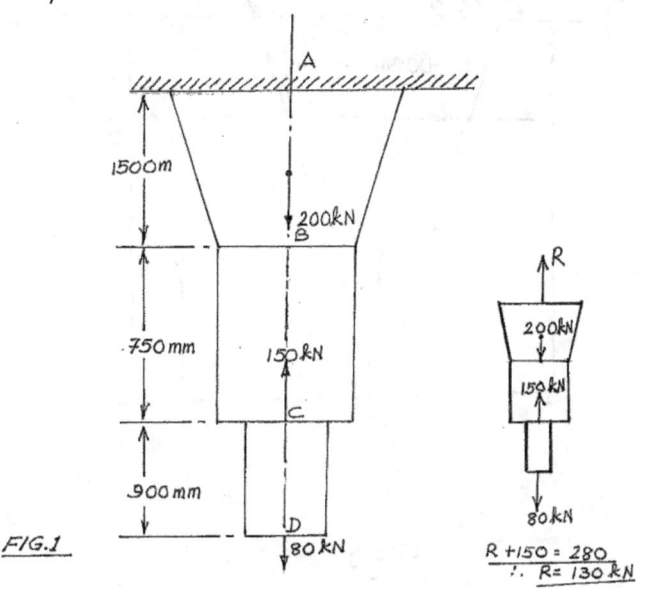

FIG.1

By inspection the reaction *'R'* at *A* necessary to maintain equilibrium is 130kN acting upwards : *R + 150 – 200 – 80 = 0* i.e.

R =130kN. The next step is to draw a Free Body Diagram. See Fig. 2

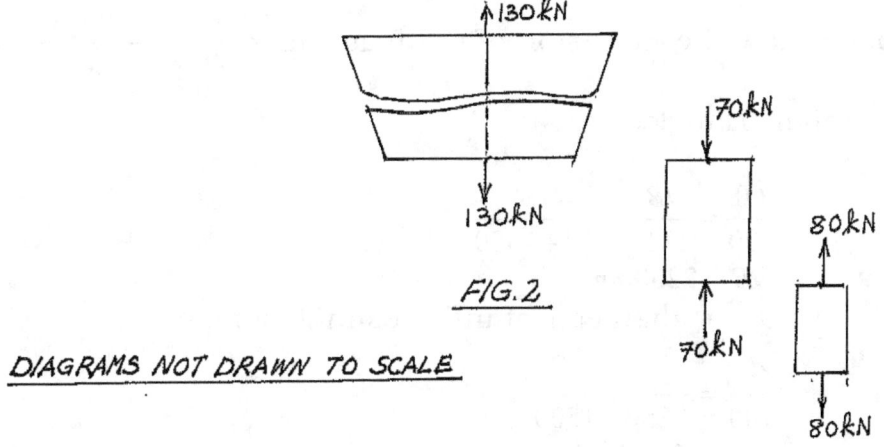

FIG.2

DIAGRAMS NOT DRAWN TO SCALE

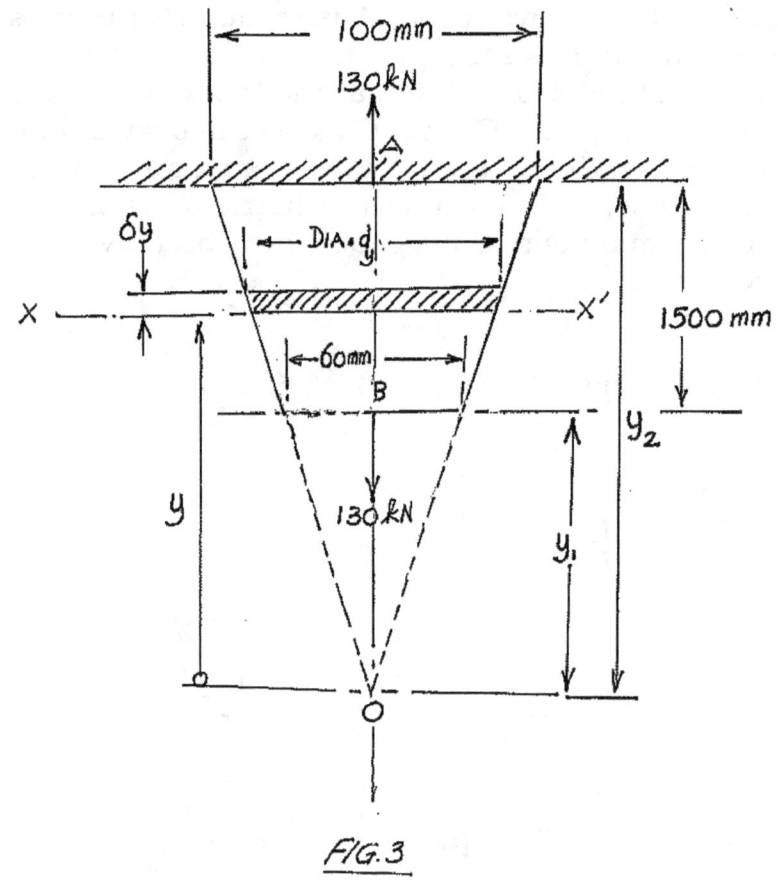

FIG. 3

Consider the top section of length 1500 mm. See Fig. 3.

Let A_y = cross-sectional area of an infinitesimal element of length 'δy'

The force at every cross section of the frustum is 130kN

By Hooke's law the extension of the frustum = $\dfrac{1}{E} \displaystyle\int_{y_1}^{y_2} \dfrac{130 \times 10^3}{A_y}\, dy$

From similar triangles :

$$\frac{60}{100} = \frac{OB}{OA} = \frac{OB}{OB + 1500}$$

$\therefore \quad OB = 2250mm$

If $\qquad d_y$ = diameter of infinitesimal element

then $\qquad \dfrac{d_y}{100} = \dfrac{y}{2250 + 1500}$

$\therefore \qquad d_y = \dfrac{100y}{3750}\,mm$

or

where y is in mm.

Cross-sectional area of infinitesimal element of diameter 'd_y' is :

$$A_y = \frac{\pi d_y^2}{4}$$

$$= \frac{\pi}{4} \cdot \left(\frac{100y}{3.75 \times 1000} \right)^2$$

$$\frac{\pi}{4} \frac{10^4 \, y^2}{14.0625 \times 10^6} \, (mm)^2$$

Accordingly by Hooke's law, the extension of the frustum 'δ_F' subjected to the tensile force of 130kN is given by:

$$\frac{1}{E} \int_{2250}^{3750} \frac{(130 \times 10^3 \times 4 \times 10^2 \times 14.0625)dy}{\pi \cdot y^2}$$

We are working in mm. so that 'E' must be expressed in N/(mm)², i.e. E = 210 x 10³ N/mm².

$$\therefore \quad \delta_F = \frac{130 \times 10^5}{210 \times 10^3} \times 56.25 \int_{2250}^{3750} \frac{dy}{y^2}$$

$$= 3482 \int_{2250}^{3750} \frac{dy}{y^2}$$

$$\delta_F = 3482 \left[-\frac{1}{y} \right]_{2250}^{3750}$$

$$= 3482 \left[+\frac{1}{y} \right]_{3750}^{2250}$$

$$= 3482 \left[\left(\frac{1}{2250} - \frac{1}{3750} \right) \right]$$

$$= 3482(0.00177)$$

$$\therefore \quad \delta_F = 0.62mm$$

Working in metres for the other sections :

Compression of section (2), δ_C.

$$= \frac{70 \times 10^3}{\frac{\pi}{4} \cdot \left(\frac{60}{1000} \right)^2} \times \frac{750}{1000} \times \frac{1}{210 \times 10^9}$$

$$= \frac{70 \times 10^3 \times 10^6 \times 1750 \times 4}{\pi \times 3600 \times 1000 \times 10^9} \times \frac{1}{210}$$

$$= \frac{7 \times 3}{\pi \times 360 \times 210} \, metres$$

$$= \frac{\frac{1000}{\pi \times 3600}}{}$$

$$\therefore \qquad \delta_C = -0.088 mm$$

Extension of Section (3), e_T

$$= \frac{80 \times 10^3 \times 900}{\frac{\pi}{4} \times \left(\frac{50}{1000}\right)^2 \times 1000} \times \frac{1}{210 \times 10^9}$$

$$= 0.000175 \ metre$$

$$= 0.175 \ mm$$

$$\therefore \qquad \text{Net deflection of free end :}$$
$$= (0.62 + 0.175 - 0.088) mm$$
$$= +0.687 mm$$

i.e.   Net deflection = 0.68mm *(extension)*

Q19. A prismatic bar of uniform composition and cross-section throughout is fitted without stress between two rigid supports as shown in Fig. 1 below.   The bar which is one metre long is subjected to vertical downward forces : $F = 1200N$ and F_2 at points on the bar 0.7m and .5m respectively from the top.   If the total compression is twice the total extension and the reaction at the lower end is four times that at the top, determine the magnitude of F_2 and the reactions R_1 and R_2 at the top and at the end of the bar respectively.

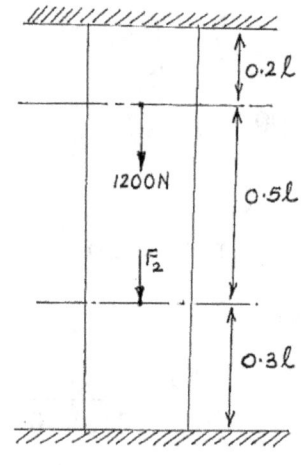

FIG 1

Ans : F_2 = -6684N (i.e. upwards); R_1 = 1828N (upwards at upper end)
R_2 = 7312N (downwards at lower end).

Q20. A uniform prismatic bar 60 cm in length is hinged at one end and supported by 2 circular cylindrical rods as shown in Fig. 1. The cross-sectional area of the rod at 'B' is twice that of the rod at 'C', but the modulus of elasticity of the rod at C is 1.5 times that of the rod at B. The weight of the bar is 20 kN. Determine the ratio of the stiffnesses of the bars at B and C i..e. S_B / S_C and also the forces in them.

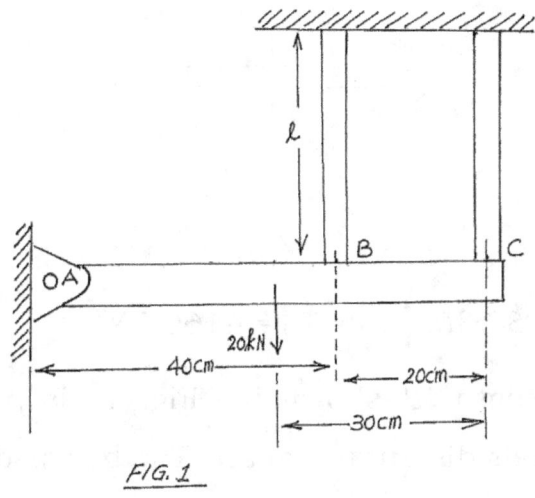

FIG. 1

Ans : $\dfrac{S_B}{S_C} = \dfrac{4}{3}$; $F_B = 5.57kN$; $F_C = 6.26kN$

Q21. Three elastic rods and a rigid column support at its four corners a rigid rectangular platform $2m \times 3m$ whose mass of 2500 kg is distributed uniformly. If the three elastic rods have the same length, cross-sectional area and are made of the same material, then determine the reactions in the rods and in the column, assuming the platform remains horizontal. See Fig. 1 below.

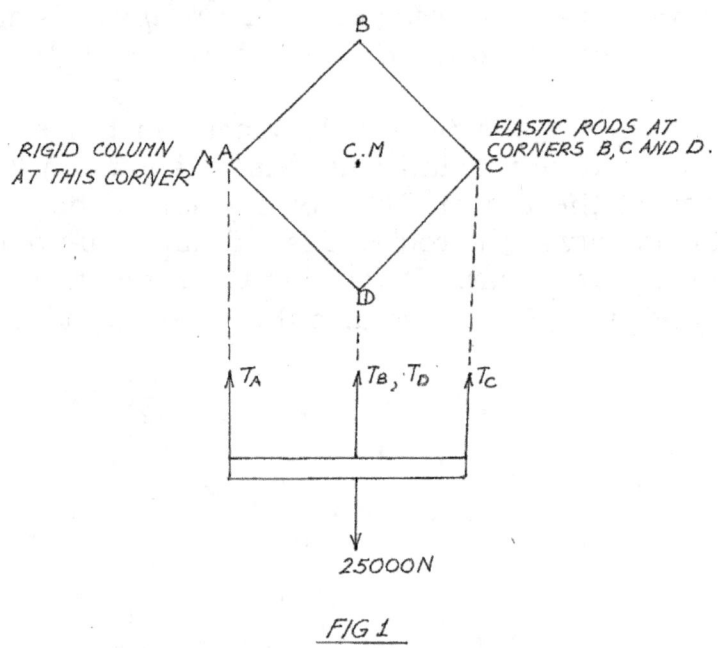

FIG 1

Ans : $T_A = T_C = 8333.3N$; $T_B = T_D = 4166.7N$.

Q22. The bar *AB* of length 'ℓ' shown in Fig. 1 is pin-jointed at B and is supported by 2 rods distant $\frac{4}{5}\ell$ apart. The bar also carries a load 'W' at a distance equal to $\frac{2}{3}$ its length from *B*. Determine the tension in each rod expressing your answer in terms of '*W*' and angles 'α' and 'θ'.

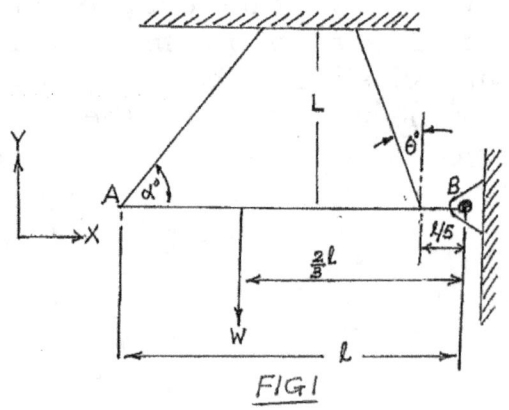

FIG I

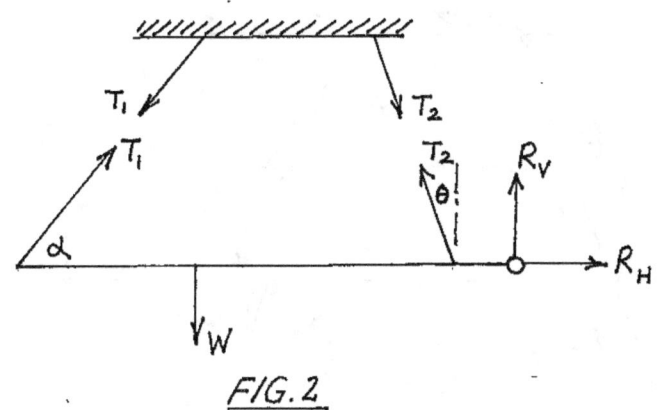

FIG. 2

The *FBD* for the arrangement in Fig. 1 is shown as Fig. 2.
Consider the conditions for static equilibrium.

For $\sum F_Y = 0$

$$T_1 Sin\alpha + T_2 Cos\theta + R_v - W = 0 \qquad \dots\dots\dots\dots\dots \text{ (i)}$$

For $\sum F_X = 0$

$$T_1 Cos\alpha - T_2 Sin\theta + R_H = 0 \qquad \dots\dots\dots\dots\dots \text{ (ii)}$$

For moments about 'B'

$$T_1 Sin\alpha(\ell) - W\left(\frac{2}{3}\ell\right) + T_2 Cos\theta\left(\frac{\ell}{5}\right) = 0$$

which may be rewritten as :

$$T_1 \ell Sin\alpha + T_2 \frac{\ell}{5} Cos\theta = \frac{2W\ell}{3}$$

$$\text{or} \quad T_1 Sin\alpha + \frac{T_2}{5} Cos\theta = \frac{2W}{3} \qquad \dots\dots\dots\dots\dots \text{ (iii)}$$

Thus we have three equations with four unknowns. The problem is indeterminate. We have to resort to means other than the laws of statics to obtain a solution.

Consider then the geometry of deformation of the rods as in Fig. 3.

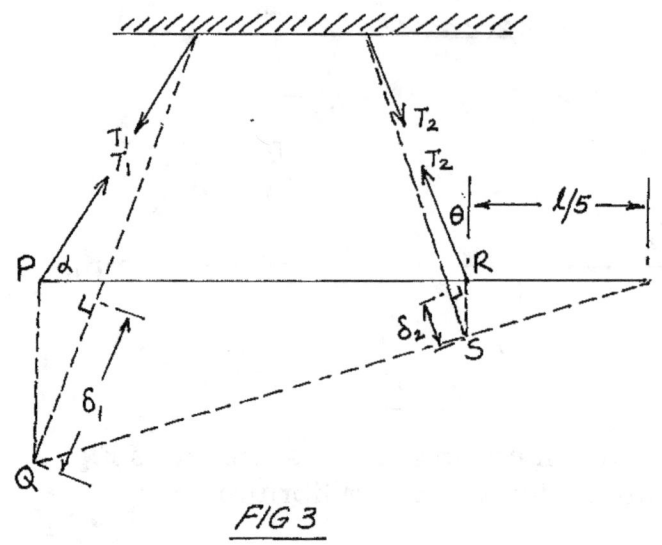

FIG 3

$$\frac{RS}{PQ} = \frac{\ell/5}{\ell} = \frac{1}{5} \qquad \text{i.e.} \qquad PQ = 5RS$$

But
$$\frac{\delta_1}{PQ} = Cos\alpha \quad ; \quad \frac{\delta_2}{RS} = Cos\theta$$

$$\therefore \quad \frac{RS}{PQ} = \frac{\delta_2}{Cos\theta} \cdot \frac{Cos\alpha}{\delta_1}$$

But
$$\frac{RS}{PQ} = \frac{1}{5}$$

$$\therefore \quad \frac{1}{5} = \frac{\delta_2}{\delta_1} \cdot \frac{Cos\alpha}{Cos\theta} \quad \text{or} \quad \frac{\delta_1}{\delta_2} = \frac{5Cos\alpha}{Cos\theta}$$

By Hooke's law
$$\delta_1 = T_1 \frac{L}{Cos\alpha} \cdot \frac{1}{EA}$$

and
$$\delta_2 = \frac{T_2 L}{Cos} \cdot \frac{1}{EA}$$

$$\therefore \quad \frac{\delta_1}{\delta_2} = \frac{T_1 L}{Cos\alpha} \cdot \frac{1}{EA} \quad \frac{Cos\theta \; EA}{T_2 L}$$

$$\frac{\delta_1}{\delta_2} = \frac{T_1 Cos\theta}{T_2 Cos\alpha}$$

But
$$\frac{\delta_1}{\delta_2} = \frac{5Cos\alpha}{Cos\theta}$$

$$\therefore \quad \frac{5Cos\alpha}{Cos\theta} = \frac{T_1 Cos\theta}{T_2 Cos\alpha}$$

i.e. $\quad \dfrac{T_1}{T_2} = \dfrac{5Cos^2\alpha}{Cos^2\theta}$

or $\quad T_1 = \left(\dfrac{5Cos^2\alpha}{Cos^2\theta}\right)T_2$

Substituting this result in (iii)

$$T_2 \cdot \dfrac{5Cos^2\alpha}{Cos^2\theta} \cdot Sin\alpha + \dfrac{T_2}{5}Cos\theta = \dfrac{2W}{3}$$

$$T_2\left(\dfrac{5Cos^2\alpha\ Sin\alpha}{Cos^2\theta} + \dfrac{1}{5}Cos\theta\right) = \dfrac{2W}{3}$$

$\therefore \quad T_2\left(\dfrac{25Cos^2\alpha\ Sin\alpha + Cos^3\theta}{5Cos^2\theta} = \dfrac{2W}{3}\right)$

$\therefore \quad T_2 = \dfrac{10W\ Cos^2\theta}{25Cos^2\alpha\ Sin\alpha + Cos^3\theta}$

and $\quad T_1 = \dfrac{5Cos^2\alpha}{Cos^2\theta}\left(\dfrac{10W\ Cos^2\theta}{25Cos^2\alpha\ Sin\alpha + Cos^3\theta}\right)$

$\therefore \quad T_1 = \dfrac{5W\ Cos^2\alpha}{25Cos^2\alpha\ Sin\alpha + Cos^3\theta}$

Q23. In the making of a pre-stressed concrete beam of cross-section 250mm x 250mm six high-strength steel wires each of 20mm diameter are stretched by a hydraulic jack which exerts a pre-stressing force causing an initial stress of 800MN/m² in each wire. After the concrete is poured in and the bond between the two materials soundly developed the pre-stressing force is removed. Calculate the residual stress in the concrete and in the reinforcement. Take E_STEEL = 210GN/m² and E_CONC = 20GN/m²

Ans : $\sigma_s = 187MN/m^2$; $\sigma_{CONC} = 18.7MN/m^2$

Q24. A vertical circular cylindrical reinforced-concrete column 300mm diameter and 2m long is subjected to a compressive load of 610kN. Assuming no buckling effects and maximum permissible working stresses in the concrete and in the rebars of 5600kN/m² and 84000kN/m² respectively, calculate the total cross-sectional area of steel required. The rebars are to be bound together by helical reinforcement. How many rebars of a specific size would you recommend? Make neat drawings each of a typical transverse and longitudinal section through a reinforced column. Take $E_c = 21 N/m^2$ and $E_s = 210 G/m^2$.

Ans : Cross-sectional area of steel = 3988(mm)^2
         Use 8 #25 rebars @ 25.4mm for a total of 4072(mm)^2.

The requested drawings are given as Figs. 1 and 2 respectively.

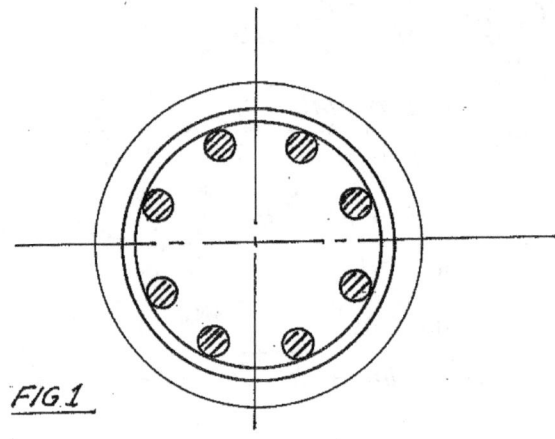

FIG 1

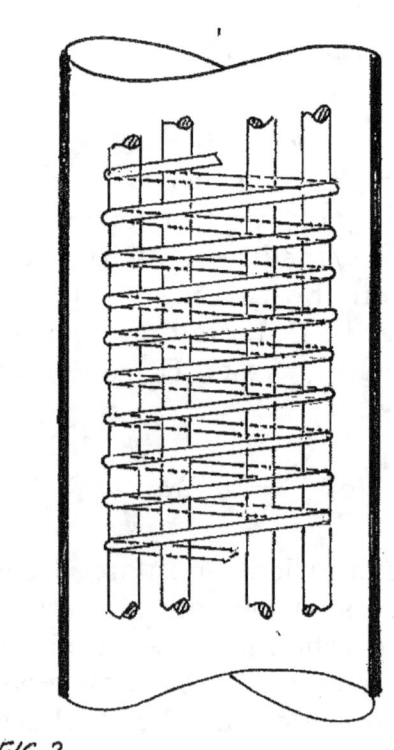

FIG.2.

Q25. Define stiffness of an elastic member in terms of the modulus of elasticity 'E', cross-sectional area 'a' and length 'ℓ'. A uniform rigid bar *ON* is hinged to a rigid wall at 'O' and is held in a horizontal position by two elastic rods both of cross-sectional area *A*, length 'L' and modulus of elasticity 'E'. The assembly is shown in Fig. 1. A force of *W* newtons is applied at the midpoint between the rods. Show that the resultant slope of the bar is given by the expression:

$$\theta = Tan^{-1}\left\{\frac{WL(2b+a)}{2AE(a^2+2ab+2b^2)}\right\}$$

Stiffness 'S' is defined as load per unit extension (or compression).

For an elastic member of cross sectional area 'a', length 'L' and modulus of elasticity 'E', the stiffness 'S' may be expressed as follows:

$$E = \frac{Load \cdot L}{a(extension)}$$

from which

$$\frac{Load}{Extension} = S = \frac{Ea}{L}$$

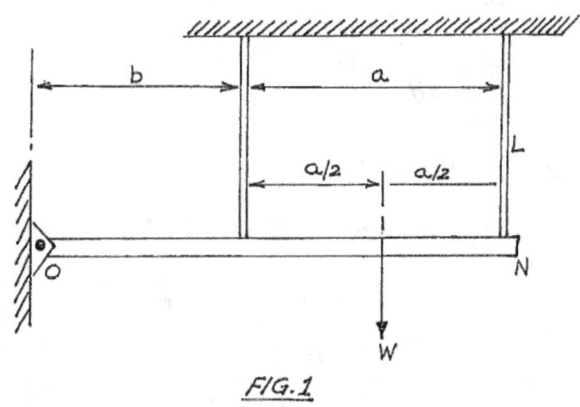

FIG.1

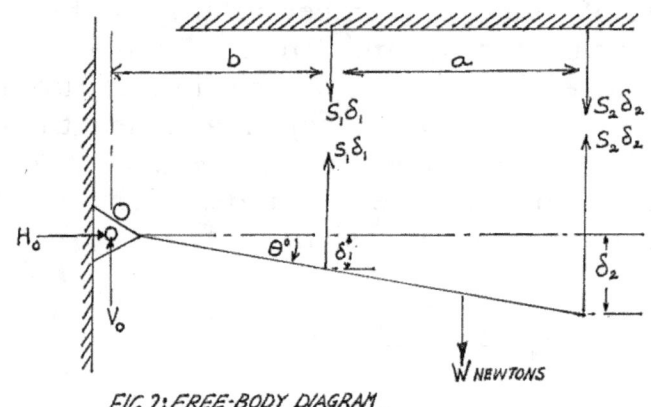

FIG.2: FREE-BODY DIAGRAM

There is no need here to evaluate H_o and V_o because we are going to take moments about 'O'. Accordingly,

$$W\left(b+\frac{a}{2}\right) = S_1\delta_1(b) + S_2\delta_2(a+b) \qquad \cdots\cdots\cdots\cdots \text{(i)}$$

Slope of bar may be expressed by the following from the geometry of the diagram :

$$Tan\theta = \frac{\delta_1}{b} = \frac{\delta_2}{b+a}$$

from which

$$\delta_1 = bTan\theta$$

and

$$\delta_2 = (b+a)Tan\theta$$

Substituting these results in (i)

$$W\left(b+\frac{a}{2}\right) = S_1\,b^2\,Tan\theta + S_2(a+b)(b+a)Tan\theta$$

$$= Tan\theta\left\{S_1\,b^2 + S_2(a+b)^2\right\}$$

or

$$Tan\theta = \frac{W(2b+a)}{2\left\{S_1 b^2 + S_2(a+b)^2\right\}}$$

Since $S_1 = EA/L = S_2$, each rod being of same material, cross-sectional area and length.

$$Tan\theta = \frac{W(2b+a)}{2S(b^2+a^2+2ab+b^2)}$$

i.e.

$$Tan\theta = \frac{WL(2b+a)}{2AE(a^2+2ab+2b^2)}$$

or

$$\theta = Tan^{-1}\left\{\frac{WL(2b+a)}{2AE(a^2+2ab+2b^2)}\right\} \qquad \text{Q.E.D}$$

Q26. The top end of the solid vertical circular cylindrical bar shown in Fig. 1 is fixed to a rigid support, the extremity of its free end being at distance 'y' from another rigid support distance 'L' from the top. A load 'P' is to be applied at a distance equal to $\frac{2}{5}ths$ the length of the bar from the bottom of the bar. If when the load 'P' is applied it stretches the bar to the extent that it must be constrained by the bottom support when the reaction there is $\frac{3}{10}ths$ the magnitude of that of the upper support, then find the value of 'y' in terms of the cross-sectional area of the bar and E = Young modulus

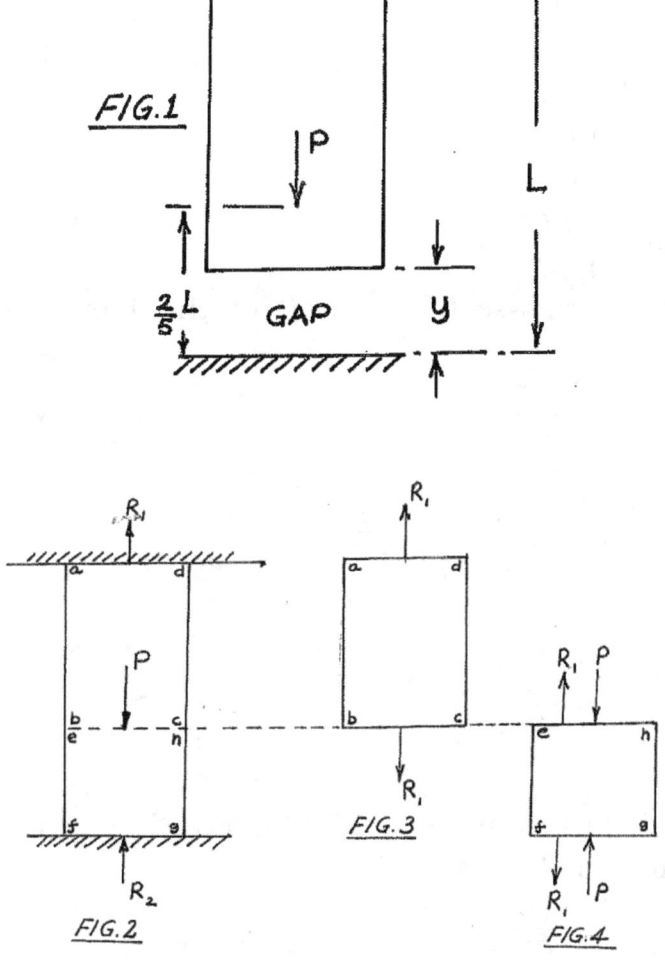

FIG.1

FIG.2

FIG.3

FIG.4

In Fig. 2 the bar at Fig. 1 is constrained by the lower rigid support. For this condition, let the reactions at top and bottom of the bar be R_1 and R_2 respectively

$$\therefore \qquad R_1 + R_2 = P \qquad \dots\dots\dots\dots\dots\dots\dots\dots\dots\dots \text{(i)}$$

Given $R_2 = \dfrac{3}{10} R_1$

$$\therefore \quad R_1 + \frac{3}{10} R_1 = P \quad \text{or} \quad \frac{13}{10} R_1 = P$$

or $\quad R_1 = \dfrac{10P}{13}$

The *FBDs* for the load configuration in Fig. 2 are given in Fig. 3 and Fig. 4 : the upper part is under a tensile load of 'R_1', and assuming 'P' > R_1, the lower part under a compressive load $P\text{-}R_1$.

Let δ_1 = deflection of upper part ; and δ_2 that of the lower.
By Hooke's law

$$\delta_1 = \frac{R_1}{A} \cdot \frac{3}{5} \frac{L}{E}$$

and because $R_1 = \dfrac{10P}{13}$

$$\delta_1 = \frac{10P}{13A} \cdot \frac{3L}{5E}$$

$$\delta_1 = \frac{6PL}{13AE}$$

'*A*' being the cross-sectional area of the bar. Noting that δ_2 is a compression

$$\delta_2 = \frac{P - R_1}{A} \cdot \frac{2}{5} \frac{L}{E}$$

$$= \left(\frac{P - \dfrac{10P}{13}}{A} \right) \cdot \frac{2L}{5E}$$

$$\delta_2 = \frac{6PL}{65AE}$$

Evidently $\quad \delta_1 + \delta_2 = y$

Noting that δ_2 is a compression it follows that a negative sign must be attached to $6PL/65AE$

$$\therefore \quad \frac{6PL}{13AE} - \frac{6PL}{65AE} = y$$

i.e. $\quad y = \dfrac{24PL}{65AE}$

Note: with connections to Q26 following

When plain reinforcement steel is ordinarily cast in concrete, any loading of the resulting composite should be such as to strictly limit, if not avoid altogether, the development of any tensile stresses in the concrete. Why? Because concrete is notoriously weak in tension. Hence any 'appreciable' tensile stress in the steel transferred to its 'married partner' in the composite, will cause cracks to develop in the concrete.

In pre-stressed concrete, ordinary reinforcement steel is replaced generally by high tensile corrugated-steel rods called rebars. The reinforcement rods are subjected to tensile stress by being stretched, commonly by means of a hydraulic jack, either before the concrete has achieved the requisite degree of hardness : pre-tensioning, or afterwards : post-tensioning.

When, as is the case of pre-tensioned or pre-stressed concrete, the source of the tensile pull is removed, the reinforcement rods contract lengthwise in an attempt to return to their original length. In the process, a compressive stress is induced in the already hardened concrete.

In the case of post-tensioning, the reinforcement rods are coated initially with a substance such as bitumen in order to facilitate their movement in the concrete after it has hardened. After hardening the rods are stretched and anchored where they emerge from the concrete.

This note is intended to whet your appetite. Consult a good book such as "Pre-stressed Concrete : Theory and Design" by Evans, R.H. and Bennett, E.W. Visit a plant that manufactures pre-stressed concrete. Ask questions. Make notes and sketches.

Now to a simple problem on pre-tensioned concrete. You might say an ordinary stress question with steel and concrete connections.

Q27. In producing a pre-tensioned, pre-stressed beam, a single high-tensile corrugated steel rod of 20mm diameter is anchored to a rigid support and stretched from the other end by a hydraulic jack which exerts a pre-stressing force 'P'. 'P' induces an initial stress of 600MN/m² in the steel rod. High quality concrete is then poured into a mould with the rod properly fitted, and allowed to set hard so that the bond between concrete and rod is soundly developed.

If the finished beam is of 250mm square section throughout then what are the final stresses in the steel and concrete given that Young's modulii for steel and concrete are respectively, $E_s = 200GN/m^2$ and $E_c = 20GN/m^2$.

Let A_s = cross-sectional area of steel rod

" A_c =   "   "   "   " concrete

$$\therefore \quad A_s = \frac{\pi}{4} \frac{(20)^2}{10^6} m^2 = 314 \times 10^{-6} m^2$$

Also,

$$A_c = \left(\frac{250 \times 250}{10^6} \right) m^2 - 314 \times 10^{-6} m^2$$

i.e. $A_c = 62186 \times 10^{-6} m^2$

In Fig 1 the pre-tensioning force 'P' when released is 'shared' by both reinforcement rod and concrete, the respective components designated P_s and P_c

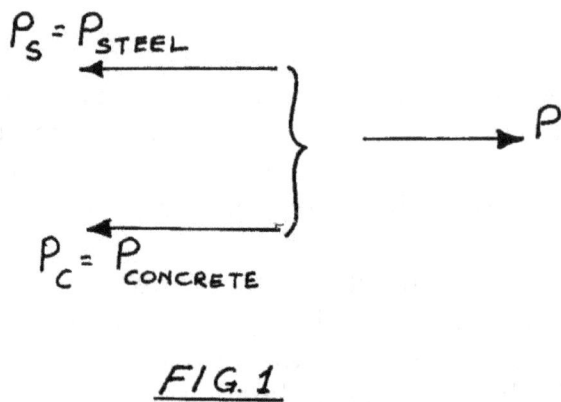

$$P_s = P_{STEEL}$$

$$\longrightarrow P$$

$$P_c = P_{CONCRETE}$$

FIG. 1

By equilibrium $P = P_s + P_c$ . . . . (i)

The lengths of the steel rod and of the concrete beam are equal being, say, L. Since the steel and concrete are now a unified component, it is evident that any contraction, say δ_s of the steel rod attempting to retain its initial configuration must also equal the contraction δ_c giving rise to the compressive stress in the concrete. That is to say the two strains, assumed elastic, are :

$$\frac{\delta_s}{L} = \frac{\delta_c}{L} \quad \text{. . . . . (ii)}$$

On this basis therefore we may write, according to Hooke's law :

$$\frac{P_s}{A_s\,E_s} = \frac{P_c}{A_c\,E_c} \qquad \dots\dots\dots\dots\dots\dots\dots\dots \text{(iii)}$$

from which

$$P_c = \frac{P_s\,A_c\,E_c}{A_s\,E_s} \qquad \dots\dots\dots\dots\dots\dots\dots\dots \text{(iv)}$$

or $\quad P_s = \dfrac{P_c\cdot A_s\,E_s}{A_c\,E_c} \qquad \dots\dots\dots\dots\dots\dots\dots\dots \text{(v)}$

We are told that the pre-tensioning of the high-tensile reinforcement rod induces a stress of 600MN/m² in it.

Therefore

$$P = 600\times10^6 \times 314\times10^{-6}$$
$$= 188400N$$

Equation (i) may now with the aid of, say, (v) be rewritten as:

$$188400 = P_s + \frac{P_s\,A_c\,E_c}{A_s\,E_s} = P_s\!\left(1 + \frac{A_c\,E_c}{A_s\,E_s}\right)$$

$$= P_s\left\{1 + \left(\frac{62186}{314}\right)\cdot\left(\frac{20\times10^9}{200\times10^9}\right)\right\}$$

$$= P_s(1 + 19.8)$$

from which $\quad P_s = 9058N$

$\therefore \quad$ by (i) $\quad P_c = 188400 - 9058N$

or $\qquad\qquad P_c = 179342N$

Accordingly residual stress 'δ_s' in steel is given by:

$$\sigma_s = \frac{9058}{314\times10^{-6}}\,N/m^2$$
$$= 28.8MN/m^2;$$

and the compressive stress 'σ_c' induced in the concrete by:

$$\sigma_c = \frac{179342}{62186\times10^{-6}}\,N/m^2$$
$$= 2.88MN/m^2$$

Q28. The uniform concrete platform in Fig. 1 has a mass of 60,000kg and is initially supported centrally by a steel column 50cm in height and of cross-sectional area 6450(mm)². Two aluminium columns each being 29.75cm in height and 19,900 (mm) cross-sectional area are placed symmetrically on each side of the column and directly below the platform. All three columns are subjected to a rise in temperature so that the weight of the platform is carried by them. At what temperature rise will this occur. Take coefficients of linear thermal expansion as follows : $\alpha_{Steel} = 14 \times 10^{-6} / ^\circ C$; $\alpha_{AL} = 23 \times 10^{-6} / ^\circ C$; Modulii of elasticity: $E_{Steel} = 210 GN/m^2$ and $E_{AL} = 67.2 GN/m^2$. Take $g = 9.81 m/s^2$.

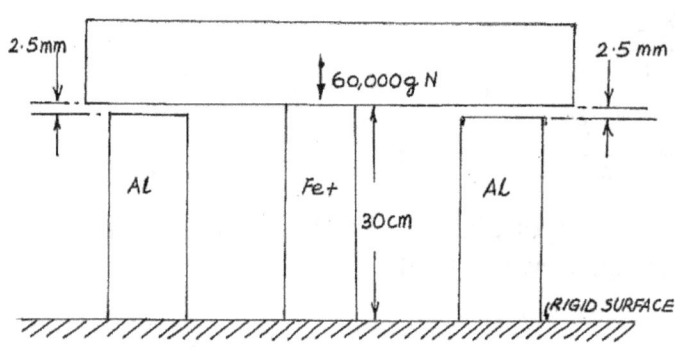

FIG. 1

Ans : 111.2°C

Q29. A solid uniform vertical monument is to be cast from steel in the form of a frustum of a cone. The design calls for a 500mm diameter base, a 200mm diameter top and 5m in height. The monument is to be erected on a solid base. Determine the shortening of the monument due to self weight. Assume the density of steel to be 7850kg/(m)³, $g \approx 10m/(sec)^2$ and Young modulus for steel = 210 GN/(mm)².

Ans : $2.8 \times 10^{-3} mm$

Q30. A temperature-activated mechanism contains a component comprising a strip of alloy steel of cross-sectional area 'a$_s$' $(= a_s)$ soldered between two strips of brass each of cross-sectional area 'a$_B$' $(= a_B)$. The length of the strips is 'ℓ'. Assuming no bending occurs, show that for a temperature rise of Δt, the stress in each of the brass strips 'σ_B' and in the steel strip σ_s are given by the expression:

$$\sigma_B = \frac{E_B (\alpha_B - \alpha_s)\Delta t}{\left(1 + 2 \dfrac{E_B a_b}{E_s a_s}\right)}$$

$$\sigma_s = \frac{2E_B a_B(\alpha_B - \alpha_s)\Delta t}{a_s\left(1 + 2\dfrac{E_B a_B}{E_s a_s}\right)}$$

in which α_B and α_s are respectively the coefficients of linear thermal expansion of brass and steel; and E_B and E_s, Young modulus for brass and steel respectively.

Q31. (a)   Explain briefly how it is possible for a body to have strain without being subjected to stress.

(b)   Two horizontal metal bars, one of brass and the other of steel are separated by a distance of 0.15mm at ambient temperature. Refer to Fig. 1. Determine (i) the increase in temperature required for the bars to make contact; and (ii) the stress in each bar when an elevated temperature of 120°C is attained. Take $\alpha_B (for\ brass) = 18.9 \times 10^{-6} /^{\circ} C$ and $\alpha_s (for\ steel) = 10.8 \times 10^{-6} /^{\circ} C$; $E_B = 105 GN/m^2$ and $E_s = 200 GN/m^2$

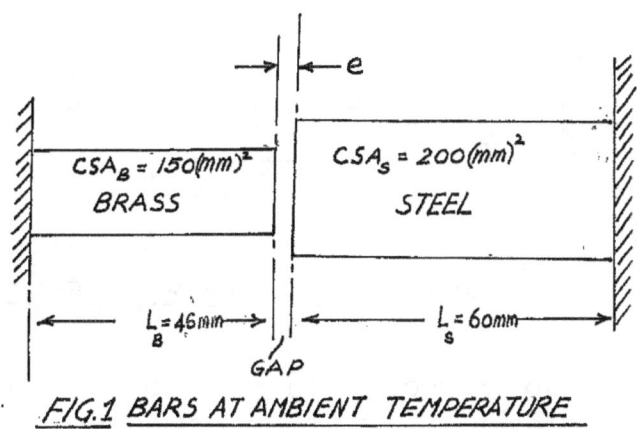

FIG.1 BARS AT AMBIENT TEMPERATURE

(a)   Strain is defined as the ratio of an increase in a dimension to the original dimension. If therefore the uniform temperature of, say, and unconstrained steel rod of initial length L_1 is T_1 and that this temperature regime be increased to another uniform temperature of T_2, then the length of the rod will increase by an amount $\alpha(T_2 - T_1)L$, where α is the coefficient of linear thermal expansion for the material of the rod. By definition therefore the strain is $\alpha(T_2 - T_1)L_1 / L_1$ or $\alpha(T_2 - T_1)$. And evidently this pure thermal strain occurred without stress.

(i)   Temperature increase to close the gap.
Let Δt = temperature increase.

Pure linear thermal expansion of brass bar for this temperature increase = $\alpha_b\ \Delta t\ L_B$

Pure linear thermal expansion of steel bar for this temperature increase $= \alpha_s \, \Delta t \, L_s$. Given that gap '$e$' $= 0.15mm$, it is evident that:

$$\alpha_B \, \Delta t \, L_B + \alpha_s \, \Delta t \, L_s = 0.15$$

Working in metres:

$$18.9 \times 10^{-6} (\Delta t) \cdot \frac{45}{1000} + 10.8 \times 10^{-6} (\Delta t) \frac{60}{1000} = \frac{0.15}{1000}$$

from which

$$850.5 \Delta t + 648 \Delta t = 0.15 (10^6)$$

or

$$1498.5 \Delta t = 150,000$$

$$\therefore \qquad \Delta t = 100.1^\circ C.$$

Therefore at an increase in temperature of 100.1°C the two bars are in contact.

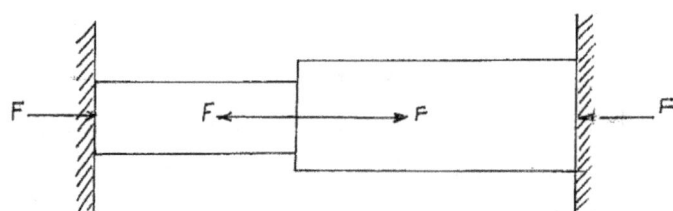

<u>FIG. 2: BARS AT ELEVATED TEMPERATURE OF 120°C</u>

(ii) With the gap closed, further heating produces further linear thermal expansion but the compressive force 'F' caused by the two bars pushing against each other as shown in Fig. 2 give rise to a compression in each bar; this compression opposes the pure linear thermal expansion in each bar.

Accordingly,

Net expansion of brass bar $= \alpha_B (\Delta T) L_B - \dfrac{F}{A_B} \dfrac{L_B}{E_B}$.

where Δt = temperature increase above 100.1°C

Also,

Net expansion of steel bar $= \alpha_S (\Delta T) L_S - \dfrac{F L_S}{A_S E}$

There is no longer any gap between the bars; gap 'e' $= 0$, so

$$\left(\alpha_B(\Delta T)L_B - \frac{FL_B}{A_B\,E_B}\right) + \left(\alpha_s(\Delta T)L_s - \frac{FL_s}{A_S\,E_S}\right) = 0$$

Note carefully that here $\Delta t = 120^\circ C - 100.1^\circ C$ i.e. 19.9°C. Substituting the numerical values:

$$\therefore \quad 18.9\times10^{-6}\times19.9\times\frac{45}{1000} + 10.8\times10^{-6}\times19.9\times\frac{60}{1000} =$$

$$F\cdot\frac{45}{1000}\cdot\frac{10^6}{105\times10^9\times150} + F\cdot\frac{60}{1000}\cdot\frac{10^6}{200\times10^9\times200}$$

from which : $\quad (16924.95 + 12895.2)\dfrac{1}{10^3} = (0.002857F)$

$\therefore \quad 4.35F = 29820.15$

i.e. $F = 6844N$ or $F = 6.84kN$

Alternatively, the problem may be solved by starting with the bars at ambient temperature, with the temperature rise being 120°C; the gap $'e'= 0.15mm$.. When the increase in temperature occurs :

Nett expansion of brass bar + Nett expansion of steel bar = 0.15mm.

Therefore

$$18.9\times10^{-6}\times120\times\frac{45}{1000} - F'\cdot\frac{45}{1000}\cdot\frac{10^6}{105\times10^9\times150} + 10.8\times10^{-6}\times120\times\frac{60}{1000}$$

$$- F'\cdot\frac{60}{1000}\cdot\frac{10^6}{200\times10^9}\cdot\frac{1}{200} = \frac{0.15}{1000}$$

which reduces to:

$$F' = \frac{18.9\times10^{-6}(120)\dfrac{45}{1000} + 10.8\times10^{-6}\times120\times\dfrac{60}{1000} - \dfrac{0.15}{1000}}{\left(\dfrac{45}{105\times10^6\times150}\right) + \left(\dfrac{60}{200\times200\times10^6}\right)}$$

i.e. $\quad F' = \dfrac{(102060 + 77760 - 150000)\cdot1}{(0.002857 + 0.0015)\ 10^3}$

$$= 29820/4.357$$

$F' = 6844N$ or $6.844kN$; Check.

Therefore the stresses σ_b, σ_s in the bars are as follows:

$$\sigma_b = \frac{6844}{150/10^6} = 45.6 MN/m^2$$

$$\sigma_b = \frac{6844}{200/10^6} = 34.2 MN/m^2$$

Q32. A bronze bolt 15mm diameter is fitted inside a circular cylindrical steel tube of outside diameter 30mm and inside diameter 25mm. Steel washers are also fitted at both ends of the assembly and the nut on the bold turned sufficiently to provide a snug fit. See Fig. 1. At an ambient temperature of 20°C there is no stress on either bolt or tube. Determine the stresses in the bolt and in the tube assuming a uniform distribution of temperature throughout the assembly when the temperature reaches 210°C. Neglect temperature increase in the washers. Take $\alpha_B = 20 \times 10^{-6}/°C$; $\alpha_s = 16.8 \times 10^{-6}/°C$; $E_B = 100 GN/m^2$ and $E_s = 210 GN/m^2$.

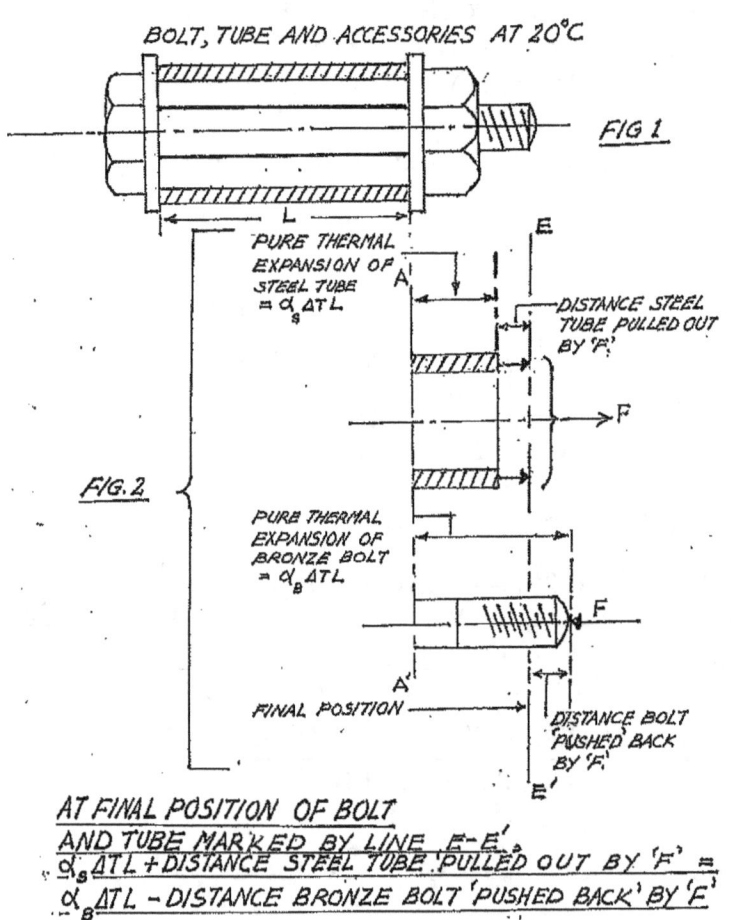

In Fig. 2 which is in 2 parts, I have shown the pure thermal expansions of the bolt and tube separately for ease of understanding and also the result of such expansion with extension and contraction. Thus with constrainment, the final position of the initial datum line $A-A'$ will be $E-E'$;

That is to say the steel tube is 'pulled-out' or extended i.e. a tensile force say, 'F' must act on it to cause it to do so. In the case of the bolt however, a compressive force of the same magnitude 'F' is necessary to push it back to the final position line $E-E'$.

Evidently, for equilibrium, the tensile force in the steel tube must equal the compressive force in the bolt.

Accordingly, we may now write

$$\alpha_s \Delta TL + \frac{PL}{A_s\ E_s} = \alpha_B \Delta tL - \frac{P}{A_B}\ \frac{L}{E_B}$$

or

$$\Delta TL(\alpha_B - \alpha_s) = PL\left(\frac{1}{A_s E_s} + \frac{1}{A_B E_B}\right)$$

$$= P\left(\frac{A_B\ E_B + A_s\ E_s}{A_s\ A_B\ E_S\ E_B}\right)$$

or

$$P = \frac{\Delta TA_s\ A_B\ E_S\ E_B(\alpha_B - \alpha_S)}{(A_B\ E_B + A_S\ E_S)}$$

$\therefore$ Stress in steel tube, $\sigma_s = \dfrac{P}{A_s}$

$\therefore$ $$\sigma_s = \frac{\Delta TA_B\ E_S\ E_B(\alpha_B - \alpha_S)}{(A_B\ E_B + A_S\ E_S)}$$

Similarly, $\sigma_B = \dfrac{P}{A_B}$

i.e. $$\sigma_B = \frac{\Delta TA_s\ E_S\ E_B(\alpha_B - \alpha_S)}{(A_B\ E_B + A_S\ E_S)}$$

The numerical data are:

$\Delta T = 190^o C;\ \ A_s = \pi(30^2 - 25^2)/4 = 216mm^2 = 216 \times 10^{-6} m^2;$

$A_B = \pi(15)^2 / 4 = 17.6.7(mm)^2 = 176.7 \times 10^{-6} m^2;$ $\qquad\qquad E_s = 210 \times 10^9 NM/m^2;$

$E_B = 100 \times 10^9 N/m^2;$

$\alpha_B = 20 \times 10^{-6}/^o C;\ \ \sigma_s = 16.8 \times 10^{-6}/^o C\ \ \alpha_s = 16.8 \times 10^{-6}/^o C$

$\therefore P = \dfrac{190 \times 216 \times 10^{-6} \times 176.7 \times 10^{-6} \times 210 \times 10^9 \times 100 \times 10^9 \times 3.2 \times 10^{-6}}{176.7 \times 10^{-6} \times 100 \times 10^0 + 216 \times 10^{-6} \times 210 \times 10^9}$

$= \dfrac{190 \times 216 \times 176.7 \times 210 \times 320}{10(62960)}$

$$= \frac{190 \times 216 \times 176.7 \times 210 \times 100 \times 3.2}{10^3 (176.7 \times 100 + 216 \times 210)}$$

$$= 7740 N \quad or \quad 7.74 kN$$

from which

$$\sigma_s = \frac{7740}{216 \times 10^{-6}} N/m^2 = 35.8 \times 10^6 N/m^2$$

i.e. $\sigma_s = 35.8 MN/m^2$

and

$$\sigma_B = \frac{7740}{176.7 \times 10^{-6}} N/m^2 = 43.8 \times 10^6 N/m^2$$

Q33. (i) Distinguish between mechanical expansion due to stress and thermal expansion.

(ii) A compressive force 'Q' acts on a composite bar assembly consisting of a solid copper cylinder of diameter 'd' snugly fitted with a circular cylindrical alloy steel tube of outside diameter 'D'. If 'Q' is applied when the temperature of the assembly is at $t_0^o C$, what must be the increase in temperature such that the entire load is carried by the copper cylinder. See Fig. 1 below.

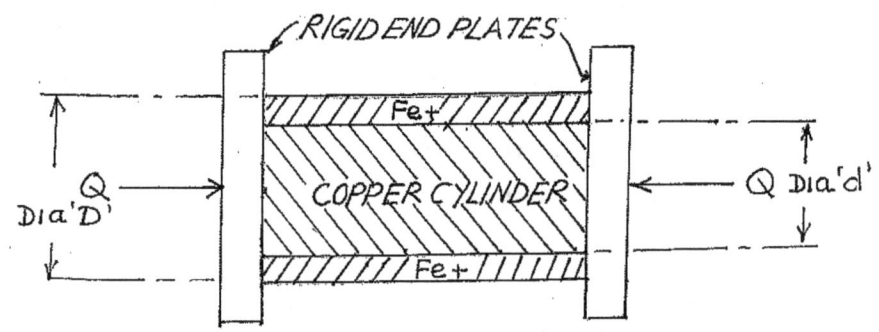

SECTION THROUGH ASSEMBLY

FIG 1

Ans : Increase in Temperature $= \frac{4Q}{\pi E_{CU}} d^2 (\alpha_{Cu} - \alpha_s)$

Q34. A compressive load 'P' is applied to a short column consisting of a circular cylindrical steel tube inside which is closely fitted a solid circular cylindrical plug of diameter 'd'. Refer to Fig. 1. If 'P' is applied when the temperature is T_0 then show that the increase in temperature such that the entire load is just carried wholly by the copper plug may be expressed by:

$$(T - T_o) = \frac{4P}{\pi d^2 E_{CU}(\alpha_{CU} - \alpha_{Fe})}$$

in which T is the higher temperature, α_{Fe} and α_{CU} the thermal coefficients of linear expansion for steel and copper respectively
and E_{CU} = modulus of elasticity for copper.

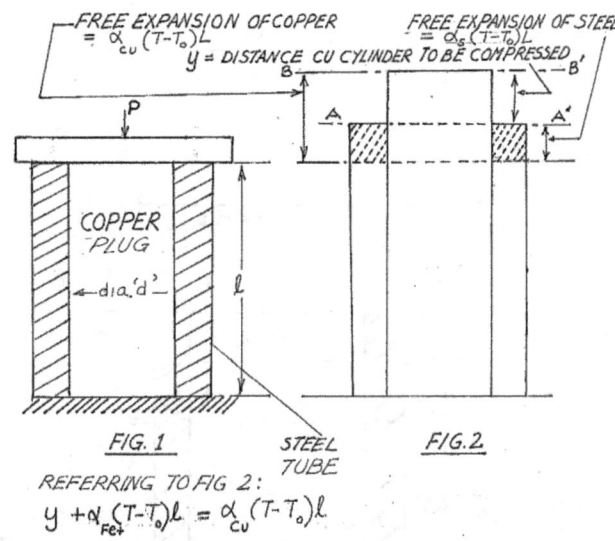

FIG. 1   FIG.2

REFERRING TO FIG 2:

$$y + \alpha_{Fe}(T-T_o)\ell = \alpha_{cu}(T-T_o)\ell$$

Assume at temperature T_o that there is no load on the composite cylinder so that both steel and copper are able to expand freely. When therefore the temperature rises form T_o to T, because the copper expands more rapidly than the steel it extends to level '$B-B$' as in Fig. 2, the free expansion being $\alpha_{CU}(T-T_0)L$. Meanwhile the steel tube extends freely also but only to level $A-A$ its expansion being $\alpha_{Fe}(T-T_o)L$. Now in order that the entire load be taken up by the copper plug, when 'P' is applied, the copper plug must be compressed down to level $A-A'$; 'P' provides the compression. Let this compression be 'y_{CU}.'

Referring once more to Fig. 2.

$$\alpha_{Fe}(T-T_o)L + y_{CU} = \alpha_{CU}(T-T_o)L \qquad \ldots\ldots\ldots\ldots \quad \text{(i)}$$

By Hooke's law, $$E_{CU} = \frac{P}{(csa)_{cu}} \cdot \frac{L}{y_{CU}}$$

$$= \frac{\dfrac{P}{\pi d^2}}{4} \cdot \frac{L}{y_{CU}} = \frac{4PL}{\pi d^2 y_{CU}}$$

from which $$y_{CU} = \frac{4PL}{\pi d^2 E_{CU}}$$

Substitution of this latter result in (i) produces

$$\alpha_{Fe}(T - T_o)L + \frac{4PL}{\pi d^2 E_{CU}} = \alpha_{cu}(T - T_o)(L)$$

from which

$$(T - T_o)(\alpha_{CU} - \alpha_{Fe}) = \frac{4P}{\pi d^2 E_{CU}}$$

i.e. $$(T - T_o) = \frac{4P}{\pi d^2 E_{CU}} \cdot \frac{1}{(\alpha_{CU} - \alpha_{Fe})}$$ .... Q.E.D

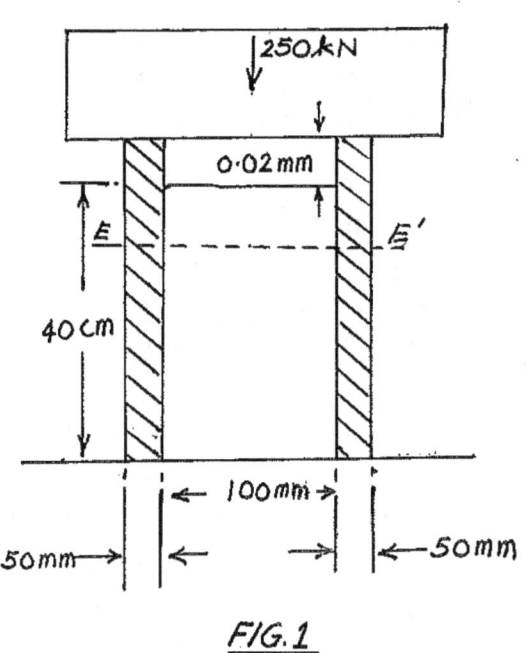

FIG.1

Q35. Fig. 1 shows a circular cylindrical steel rod encased in a copper tube over its entire length, the tube being a close fit with the rod. The assembly rests on a rigid foundation and supports a rigid slab of weight equal 500kN. Because of an error originating in the manufacturing process, the copper tube's length was 0.02mm more than the dimension shown on the workshop drawing. Given that the diameter of the steel rod is 100mm and the outside diameter of the copper tube is 200mm, determine the stress in each component as a result of the imprecision. By what percent is the magnitude of each stress altered assuming the copper tube had its correct length? Take $E_{STEEL} = 210GN/m^2$ and $E_{CU} = 103 GN/m^2$.

~ 528 ~

Let us work in metres :

Cross-sectional area of steel rod, $(csa)_s = \dfrac{\pi}{4}\left(\dfrac{100}{1000}\right)m^2$

i.e. $(csa)_s = \dfrac{\pi}{4}\left(\dfrac{1}{100}\right) = \dfrac{\pi}{400}m^2$

Cross-sectional area of copper tube

$$(csa)_{CU} = \dfrac{\pi}{4}\left\{(200)^2 - (100)^2\right\} \cdot \dfrac{1}{10^6}$$

$$= \dfrac{\pi}{4}\left(\dfrac{30000}{10^6}\right)$$

$$(csa)_{CU} = \dfrac{3\pi}{400}m^2$$

For vertical equilibrium, $P_s + P_{CU} = 250kN$ . . . . (i)

in which P_s = load carried by steel rod, and P_{CU} = load carried by copper tube. Considerations of statical equilibrium can provide no further assistance. To resolve this indeterminacy we must resort to a consideration of deformations. Now if the copper tube had been made 40cm long according to the workshop drawing then by Hooke's law:

$$E_{CU} = \dfrac{P_{CU}}{(csa)_{CU}} \cdot \dfrac{\ell}{\delta_{CU}} = \dfrac{P_{CU}}{(csa)_{CU}} \cdot \dfrac{40/100}{\delta_{CU}}$$

Further,

$$E_s = \dfrac{P_s}{(csa)_s} \cdot \dfrac{\ell}{\delta_s} = \dfrac{P_s}{(csa)_s} \cdot \dfrac{40/100}{\delta_s}$$

in which δ_{CU}, δ_s represent the respective deformation in the copper and steel.

Evidently $\delta_{cu} = \delta_s$. Expressing this equality using the latter two expression results in the following :

$$\dfrac{P_s}{P_{cu}} = \dfrac{E_s(csa)_s}{E_{CU}(csa)_{CU}}$$

Substituting the values of cross-sectional areas obtained previously we obtain,

$$\dfrac{P_s}{P_{cu}} = \dfrac{210\times10}{103\times10^9} \cdot \dfrac{\pi}{400} \cdot \dfrac{400}{3\pi}$$

i.e. $\dfrac{P_s}{P_{CU}} = \dfrac{210}{309} = 0.67$

which may be referred as $P_s = 0.67P_{CU}$ . . . . (ii)

Substituting this result in (i), we obtain
$$0.67P_{cu} + P_{cu} = 250kN$$
from which $P_{cu} \approx 149.7$
$$\therefore \qquad p_S \approx 100.3$$
Consider next the assembly with the imprecision.

Look again at Fig. 1. Assuming the equilibrium position of the block is at, say, level $E - E'$ it means that the distance travelled by the block is equal to this error in manufacture plus the deformation due to compression of the steel rod 'δ_s' to the level $E - E'$. Evidently, this total distance must be equal to the entire compression of the block of the upper tube. This is depicted in Fig. 2

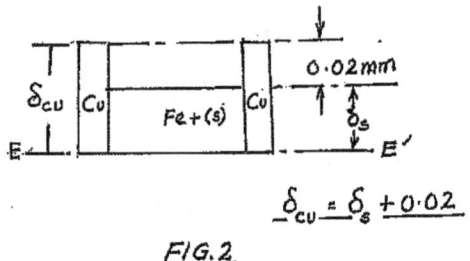

$$\delta_{cu} = \delta_s + 0.02$$

FIG.2

Accordingly, $\delta_{cu} = \delta_s + 0.02mm$ . . . . (iii)

But
$$\delta_{cu} = \frac{P_{cu}}{E_{cu}} \cdot \frac{1}{(csa)_{cu}} \, 40/100$$

and,
$$\delta_s = \frac{P_s}{E_s} \cdot \frac{1}{(csa)_s} \cdot 40/100$$

$\therefore$ By (ii)
$$\frac{P_{cu}}{E_{cu}} \cdot \frac{1}{(csa)_{cu}} \cdot \frac{40}{100} = \frac{P_s}{E_s} \cdot \frac{1}{(csa)_s} \cdot \frac{40}{100} + \frac{0.02}{1000}$$

or,
$$\frac{P_{cu}}{103 \times 10^9} \cdot \frac{400}{3\pi} \cdot \frac{40}{100} = \frac{P_s}{210 \times 10^9} \cdot \frac{400}{\pi} \cdot \frac{40}{100} + \frac{0.02}{1000}$$

or,
$$\frac{P_{cu}}{103 \times 10^9} \cdot \frac{160}{3\pi} = \frac{P_s}{210 \times 10^9} \cdot \frac{160}{\pi} + \frac{0.02}{1000}$$

i.e.
$$\frac{160P_{cu}}{3\pi \times 103} = \frac{160P_s}{210\pi} + 0.02(10^6)$$
$$0.164P_{CU} = 0.242P_s + 20000$$

i.e. $P_{cu} = 1.475P_s + 121951$ . . . . (iv)

By (i), with load in newtons, $P_{cu} = 250000 - P_s$

$$\therefore \qquad 250000 - P_s = 1.475 P_s + 121951$$

$$\text{or} \qquad 2.475 P_s = 128049$$

$$\text{i.e.} \qquad P_s = 51737 N \text{ or } 51.7 kN$$

$$\text{so that } P_{cu} = 198.3 kN$$

Summarising results:

(a)   Stresses with imprecise fit:

$$\sigma_{cu} = \frac{198.3 kN}{(3\pi / 400) m^2} = 8415 kN / m^2$$

$$\sigma_{cu} = 8.4 MN / m^2$$

$$\sigma_s = \frac{51.7 kN}{(\pi / 400)}$$

$$= \frac{517 \times 40}{\pi} = 6583.5 kN / m^2$$

$$\sigma_s = 6.6 MN / m^2$$

(b)   Stresses with precise fit :
For copper : $P_{Cu} = 149.7 kN$

$$\therefore \qquad \sigma'_{cu} = \frac{149.7}{198.3} \times 8.4 MN / m^2$$

$$\approx 6.3 MN / m^2$$

$$\text{and} \quad \sigma_s = \frac{100.3}{51.7} \times 6.6 MN / m^2$$

$$= 12.8 MN / m^2$$

$\therefore$ For copper tube <u>increase</u> in stress due to imprecise fit

$$= \frac{(8.4 - 6.3)100}{6.3} = 33.3\%$$

For steel tube <u>decrease</u> in stress due to imprecise fit

$$= \frac{(12.8 - 6.6)100}{12.8} = 48.4\%$$

Q36. Make a neat drawing of a typical "expansion bend" for a long straight length of pipe carrying, say, superheated steam.

Explain how it functions; and the method of its assembly in the pipeline

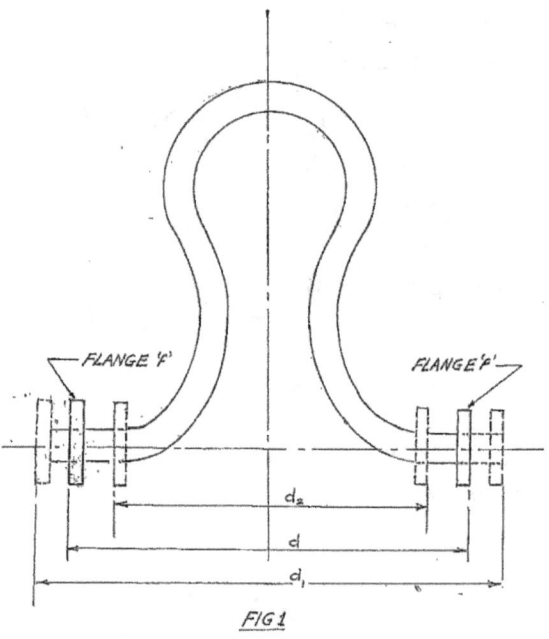

FIG 1

The drawing is given as Fig. 1

Before explaining how the bend functions it is necessary to deal with the method of assembly.

The bend is first produced in the engineering workshop with the distance between the flanges 'F' being, say 'd'. The distance between the flanges at the free ends of the pipeline to which the expansion bend is to be connected must be slightly greater than 'd', say, 'd_1'. Therefore the bend has to be strained in order that bolting up to the pipeline to the bend would occur.

The bend functions by 'absorbing' thermal expansion of the pipeline thereby reducing thermal stress in its walls. For example, in the present illustration, the flanges of the bend expand for a distance $d_1 - d_2$. In practice, the two extreme ends of the pipeline are anchored to rigid structures as for example a steam manifold or a boiler or some solid stationary object such as a reciprocating steam engine or a steam turbine.

Q37. A brass tube without end restraints is subjected to a linear temperature gradient, the temperature at one end being 35ºC and 85ºC at the other. The tube is 85cm long and the coefficient of linear thermal expansion of brass is $18.9 \times 10^{-6} /^{\circ} C$. Calculate the total extension of the tube at 85ºC, assuming thermal strain is uniform throughout.

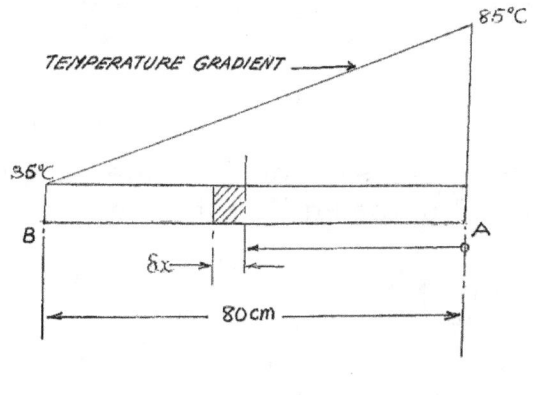

FIG. 1 .

In Fig. 1 above *AB* is the tube and the sloping line above it represents the linear temperature gradient. Consider an infinitesimal length of tube of length δx, distant 'x' from 'A', the end at 'A' being at 85°C. Temperature

at this section $= 85 - \dfrac{x}{80}(85-35)°C$

$$= \left(85 - \frac{5x}{8}\right)° C$$

Thermal strain as opposed to extension in length

$$x = \frac{(\alpha)(Temp.\ increase\ at\ section\ at\ x)(x)}{x}$$

$= \alpha(Temp.\ increase\ at\ section\ at\ x)$. But temperature increase at section at

$$x = \left\{\left(8.5 - \frac{5x}{8}\right) - 35\right\} = \left(50 - \frac{5x}{8}\right)° C. \quad \text{Accordingly, thermal strain in length}$$

$$'x' = 18.9 \times 10^{-6}\left(50 - \frac{5x}{8}\right)$$

$$= 94.5 \times 10^{-6}\left(10 - \frac{x}{8}\right)$$

Assuming as is given that this strain is uniform throughout the bar it follows that it must also be the strain in the infinitesimal length 'δx' as well.

Accordingly, strain in length 'δx' by

$$94.5 \times 10^{6}\left(10 - \frac{x}{8}\right) = \frac{Extension\ of\ '\delta x'}{\delta x}$$

~ 533 ~

∴   Linear   thermal   expansion   of   infinitesimal   length

$$'\delta x' = 94.5 \times 10^{-6} \left(10 - \frac{x}{8}\right) \cdot \delta x$$

The integral of this expansion of the infinitesimal length δx of course gives the extension 'δ' of the entire tube, i.e.

$$\delta = \int_0^{80} 94.5 \times 10^{-6} \left(10 - \frac{x}{8}\right) dx$$

$$= 94.5 \times 10^{-6} \left[10x - \frac{x^2}{16}\right]_0^{80}$$

$$= 94.5 \times 10^{-6} \left[800 - \frac{6400}{16}\right] = 94.5 \times 10^{-6} \times 400$$

$$= 37800 \times 10^{-6} cm$$

$$= 0.0378 cm$$

i.e. total extension of bar $= 0.0378cm$

approx 0.38mm

APPENDIX I

TABLES OF DIMENSIONS AND PROPERTIES OF SOME STRUCTURAL SECTIONS

APPENDIX 1 – TABLE 1

TABLE OF DIMENSIONS AND PROPERTIES
(EXTRACTED FROM BS EN 10219: 1997)

SQUARE HOLLOW SECTIONS

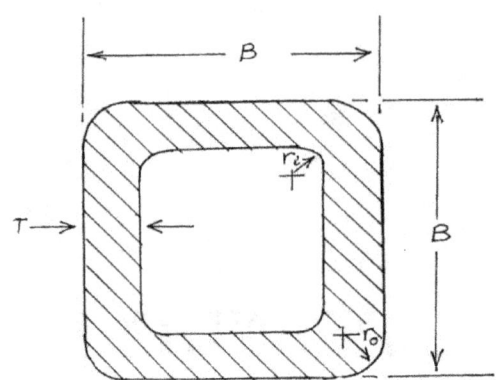

Size BxB	Thickness T	Corner Radii mm		Area of Section (cm)^2	Mass per Metre kg/m	I_{xx}, I_{yy}	Radius of Gyration	Section Modulus
		Ext	Int					
mm x mm		r_o	r_i			(cm)^4	cm	(cm)^3
20 x 20	2	4	2	1.34	1.05	0.692	0.72	0.692
25 x 25	2	4	2	1.74	1.36	1.48	0.924	1.19
30 x 30	2	4	2	2.14	1.68	2.72	1.1	2.1
30 x 30	3	6	3	3.01	2.36	3.5	1.08	2.34
40 x 40	2	4	2	2.94	2.31	6.94	1.54	3.47
40 x 40	2.5	5	2.5	3.59	2.82	8.22	1.51	4.11
40 x 40	4	8	4	5.35	4.2	11.1	1.44	5.54
50 x 50	2	4	2	3.74	2.93	14.1	1.95	5.66
50 x 50	3	6	3	5.41	4.25	19.5	1.9	7.79
50 x 50	4	8	4	6.95	5.45	23.7	1.85	9.49
60 x 60	2	4	2	4.54	3.56	25.1	2.35	8.38
60 x 60	3	6	3	6.61	5.19	35.1	2.31	11.7
70 x 70	4	8	4	10.1	7.97	72.1	2.67	20.6
80 x 80	8	20	12	20.8	16.4	168	2.84	42.1
90 x 90	6	12	6	19.2	15.1	220	3.39	49
100 x 100	12	36	24	36.1	28.3	408	3.36	81.6
120 x 120	8	20	12	33.6	26.4	677	4.49	113
140 x 140	12.5	37.5	25	57	44.8	1425	5	204
150 X 150	16	48	32	74.8	58.7	2009	5.18	268
180 X 180	5	10	5	34.4	27	1737	7.11	193

APPENDIX 1 – TABLE 2

TABLE OF DIMENSIONS AND PROPERTIES
(EXTRACTED FROM BS EN 10056: 1999)

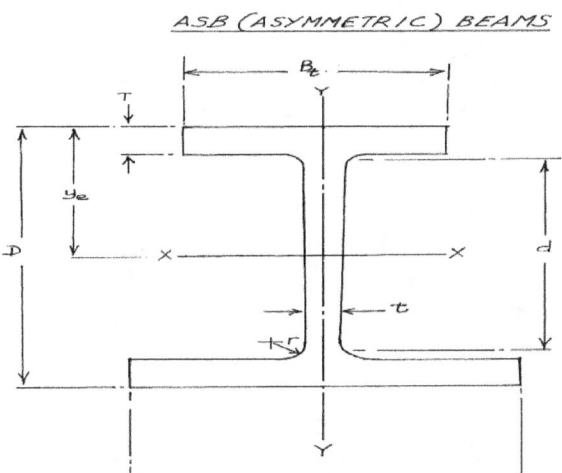

Designation	Mass per metre	Depth of Section	Width of top flange	Width Of bottom flange	Thickness of Web	Thickness of Flange	Root Radius	Depth Between Fillets	I_{xx}	I_{yy}	k_{xx}	k_{yy}	Elastic Neutral Axis Position, Y_e
Serial Size	kg/m	d mm	B_t mm	B_b mm	t mm	T mm	r mm	D mm	$(cm)^4$	$(cm)^4$			cm
300 ASB (FE) 249	249.2	342	203	313	40	40	27	208	52920	13190	12.9	6.45	19.2
300 ASB 196	195.5	342	183	293	20	40	27	208	45870	10460	13.6	6.48	19.8
300 ASB (FE) 185	184.6	320	195	305	32	29	27	208	35660	8752	12.3	6.1	18
300 ASB 155	155.4	326	179	289	16	32	27	208	34510	7989	13.2	6.35	18.9
300 ASB (FE) 153	152.8	310	190	300	27	24	27	208	28400	6840	12.1	5.93	17.4
280 ASB (FE) 136	136.4	288	190	300	25	22	24	196	22220	6256	11.3	6	16.3
280 ASB 124	123.0	296	178	288	13	26	24	196	23450	6410	12.2	6.37	17.2
280 ASB 105	104.7	288	176	286	11	22	24	196	19250	5298	12	6.3	16.8
280 ASB (FE) 100	100.3	276	184	294	19	16	24	196	15510	4245	11	5.76	15.6
280 ASB 74	73.6	272	175	285	10	14	24	196	12190	3334	11.4	5.96	15.7

APPENDIX 1 – TABLE 3

TABLE OF DIMENSIONS AND PROPERTIES
(EXTRACTED FROM BS EN 10056: 1999)

STRUCTURAL TEES SPLIT FROM UNIVERSAL BEAMS

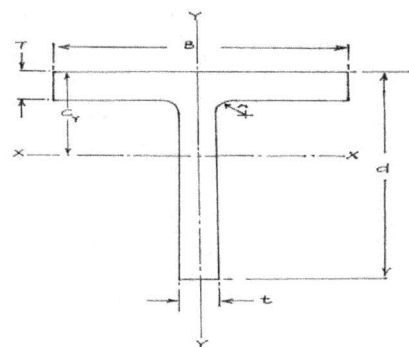

Serial Size	Mass per metre	Width of Section mm	Depth of Section mm	Thickness		Root Radius mm	Position of Centre of Mass From x-axis	Second Moment of Area (cm)^4	
				Web t	Flange T			I_{xx}	I_{yy}
B x d x kg/m	kg/m	B	d	t	T	r	C_y		
254 x 343 x 63	62.6	253	338.9	11.70	16.2	15.2	8.85	8980	2190
305 x 305 x 75	74.6	304.8	306.1	11.8	19.7	16.5	6.45	7410	4650
229 x 305 x 70	69.9	230.2	308.5	13.1	22.1	12.7	7.61	7740	2250
229 x 305 x 63	62.5	229	306	11.9	19.6	12.7	7.54	6900	1970
229 x 305 x 57	56.5	228.2	303.7	11.1	17.3	12.7	7.58	6270	1720
229 x 305 x 51	50.6	227.6	301.2	10.5	14.8	12.7	7.78	5690	1460
210 x 267 x 61	61	211.9	272.2	12.7	21.3	12.7	6.66	5160	1690
210 x 267 x 55	54.5	210.8	269.7	11.6	18.8	12.7	6.61	4600	1470
210 x 267 x 51	50.5	210	268.3	10.8	17.4	12.7	6.53	4250	1350
210 x 267 x 46	46	209.3	266.5	10.1	15.6	12.7	6.55	3880	1190
210 x 267 x 41	41.1	208.8	264.1	9.6	13.2	12.7	6.75	3530	1000
191 x 229 x 49	49.1	192.8	233.5	11.4	19.6	10.2	5.53	2970	1170
191 x 229 x 45	44.6	191.9	231.6	10.5	17.7	10.2	5.47	2680	1040
191 x 229 x 41	41	191.3	229.9	9.9	16	10.2	5.47	2470	935
191 x 229 x 37	37.1	190.4	228.4	9	14.5	10.2	5.38	2220	836
191 x 229 x 34	33.5	189.9	226.6	8.5	12.7	10.2	5.46	2030	726
152 x 229 x 41	41	155.3	232.8	10.5	18.9	10.2	5.96	2600	592
152 x 229 x 37	37.1	154.4	230.9	9.6	17	10.2	5.88	2330	523
152 x 229 x 34	33.6	153.8	228.9	9	15	10.2	5.91	2120	456
152 x 229 x 30	29.9	152.9	227.2	8.1	13.3	10.2	5.84	1880	397
152 x 229 x 26	26.1	152.4	224.8	7.6	10.9	10.2	6.04	1670	322
127 x 152 x 24	24	125.3	155.4	9	14	8.90	3.94	662	231
127 x 152 x 19	18.5	123.4	152.1	7.1	10.7	8.9	3.78	501	168
102 x 152 x 17	16.4	102.4	156.3	6.6	10.8	7.6	4.14	487	97.1
102 x 152 x 13	12.4	101.6	152.5	5.8	7	7.6	4.43	377	61.5
102 x 127 x 14	14.1	102.2	130.1	6.3	10	7.6	3.24	277	89.3
102 x 127 x 13	12.6	101.9	128.5	6	8.4	7.6	3.32	250	74.3
102 x 127 x 11	11	101.6	126.9	5.7	6.8	7.6	3.45	223	59.7
133 x 102 x 15	15	133.9	103.3	6.4	9.6	7.6	2.11	154	192
133 x 102 x 13	12.5	133.2	101.5	5.7	7.8	7.6	2.1	131	154

APPENDIX 1 – TABLE 4

TABLE OF DIMENSIONS AND PROPERTIES
(EXTRACTED FROM BS EN 10219: 1997)

RECTANGULAR HOLLOW SECTION

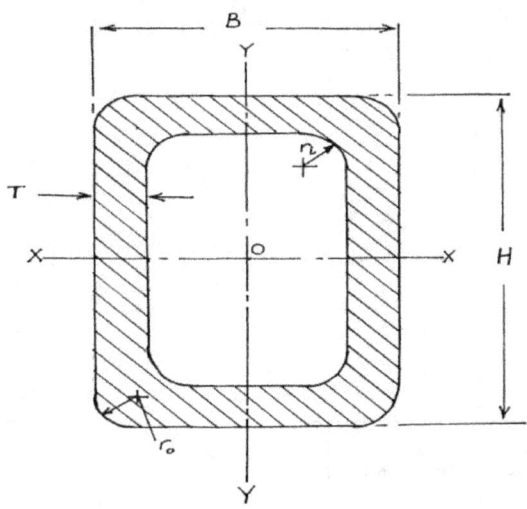

Size H x B mm x mm	Th'k's T mm	Corner Radius mm		Mass/m kg/m	Cross Sectional Area (cm)^2	Second Moment of Area (cm)^4		Radius of Gyration cm		Section Modulus (Elastic) (cm)^3	
		Ext r_o	Int r_i			I_{xx}	I_{yy}	k_{xx}	k_{yy}	Axis X-X	Axis Y-Y
40 x 20	2	4	2	1.68	2.14	4.05	1.34	1.38	0.793	2.02	1.34
40 x 20	3	6	3	2.36	3.01	5.21	1.68	1.32	0.748	2.6	1.68
50 x 25	2	4	2	2.15	2.74	8.38	2.81	1.75	1.01	3.35	2.25
50 x 25	3	6	3	3.07	3.91	11.2	3.67	1.69	0.969	4.47	2.93
50 x 30	2.5	5	2.5	2.82	3.59	11.3	5.05	1.77	1.19	4.52	3.37
50 x 30	3	6	3	3.3	4.21	12.8	5.7	1.75	1.16	5.13	3.8
60 x 30	4	8	4	4.2	5.35	15.3	6.69	1.69	1.12	6.1	4.46
60 x 40	2	4	2	2.93	3.74	18.4	9.83	2.22	1.62	6.14	4.92
60 x 40	2.5	5	2.5	3.6	4.59	22.1	11.7	2.19	1.6	7.36	5.87
60 x 40	5	10	5	6.56	8.36	35.3	18.4	2.06	1.48	11.8	9.21
70 x 50	2	4	2	3.56	4.54	31.5	18.8	2.63	2.03	8.99	7.5
70 x 50	4	8	4	6.71	8.55	54.7	32.2	2.53	1.94	15.6	12.9
70 x 50	6	10	5	8.13	10.4	63.5	37.2	2.48	1.9	15.6	12.9
80 x 40	2	4	2	3.56	4.54	37.4	12.7	2.87	1.67	9.34	6.36
80 x 40	3	6	3	5.19	6.61	52.3	17.6	2.81	1.63	13.1	8.78
80 x 60	4	8	4	7.79	10.1	87.9	56.1	2.94	2.35	22	18.7
90 x 50	2	4	2	4.19	5.34	57.9	23.4	3.29	2.09	12.9	9.35
90 x 50	4	8	4	7.97	10.1	103	40.7	3.18	2	22.8	16.3
10 x 100 x 40	3	6	3	6.13	7.81	92.3	21.7	3.44	1.67	18.5	10.8
10 x 100 x 50	6	12	6	12.3	15.6	179	58.7	3.38	1.94	35.8	23.5
100 x 60	6	12	6	13.2	16.8	205	91.2	3.49	2.33	41.1	30.4
100 x 80	4	8	4	10.5	13.3	189	134	3.77	3.17	37.9	33.5
120 x 60	2.5	5	2.5	6.74	8.59	161	55.2	4.33	2.53	26.9	18.4
120 x 80	6.3	15.75	9.45	17.5	22.2	408	217	4.28	3.12	68.1	54.3
140 x 80	4	8	4	13	16.5	430	180	5.1	3.3	61.4	45.1
15 x 150 x 100	4	8	4	14.9	18.9	595	319	5.6	4.1	79.3	63.7
150 x 100	8	20	12	27.7	35.2	1008	536	5.35	3.9	134	107
160 x 80	4	8	4	14.2	18.1	598	204	5.74	3.35	74.7	50.9
160 x 80	5	10	5	17.5	22.4	722	244	5.68	3.3	90.2	61

APPENDIX 1 – TABLE 5

TABLE OF DIMENSIONS AND PROPERTIES
(EXTRACTED FROM BS EN 10056: 1999)

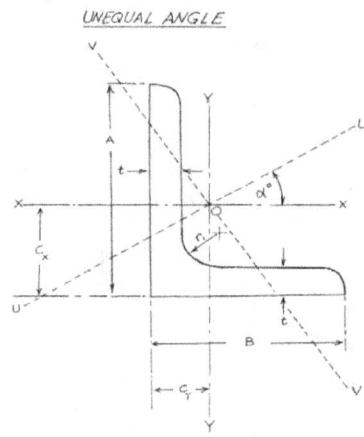

Designation A x B x t	Mass per Metre	Root Radius r_1	Toe Radius r_2	Area of Section	Distance of Centre of Mass		I_{xx}	I_{yy}	I_{uu}	I_{vv}	k_{xx}	k_{yy}	k_{uu}	k_{vv}	Angle between Axes X-X and U-U
mm x mm x mm	kg/m	mm	mm	(cm)^2	c_x mm	c_y mm	(cm)^4	(cm)^4	(cm)^4	(cm)^4	cm	cm			tan α°
200 x 150 x 18'	47.1	15	7.5	60	6.33	3.85	2376	1146	2920	623	6.29	4.37	6.97	3.22	0.549
200 x 150 x 15'	39.6	15	7.5	50.5	6.21	3.73	2023	979	2480	526	6.33	4.4	7	3.23	0.551
200 x 150 x 12'	32.0	15	7.5	40.8	6.08	3.61	1653	803	2030	430	6.36	4.44	7.04	3.25	0.552
200 x 100 x 15'	33.7	15	7.5	43	7.16	2.22	1759	299	1860	193	6.4	2.64	6.59	2.12	0.260
200 x 100 x 12'	27.3	15	7.5	34.8	7.03	2.10	1441	247	1530	159	6.43	2.67	6.63	2.14	0.262
200 x 100 x 10'	23.0	15	7.5	29.2	6.93	2.01	1219	210	1290	135	6.46	2.68	6.65	2.15	0.263
150 x 90 x 15'	26.6	12	6	33.9	5.21	2.23	761	205	841	126	4.74	2.46	4.98	1.93	0.354
150 x 90 x 12'	21.6	12	6	27.5	5.08	2.12	627	171	694	104	4.78	2.49	5.02	1.94	0.358
150 x 90 x 10'	18.2	12	6	23.2	5	2.04	533	146	591	88.3	4.8	2.51	5.05	1.95	0.360
150 x 75 x 15'	24.8	12	6	31.7	5.52	1.81	713	119	753	78.6	4.75	1.94	4.88	1.58	0.253
150 x 75 x 12'	20.2	12	6	25.7	5.4	1.69	589	99.6	623	64.7	4.78	1.97	4.92	1.59	0.258
150 x 75 x 10'	17	12	6	21.7	5.31	1.61	501	85.4	531	55.1	4.81	1.99	4.95	1.6	0.261
125 x 75 x 12'	17.8	11	5.5	22.7	4.31	1.84	354	95.5	391	58.5	3.95	2.05	4.15	1.61	0.354
125 x 75 x 10'	15	11	5.5	19.1	4.23	1.76	302	82.1	334	49.9	3.97	2.07	4.18	1.61	0.357
125 x 75 x 8'	12.2	11	5.5	15.5	4.14	1.68	247	67.6	274	40.9	4	2.09	4.21	1.63	0.360
100 x 75 x 12'	15.4	10	5	19.7	3.27	2.03	189	90.2	230	49.5	3.1	2.14	3.42	1.59	0.540
100 x 75 x 10'	13	10	5	16.6	3.19	1.95	162	77.6	197	42.2	3.12	2.16	3.45	1.59	0.544
100 x 75 x 8'	10.6	10	5	13.5	3.1	1.87	133	64.1	162	34.6	3.14	2.18	3.47	1.6	0.547
100 x 65 x 10'	12.3	10	5	15.6	3.36	1.63	154	51	175	30.1	3.14	1.81	3.35	1.39	0.410
100 x 65 x 8'	9.9	10	5	12.7	3.27	1.55	127	42.2	144	24.8	3.16	1.83	3.37	1.4	0.413
100 x 65 x 7'	8.8	10	5	11.2	3.23	1.51	113	37.6	128	22	3.17	1.83	3.39	1.4	0.415

APPENDIX 1 – TABLE 6

TABLE OF DIMENSIONS AND PROPERTIES
(EXTRACTED FROM BS EN 10219: 1997)

CHANNELS

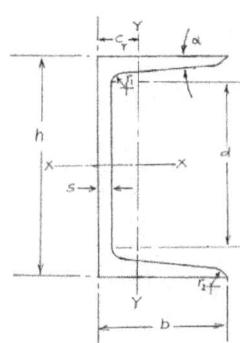

Designation h x b x kg/m	Mass/m kg/m	Depth of Section h	Width of Section b	Thickness		Radii		Flange Taper x Degrees	Depth Between Fillets d	Cross-Sectional Area	Second Moment of Area (cm)⁴		Radius of Gyration (cm)		Section Modulus (Elastic) (cm)³		Elastic Neutral Axis C_y
				Web s	Flange t	Root r_1	Toe r_2				I_{xx}	I_{yy}	Axis X-X	Axis Y-Y	Axis X-X	Axis Y-Y	
		mm	mm	mm	mm	mm	mm		mm	(cm)²							cm
432 x 102 x 65	65.5	431.8	101.6	12.2	16.8	15.2	4.8	5	362.5	83.4	21373	627	16	2.74	990	80	2.31
381 x 102 x 55	55.01	381	101.6	10.4	16.3	15.2	4.8	5	312.6	70.1	14869	579	14.6	2.87	781	75.7	2.52
305 x 102 x 46	46.21	304.8	101.6	10.2	14.8	15.2	4.8	5	239.3	58.9	8208	499	11.8	2.91	539	66.5	2.65
305 x 89 x 42	41.81	304.8	88.9	10.2	13.7	13.7	3.2	5	245.4	53.3	7078	326	11.5	2.48	464	48.6	2.18
254 x 89 x 36	35.66	254	88.9	9.1	13.6	13.7	3.2	5	194.7	45.4	4445	302	9.89	2.58	350	46.7	2.42
254 x 76 x 28	28.18	254	76.2	8.1	10.9	12.2	3.2	5	203.9	35.9	3355	162	9.67	2.12	264	28.1	1.85
229 x 89 x 33	32.68	228.6	88.9	8.6	13.3	13.7	3.2	5	169.9	41.6	3383	285	9.01	2.61	296	44.8	2.53
229 x 76 x 26	26.08	228.6	76.2	7.6	11.2	12.2	3.2	5	177.8	33.2	2615	159	8.87	2.19	229	28.4	2
203 x 89 x 30	29.77	203.2	88.9	8.1	12.9	13.7	3.2	5	145.2	37.9	2492	265	8.11	2.64	245	42.4	2.65
203 x 76 x 24	23.85	203.2	76.2	7.1	11.2	12.2	3.2	5	152.4	30.4	1955	152	8.02	2.24	192	27.7	2.14
178 x 89 x 27	26.79	177.8	88.9	7.6	12.3	13.7	3.2	5	121	34.1	1753	241	7.17	2.66	197	39.3	2.76
178 x 26 x 21	20.84	177.8	76.2	6.6	10.3	12.2	3.2	5	128.8	26.6	1338	134	7.1	2.25	151	24.8	2.2
152 x 89 x 24	23.87	152.4	88.9	7.1	11.6	13.7	3.2	5	96.9	30.4	1168	216	6.2	2.66	153	35.8	2.87
152 x 76 x 18	17.91	152.4	76.2	6.4	9	12.2	2.4	5	105.9	22.8	852	114	6.11	2.23	112	21	2.21
127 x 64 x 15	14.92	127	63.5	6.4	9.2	10.7	2.4	5	84	19	482	67.2	5.04	1.88	76	15.2	1.94
102 x 51 10	10.4	101.6	50.8	6.1	7.6	9.1	2.4	5	65.8	13.3	207	29.1	3.95	1.48	40.8	8.14	1.51
76 x 38 x 7	6.71	76.2	38.1	5.1	6.8	7.6	2.4	5	45.8	8.56	74.3	10.7	2.95	1.12	19.5	4.09	1.19

APPENDIX 1 – TABLE 7

TABLE OF DIMENSIONS AND PROPERTIES
(EXTRACTED FROM BS EN 10056: 1999)

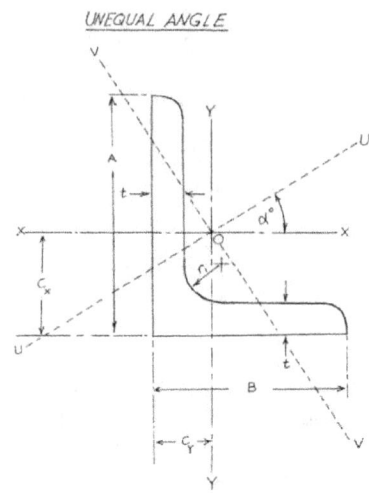

UNEQUAL ANGLE

Designation A x B x t	Mass per Metre	Root Radius r_1	Toe Radius r_2	Area of Section	Distance of Centre of Mass		I_{xx}	I_{yy}	I_{uu}	I_{vv}	k_{xx}	k_{yy}	k_{uu}	k_{vv}	Angle between Axes X-X and U-U
					c_x	c_y									
mm x mm x mm	kg/m	mm	mm	$(cm)^2$	mm	mm	$(cm)^4$	$(cm)^4$	$(cm)^4$	$(cm)^4$	cm	cm			tan α°
200 x 150 x 18'	47.1	15	7.5	60	6.33	3.85	2376	1146	2920	623	6.29	4.37	6.97	3.22	0.549
200 x 150 x 15'	39.6	15	7.5	50.5	6.21	3.73	2023	979	2480	526	6.33	4.4	7	3.23	0.551
200 x 150 x 12'	32.0	15	7.5	40.8	6.08	3.61	1653	803	2030	430	6.36	4.44	7.04	3.25	0.552
200 x 100 x 15'	33.7	15	7.5	43	7.16	2.22	1759	299	1860	193	6.4	2.64	6.59	2.12	0.260
200 x 100 x 12'	27.3	15	7.5	34.8	7.03	2.10	1441	247	1530	159	6.43	2.67	6.63	2.14	0.262
200 x 100 x 10'	23.0	15	7.5	29.2	6.93	2.01	1219	210	1290	135	6.46	2.68	6.65	2.15	0.263
150 x 90 x 15'	26.6	12	6	33.9	5.21	2.23	761	205	841	126	4.74	2.46	4.98	1.93	0.354
150 x 90 x 12'	21.6	12	6	27.5	5.08	2.12	627	171	694	104	4.78	2.49	5.02	1.94	0.358
150 x 90 x 10'	18.2	12	6	23.2	5	2.04	533	146	591	88.3	4.8	2.51	5.05	1.95	0.360
150 x 75 x 15'	24.8	12	6	31.7	5.52	1.81	713	119	753	78.6	4.75	1.94	4.88	1.58	0.253
150 x 75 x 12'	20.2	12	6	25.7	5.4	1.69	589	99.6	623	64.7	4.78	1.97	4.92	1.59	0.258
150 x 75 x 10'	17	12	6	21.7	5.31	1.61	501	85.4	531	55.1	4.81	1.99	4.95	1.6	0.261
125 x 75 x 12'	17.8	11	5.5	22.7	4.31	1.84	354	95.5	391	58.5	3.95	2.05	4.15	1.61	0.354
125 x 75 x 10'	15	11	5.5	19.1	4.23	1.76	302	82.1	334	49.9	3.97	2.07	4.18	1.61	0.357
125 x 75 x 8'	12.2	11	5.5	15.5	4.14	1.68	247	67.6	274	40.9	4	2.09	4.21	1.63	0.360
100 x 75 x 12'	15.4	10	5	19.7	3.27	2.03	189	90.2	230	49.5	3.1	2.14	3.42	1.59	0.540
100 x 75 x 10'	13	10	5	16.6	3.19	1.95	162	77.6	197	42.2	3.12	2.16	3.45	1.59	0.544
100 x 75 x 8'	10.6	10	5	13.5	3.1	1.87	133	64.1	162	34.6	3.14	2.18	3.47	1.6	0.547
100 x 65 x 10'	12.3	10	5	15.6	3.36	1.63	154	51	175	30.1	3.14	1.81	3.35	1.39	0.410
100 x 65 x 8'	9.9	10	5	12.7	3.27	1.55	127	42.2	144	24.8	3.16	1.83	3.37	1.4	0.413
100 x 65 x 7'	8.8	10	5	11.2	3.23	1.51	113	37.6	128	22	3.17	1.83	3.39	1.4	0.415

APPENDIX 1 – TABLE 8

TABLE OF DIMENSIONS AND PROPERTIES
(EXTRACTED FROM BS EN 10056: 1999)

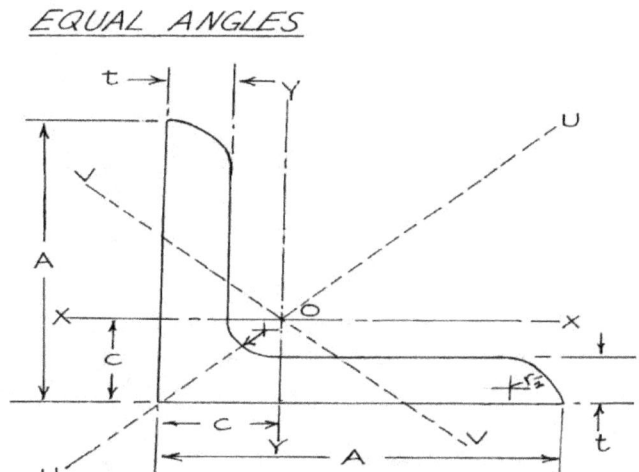

EQUAL ANGLES

Designation Serial Size A x A x t	Mass per metre	Root Radius r_i	Toe Radius r_2	Area of Section	Distance of Centre of Mass c	I_{xx}	I_{yy}	I_{uu}	I_{yv}	k_{xx}, k_{yy}	k_{uu}	k_{vv}
mm x mm x mm	kg/m	mm	mm	(cm)^2	cm					cm		
200 x 200 x 24	71.1	18	9	90.6	5.84	3331	3331	5280	1380	6.06	7.64	3.9
200 x 200 x 20	59.9	18	9	76.3	5.68	2851	2851	4530	1170	6.11	7.70	3.92
200 x 200 x 18	54.2	18	9	69.1	5.6	2600	2600	4150	1050	6.13	7.75	3.9
200 x 200 x 16	48.5	18	9	61.8	5.52	2342	2342	3720	960	6.16	7.76	3.94
150 x 150 x 18	40.1	16	8	51	4.37	1050	1050	1680	440	4.54	5.73	2.92
150 x 150 x 15	33.8	16	8	43	4.25	898	898	1430	370	4.57	5.76	2.93
150 x 150 x 12	27.3	16	8	34.8	4.12	737	737	1170	303	4.6	5.8	2.95
150 x 150 x 10	23	16	8	29.3	4.03	624	624	990	258	4.62	5.82	2.97
120 x 120 x 15	26.6	13	6.5	33.9	3.51	445	445	710	186	3.62	4.57	2.34
120 x 120 x 12	21.6	13	6.5	27.5	3.4	368	368	584	152	3.65	4.6	2.35
120 x 120 x 10	18.2	13	6.5	23.2	3.31	313	313	497	129	3.67	4.63	2.36
120 x 120 x 8	14.7	13	6.5	18.7	3.23	256	256	411	107	3.69	4.67	2.38
100 x 100 x 15	21.9	12	6	27.9	3.02	249	249	395	105	2.98	3.76	1.94
100 x 100 x 12	17.8	12	6	22.7	2.9	207	207	328	85.7	3.02	3.8	1.94
100 x 100 x 10	15	12	6	19.2	2.82	177	177	280	73	3.04	3.83	1.95
100 x 100 x 8	12.2	12	6	15.5	2.74	145	145	230	59.9	3.06	3.85	1.96
90 x 90 x 12	15.9	11	5.5	20.3	2.66	148	148	235	62	2.7	3.4	1.75
90 x 90 x 10	13.4	11	5.5	17.1	2.58	127	127	201	52.6	2.72	3.42	1.75
90 x 90 x 8	10.9	11	5.5	13.9	2.5	104	104	166	43.1	2074	3.45	1.76
90 x 90 x 7	9.6	11	5.5	12.2	2.45	92.6	92.6	147	38.3	2075	3.46	1.77

This page is intentionally left blank.

APPENDIX II

TENSILE FRACTURE TOUGHNESS

APPENDIX II

TENSILE AND FRACTURE TOUGHNESS
PROPERTIES OF SOME MATERIALS
AT ROOM TEMPERATURE

Material	Yield Stress MN/m^2 or MPa	Tensile or Ultimate Strength MN/m^2 or MPa	Elongation at Fracture based on Original Length %	% Reduction in Area %	Fracture Toughness $(MN/m^2)\sqrt{m}$ or MPa$\sqrt{m}$ or $MN/m^{3/2}$
AISI 1144 Steel	540	840	5	7	66
AISI 4130 Steel	1090	1150	14	49	110
Titanium Alloy	925	1000	16	34	66
Aluminium Alloy (7075)	415	485	13	N/A	24
ASTM A517-F Steel	760	830	20	66	187

APPENDIX III

BIBLIOGRAPHY

BIBLIOGRAPHY

Asimov, Isaac. *Gold*. Harper-Collins Publisher, 1995

Bright, J. *Freemasonry: Some Deeper Considerations*. England: Toye, Kenning & Spencer (Butterworth) Limited, Gartree Press Limited, 1992

Camm, F. J. *The Elements of Mechanics and Mechanisms*. George Newness Limited Tower House

Dadourian, H. M. *How to Study How to Solve*. Addison-Wesley Press, Inc, USA

Einstein, Albert. *Ideas and Opinions Lecture on "Geometry and Experience" before the Russian Academy of Sciences, January 27, 1921*. Based on Mein Weltbild, edited by Carl Seelig and other sources. New translations and revisions by Bargmann, Sonja, New York: Wings Books, 1954

Ewalds, H. L. and Wanhill, R. J. H. *Fracture Mechanics*. 1st Edition 1984, Reprinted 1993. Edward Arnold, 1993

Geary, A., Lowry, H. V., Hayden, H. A. *Advanced Mathematics for Technical Students, Part I*, 1st Edition, Reprinted 1954. London: Longmans, Green and Company, 1954

Gillings, R. J. *Graphs*. 3rd Edition reprinted 1961. Australian Publishing Company, Sydney

Levinson, J. J. *Introduction to Mechanics*. 1st Edition. Prentice-Hall Incorporated, 1961

Merchant, W. and Bolton, A. *An Introduction to the Theory of Structures*. Blackie and Sons Limited, 1956

Newton Friend, J. *Still More Numbers: Fun and Facts*. 1st Edition. New York: Charles Scribner's Sons, 1964

Parker, H. *Simplified Design of Roof Structures for Architects and Builders*. 2nd Edition. John Wiley and Sons, 1962

Parry, His Honour Edward Abbott. *The Seven Lamps of Advocacy*. London W.C.2.: T. Fisher Unwin, Adelphi Terrace, 1923

Petroski, H. Design Paradigms: *Case Histories of Error and Judgment in Engineering*. 1st Edition. Cambridge University Press, 1994

Pippard, A. J. S and Baker, J. F. *The Analysis of Engineering Structures.* 3rd Edition. London: Edward Arnold (Publishers) Limited, 1957

Quiller-Couch, Arthur. *On the Art of Writing.* Cambridge at the University Press

Timoshenko, S. *Strength of Materials Parts I and II.* 3rd Edition. London: D. Van Nostrand, 1957

Timoshenko, S. and Goodier, J. N. *Theory of Elasticity.* 2nd Edition. Mc Graw-Hill Book Company Incorporated, 1951

Timoshenko, S. and Mac Cullough, G. H. *Elements of Strength of Materials.* 3rd Edition. D. Van Nostrand Company Incorporated, 1956

Wolfe, Tom. *The Bonfire of the Vanities.* Bantam Books, 1988

Zienkiewicz, O. C. and Taylor, R. L. *The Finite Element Method.* 4th Edition. Mc Graw-Hill Book Company, 1994

INDEX

This page is intentionally left blank.